VOLUME SIXTY

Progress in
MEDICINAL CHEMISTRY

VOLUME SIXTY

Progress in
MEDICINAL CHEMISTRY

Edited by

DAVID R. WITTY

Director Witnet Ltd,
Platinum Building,
St John's Innovation Park,
Cowley Road, Cambridge CB4 0DS,
United Kingdom

BRIAN COX

Emeritus Professor of Pharmaceutical Chemistry,
Chichester 3 3R341,
School of Life Sciences,
University of Sussex,
Brighton, BN1 9RH,
United Kingdom

ELSEVIER

Elsevier
Radarweg 29, PO Box 211, 1000 AE Amsterdam, Netherlands
The Boulevard, Langford Lane, Kidlington, Oxford OX5 1GB, United Kingdom
50 Hampshire Street, 5th Floor, Cambridge, MA 02139, United States

First edition 2021

ISBN: 978-0-323-85056-8
ISSN: 0079-6468

For information on all Elsevier publications
visit our website at https://www.elsevier.com/books-and-journals

Publisher: Zoe Kruze
Acquisitions Editor: Jason Mitchell
Developmental Editor: Tara Nadera
Production Project Manager: Abdulla Sait
Cover Designer: Miles Hitchen

Typeset by SPi Global, India

Contents

Contributors

Thierry Bonnaud
Johnson Matthey, Pharmorphix, Cambridge, United Kingdom

Lee J. Byrne
Faculty of Science, Engineering and Social Sciences, Natural and Applied Sciences, School of Psychology and Life Sciences, Canterbury Christ Church University, Canterbury, Kent, United Kingdom

Jason C. Cole
Cambridge Crystallographic Data Centre, Cambridge, United Kingdom

D. Conole
Department of Chemistry, Molecular Sciences Research Hub, Imperial College London, London, United Kingdom

Dyanne Cruickshank
Johnson Matthey, Pharmorphix, Cambridge, United Kingdom

E. De Vita
Department of Chemistry, Molecular Sciences Research Hub, Imperial College London, London, United Kingdom

Ilenia Giangreco
Cambridge Crystallographic Data Centre, Cambridge, United Kingdom

Thomas Kendall
Technobis Crystallization Systems, Alkmaar, The Netherlands

C.S. Kounde
Department of Chemistry, Molecular Sciences Research Hub, Imperial College London, London, United Kingdom

Ruth A. Lunt
Johnson Matthey, Pharmorphix, Cambridge, United Kingdom

M. Maneiro
Department of Chemistry, Molecular Sciences Research Hub, Imperial College London, London, United Kingdom

Paula L. Nichols
Synple Chem AG, Kemptthal; ETH, Zurich, Switzerland

Adam R. Patterson
Johnson Matthey, Pharmorphix, Cambridge, United Kingdom

C. Daniel Scott
Johnson Matthey, Pharmorphix, Cambridge, United Kingdom

Francesca Stanzione
Cambridge Crystallographic Data Centre, Cambridge, United Kingdom

Sam Stratford
Johnson Matthey, Pharmorphix, Cambridge, United Kingdom

E.W. Tate
Department of Chemistry, Molecular Sciences Research Hub, Imperial College London, London, United Kingdom

Carol M. Trim
Faculty of Science, Engineering and Social Sciences, Natural and Applied Sciences, School of Psychology and Life Sciences, Canterbury Christ Church University, Canterbury, Kent, United Kingdom

Steven A. Trim
Venomtech Ltd., Sandwich, Kent, United Kingdom

Q. Zhang
Department of Chemistry, Molecular Sciences Research Hub, Imperial College London, London, United Kingdom

Preface

The focus of the 60th edition of *Progress in Medicinal Chemistry* is on the tools, technologies and strategies that underpin the science. We start with an assessment of the applications in drug discovery of nature's most active biomolecules: the venoms used for attack and defence in a wide range of organisms. Next, though targets for drug discovery programmes have typically been inhibitors, agonists and antagonists of proteins in physiological pathways, PROTACs represent an exciting alternative catalytic strategy for the design of drugs to modulate a specific protein's function by changing its concentration through degradation. Another chapter reviews the practices of the medicinal chemist—past, present and future—with a focus on the impact of automation and how that is changing both the capabilities and our expectations of experimentalists. Molecular docking is now used in almost all lead optimisation programmes, and in a further chapter, we assess the scope of docking software and the newest applications of this key technology. Finally, chemotypes for many drug targets present challenges for formulation due to their necessary physicochemical properties, such as lipophilicity, solubility or crystallinity. However, cocrystal formation by design presents new opportunities for the chemist to create deliverable and patentable drugs.

Venoms are used for defensive or predatory purposes in a surprising range of animals and are typically composed of a huge variety of active biomolecules, even within a single species. Their biological evolution has made venom compounds both stable in vivo and exquisitely potent ligands for their intended target proteins. As these include a wide range of disease-associated enzymes, they can be effective starting points for drug design and valuable tools to elucidate pathways. In Chapter 1, Trim and colleagues describe how an improved knowledge of venoms is making them increasingly applicable to drug discovery.

No strategy has more radically changed the way in which we think about designing drugs than the PROTAC hypothesis. In its simplest form, with this approach drug molecules are designed to catalytically cause the degradation of target proteins using the cell's own protein recycling mechanisms, and thereby affect the activity of a specific enzyme function. The first molecules designed using this strategy are now in clinical evaluation, but significant challenges remain for medicinal chemists working to create them.

Tate and coworkers review the great advances that have been made since the inception of this field less than two decades ago, in Chapter 2.

Chapter 3 reviews the gradual adoption of automated technologies within the medicinal chemistry lab as tools for drug design, synthesis and analysis. Much of the key equipment in today's labs would be recognisable from half a century ago, but the application of increasingly sophisticated robotic technologies, underpinned by advances in computing power, is changing the laboratory. Nichols argues that requirements for the medicinal chemistry practitioner of the future may be very different from what we have expected of the laboratory chemist hitherto.

The ability to model the interaction between ligand molecules and their biological targets has become a cornerstone in modern drug discovery and lead optimisation. As the power of computing has grown, we have gained ever more confidence in the predictions of binding models. Likewise the application software has become more accessible to the nonexpert user. In Chapter 4, Stanzione and Cole evaluate the current state of technology, how it has been used to help drug discovery and what capabilities we can expect to be able to apply in the future.

The final chapter focuses on cocrystal technology and its utility in the production of developable solid dose forms. Once conventional salt form options have been examined and eliminated, there is a tendency in the industry to move directly to higher cost and higher risk particle size engineering and/or stabilised amorphous form methodologies, without considering cocrystals as a viable option. Methods for identifying cocrystals using a variety of analytical techniques and their scale-up by techniques such as mechanochemistry and solution-based crystallisation are reviewed. The benefits of utilising cocrystal technologies, in particular as a remedy for dissolution rate/solubility-limited absorption issues, are illustrated with a selection of literature and in-house examples. By improving physicochemical performance, this technology potentially extends the range of compounds designed though medicinal chemistry that can be considered drug-like.

David R. Witty
Brian Cox
November 2020

CHAPTER ONE

Utilisation of compounds from venoms in drug discovery

Carol M. Trim[a], Lee J. Byrne[a], and Steven A. Trim[b,*]
[a]Faculty of Science, Engineering and Social Sciences, Natural and Applied Sciences, School of Psychology and Life Sciences, Canterbury Christ Church University, Canterbury, Kent, United Kingdom
[b]Venomtech Ltd., Sandwich, Kent, United Kingdom
*Corresponding author: e-mail address: s.trim@venomtech.co.uk

Contents

Progress in Medicinal Chemistry, Volume 60
ISSN 0079-6468
https://doi.org/10.1016/bs.pmch.2021.01.001

Keywords: Macrocycle, Venom, Toxin, Peptide, Polyamine, Drug discovery

1. Introduction to venoms and historical perspective

1.1 Evolution

Evolution of a venom system presents a species with a powerful advantage in terms of ability to catch prey, evade predators and even compete for a mate. This has led to many independent, unrelated venom systems developing [1,2]. Venoms evolved from other somatic proteins through co-option of genes [3,4], protein neofunctionalisation [5,6] and gene duplication [7–9]. This means that they already possess, inherited from the ancestral proteins, functionally active groups for engaging other biological proteins.

Venoms have been mostly studied from an evolutionary perspective and because they are of great medical importance. Snakebite envenomation alone kills >100,000 people per year and causes morbidity in >400,000 people each year [10]. It is estimated that scorpion stings cause 3000 deaths a year worldwide [11] but it is rare for a human to die from a spider bite [12]. Morbidity and mortality are essentially due to large doses of pharmacologically active components delivered together. Here is where the interest to drug discovery lies.

1.2 Venom sources and pharmacology

Physiologically, venom components are expressed and stored in the venom gland and reside there until delivered by the organism into another, via an injection system, when required. This has led to selection of stable folds to maintain peptide components in active conformation [13–15] ready for use. Thus, venom orthologues are more stable than their somatic orthologues. For example, natriuretic peptides from mambas (*Dendroaspis* spp.) venom are stable for many hours in rat plasma whereas brain and atrial natriuretic peptides are mostly degraded in under an hour [16]. The glucagon like peptide, exendin-4, from the Gila monster *Heloderma suspectum* has a 10-fold greater plasma half-life than human GLP1 [17] making this peptide extremely useful for drug discovery.

Fig. 1 Examples of the range of species that utilise venom for defence/feeding/competition. Examples shown are (A) Blue ringed octopus *Hapalochlaena fasciata*, (B) Southern Copperhead snake *Agkistrodon contortrix*, (C) *Pterois volitans* Lion Fish, (D) Indian ornamental theraphosid *Poecilotheria regalis*, (E) Monocled cobra *Naja kaouthia*, (F) European wasp *Vespula germanica*, (G) White spot jellyfish *Phyllorhiza punctata*, (H) Giant house spider *Eratigena atrica*, (I) Giant forest scorpion *Heterometrus mysorensis*, (J) Gila monster *Heloderma suspectum*, (K) Chinese red-headed centipede *Scolopendra subspinipes*, (L) Solenodon *Solenodon paradoxus*.

Animals from a wide range of species utilise venoms for hunting and as a defence mechanism, and examples of the evolved diversity are shown in Fig. 1. These venoms can have a varied range of consequences on the victim such as a cytotoxic, neurotoxic, myonecrotic, haemolytic, cardiotoxic, necrotic, nephrotoxic, hypotensive, haemorrhagic or algogenic effects based on their biological activities and unique origins.

1.2.1 Terrestrial reptiles

The venom systems of reptiles, now correctly known as non-avian reptiles due to the genetic inability to separate them from birds (the direct decedents of the dinosaurs), were the first to be characterised. The majority of snake species are non-venomous, however those that are, include vipers, rattlesnakes, mambas, cobras [18] and many species of sea snakes, which will be discussed later. The helodermatid lizards are also venomous, these are the Beaded lizard (*Heloderma horridum*) and the Gila Monster (*Heloderma suspectum*). The venom system in these lizards may have evolved independently from the snake system as initially proposed [19]. However later work by Fry and colleagues in 2012 proposed the Toxicofera clade, which has largely been adopted, suggesting a single venomous reptile ancestor and loss of venom from the other suborders [20]. The debate continues on as to the true evolution of venom systems in non-avian reptiles as either a single event [20] or multiple convergent events as restated [21].

Snake venoms are frequently described as being composed of a complex mixture of proteins and peptides which typically make up 90–95% of their dry weight. By function these include three-finger toxins, disintegrins, phospholipase A2s (PLA2s), metalloproteinases, hyaluronidases, cholinesterases, L-amino acid oxidases, serine proteases, lectins, and growth factors. Other components include inorganic cations, such as calcium, zinc, magnesium, sodium, and potassium [22]. Envenomation can cause the recipient of a snakebite to be exposed to a wide range of effects from pain and necrosis to haemorrhage or coagulation. Despite this complexity many snake venoms contain a few major protein classes with lots of minor components [22] and the evidence for this can be seen in the ion exchange chromatogram Fig. 2A. In snakes, venom proteins have also been found in extracellular vesicles (SVEVs) within the venom of four different species of snake (*Agkistrodon contortrix contortrix*, *Crotalus atrox*, *Crotalus viridis* and *Crotalus cerberus oreganus*) [23] but this has not yet been reported in species other than snakes.

1.2.2 Arthropods

A wide range of arthropods are venomous [24]. This includes the major groups of Arachnida including spiders (Araneae) [25]. Spiders are one of the largest venomous clades [26] but not all of these species produce venoms that are harmful to humans. Typical spider venom is estimated to comprise over 400 venom peptides [27] and the majority of these fall within a small size range of 8–11 kDa [28]. This similarity is demonstrated through reverse phase HPLC in Fig. 2. The vast majority of spiders use their venom in a predatory manner but the neotropical jumping spider *Bagheera kiplingi* actually feeds mainly on plants [29,30]. Spider venoms have different levels of toxicity to humans and other animals. Funnel web and widow spiders (*Latrodectus hasselti and L. mactans*), can kill a human, although a successful antivenom means there are few fatalities [31], whereas British native spiders are of no risk to mammals at all but are very effective at killing their prey item, which could be an insect or another arthropod. Examples of the latter would be the giant house spider *Eratigena atrica* (Fig. 1H) and the European spider *Tegenaria agrestis* [32].

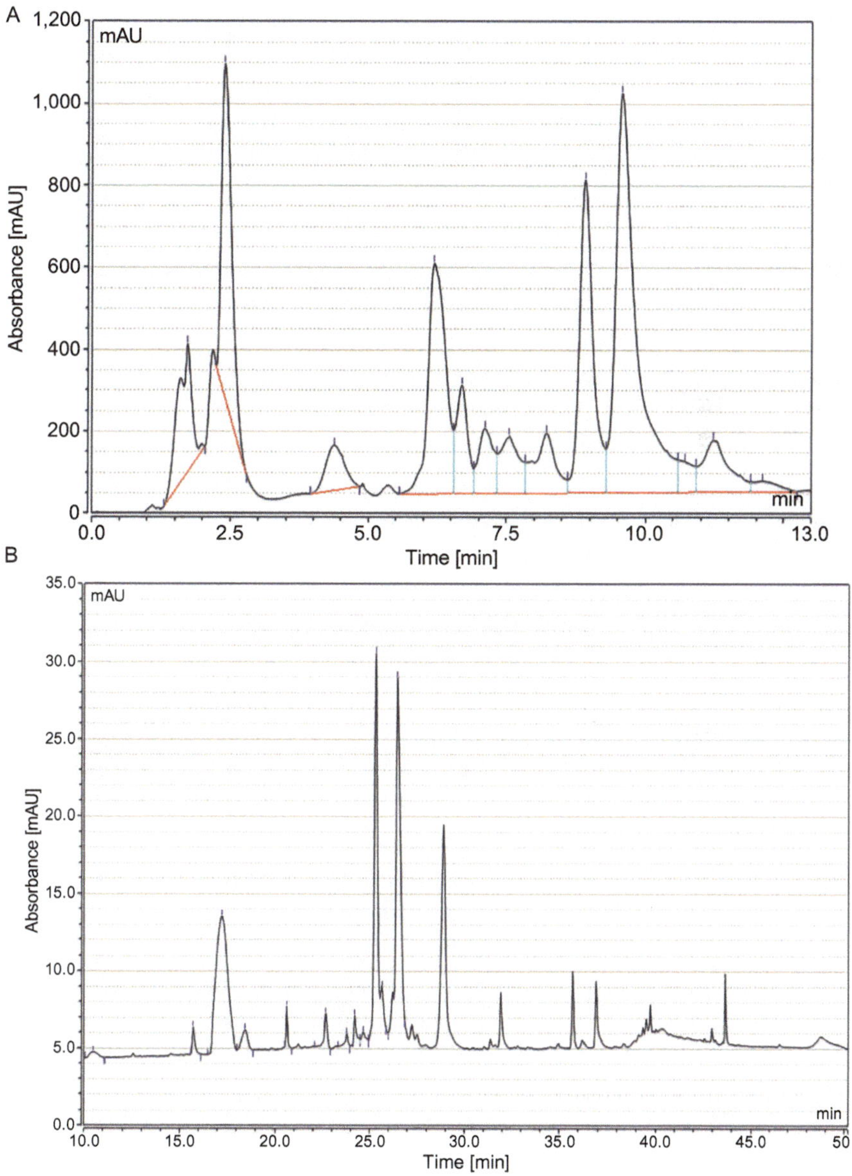

Fig. 2 HPLC separation of the whole venom demonstrates the standard techniques and example chromatograms. (A) Ion exchange chromatography venom from the Indochinese spitting cobra *Naja* siamensis performed on the ThermoFisher (Dionex) UltiMate3000, demonstrating dominant peaks of few protein families and many minor components. (B) Reverse phase chromatography of whole venom from Lamberton's Tarantula (Theraphosid), *Monocentropus lambertoni* demonstrating the range of hydrophobicities in theraphosid venoms.

The main active components in brown recluse spider (Loxosceles) venom are inhibitor cystine knot (ICK) peptides, phospholipases-D and astacin-like metalloproteases. Also found in the venom are serine proteases, serpins, hyaluronidases, venom allergens, and a translationally controlled tumour protein (TCTP) [33,34]. Spiders of the genus *Latrodectus* (widow and red back spiders) have venom that contains latrotoxins which are neurotoxic and are the main components which cause intense pain to the recipient of a bite [26]. There are insect-selective neurotoxins from the venom of the Blue mountains funnel web spider *Hadronyche versuta* called Janus-faced atracotoxins (J-ACTXs) [35] of use to crop science. There are many other neurotoxic peptides such as; hainantoxin-III and hainantoxin-IV from Theraphosidae spiders. These toxins were initially isolated from the venom of the Chinese bird spider *Seleconosmia hainana* [36]. Later confirmed through genomics and proteomics although taxonomic changes meant they were assigned to *Haplopelma (Ornithoctonus) hainana* [9]. At the time of writing the official *toxin nomenclature is* μ-theraphotoxin-Hhn2a and μ-theraphotoxin-Hhn1b reflecting the Haploplema genera, however recent mitochondrial genome sequencing confirms this spider as *Cyripagopus hainanus* [37].

Other arachnids that are also venomous including the Scorpions (Scorpiones) [38] and Pseudoscorpions (Pseudoscorpiones); the latter have an average venom yield of 5 nL [39]. It was also once questioned whether the ticks (Parasitiformes) were venomous but they are, according to the latest definitions of a venomous animal, i.e., the animal actively injects its toxins [40].

Centipedes (Chilopoda) [41] and Remipedes (Remipedia: *Speleonectes tulumensis*) [42] are also venomous and so are blood worms, such as *Glycera dibranchiata, Glycera fallax* and *Glycera tridactyla* [43]. A wide range of Insects (Insecta) are also venomous including Hemiptera or the true bugs and an example of these are the assassin bugs [44]. Hymenoptera (bees, wasps, and ants) produce venom and while these are the best known, there are also many other insects that possess venom [44]. Adult moths (Lepidoptera) are not venomous themselves, but moth caterpillars can be venomous [45]. An example of Dipterans (true flies) that produce complex venom, rich in peptides, (in particular disulphide rich peptides) is the superfamily Asilidae, known as Assassin flies, such as *Dolopus genitalis* [14].

1.2.3 Mammals

There are a number of venomous mammals including the slow loris species (*Nyciticebus* spp.) [46] and the male platypus (*Ornithorhynchus anatinus*) [47].

The platypus uses its venom in competition as well as in defence [47], and has paired venom glands that are connected to spurs on each hind leg [48]. Echidna were once thought to be venomous but they do not fit the definition of a venomous animal [47], however some other insectivores are truly venomous. From the order Eulipotyphla the Hispaniolan Solenodon *Solenodon paradoxus* [49] Fig. 1 and a number of shrews including *Neomys fodiens* the Eurasian water shrew are a venomous [50]. Blarina toxin (BLTX) has been isolated from the venom of the short tailed shrew *Blarina brevicauda* and it is suggested that its toxicity is due to a kallikrein-like activity of the venom [51]. Blarinasin is also a tissue kallikrein-like protease found in the same shrew [52].

1.2.4 Cone snails (Conoidea)

Like scorpions, cone snails also use their venom for both defensive and predatory purposes. It contains peptides called conotoxins that have neurotoxic effects on the victim. These peptides have evolved to adapt to new prey [53] The geography cone (*Gastridium geographus*) and the piscivorous cone snail (*Conus tulipa*) are thought to be the most deadly species of cone snail [54] to humans, probably due to their diet of vertebrates, compared with other cones. There is also another lineage of less well-studied cone snails belonging to the *Gemmula* genera [54].

1.2.5 Sea snakes

Many other venomous creatures inhabit the ocean, including sea snakes. An example being the spine-bellied sea snake (*Hydrophis curtus*) whose venom shows myotoxicity against a human skeletal muscle cell line and is selective for skeletal muscle cells compared to cardiac muscle [55]. The elegant sea snake (*Hydrophis elegans*), olive sea snake (*Aipysurus laevis*) and the Australian beaked sea snake (*Hydrophis zweifeli*) possess venoms which have a higher myotoxic effect compared to the spine-belled sea snake. These snakes are passive in their natural ocean habitat and only really present a hazard to humans when caught in nets or washed up.

1.2.6 Fish

There have been fewer studies on fish venoms compared with other venomous animals even though >50% of venomous vertebrates are fish [56]. Fish have convergently evolved venom systems multiple times [57,58] leading to multiple lineages. The scorpionfish (*Scorpaena plumieri*) has venom-containing cytolysins which cause cardiotoxic, neurotoxic and inflammatory effects [59].

The lionfish (*Pterois volitans*) (Fig. 1C), has venom which causes algogenic and inflammatory effects [60]. One of the most dangerous venomous fish is the reef stonefish (*Synanceia verrucosa*). The venom glands are in the dorsal spines and cause necrosis, haemorrhage and inflammatory cell infiltration when tested in rats [61]. Humans fall victim to this fish when walking in shallow reef waters and their sting has been shown to even penetrate flipflops by several television presenters. Stingrays, unrelated to the other venomous fish as they are cartilaginous (Chondrichthyes) and more closely related to sharks, deliver envenomation, typically to the lower limbs which causes pain, swelling and necrosis [62,63]. Also, several species of sharks are venomous [58], although this property has been poorly studied compared with other taxa.

The greater (*Trachinus draco*) and lesser (*Echiichthys vipera*) weever fish have some unique venom components but these have not been studied in detail [64]. Some have been shown to cause specific effects. For example Dracotoxin in the greater weever fish causes membrane depolarising activity and haemolysis [65]. Components of the venom of the lesser weever fish venom cause cell lysis in erythrocytes, cell cycle arrest and tissue necrosis [66].

1.2.7 Cnidaria

The Cnidaria are a group of aquatic animals all containing subcellular stinging capsules called nematocysts. Coral and jellyfish venom is much understudied [67] but this field is growing as techniques improve. Members of this phylum include corals, which have interesting bioactive compounds [68], hydras, which have pore forming toxins called cytolysins [69], sea anemones [70] and jellyfish [71]. Sea anemones produce a number of peptide toxins that modulate ion channels [72]. The box jellyfish *Chironex fleckeri* venom is neurotoxic and causes tissue necrosis [71] but also has haemolytic effects [73]. The Portuguese man-of-war (*Physalia physalis*) is actually a colony of polyps rather than a single organism, however, together they maintain stinging tentacles that cause cardiotoxic, necrotic and neurotoxic effects [74] in humans, even after death.

1.2.8 Other venomous species

The blue-ringed octopus (*Hapalochlaena fasciata*) (Fig. 1A) has tetrodotoxin (TTX) in its venom which causes pain and neurotoxic effects (paralysis) by inhibiting signal transduction by nerve cells through sodium channel blockade [75]. This toxin is synthesised by bacteria and is also found in other several species, for example in pufferfish and salamanders. Even sponges can cause pain and swelling, fire sponges are an example (*Tedania* spp.) [76]. Sea urchins trigger an inflammatory reaction and can cause tissue necrosis and muscular paralysis in some and the crown-of thorns starfish (*Acanthaster planci*) venom has strong haemolytic activity [77].

1.2.9 Application to drug discovery

Convergent evolution from unrelated species has led to extremely varied compositions of venom across a range of species lineages. For drug discovery, this means there is a very diverse compound library of potentially millions of compounds from an estimated 220,000 venomous species currently known [1]. The Australian funnel web spiders have the most complex venoms venom discovered to date, the venom of the Darling Downs funnel web spider (*Hadronyche infensa*) contains over 3000 different peptides [6]. Cone snails have over 1000 venom peptides [78]. Platypus venom has 83 peptides from 13 different families [48]. In contrast the Northern short-tailed shrew has much less complex venom with only seven proteins isolated so far [5].

So, if venoms cause high mortality and morbidity in so many people why would venom components be considered as potential drugs? Looking at the problem from a pharmacological standpoint the opportunities are obvious, and it is not a surprise that receipt of a random dose of a complex mixture of pharmacologically active compounds is dangerous. The clear action of these compounds led to early experimentation and application as traditional medicines with variable results [79] but now, in the hands of drug discovery scientists, their true potential is being realised.

2. Approved drugs from venoms

There are a number of currently approved drugs that have their origins in venom components, and these range from small molecules to peptides and proteins. Looking specifically at peptide therapeutics there are a number that have been derived from venoms which have been approved for clinical use [80] with origins ranging from snakes and cone snails to Gila monsters (Fig. 1J).

2.1 Hypertension

Capoten® (Captopril) an angiotensinogen converting enzyme (ACE) inhibitor was the first successful therapeutic to have been derived from a venom component [18,81,82]. It is a synthetic small molecule that was derived from work investigating the short, proline-rich bradykinin potentiating peptides isolated from the venom of the Brazilian or arrowhead pit viper *Bothrops jararaca*. Bradykinin potentiating peptides act by blocking the production of angiontensin II by inhibiting somatic ACE and are important in blood circulation physiology; their effects originally discovered from studying patients envenomed by *Bothrops jararaca*. Investigative work has examined the possibility of using these peptides as direct therapeutics to control blood pressure but they were not developed due to poor oral bioavailability. Instead, the key features of the peptides were used to screen small organic molecule libraries leading to the identification of 1-((2s)-3-mercapto-2-methylpropionyl)-1-proline (Captopril) Fig. 3. The significantly increased potency of this molecule attributed to the orientation of the molecule's proline residue within the substrate-binding pocket of ACE together with the thiol moiety coordinating with the active site Zn. Improved versions of the original drug have been developed, including Ramipril which is currently prescribed to treat hypertension [83].

2.2 Angina

A number of venom-derived drugs have been approved to treat unstable angina, in patients that are at high risk of death from myocardial infarction. Angina drugs Aggrestat® (tirofiban) and Integrilin® (eptifibatide) [18] are disintegrins that block the aggregation of platelets by binding to the aIIbb3 receptor thus preventing fibrinogen binding. These were developed from the small disulphide-rich venom proteins echistatin, isolated from *Echis carinatus* (the saw-scale viper), and barbourin, from *Sistrusus miliarius barbouri* (the south-eastern pygmy rattlesnake) (Fig. 3). These mimic the short amino acid sequence Arg-Gly-Asp (RGD) and Lys-Gly-Asp (KGD) respectively, that are key to binding to the receptor, and it has been found that the substitution of the lysine for arginine in the barbourin pharmacophore increases specificity for the aIIbb3 receptor [83]. The RGD motif within echistatin is located away from the disulphide-constrained core of the protein in a loop region. The development of tirofiban from echistatin involved altering the spacing of the amino acid RGD pharmacophore side chains within the venom pharmacophore. In contrast eptifibatide was developed from barbourin as a small cyclic disulphide-constrained peptide containing the barbourin KGD motif.

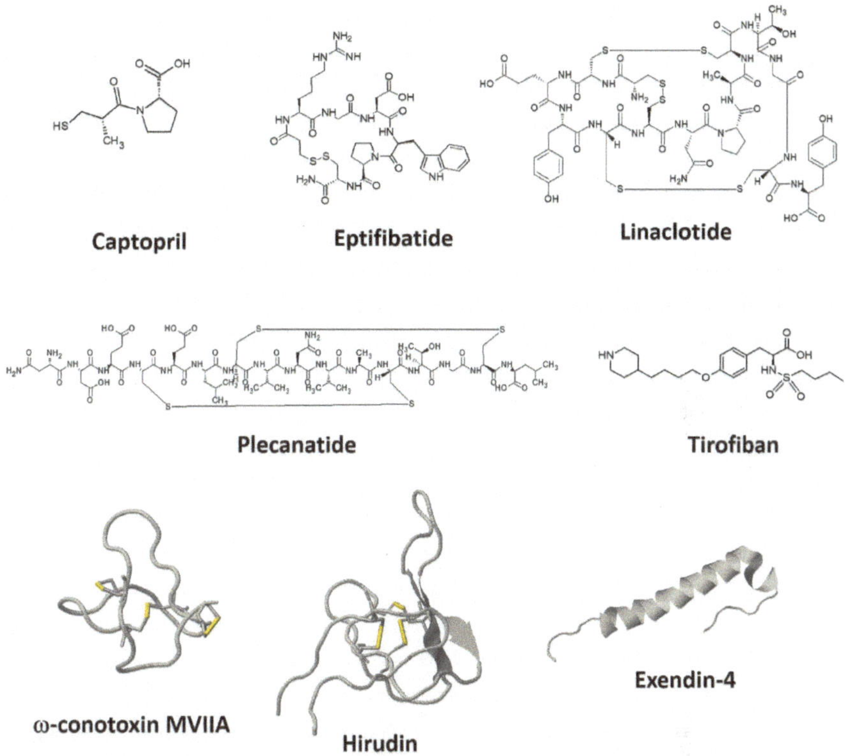

Fig. 3 Chemical and protein structures of toxin-derived compounds either used as clinically approved drugs or used as the basis of clinically approved drugs. Captopril was derived from a nonapeptide isolated from the venom of *Bothrops jararaca*. It is an angiotensin converting enzyme (ACE) inhibitor used to regulate blood pressure. Eptifibatide was derived from barbourin, a disintegrin protein isolated from *Sistrurus m. barbouri*. Tirofiban is based on echistatin disintegrin isolated from *Echis carinatus* and a potent inhibitor of platelet aggregation. ω-Conotoxin MVIIA isolated from *Conus magnus* is used as the basis for the synthetic peptide Ziconotide, a N-type calcium channel blocker. Recombinantly produced hirudin (lepirudin) is a potent thrombin inhibitor. Synthetically produced exanatide (exendin-4) is a glucagon-like peptide-1 receptor agonist and is used to treat diabetes mellitus type 2.

2.3 Anticoagulants

Viprinex™ (ancrod) is derived from the Malayan pit viper (*Calloselasma rhodostoma*) and was developed as an anticoagulant and for use in the treatment of acute ischaemic stroke [84,85]. It is not currently approved for clinical use in any country although its effects on stroke have been studied in clinical trials [86,87].

Exanta™ (ximelagatran) from cobra venom was a promising anticoagulant peptide [88] developed by Astra Zeneca to try to replace warfarin as an anticoagulant. The problem with warfarin is that it carries a high risk of haemorrhage at therapeutic concentrations and its therapeutic window is narrow. Exanta™ was hailed as a wonderful new drug and was already being used in a number of different countries when unfortunately it had to be withdrawn in 2006 because of serious liver injury displayed by raised levels of liver enzymes. However development of this drug demonstrated that a specific oral thrombin inhibitor was a possibility and led to the development of non-vitamin K antagonist oral anticoagulants (NOACs) [89].

Defibrase® has been developed from the enzyme component batroxobin found in toxin from another Brazilian Lancehead snake (*Bothrops moojeni*) which acts as a defibrinogenating agent and is used to treat thrombosis [90]. It is registered for use in China and Japan. Reptilase® is the trade name for a thrombin-like serine protease that has been similarly developed but from the Common lancehead (*Bothrops atrox*). This work was first published in 1958 [91] and reptilase® was later used in patients to breakdown fibrin clots to restore blood flow [92] and reduce bleeding post cardiac surgery [92]. It is also used in a blood test used in the UK's National Heath Service to investigate normal clotting time even in the presence of heparin [93]. Hemocoagulase (Botroclot®), used in the prevention of bleeding and treatment of surgical bleeding (haemorrhage), is derived from *Botrops jararaca* [94].

Alfimeprase, a fibrolase, is a fibrinolytic enzyme isolated from southern copperhead snake (*Agkistrodon contortrix contortrix)* venom which has very effective thrombolytic activity. It looked very promising in vivo but this enzyme did not get past Phase III trials as needed endpoints of the trial were not met [95].

Refludan® (lepirudin) and Hirulog® (bivalirudin) are anticoagulant drugs for patients who have developed a reaction to other anticoagulants. Their structures are based on hirudin (Fig. 3), a direct thrombin inhibitor (DTI), isolated from the saliva of leech *Hirudo medicinalis* [96]. Refludan® was discontinued in 2012, not due to safety concerns but because of discontinuation of manufacturing [97]. Bivalirudin was developed as a synthetic peptide and was approved for use in 2000 as an anticoagulant in patients with unstable angina undergoing percutaneous transluminal coronary angioplasty, and in patients at risk of heparin-induced thrombocytopenia (HIT) undergoing percutaneous coronary intervention (PCI) [79].

2.4 Analgesics

Chronic pain is another area in which successful drugs have been derived from venoms. Prialt® (ziconotide) has been developed from a peptide from the magical cone snail (*Conus magus*) venom ω-conotoxin MVIIA [98] (Fig. 3) and is a blocker of voltage gated calcium channel $Ca_V2.2$. Leconotide, an even more selective drug based on the same toxin, has shown superior efficacy in rat models of diabetic neuropathic pain compared to Ziconotide [99]. However this has been misrepresented in a review by Pennington et al. suggesting Leconotide displayed side effects in clinical trials and was not developed further [79].

Byetta® (Exanatide) was developed to treat type 2 diabetes from a peptide called exendin-4 (Fig. 3) from *Heloderma suspectum* (Gila Monster) (Fig. 1J). This peptide mimics glucagon-like peptide 1 (GLP-1) by stimulating the glucagon-like peptide 1 receptor [100]. Later it was found that there is an increased risk of having pancreatitis with this drug. Further developments in this area have led to improved drugs being produced and these are Bydureon® (exenatide extended-release) and Lyxumia® (Lixisenatide) although Bydureon® may increase the risk of developing thyroid cancer.

3. Composition of venoms: An overview of evolutionary complexity

Unlike synthetic compound libraries the nomenclature of biological compounds from venoms relies on taxonomic descriptions of the species. This work has focused on the protein components but the growing adoption of rational nomenclature pushes the focus towards functional activity [22,101–103]. Universal adoption of this nomenclature with its Greek prefix denoting activity will greatly aid in the uptake of venom proteins in drug discovery. Further work is on going on reinforcing uptake of rational nomenclature to describe diverse venomous groups; this will further open up the utility of venom compounds to drug discovery scientists not familiar with the animals and systems the toxins come from.

3.1 Proteinaceous venom components

3.1.1 Enzymes

Large enzymes predominate in viper venoms (Serpentes, Viperidae) [104] and on the face of it are far from drug like molecules; however some have made it as successful drugs. Reptilase®, mentioned above, is an example of a thrombin like serine protease from the lance head viper (*Bothrops* spp.). Snake venom serine proteases act as pro-coagulant through conversion of soluble fibrinogen to insoluble fibrin [105], but also act to breakdown clots through continued catalysis of fibrin into fibrinopeptides which are uncoagulable resulting in afibrinogenemia [106]. Those deposited in UniProt have a molecular mass of 27.7 (±2.5) kDa. Serine proteases are common in viper venoms [104–106]. They are also commonly reported in Cnidarian venoms [69,107–110], which is a clear case of convergent evolution as there is no venomous ancestor between Cnidaria and Serpentes taxa. Other marine venomous species containing serine proteases include Stingrays (Elasmobranch) [111], and the Squid family (Coleoidae) [112,113]. Hemodynamic serine proteases are also reported in wasps and ants (Hymenoptera) [113–115], theraphosids (Theraphosidae) [116], true spiders (Araneidae) [117] and scorpions (Buthidae) [118] but have not been used therapeutically. Evolution by natural selection is the famous blind watch maker and thus mutations in some serine proteases have also led to them inhibiting ion channels [119] a feature also common in phospholipase A2 enzymes [120,121].

Phospholipase enzymes are almost ubiquitous in venoms and have been described from a wide diversity of taxa, including; snakes [121], true spiders [122], scorpions [123], jellyfish [109], fish [124], snails [125], bees and wasps [126]. They have only been described from a couple of Theraphosidae genera *Pamphobeteus* [127] and *Lasiodora* [128,129]. However, Phospholipase A2 activity is present in a wider diversity of species than currently published or entered in databases and this makes them ideal for checking the activity of diverse libraries of venoms (Fig. 4) and venom fractions. Phospholipase A2 activity is highest (per μg venom protein) in snakes but also many scorpion and theraphosid species (Fig. 4). Phospholipase enzymes such as PLA2 act as spreading factors [129], but are also cytotoxic, especially PLD [130], and are responsible for neuro- and hemo-toxic effects [121]. The catalytic site of action of PLA2 enzymes is at the *sn*-2 position of sphingolipids such as those found in cell membranes.

Hyaluronidase is another enzymes frequently recruited into venom gland expression and is found in such diverse species as cobras [131], spiders [130], centipedes [132] and even the duck-billed platypus [133], which have all evolved venom independently.

Phosphodiesterase (PDE) enzymes as drug targets gained popularity in 1996 with the publication of a potent and selective inhibitor of PDE5, sildenafil (Viagra™), for male erectile dysfunction [134]. These enzymes were first purified from snake venom in 1963 [135] and have since been used for

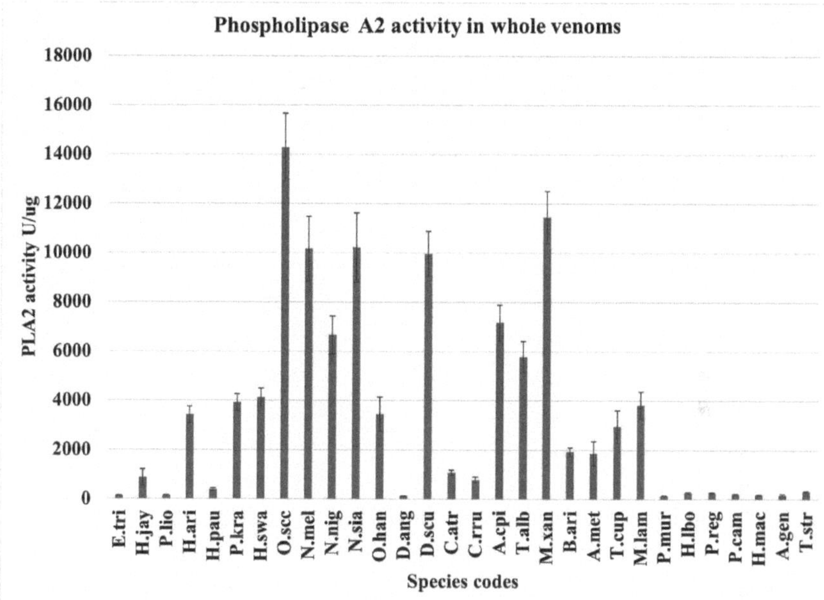

Fig. 4 Phospholipase A2 activity measured with the EnzCheck Phospholipase A2 activity assay kit (Invitrogen) using a whole venom diversity set (Venomtech Ltd). PLA2 activity is greatest in snake venoms, specifically Elapidae > Viperidae, then Scorpion > theraphosidae. Species codes Centipede- E. tri (*Ethmostigmus trigonopodus*). Scorpions—H. jay (*Hottentotta jayakari),* P. lio (*Parabuthus liosoma*), H. ari (*Hadrurus arizonensis*), H. pau (*Hadogenes paucidens*), P. kra (*Protoiurus Kraepelini*), H. swa (*Heterometrus swammerdami*). Elapidae; O. scc (*Oxyuranus scutellatus scutellatus*), N. mel (*Naja melanoleuca*), N. nig (*Naja nigricollis*), N. sia (*Naja siamensis*), O. han (*Ophiophagus Hannah*), D. ang (*Dendroaspis angusticeps*). Viperidae; D. scu (*Deinagkistrodon acutus*), C. atr (*Crotalus atrox*), C. rru (*Crotalus ruber*), A. cpi (*Akistrodon contortrix pictigaster*), T. alb (*Trimeresurus albolabris*), M. xan (*Montivipera xanthina*), B. ari (*Bitis arietans*). Theraphosidae; A. met (*Avicularia metallica*), T. cup (*Tapinauchenius cupreus*), M. lam (*Monocentropus lambertoni*), P. mur (*Pterinochilus murinus*), H. lbo (*Cyriopagopus albostriatum),* P. reg (*Poecilotheria regalis*), P. cam (*Psalmopoeus cambridgei*), H. mac (*Heteroscodra maculata*), A. gen (*Acanthoscurria geniculate*), T. str (*Theraphosa stirmi*).

identification of sildenafil analogues [136]. Priapism has been noted in patients envenomed by Brazilian wandering spiders (*Phoneutria* spp.) [137] and scorpions [138], however these effects are independent of PDE enzymes. More recently the first ever case of priapism from snake venom was reported from an envenomation case involving a juvenile Russel's Viper (*Daboia russelii*) [139]. The mechanism is currently unknown but could be through PDE inhibition. Phosphodiesterases have also been discovered in scorpion venoms [123] and stonefish venom [140].

ADPase, ATPase and 5′-nucelotidase activities are commonly found in viper venoms containing phosphodiesterase components [141]. These enzymes are considered one of the minor protein families in venoms [22], but at this moment are also poorly studied. DNase and 5′-nucleotidase enzymes are common in viper venoms [141] and stone fish venom [140] where they act much like other enzymes of this class degrading polynucleotide chains and free nucleosides. However, these venom proteins have had a critical role in phosphodiesterase drug discovery as the PDE assays rely on 5′-nucleotidase from (*Crotalus atrox*) Western Diamondback rattlesnake venom to function. These enzymes appear to be restricted to snake venoms, however. There is one report of nuclease activity in lizard venom [142] and one in the assassin fly [143], however both of these are transcriptomic and proteomic studies rather than confirmation of functional activity. The most interesting nuclease activity is from the crown of thorns starfish (*Acanthaster planci*) venom as it is cell and nuclear penetrant where it causes hepatocyte apoptosis through DNA degradation [77]. Phosphatases, both alkaline and acid, are also known from snake venoms.

Metalloproteases are a major component of viper venoms and are particularly common in rattlesnake venoms where they compose up to 43% of the venom and are also responsible for major muscle damage after envenomation [144]. These enzymes are also inhibited by cation chelation and non-selective metalloprotease inhibitors which are being investigated as potential snakebite envenomation treatments [144,145]. Previously thought only to be present in viper venoms, these enzymes have more recently been discovered in recluse spiders (*Loxosceles* spp.) [146], bark scorpions (*Tityus* spp.) [118], jellyfish [69] and even centipedes [147]. Many venom metalloprotease transcripts are translated with their associated disintegrin domains, but these are not as well represented in the venom proteomes [22].

L-Amino Acid Oxidase enzymes (LAAO) are a major component of snake venoms and are particularly abundant in pit vipers, including rattlesnakes (Crotalinae) and contribute to the yellow colouration of such venoms as a result of the isoalloxazine ring system contained within the flavin cofactor [148]. Though thought to be restricted to snake venoms, there is a recent report of an LAAO found within a scorpion (*Tityus serrulatus*) venom [123] and, if the yellow colouration is indicative of the presence of LAAO, then it is likely that these enzymes will be discovered in the venoms of the Indian tiger or ornamental spiders (Genus *Poecilotheria*) (Fig. 1D). Although there are numerous reports of the antimicrobial action of snake venom LAAO, it is highly unlikely that these will yield any useful compounds in the fight against infection due to the probable mode of action being release of H_2O_2 [149]. The presence of LAAO activity in venom suggests that there could be selection pressure either to inhibit this enzyme in the venom gland, or for the presence of D-amino acids that are resistant to it. Peptide isomerases have so far only been reported from a few species including duck-billed platypus [150] and funnel web spiders [151]. D-amino acids are well known in cone snail venoms [152] and reported from funnel web spiders [153]. It is therefore possible that isomerisation is underreported in venoms and could potentially account for differences in synthetic versions of natural peptides.

3.1.2 Cysteine-rich secreted proteins

Cysteine rich secretory proteins (CRiSP) are predominantly found in snake venoms and some arthropods, but also in the venomous saliva of monitor lizards [154]. In mammals these proteins play a major role in the mammalian reproductive system, such that the venom versions may have utility in reproductive medicine. These 20–30 kDa proteins consist of over 200 amino acids, with 16 cysteines, of which 10 are present in the C-terminal producing the characteristic cysteine-rich motif [155]. CRiSPs are highly conserved throughout evolution with proteins deposited in UniProt from tomatoes to turkey vultures and drosophila to dogs, so it is no surprise that they are also present in a wide range of unrelated venom systems. Of these, they have been predominantly studied in non-avian reptiles [155]. Snake venom CRiSPs are some of the largest ion channel blocking proteins and act on cyclic nucleotide gated ion channels [156] and L-type calcium channels [155]. L-type calcium channel block results in blockade of smooth muscle contraction. The diversity of these proteins in the natural world and their subsequent neofunctionalisation in venom systems presents a great opportunity to the drug discovery scientist to find structurally improved stability in a molecule by studying the venom orthologue of the CRiSP of interest.

3.1.3 Other large proteins

Cobra venom factor (CVF) is one of the largest molecules present in venoms at approximately 180 kDa, depending on species. It is a structural and functional mimic of mammalian complement C3 but with a significantly longer half-life [157]. Through persistent activation of the complement system the net result is complement depletion and thus it is used as a tool for this purpose in vivo studies such as ischemic reperfusion injury [158].

3.1.4 Membrane attack complexes pore forming (MACPF) proteins

These large proteins (>35 kDa) are part of the membrane attack complex pore forming (MACPF) proteins family and are related to perforins. These genes are present throughout the natural world, being predominantly expressed in the immune system. MACPF proteins bind to cellular membranes and form large pores for cellular destruction [159]; this has clearly made them ideal as weapons and thus they have been repurposed in venom systems. They are a major component of aquatic venoms, in Cnidaria they have evolved into a gene family of venom MACPF proteins called actinoporins which are toxic to fish and crustaceans [72]. Other aquatic venoms including those from fish and cone snails that have also repurposed these toxins [160]. With the high risk of internal damage from such toxins when they are stored in the venom glands, there is a very high likelihood of important inhibitors of the immune system membrane attack complex being present within venoms and these could potentially dramatically improve the lives of patients with overactivity disorders of the immune system.

3.1.5 Cystatins

Cystatins are a group of poorly studied C1 Cysteine protease inhibitors of around 140 amino acids, predominantly found in snake venoms [161] and some others where they have no known toxic activity. However they have potential as anti-metastatic agents for cancer [162].

3.1.6 Peptide toxins

Unlike the large enzymes the smaller protein and peptide components of venoms present the drug discovery scientist with a wealth of biotherapeutic potential. The delineation of peptide from polypeptide and protein has not been formally decided. Typically in venom research, and in this chapter, non-catalytic polypeptide chains of approximately <100 amino acids are referred to as peptides, and those larger as proteins. The largest peptides are the three-finger neurotoxins and natriuretic peptides which have many members around and slightly over the 100 amino acids. However, sizes range down to small cationic and other linear peptides of <10 amino acids.

Inhibitor cysteine knot (ICK) peptides, also called knottins, dominate the venom drug discovery literature as these amino acid chains are stably folded with typically three to six disulphide bonds and inhibit a wide range of ion channels across a wide range of species including mammals and insects. Thus, they have utility in crop protection as insecticides as well as in human and veterinary drug discovery. They are predominantly found in arachnid and cone snail venoms, but such is the taxonomic distance between these that it is unsurprising to discover recruitment in other species, such as robber flies [14]. Although initially discovered as an ion channel inhibiting motif, the knottin family of peptides have a wide range of specific activities including GPCR agonism [163], intracellular stimulation of ryanodine receptors [164] and ion channel activation [165]. The stable confirmation of disulphide bonds allows mutation of the loops for fine tuning pharmacology, both by evolution and human intervention, without significantly altering the tertiary structure. This makes the knottin superfamily probably the most useful class of venom peptides for drug discovery. ICK peptides have also been found to inhibit venom proteases [166] and are expected, to have utility at many other targets.

Another class of enzyme inhibitors from venoms contain the Kunitz fold typical of mammalian trypsin inhibitors. Kunitz peptides are among the most common peptides in snake venoms [22] but are also found in arachnids including the tiny Pseudoscorpions [39,167] whose venom, like those of snakes, is rich in proteases. It is likely that these serine protease inhibitors evolved due to the need to protect the venom glands from internal damage from venom enzymes and later evolved further for action in prey such as clot prolongation from plasmin or kallikrein inhibition [168].

To date no phosphodiesterase inhibitors found in venoms have been published, but their presence is postulated due to the prevalence of PDEs in venoms.

Other enzyme inhibitors include the well-characterised bradykinin potentiating peptides from the lancehead vipers (*Bothrops* spp.) which exert their hypotensive effect though inhibition of angiotensin converting enzyme (ACE). Discovery of this action inspired the ACE inhibitor class of small molecule drugs on the market today. These peptides have been discovered in diverse classes of pit vipers and some scorpions. They are very short (5–13 amino acids) and possess a post-translationally modified pyroglutamic acid. The peptides have already shown great utility in drug discovery but the smallest, at five amino acids and 611.7 Da, could be useful in its own right. There is also a potential for cyclisation through the N-terminal cleavage of the pyroglutamic acid [169].

Bothrops natriuretic peptides are located on the bradykinin potentiating peptide transcripts and post-translationally cleaved into the active forms along with the many isoforms of BPPs [170]. However this does not appear to be the case for the elapid, lizard and scorpion venom natriuretic peptides, but could be the case for those from the duck-billed platypus [171].

Three-finger toxins (3FTX) are almost exclusively restricted to elapid snakes, where they make up a large percentage of the venom proteome and are found across all the Elapid taxa. There are a few exceptions found in other snakes such as vipers and rear-fanged colubrids. Three disulphide bonds hold these peptides in a stable structure resembling a thumb and first two fingers of a human hand when held apart. Three-finger toxins are responsible for the majority of the neurotoxic and cardiotoxic actions of these snakes and the cytotoxic actions of the spitting cobras. These incredibly diverse toxins represent another class of venom peptides with great drug potential due to their extreme pH and thermal stability, coupled with selective target engagement such as beta adrenergic receptor antagonism [171], muscarinic receptor antagonism [172] and vasopressin antagonism [173].

Snaclecs are a class of C-type lectins found specifically in snake venoms where they modulate prey and predator haemodynamics, but they are also very useful as drug discovery tools [174] and have made it to licenced drugs [22]. Lectins are also found in Cnidarian venoms and a few other unrelated species where they act in a similar way.

Cationic peptides are commonly found within venom systems, both short and long forms, however, the long forms are still only about 40 amino acids. These short non-disulphide linked peptides exist structurally as cationic helices that interact with membrane lipids, some form pores causing lysis which can be specific for bacterial cells [175]. Others, like crotamine, are cell penetrant and are being investigated as potential cancer therapeutics [176]. These alpha-helical cationic peptides are very common in scorpion venoms where there main target appears to be intracellular calcium signalling through ryanodine receptors but their pharmacology can be engineered out making them useful intracellular shuttle vectors [177]. Probably the most well characterised alpha helical peptide is melittin from honeybees (*Apis mellifera*) but similar peptides are found in many hymenoptera where they have similar antimicrobial, cell penetrant properties [178,179] and antiviral actions [180].

3.1.7 Growth factors

Nerve growth factor (NGF) and other growth factors have been discovered to be components of venoms. Snake venom NGF is made up of a homodimeric complex of beta-subunits. The activity of NGF was first described using snake venom from the moccasin snake *Agkistrodon piscivorus*, the first source used to isolate NGF [181] and was later found in Viperidae, Crotalidae and Elapidae family venoms. The structures of these vary between species of snake. Vascular endothelial growth factor (VEGF) is key in the formation of blood vessels but has also been found in snake venom. It varies in structure and function in different snake species and in particular around the C-terminal putative coreceptor-binding and receptor-binding loop regions of the proteins. The puff adder *Bitis arietans* has venom type VEGFs unique from other VEGFs [182]. Epidermal Growth Factor (EGF) repeat-containing transcripts were discovered in the rear fanged snake *Thamnodynastes strigatus* [183]. In the venom of the giant red bull ant, *Myrmecia gulosa*, a peptide has been discovered which shares the same six cysteine containing structural domain as EGF. This peptide has been named MIITX2-Mg1a [184]. EGF peptide domains have also been found in sea anemones including *Stichodactyla gigantean* [185]. Some of these venom proteins are likely to be responsible for some of the effects seen on tyrosine kinase receptors [186].

3.1.8 Nonproteins

Tetrodotoxin is probably the most well known of the non-protein venom components, although its presence in pufferfish (for which its most associated) is not associated with a venom system, it is a poison as the aggressor has to do the biting. Tetrodotoxin (TTX), palytoxin and saxitoxin are produced by zoanthids, dinoflagellates and cyanobacteria in aquatic environments and thus when found within venom systems the host must have acquired mutations to protect itself as in the blue ringed octopus [75] (Fig. 1A). It is extremely useful in neuroscience research as it only binds to a select group of voltage gated sodium channels, sparing the pain relevant $Na_V1.8$ and $Na_V1.9$ along with the cardiac $Na_V1.5$ channel [75]. Thus, it is used frequently at the end of electrophysiology studies to confirm whether the channels being studied are TTX sensitive or resistant.

Scorpion venoms are unique in potentially containing mucopolysaccharides and mucoproteins [187] in their venom and also have an unusual prevenom which is rich in potassium ions [188]. Neither of these are presently considered of interest to the drug discovery world, and the high viscosity of scorpion venom is a constant challenge when collecting and purifying the components from these venoms.

Many venoms, especially those from true spiders (Araneomorphae) and insects contain a host of small non-peptide components such as acylpolyamines, free amino acids and biogenic amines which, until recently, have been difficult to study. A team from the University of Zurich, Switzerland, has produced a database of such toxins from spiders and a rational nomenclature to make the research of these toxins much easier [189]. The VenoMS database (www.venoms.ch) contains a host of identified small molecules from spider and wasp venoms that need to be taken into account when screening. Glutamate is also likely to be added to the list soon although not currently in the database. Snake venoms, and probably many other venoms, also contain polyamines and other metabolites but at this stage it is not clear if these are produced by the venomous animal or the newly discovered venom microbiome [190,191]. The other class of small molecules known from venoms are the alkaloids, principally restricted to ant venoms [192]. However, there are many erroneous references to toad and millipede venom, where in fact these are sequestered toxins acting as a poison as they lack the delivery system characteristic of a venom.

4. Drug like structures within venoms

Determination of the three dimensional structure of a venom component provides an insight into the molecule's interaction with its target, allowing the researcher to identify critical parts of the molecule essential for activity and permitting the development of the compound to generate leads for potential therapeutics. Knowledge of the structure identifies moieties that can be modified to increase potency, selectivity or to introduce modifications to improve desirable drug-like properties, such as increased stability, solubility etc., just as in those studies performed with synthetic small molecules.

4.1 Structures found in venom components

Recent work that utilised a combined proteomic/transcriptomic approach to characterise venom from 191 different animal species has revealed over 20,000 venom toxin sequences [162]. From these some 3600 physical peptides were recombinantly produced for further characterisation, with a remarkable 8% found to possess activity against the human melanocortin receptor (hMCR4). Half of these peptides contain at least three disulphide bonds, with 23% having four. From a size perspective 28% of the peptides are around 31–40 amino acids in length, with 15% being 41–50 amino acids and 20% of peptides are between 10 and 30 amino acids in length. Of the possible post-translation modifications (PTMs), 38% of peptides are hydroxylated and 34% amidated. Other PTMs include pyroglutamic acid addition (in 21% of cases) or show addition of 4-carboxyglutamic acid [163].

Peptide toxins can be broadly separated into two classes based upon their structure: disulphide-rich peptides and linear peptides. Disulphide-rich peptides have significant structural and functional variation while linear, non-disulphide bridged peptides tend to be helical in structure.

4.1.1 Disulphide-rich peptides

Within the disulphide-rich peptides class is the inhibitor cysteine knot (ICK) [193]. The ICK is a structural motif characterised by three disulphide bonds and arranged such that two of the disulphide bonds form a ring through which the third threads (Fig. 5), while the main secondary structural element consists of a beta hairpin loop near the C-terminus. The cysteine knot is a very stable arrangement, and this structure is prevalent in many peptides from venomous organisms from all kingdoms. Further structural variations of this basic motif are also known that possess additional disulphide bonds, presumably to increase stability further.

A representative ICK peptide is the ProTX-III peptide from the green-velvet theraphosid *Thrixopelma pruriens*, which has been found to inhibit sodium channels and has potential for use in treating chronic pain. Structure/function relationships of this peptide reveal that the C-terminal region is responsible for interacting with the human $Na_V1.7$ sodium channel without affecting voltage sensing, and that possibly side chain orientations

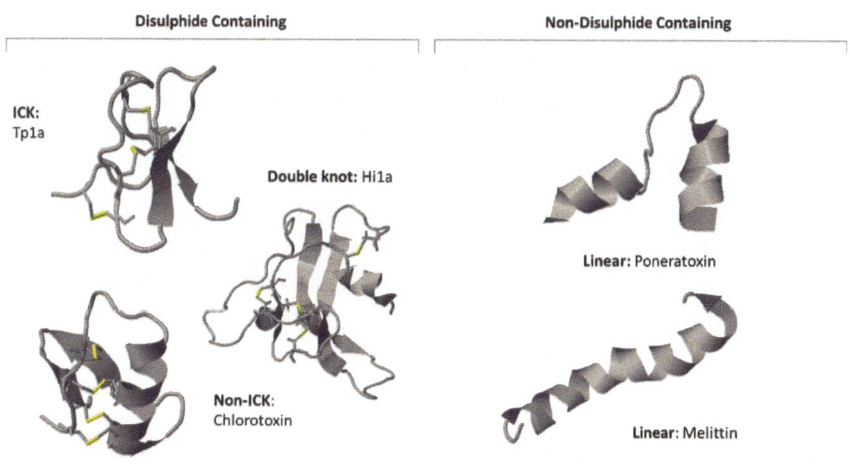

Fig. 5 Basic structures of disulphide bond containing and non-disulphide bond containing (linear) venom peptides. ICK inhibitor cysteine knot ProTx Tp1a from *Thrixopelma pruriens*; double knot peptide Hi1a from *Hadronyche infensa*; non-ICK disulphide bonded peptide chlorotoxin from *Leiurus quinquestriatus*; linear toxin poneratoxin from *Paraponera clavata*; linear toxin melittin from *Apis mellifera*.

within this peptide are responsible for the distinct interactions and mechanism of action when compared to ICK peptides from other spiders [193,194]. It should be referred to as μ-theraphotoxin-Tp1a using the rational nomenclature system [102].

Double knot peptides, where there are two independently-folded ICK motifs, were predicted to be found from amino acid sequences isolated from the Chinese bird spider or earth tiger tarantula, *Ornithoctonus huwena* [195] and were subsequently isolated from the funnel web spider *Hadronyche infensa*, (Hi1a peptide, Fig. 5) [196]. The Hi1a peptide has been found to target the acid-sensing ion channel 1a (ASIC1a) and has potential for development as a therapy for ischemic brain injury. Structure/activity studies revealed that both ICK domains need to be present for full potency of the peptide revealing the importance of the ICK domains for activity. Other tools for ASIC channels come from unrelated venoms including sea anemones, black mambas and another the Trinidad chevron spider (*Psalmopoeus cambridgei*) [28].

A derivative of the ICK motif has also been identified in the form of a disulphide-directed beta hairpin (DDH) and consists of a two disulphide bonds (connectivity 1–4 and 2–3) which may represent either an evolutionary precursor to the ICK motif or have arisen via a convergent evolutionary process [35,197]. A further structure identified within the ISTX-1 sodium channel inhibitor from the tick *Ixodes scapularis* [198] has the disulphide arrangement of the ICK motif but a spatial arrangement distinct from ICK or DDH.

There are many other disulphide-containing peptides present in arthropod venoms that are distinct from the ICK motif and its derivatives. A structurally diverse disulphide-containing peptide is found in chlorotoxin from scorpion venom. It possesses a complex CS alpha/beta structural motif which has been found to vary in the number of disulphide bonds and their respective connectivity's (Fig. 5). This toxin binds to chloride channels in the brain.

4.1.2 Three-finger toxins

Elapid snake venoms are rich in three-finger toxins (3FTXs) which range from 58 to 81 amino acid residues, and are folded into a three-finger structure stabilised by disulphide bridges. These non-enzymatic peptides bind post-synaptically at neuromuscular junctions to induce paralysis in victims [199]. This diverse group of peptides differ in length and the number of disulphide bridges and can be present as monomers, covalent- or non-covalent homo and heterodimers [198,200].

4.1.3 Linear peptides

Linear, non-disulphide bridged poly-cationic peptides predominantly occur in the venoms of hymenoptera (ants, bees and wasps) [201]. Poneratoxin from *Paraponera clavata* is a linear 25 amino acid peptide with an apolar *N*-terminal and a polar charged C-terminal helix connected by a beta turn resulting in a tertiary bent-conformation (Fig. 6). In frog skeletal muscle, application of this peptide causes paralysis and appears to modulate sodium channel activity [202]. Melittin from the venom of the honey bee (*Aris mellifera*) is one of the most characterised examples of these linear peptides. Melittin itself is a 26 amino acid peptide with a C-terminal amidation [203]. It forms an alpha helix that contains a hydrophobic N-terminal region, a central region with both hydrophobic and hydrophilic faces and a C-terminal region that is hydrophilic (Fig. 5). The peptide can bind to three other peptide units to form an antiparallel tetramer in a bent conformation [178].

4.1.4 Large protein structures

In contrast to venoms from arthropods, snake venoms are more complex, exhibiting a broader range of pharmacological effects. They consist of a diverse array of larger proteins and peptides than the predominantly disulphide bonded peptides found in spiders, scorpions and cone snails [202,203]. Snake venom metalloproteinases (SVMP), snake venom serine proteases, phospholipases A2 (PLA2s) and three-finger peptides (3FTX) represent the major protein families present in snake venom. Additional protein components include cysteine-rich secretory proteins, L-amino acid oxidases, Kunitz peptides, C-types lectins, disintegrins and natriuretic peptides [22,204,205].

Snake venom metalloproteinases are a major component of viper venoms. These zinc-dependent proteases range from 20 to 110 kDa in size and can be differentiated according to structure into three different classes; P-I, P-II and P-III. P-III proteins consist of a metalloproteinase and disintegrin-like domain together with cysteine-rich domains; P-IIs contain a metalloprotein and disintegrin domain, while P-Is contain only a metalloproteinase domain. The pharmacological effects of these multifunctional proteins are in influencing various steps of the blood-clotting cascade causing systemic haemorrhage and incoagulable blood in prey/victims. The recombinant P-I metalloproteinase fibrolase with a modified N terminus was called

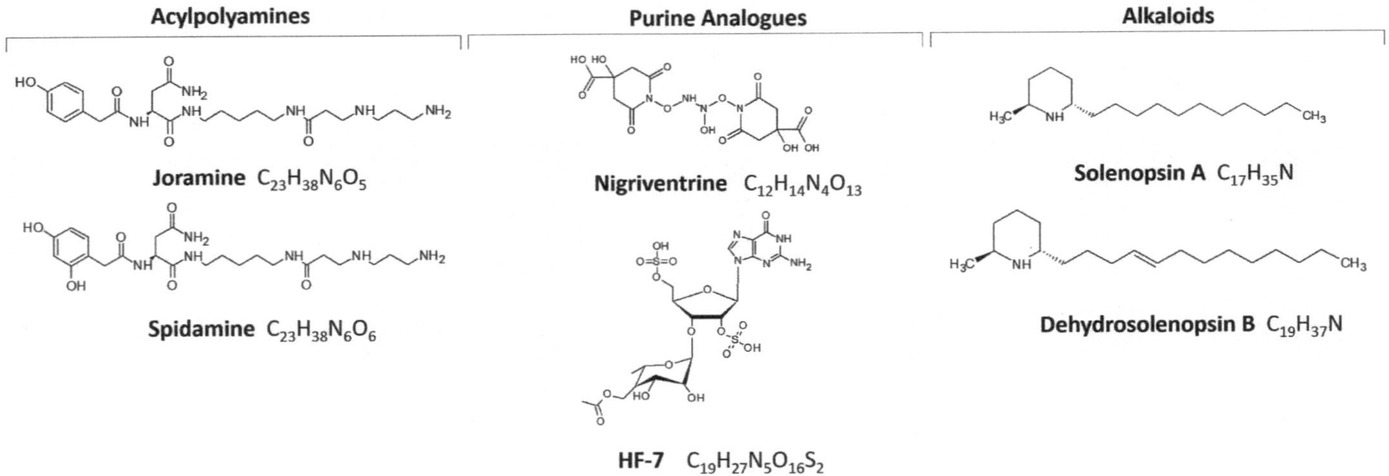

Fig. 6 Examples of small molecule components of venoms. Acylpolyamines joramine and spidamine from *Nephila clavata*; nigriventrine from *Phoneutria nigriventer*; HF-7 from *Hololena curta*; solenopsin A from *Solenopsis saevissima*; dehydrosolenopsin B from *Solenopsis invicta*.

Alfimeprase (Amgen) passed phase II for clearing peripheral artery occlusions, but failed to reach endpoints in phase III [95].

Snake venom serine proteases (SVSPs) range from 26 to 67kDa and possess two distinct structural domains. SVSPs act by cleaving polypeptide chains after positively charged or hydrophobic amino acid residues resulting in protein degradation. Within the prey or victim, SVSPs cause a variety of effects, although the molecular mechanisms remain unclear. These include rupturing of capillary blood vessels, inducing edema and hyperalgesia, and hemotoxic effects. The thrombin-like serine protease batroxobin discussed above converts fibrinogen into fibrin through the cleavage and release of fibrinopeptide. Ancrod is also a serine protease that acts similarly to batroxobin in cleaving fibrinogen.

Phospholipase A2 cleaves the fatty acid tail from the glycerol backbone of phospholipids releasing the cleaved fatty acid as arachidonic acid. Subsequent downstream modification of the arachidonic acid produces eicosinoids which in turn mediate the inflammation response. Venom PLA2s range from 13 to 15kDa and are classified into four major sub-types of which Types 1 and 2 are a major component of venoms [206]. The single chain polypeptides typically contain 7 (sometimes 6 or 8) disulphide bridges and possess more alpha-helical secondary structure (50%) than beta-sheet (10%). The active site also acts as the binding site for the Ca^{2+} ion, an obligatory co-factor for catalytic activity. This basic structure is highly conserved between species but there are variations between isozymes in the number and arrangement of disulphide bonds, addition of a C-terminal extension or in substitutions of particular amino acids within the active site that affect catalytic activity.

4.2 Venom small molecule structures

Small molecule or low molecular mass component toxins are significant components within a number of arthropod venoms. They are active against several biological targets including glutamate receptors, GABA receptors, NMDA receptors and monoamine oxidase [207–209]. They vary structurally but the acylpolyamines dominate (Fig. 6). Acylpolyamines, which are primarily antagonists of ionotropic glutamate receptors, are composed of an aromatic acyl headgroup linked via an amide to a polyamide backbone and ending with either a basic amino acid tail (the amino acid-containing polyamines) or an amine tail (the non-amino acid containing acylpolyamines) [210,211]. In spiders, both types exist [212], with the non-amino acid acylpolyamines being more widely distributed across arthropod species. They consist of an acyl head group attached to the polyamine backbone via a three to five carbon linkage composed of methylene units with hydroxylamines separating them. The amino acid acylpolyamines contain either one or two of the basic residues asparagine, ornithine or ω-*N*-methyllysine between the acyl headgroup and the polyamine backbone, and possibly up to three terminal amino acids. The polyamine backbone typically consists of methylene groups interspersed with amine (NH), methyl amine (NCH_3) or dimethyl $N(CH_3)_2$ amines and can vary between 7 and 43 unit in length [212].

Small structural differences can affect the mode of action of the acylpolyamines and therefore provide possibilities for development. As an example, Chiba et al. purified two acylpolyamines (spidamine and joramine) from the orb weaver spider *Nephila clavata* and determined their structures by NMR spectroscopy [213]. On comparing the two molecules, joramine was found to differ from spidamine by the absence of an hydroxyl group on the C2 carbon of the aromatic headgroup. From an assessment of their activity, joramine was shown to reversibly block lobster neuromuscular synapses whereas spidamine was found to block them irreversibly. Structural studies suggest this hydroxyl group causes a conformation change in the molecule that may be responsible for this difference [212,213].

Other low molecular weight molecules found in venoms with activity and potential for therapeutic development include purine derivatives of adenosine, inosine, guanosine and cytidine. HF-7, isolated from the corner funnel weaver or funnel web spider *Hololena curta* is a guanosine derivative that possesses an acetate ester attached to the C4 position of an alpha-linked guanosine molecule with two sulphate groups attached to the ribose sugar (Fig. 6). HF-7 has been demonstrated to block kainite receptors and L-type calcium channels, albeit weakly [214]. Other molecules include the neuroactive nigriventrine from the wandering spider *Phoneutria nigriventer* which causes convulsions in rats, while alkaloids found in a variety of venoms have been shown to inhibit nitric oxide synthases (NOS). Solenopsins, a heterogenous group of alkaloids isolated from fire ants, have a variety of biological activities, ranging from antibacterial and antibiofilm activity to interfering in neuromuscular transmission, mast cell histamine release, ATPase/sodium pump inhibition, phosphatidylinositol 3-kinase signalling and angiogenesis [215]. Two example solenopsins isolated from the venom of the fire ant *Solenopsis invicta* and the red ant *Solenopsis saevissima* are shown in Fig. 6.

5. Drug like properties of venom components

Due to their extremely tight binding to cellular targets, peptide-based therapeutics can offer high specificity and low toxicity with this high specificity resulting from the large chemical space created by native amino acid side chain variation. By exploring this space with backbone and side chain combinations, peptide chemists can expect to select binders, agonists and antagonists of almost any biologically important target. Development of peptides as therapeutics requires a compromise between peptide length and pharmacologically useful levels of activity. There are numerous variables including size and accessibility of the ligand binding surface, the possible induced fit, ligand stability, and receptor residency time. Native peptide ligands are susceptible to proteolysis which can be reduced by chemically altering the amide bond and side chains [216]. However, even if cleaved, small fragments may still retain biological activity [217]. As many venom systems contain proteases or have evolved to be injected into other animals where proteases exist, they contain many modifications to greatly increase stability [218]. Mamba natriuretic peptides have a naturally longer plasma half-life to that of the mammalian peptides [16]. An additional problem remains that peptide ligands and their fragments are rapidly excreted through the kidneys giving short turnover times.

Many of the toxin-derived drugs currently approved for clinical use have small volumes of distribution and exert their action at the site of administration. A small volume of drug distribution of current toxin-derived peptide drugs also indicates efficient clearance by the liver and kidney or by fairly rapid proteolysis, despite toxin adaptations. Together these clearance processes result in a short half-life value of peptide therapeutics. Depending on the therapy required, this may be advantageous or disadvantageous. For short acting drugs such as anti-clotting agents, rapid clearance may be desirable, but for therapies requiring prolonged exposure to the peptide, repeated administration or continuous drug infusion may be required.

Elimination of the peptide can be avoided by covalent modification using a wide variety of natural and synthetic polymeric conjugates that reduce the ease of proteolytic degradation and renal clearance. These conjugates include polyethylene glycol (PEG), hydroxyethyl starch (HES), human serum albumin (HAS), XTEN (a recombinant polyamide), PAS (a recombinant polypeptide made of proline, alanine and serine), polyglutamic acid and monoclonal antibodies.

Improving the access of therapeutic peptide to their targets across biological barriers can be achieved by adding moieties that are recognised by a cell's passive or active transport mechanisms. It has been found that the incorporation of positively charged amino acids at the termini of peptides improves tissue and cellular access [219,220] although large numbers of positively charged residues may destroy mammalian membranes [175]. An alternative is to conjugate the therapeutic peptide to ligands of cell surface receptors for receptor-mediated uptake. The most frequently used of these are sugar moieties coupled to peptides which can increase tissue penetration, including transport across the blood/brain barrier and/or intestinal absorption while still maintaining activity [221–224].

Typically, covalent modifications reduce potency but this is compensated for by a concomitant extension in half-life. The new covalently modified peptide would become a new chemical entity (NCE) and as such may have different pharmacokinetic, pharmacodynamics and immunogenic properties [225]. If the covalent modification is reversible then the NCE would take the form of a pro-drug, releasing the peptide as the active pharmaceutical ingredient (API). In this instance, the NCE would not need to interact with the target. However, if the covalent modification is irreversible, then the NCE must interact with the target.

The oral bioavailability of peptides remains an obstacle to their development as drugs outside a clinical setting. Peptides are rarely absorbed across the intestinal mucosa and various chemical modification strategies have been employed to try to address this, including the conjugation to passive and active transport enhances as discussed previously. A successful insecticide orally active to insects on crops has been developed from spider venom [13] engineering stability in the hydrogen bond structure of analgesic α-conotoxin Vc1.1 delivers oral bioavailability and target engagement [226]. However, peptides do not necessarily need to be administered orally and can be administered subcutaneously thereby avoiding potentially degradative intestinal mucosal enzymes and first passage through the liver.

Determination of the important pharmacological parameters for peptides including pharmacodynamic and pharmacokinetic properties (adsorption, distribution, metabolism and excretion, ADME) is challenging as they usually exhibit different patterns to those seen for more traditional small molecule pharmaceuticals. Due to profound differences in structure and conformation, toxin-derived peptide drugs vary in their physicochemical properties and subsequently in their pharmacokinetics. This provides a wide substrate for chimeric peptides to maintain target engagement and improve PK.

The structural differences between peptides and peptide-derived drugs have a marked effect on their physicochemical properties. In particular the molecular weight of the proposed peptide drug is important as it directly affects its suitability for oral administration. Depending upon the site of action of the drug this may either be an advantage or disadvantage. For example the relatively high molecular weight peptide drugs linaclotide and plecanatide (Fig. 3) act within the GI lumen and this together with their low water solubility ensures limited systemic absorption after oral administration, contributing to their selectivity and safety [227]. Molecular weight is not an issue if the drug is administered intravenously or locally, e.g., eptifibatide, and indeed could be advantageous in limiting the distribution of the drug to other locations in the body. Some examples of the pharmacokinetic properties of toxin-derived peptide drugs that are clinically approved for use are discussed in the following paragraphs.

As described previously, captopril is a small molecular weight drug that mimics a bradykinin-potentiating peptide from *Bothrops jararaca*. Its chemical structure and physicochemical properties ensure that is soluble in the aqueous environment, however absorbance is reduced across the GI tract (and other biological membranes). This means that around 60–70% of an oral dose of captopril is absorbed with peak blood concentration reached 45–60 min after oral administration although this reduces to 30–40% if co-administered with food [227,228]. Once in body fluids and plasma, captopril's sulfhydryl group binds to albumin and other thiol-containing molecules (such as cysteine and glutathione). A disulphide bond can also form between two adjacent captopril molecules to form a dimer. Together, these compounds act as a reservoir of the pharmacologically active drug and prolong its pharmacological effects. Pharmacokinetic data for captopril and other drugs developed from venom toxins are shown in Table 1. Approximately 30% of captopril binds to plasma proteins in central circulation with a volume of distribution of 0.8 L (corresponding to a 56 L in a 70 kg patient). Captopril is cleared at a rate of 49 L/h primarily through the kidneys where it undergoes active secretion in the tubuli of the nephron. This results in an elimination half-life of approximately 2 h [228].

Eptifibatide, derived from a disintegrin peptide from the pygmy rattlesnake (see previously) is administered as an intravenous bolus followed by infusion with doses adjusted to patient's weight (Table 1). The volume of distribution in patients with coronary disease is 185–260 mL, being somewhat higher than that found in healthy individuals (220–270 mL). Unlike captopril, eptifibatide remains largely unbound within the plasma and body fluids

Table 1 The major pharmacokinetic parameters of some of the clinically approved toxin-derived peptide drugs (in a typical adult 70-kg patient with normal renal function).

Drug	Absolute bioavailability F	Volume of distribtution V or apparent V (V/F) L	f_u%	Clearence CL or apparent CL (CL/F) L/h	$t_{1/2}$ h	T_{max} h
Captopril	60 – 75% (PO)	56	65–70	49	2	0.75–1
Eptifibatide	10 – 89% (variable, PO)	210–350	10	21–29	5–27	1.5–2
Ziconotide	50% (intrathecal)	0.155	–	–	2.9–6.5	–
Lepirudin	100% (IV)	12.2	–	9.8	–	1.3
Linaclotide	0% (PO, local activity in GI – not absorbed systemically)	–	–	–	–	–
Plecanatide	1% (PO, local activity in GI – not absorbed systemically)	–	–	–	–	–

Where Fu, fraction of drug unbound in plasma; $t_{1/2}$, elimination half-life; T_{max}, time taken to achieve maximum plasma concentration post administration; L, litres; h, hours.

Adapted from Stepensky D. Pharmacokinetics of toxin-derived peptide drugs. Toxins (Basel). 2018;10 (11):483.

and has an elimination half-life of 2.5 h. Renal clearance accounts for 50% of the total clearance with eptifibatide where the majority of the drug excreted is in the form of the original drug or its metabolites [229].

Ziconotide derived from the ω-conotoxin MVIIA peptide (Fig. 3) is administered intrathecally within an isotonic solution via a micro infusion device [230]. This special route of administration is not dependent upon patient related factors such as weight, gender or renal function. The volume of distribution within the cerebrospinal fluid is 155 mL similar to the estimate total cerebrospinal fluid volume [231] with clearance within the body via cleavage by protease enzymes. Any ziconotide that leaves the brain and spinal fluid exerts low systemic exposure due to extensive dilution in body fluids. Ziconotide has a terminal half-life of 4.6 h [227].

6. Target engagement

The majority of characterised venom peptides and proteins target interactions with extra cellular or cell surface proteins and lipids, however there are also cell penetrant peptides interacting with intracellular targets such as Ryanodine receptors. Which will be discussed here [232]. Natural selection is a strong force for shaping pharmacologically active proteins and selecting those that optimally engage with the target proteins, as those that do not reach or engage with the target are of no benefit to the host.

6.1 GPCRs

G-protein coupled receptors are a well characterised family of cell surface receptors and represent one of the key target classes for drug discovery and modern pharmaceuticals. Published knowledge of venom components acting at these receptors has, so far, predominantly focused on glucagon like peptide receptor and muscarinic GPCRs. However, there are promising signs for other GPCR targets including adrenoceptors, vasopressin and somatostatin. Binding to lysophosphatidic acid receptor (LPAR6) has also been hypothesised through molecular dynamics studies [233]. Conference posters have also been presented such as those from AstraZeneca in 2017 which also demonstrated hits to orphan receptors GP120, GPR39 and vasoactive intestinal polypeptide receptor 1 (VIPR1) [234]. Although many interactions of venoms with drug discovery targets go beyond what we know about their evolutionary purpose, they are very useful for drug discovery.

Exendin-4 has been extensively mentioned due to the success of Byetta® and this is in a large part due to the peptide binding the glucagon like peptide receptor with a Ki 0.46 nM. Another subnanomolar GPCR ligand is AdTx1 from the green mamba (*Dendroaspis angusticeps*) which generates insurmountable inhibition of α1a adrenoceptor at sub-nanomolar concentrations K_i 0.35 nM. The slow on-rate of $6 \times 10^6\,M^{-1}\,min^{-1}$ is balanced by the long half-life of the receptor–toxin complex $t_{1/2}$ of 3.6 h [235]. The highly selective inhibition of α1a adrenoceptor occurs at an orthosteric binding site [234].

Muscarinic acetylcholine receptors are important GPCR drug targets expressed in neuronal cells (M1 and M4) and smooth muscle (M2 and M3). Several toxins from African and Asian Elapid snakes have been discovered with low nanomolar Ki, good selectivity and slow dissociation

[172]. Muscarinic toxins are another example of why a universally adopted toxin nomenclature is required [103] for snake venoms as for other taxa, with authors using m1 toxin and MT1 to identify separate muscarinic toxins from *Dendroaspis angusticeps* [236,237]. This has led to the UniProtKB entry (Q8QGR0) for muscarinic toxin 7 to list muscarinic toxin 1 as an alternative name. As previously discussed, venom genes are subject to multiple gene duplication events throughout their evolution and this gives rise to multiple, closely related toxins, which is the perfect substrate for structure activity relationship studies like those performed with small molecules. This is due to individual amino acid mutations which can lead to changes in selectivity, potency and/or binding kinetics, thus producing good substrates for SAR analysis. Reactive groups can be elaborated at the amino acid level meaning synthesis of modifications is very straightforward. Site-directed mutagenesis or substitutions in solid-state synthesis can be used to test hypotheses from such SAR assessments. With the muscarinic toxins these studies have identified that key residues at the tip of the middle finger of these three-finger toxins reach down into the orthosteric binding cleft of the receptor [172], whereas those amino acids distal to the tip are responsible for the irreversibility of the binding [236].

Venom from the iconic mamba snakes of Africa also contain GPCR-specific tool molecules such as mambaquaretin-1 which inhibits vasopressin 2 receptor with a K_i of 2.81 nM, but which is inactive on other GPCRs and cardiac ion channels at 1 μM [173].

It is not just non-avian reptile venoms that have useful interaction with drug-relevant GPCRs; the cone snail cysteine knot peptides also possess GPCR activity and a good example of this is Conotoxin τ-CnVA which selectively blocks the Somatostatin receptor SST3 with K_i of 1.5 μM [238]. Another α-conotoxin is a competitive antagonist at LPAR6, blocking binding of oleoyl-L-α-lysophosphatidic acid [233].

6.2 Ion channels

Ion channels represent promising yet challenging drug targets, these challenges are often down to highly conserved regions within each family and thus limited substrate for discovery or design of potent and selective small molecule drugs. This is where the larger interaction surface of venom peptides provides key selectivity opportunities while maintaining drug like potency. Venom peptides bind to ion channel targets in three main locations; pore blocking [239], voltage sensor [240] and ligand binding site

[199]. α-Toxins block inactivation and β-toxins block activation. However CRiSP toxins with their large size are still able to act as pore blockers but interact with around 14 amino acids present in the major extra cellular loops of cyclic nucleotide gated ion channels [156].

6.3 Ionotropic receptors

The group of α-neurotoxins that inhibit postsynaptic nicotinic acetylcholine receptors on smooth muscle were some of the first venom actions to be discovered. Initially discovered from elapid snakes where they represent a dominant toxin class, these neurotoxins are poorly reversible, and sometimes irreversible, thus causing life threatening skeletal muscle paralysis which is a hall mark of envenomation from these species [199]. α-Bungarotoxin binding is now part of a standard assay when studying nicotinic acetylcholine receptors as it is a potent, high affinity (K_d 0.35 nM) and irreversible ligand at α7-nicotinic acetylcholine receptors [241]. Highly charged amino acids lysine and arginine in loop II of these three finger toxins are critical for target engagement [199]. There are also a wealth of nicotinic acetylcholine receptor tools from the cone snails which independently potently bind to different receptors producing a range of useful pharmacological tools [242]. Elapid three-finger neurotoxins are also able to potently inhibit GABA A binding with a K_i of 0.8 nM [243].

7. Therapeutic areas and discovery phase leads from venoms

Discovery of new components in venom that have an effect on a range of disease targets continues and there are molecules from a various species currently being tested in vivo. There is also promising emerging data in recent literature demonstrating the effect of venoms on a wider range of disease targets at the very early stages of drug discovery.

7.1 Cancer

There have been several developments in cancer research using whole and purified components of venom. A number of studies have been undertaken utilising scorpion venom. Iranian scorpion *Hemiscorpius lepturus* whole venom has been shown to inhibit tumour growth displaying an anti-proliferative effect via the Trp53/Bcl2/Casp3 pathway in the colon cancer cell line CT26, while showing low toxicity on a non-cancerous cell line [244]. Whole venoms from the scorpions *Androctonus crassicauda*, *Messobuthus eupeus and Hemiscorpius lepturus* have been used in a CT26 tumour mouse model to show that a strong in vivo anticancer immune response [245]. A protein component of the venom from the Indian black scorpion *Heterometrus bengalensis* (named Bengalin by authors) has been shown to have involvement in the MAPK pathway in human leukaemic U937 cells [246].

Contortrostatin, a snake venom disintegrin from the southern copperhead viper *Agkistrodon contortrix contortrix* (Fig. 1B), has been shown to reduce metastasis in breast carcinomas xenograft models in nude mice and also displayed antiangiogenic activity [247]. This contortrostatin has had the three amino acids from its C-terminal end replaced with part of another disintegrin, echistatin, to produce Vicrostatin and a delivery method for this drug has been established [248]. Vicrostatin can be recombinantly produced in an *Escherichia coli* expression system. In vitro and in vivo studies are ongoing in breast, prostate, ovarian and brain tumours [249]. Disintegrins in snake venom block integrin signalling on platelets and during metastasis leading to their investigation in cancer therapeutic discovery [250].

Venoms and toxins have been shown to have effects on many of the hallmarks of cancer [251]. In breast cancer cell lines, it has been shown that whole venoms from a range of species can affect the phosphorylation of receptor tyrosine kinases which are targets of cancer treatments [186]. These targets include the epidermal growth factor Receptor (EGFR) which is overexpressed or mutated in a number of cancers. Theraphosidae and Elapidae venoms have also been shown to reduce phosphorylation of the Ephrin A6 receptor which is another anti-cancer target. Anti-oncogenic activity has been shown in toad secretions (not venoms) from the following species *Rhinella schneideri* and *Rhinella marina* [252].

Samsum ant (*Pachycondyla sennaarensis*) venom has shown anticancer activities in cancer cell lines through a reduction in insulin growth factor 1 (IGF-1) mediated cell proliferation and reduction in phosphorylation of the kinases AKT and extracellular-signal-regulated kinase (ERK) which are also pathways involved in cancer [253]. This effect has also been shown in in vivo models in rats [254]. Cardiotoxin 3 (CTX-III) from the Taiwan or Chinese cobra *Naja atra* induces apoptosis in the lung cancer cell line A549 by inactivating EGFR, PI3K and JAK/STAT3 kinase pathways [255].

The giant jellyfish *Nemopilema nomurai* venom has been shown to have anticancer properties in the human hepatocellular carcinoma cell line HepG2 and in the human breast adenocarcinoma cell line MDA-MB-231 by targeting the AKT and mTOR kinase signalling pathways [256,257].

The peptide melittin mentioned earlier in this chapter is the main component found in the European honeybee *Apis mellifera* venom and this peptide exhibits anticancer affects in renal, lung, liver, prostate, bladder and leukaemia [179]. As there are issues with the cytotoxicity, degradation and haemolytic activity of melittin much work has been done to develop more targeted therapies with this peptide to produce anticancer agents. This has been undertaken utilising conjugation with hormone receptors, gene therapy or nanoparticle based delivery of melittin in tumours [258].

In the venom of the lesser weaver fish (*Echiichthys vipera)* novel cytotoxic components have been discovered that have an effect on human colon cancer cells, but not yet identified due to limited information about this venom [64,259].

7.1.1 Imaging tumours in surgery

The peptide chlorotoxin from the venom of the Israeli deathstalker scorpion (*Leiurus quinquestriatus*) has been widely studied due to its ability to bind selectively to tumour cells. It is being used as an imaging agent to label gliomas in surgery so they can be safely removed from the brain without removing as much healthy tissue [260]. This new drug diagnostic tool, currently in Phase II clinical trials is called Tozuleristide [78].

7.2 Antimicrobial effects

Antibiotic resistance is a worldwide health problem [261] and to add to that the pharmaceutical development pipelines are lacking new antibiotics. A range of venoms from a number of different species have been shown to have an antimicrobial effect [252,262]. Venom peptides, lycotoxin (from the wolf spider) and cathelicidin (from the king cobra and banded krait) cause strong reversible inhibition of ATP synthase in *Escherichia coli* [263]. Cathelicidin, which is a synthetic peptide derived from snake venom also has also been shown to have wound-repair activity [262]. Ant venom has been shown to have antimicrobial, antiparasitic, anti-inflammatory and antioxidant activity [264]. As well as anti-cancer effects melittin has been shown to have antiviral effects on a range of viruses including a number of studies with human immunodeficiency virus (HIV) [180]. It has also shown antifungal and antiparasitic activity [258]. A modified cationic peptide from the Chinese swimming scorpion (*Lychas mucronatus*) called Mucroporin-M1 has shown effective inhibition of viruses such as measles (EC_{50} 7.15 μg/mL) and SARS CoV (EC_{50} 14.5 μg/mL) [265]. There are also preliminary studies suggesting that venom peptides could block binding of SARS CoV2 to the human ACE2 receptor, although nothing has been published to date with peer review.

Melittin also has anti-bacterial effects [266]. A melittin related anti-bacterial peptide, RV-23 from the frog skin of *Rana draytonii*, which has lower cytotoxic effects than bee melittin, and another related peptide AR-23 which is from the skin of the frog *Rana tagoi* have also been studied. RV-23 was shown to have a higher lytic activity [267].

7.3 Antiparastic effects

Snake venom has been shown to have an antiparasitic effect. The venom from the cobra *Naja sumatrana* showed antiprotozoan activity towards the ciliate *Tetrahymena pyriformis* and the effect of these cytotoxic peptides from the three-finger toxin family caused immobilisation and death in these organisms [268].

Antileishmanial activity of venom from the common krait *Bungarus caeruleus* snake venom has been displayed in a mouse model [269]. Spider venoms have been also shown in mice to have an anti-parasitic effect when these mice were infected with tachyzoites from *Toxoplasma gondii*. Using spider venom from *Ornitoctonus huwena* showed a better survival rate in the mice than the control when treated. This has potential in the treatment of toxoplasmosis and new drugs are urgently needed to the serve side effects of current drugs [270].

7.4 Analgesics

Venoms can have an effect on a range of ion channel targets and act as an analgesic [28]. Spider venom-derived cysteine knot peptides have therapeutic value selectively targeting voltage gated sodium channels which could be used in epilepsy and chronic pain treatments [271]. This is a very active area of venom drug discovery.

7.5 Diabetes

The FDA approved drug Byetta® (exanatide) was mentioned in the earlier section in the treatment of diabetes but since the development of this drug many more promising peptides have been discovered that act as pharmacological tools and therapeutics for this disease. These include platypus glucagon-like peptide-1 (pGLP-1) from the male platypus (*Ornithorhynchus anatinus*) which acts as an GLP-1 receptor agonist, a fast acting insulin analogue from the geography cone snail (*Conus geographus*), a blocker of voltage gated potassium channel $K_V2.1$, hanatoxin, from the Chilean rose tarantula (*Grammostola rosea*) and guangxitoxin-1 from the Chinese earth tiger tarantula (*Chilobrachys guangxiensis*). There is also a peptide called conkunitzin-S1 which is a blocker of voltage gated potassium channel $K_V1.7$ which has been isolated from the striated cone snail (*Conus striaus*), and, from the Eastern Indian red scorpion (*Buthus tamulus*), the peptide Iberiotoxin which is a large-conductance calcium-activated potassium (BK) channel blocker [272].

An Echidna (*Tachyglossus aculeatus*) peptide Echidna GLP-1 (eGLP-1) is also a GLP-1 receptor agonist but this is outside the scope of this article as echidna are not defined as venomous [47].

7.6 Anti-coagulants and procoagulants

As shown by reptilase™ from *Bothrops atrox,* discussed earlier, snake venoms have procoagulant properties but some have anticoagulant properties as well. In a study of 28 different snake venoms from a range of families [273] the following snakes showed procoagulant activity: Russell's viper (*Daboia russellii siamensis*); fer-de-lance pit viper (*Bothrops asper)*; Brazilian lancehead pit viper (*Bothrops moojeni)*; and the Southern Pacific Rattlesnake *Crotalus oreganus helleri*. The Western diamond-backed rattlesnake (*Crotalus atrox*) and the red spitting cobra (*Naja Pallida*) demonstrated anticoagulant activity. Both activities can have useful medical applications.

7.7 Insecticides

Janus-faced atracotoxins (J-ACTXs) neurotoxins that could be developed into new insecticides, are found in the venom of the Australian funnel web spider [35] targeting voltage-gated sodium channels. Sea anemones have also been shown to be a source of insecticidal peptides [72].

7.8 Venoms that modulate ion channels

A range of venom from species including snakes, arachnids [274] and fish hunting molluscs have been shown to modulate voltage-gated calcium channels by peptide toxins [239]. These components could be used as tools to investigate new pathways or as potential new drugs. Venom from *Ornithorhynchus anatinus* (the duck billed platypus) potently induces calcium ion influx in human neuroblastoma cells [133]. A number of sea anemone venoms also have an effect on voltage gated sodium channels targeting receptor site-3 [275]. Centipede venoms are also a source of drug leads as they have an effect on neuronal ion channels and receptors [41].

7.9 Autoimmune disease

A key therapeutic target for a range of autoimmune diseases is the voltage-gated potassium channel $K_V1.3$ [276]. The peptide toxin ShK from the sea anemone (*Stichodactyla helianthus*) blocks this channel potently. It also blocks $K_V1.1$, $K_V1.4$ and $K_V1.6$ weakly [277] but has been synthesised and modified by just one amino acid (K18A mutation) to increase its selectivity for $K_V1.3$ and $K_V1.1$ which makes it a better lead for autoimmune disease targeting [278].

Dalazatide targets potassium channels and is a treatment for psoriasis and other autoimmune diseases. It was recently trialled in humans, but further larger studies are needed [279]. It is a 37-amino acid synthetic peptide which was originally derived from ShK and is a specific inhibitor of potassium channel $K_V1.3$ [280] and blocks the activation of the effector memory T-cells. Other peptides with $K_V1.2$ activity with similar sequences to Shk and BgK from *Bunodosoma granulifera* have been discovered from *Oulactis* sp. sea anemones [281], this produces good substrate for SAR.

8. Strategies for discovering novel actives within venoms

The initial investigations into venoms dealt with the whole mixture as a starting point to discover why they were dangerous to humans and what could be done about it. Cobras were thought to produce alkaloids and so called cobric acid in their venoms as proteins were not thought then to be so toxic. The conclusion that toxic proteins were responsible and not alkaloids was discussed in the Indian Medical Gazette of 1887 [282]. This is likely to be the first mention of proteins being potent and therefore toxic. At the turn of the 20th century the only published research on venoms was concerned with those of medical significance and how to limit their effects on human health. It was only in 1935 when the use of venom in drug discovery was first demonstrated in the scientific literature [283]. This was the use of whole Russell's viper (*Daboia russelii*) venom as procoagulant for haemophiliac patients during tooth extraction [283]. Whole venom is still used in clinical practice, for example the dilute whole Russell's viper venom clotting time (RVVT) is still the mandated test for lupus [284] and the venom was patented in 1992 as a procoagulant in vacutainers for blood collection [285]. So even though whole venoms are complex mixtures with many synergistic and even potentially antagonistic properties they are still used by many in early drug discovery research both in vitro [168] and in vivo [254]. This can yield good results if it is followed up through subsequent purification and identification of the active compounds [173]. Use of whole venoms in the initial screening is also useful in situations where there is no prior evidence for venom activity such as P2X receptors [286] and receptor tyrosine kinases [186]. However there are many drawbacks with the use of whole venom due to the complexity of the mixtures, such that there is a high probability of confounding actions. The use of binary biochemical assays where the experimental system is a single binding event, reduces the problems associated with assessing complex mixtures, however these assays are also amenable to High Throughput Screening (HTS) and therefore could start with the purified components. The rise of automated biochemical and cell based screening, in particular automated ion channel screening, has revolutionised drug discovery biology. These advances came into wider use at the turn of the 21st centurary to keep up with the explosion of synthetic chemistry and ever larger compound libraries. Cell based assays are particularly sensitive to cytotoxins masking possible active molecules and thus screening of separated venom components is preferable.

As with all screening campaigns and thus compound libraries, the most efficient start point is a targeted library. Preparation of venom libraries begins with collection. The method of collection depends on the organism and ranges from a snake biting onto a paraflm topped container through to electrical stimulation of venom glands. The venom is then quantified and fractionated using HPLC purification to separate it into its component parts. Fig. 2A shows an example of an ion exchange chromatogram from spitting cobra *Naja siamensis* venom which has a few dominant protein families. This is compared with a second reverse phase chromatogram from the theraphosid spider *Monocentropus lambertoni* which has many peaks of similar intensity (Fig. 2B). The complexity of the venoms often requires two orthogonal separation methods such as reverse phase and size exclusion or ion exchange to get the purest fraction library. This is a very practical approach as the most useful drug-like molecules in the venoms are very stable when treated this way. Once lyophilised into society for biomolecular screening (SBS) standard assay plates of a chosen format they can be handled in a similar way to other compound libraries. Fig. 7 shows an example workflow. Despite the trend of assay miniaturisation leading to

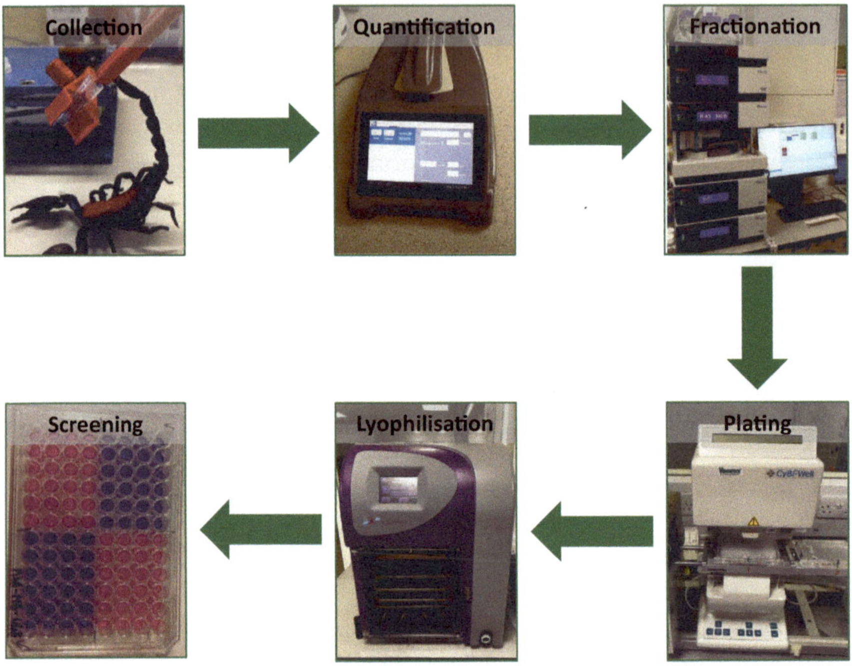

Fig. 7 Example workflow from venom extraction to assay. Venom is extracted and quantified using microspectrophotometry. Venom is fractionated using high pressure liquid chromatography (HPLC) and this could be by a number of dimensions utilising reverse phase and size exclusion or ion exchange chromatography. The venom fractions are then standardised, plated into SBS standard assay plates, lyophilised ready for screening against drug targets or cell lines.

multi-thousand well plates, the dominant work-horse sceening plates are still 384-well and, to a lesser extent, 1536-wells; this was nicely predicted by Mayr and Fuerst in 2008 [287]. Thus it is not surprising that the 384-well and 384-well low dead volume (LDV) assay format is currently dominant in venom peptide screening, although several labs are still using 96-well formats. Assay format and thus well size, is key to the ability to utilise the venoms. Large snakes produce many milligrams of protein with each venom yield, whereas the yield is many orders of magnitude less when considering small scorpions, wasps or cone snails. Even more extreme are the negected venoms of many invertebrates such as pseudoscorpions, flies and gnathiid isopods, due to their small size of a few millimetres body length [288]. The first recoverable venom collection from a pseudoscorpion (*Synsphyronus apimelus*) produced 5 nL of venom on average [39]. Thus miniaturised wells allow for the most use from the smaller species. There are a few suppliers around the world who produce assay ready libraries of venom fractions which make them easily accessible to drug discovery.

While the native venom, collected recoverably from the source animals, represents the true power of the evolved molecules, there are many technical challenges to producing compound libraries this way, as discussed. The rise in gene sequencing throughput has allowed other methods of natural product library production, such as recombinant expression in model organisms or solid state peptide synthesis from known sequences [163,289]. Typically venom glands are terminally dissected from euthanised specimens [163,288] so that the RNA can be extracted for reverse transcription and subsequent recombinant expression. However one group do report surgically removing one of the pair of venom glands from a Heloderma lizard such that the animal could recover and still have a working venom gland, while the scientists could access the RNA [163]. Transcriptomics is a very powerful tool to characterise venoms through recombinant production or synthesis from gene sequence data, but it usually comes at the cost of the animal. However a seminal paper in 2012 identified the unusual stability of messenger RNA in whole venom [290] thus negating the need for sacrificing the animal or invasive surgery. But, to date, this ethical and practical approach to obtaining venom transcripts represents a minority, with only one paper on comparison of PLA2, readily found [291]. Recombinant and synthetic libraries have their benefits in hit follow up as the compounds have already been identified and are ready for scale-up. Also the toxins can be expressed with novel tethers to enable use of particular screening technologies such as for difficult ion channels like TRPA1 [292]. However a major drawback is that the native post translational modification of the venom proteins, often important in their function and stability, are not replicated and thus the starting point may not be as good [8,293,294].

However the venom component library is prepared, for the most part it can be screened with a similar methodology to standard compound libraries, thus enabling screening of the full range of targets used with other libraries. This process has been used to identify novel leads from venoms on difficult targets such as ion channels, GPCRs, kinases and protein-protein interactions. Automated patch clamp methods have proven useful for screening venom fractions to discover novel, selective, inhibitors of the pain relevant sodium channel $Na_V1.7$ [295]. Another method used for high throughput screening of $Na_V1.7$ has been using a chimeric protein containing the $Na_V1.7$ voltage sensor domain fused to the more amenable $K_V2.2$ channel such that fluorescence can be measured [240]. One major benefit of the venom libraries is that they are soluble in assay buffers, negating the need for DMSO which is particularly problematic in cell based assays. Also, DMSO at standard assay concentrations of between 1% and 0.1% can alter the stability and other physical properties of the venoms [296], thus leading to potential complexities later on.

Once hits have been identified from the screen and confirmed in the usual way, hit follow up methods differ depending on the type of venom library used. Those from native venoms can be identified by intact mass and digested peptide high resolution MS/MS techniques when the starting material has not been previously identified. These libraries are very efficient as only interesting hits need to be identified and produced. In contrast, those from synthetic or recombinant libraries are already known, due to the extra steps in the production. Either way the resulting information can easily be used to take into lead development to tune the hit into more drug like molecules such as by reducing the size, or changing selectivity or PK parameters by comparing with other venom compounds and developing structure activity relationships (SAR). Such approaches include screening venoms of closely related species where the combination of gene duplication and natural selection has produced a library of similar compounds with minor mutations providing good substrates for SAR work.

9. Utilising SAR with venom peptides

The classification of venom peptide toxins into groups permits study and understanding of the structure-function relationship of each individual group. Classification could be based on a number of different determinants: receptor or ion channel specificity or cellular target, peptide length, structural scaffold, number of disulphide bonds, mechanism of action etc. By using amino acid sequence similarities, positioning of key amino acid residues (i.e. cysteines in disulphide bonds) and comparing key structural folds, peptide venoms can be classified into each group and the knowledge gained used to engineer peptides with desirable pharmacology. In this regard, the approach is to use positional substitution of amino acids within the sequence (positional scanning) by producing the peptide analogue and measuring the resulting activity/binding specificity against the target. For example, to determine the contribution of a particular residue within a peptide, alanine scanning can be performed whereby peptide analogues of the original peptide are produced in which alanine systematically replaces sequential amino acids within the peptide. Due to alanine's non-bulky methyl side chain, any structure within the peptide is likely to be preserved while removing any potential chemical activity (and hence potential functionality) of the original amino acid. Further positional substitutions with other natural and non-natural amino acids can also be used to determine the impact of changing charge, hydrophobicity, bulkiness, etc at key positions within the peptide sequence as part of a rational design process.

As an example, extensive structure activity relationship studies have been performed on inhibitory cysteine knot (ICK) peptides from spider venoms that target voltage gated sodium channels (Na_V) and on binding act allosterically to modulate the channels [271]. These peptides have been classified into groups based on the sequence similarity and the position of the key cysteine amino acids that, when participating in the disulphide bond, make up the characteristic cysteine knot. The conserved ICK is composed of three disulphide bonds between cysteine residues C1-C4, C2-C5 and C3-C6 resulting in a peptide with a globular structure with four loops and a C-terminal tail. There are 12 distinct families of Na_V modulators (NaSpTx 1-12) and extensive SAR studies have resulted in a detailed study of the potency, selectivity and pharmacology of peptides from families NaSpTx 1, 3 and 7 [271].

SAR investigations using alanine scanning and substitutions of other amino acids have been carried out on several members of the NaSpTx 1 family of proteins. GpTx-1 from the theraphosid *Grammostola porteri* is a nanomolar inhibitor of $Na_V1.7$ (IC_{50} 10 nM) and alanine scanning revealed that amino acid residues Trp29, Lys31 and Phe34 were essential to this inhibition. As part of a rational design to create an optimised GpTx-1 peptide, additional positional substitutions Phe5Ala, Met6Phe, Thr26Leu and Lys28Arg produced a peptide that possessed a 6-fold enhanced potency for $Na_V1.7$ and a 1000-fold selectivity over $Na_V1.4$ and $Na_V1.5$ [297]. Later scans with amino acids other than alanine revealed further residues critical to activity while structural models of GpTx-1 bound to the $Na_V1.7$ protein predicted interactions between critical residues in the peptide and those within the voltage sensor region of $Na_V1.7$. A summary of the structure activity relationships undertaken for this peptide is shown in Table 2.

By aligning the amino acid sequences of members of the NaSpTx family, the positions of the cysteine residues and regions of conservation can be identified [298]. Pro12, Trp31 and Lys33 are not buried and appear to be critical for activity as is the serine residue in position 26 [240,297]. Zhang et al. used this sequence conservation to engineer analogues of HNTX-1, a member of the NaSpTx 1 family, to engineer a gain-of-function analogue

Table 2 Structure-activity relationships analysis of GpTx-1 from *Grammastola porteri* selectively blocks human $Na_V1.7$, and has been improved through four mutations.

Name	1	2	3	4	5	6	7	8	9	10	11	12	13	14	15	16	17	18	19	20	21	22	23	24	25	26	27	28	29	30	31	32	33	34	NaV1.7	NaV1.5	NaV1.4
Wild type	D	C	L	G	F	M	R	K	C	I	P	D	N	D	K	C	C	R	P	N	L	V	C	S	R	T	H	K	W	C	K	Y	V	F	0.09 ± 0.01	>5	2.7 ± 1.2
F5A	D	C	L	G	A	M	R	K	C	I	P	D	N	D	K	C	C	R	P	N	L	V	C	S	R	T	H	K	W	C	K	Y	V	F	0.63 ± 0.12	27 ± 10	45 ± 3
M6A	D	C	L	G	F	A	R	K	C	I	P	D	N	D	K	C	C	R	P	N	L	V	C	S	R	T	H	K	W	C	K	Y	V	F	0.43 ± 0.05	>5	>4.6
T26A	D	C	L	G	F	M	R	K	C	I	P	D	N	D	K	C	C	R	P	N	L	V	C	S	R	A	H	K	W	C	K	Y	V	F	0.26 ± 0.05	>5	3.2 ± 0.1
K28A	D	C	L	G	F	M	R	K	C	I	P	D	N	D	K	C	C	R	P	N	L	V	C	S	R	T	H	A	W	C	K	Y	V	F	0.47 ± 0.05	>5	>4.9
Optimised	D	C	L	G	A	F	R	K	C	I	P	D	N	D	K	C	C	R	P	N	L	V	C	S	R	L	H	R	W	C	K	Y	V	F	0.0016	>10	1.9

From data published by Gonçalves TC, Benoit E, Kurz M, Lucarain L, Fouconnier S, Combemale S, et al. From identification to functional characterization of cyriotoxin-1a, an antinociceptive toxin from the spider Cyriopagopus schioedtei. Br J Pharmacol 2019;176(9):1298–314 with amino acids shown by single letter codes.

that was a potent blocker of the human $Na_V1.7$ channel [298]. HNTX-1 itself displays weak affinity for insect sodium channels and has no effect on mammalian channels but was used as a framework to engineer analogues which were subsequently tested for activity. Using the conserved sequence within the NaSpTx 1 family, amino acids were substituted at various positions within the peptide leading to the eventual analogue (HNTX-1: E1G-N23S-D26H-L32W) with potent human $Na_V1.7$ blocking activity and an I_{C50} of $0.036 \pm 0.007\,\mu M$. As part of the development of this analogue, a consensus sequence motif (X1X2SWCKX3) was identified and shown to be critical for suppression of the human $Na_V1.7$ channel.

The HNTX-1 three dimensional protein structure (PDB code 1NIX) revealed that the basic amino acid residues Arg25, Lys27 and Lys30 came together to form a positively charged surface that would interact with Na_Vs by electrostatic interactions. Nearby residues Phe5, Tyr20, Trp28, Val31, Leu31 and Leu32 clustered to form a hydrophobic patch. Both these surfaces are highly conserved within the NaSpTx 1 family. Upon modelling the potent HNTX-1:E1G-N23S-D26H-L32W peptide onto the HNTX-1 template it was found that these amino acid substitutions strengthened the interactions: Asp26His increased the positive electrostatic interaction with the Na_V; Glu1Gly reduced the negative charge, while Leu32Trp strengthened the hydrophobic interaction and the Asn23Ser substitution was required to better orient the critical Trp29 and Lys31 residues correctly [298].

A similar approach has been used on other members of the NaSpTx 3 and NaSpTx 7 families, revealing key motifs and residues important for activity. ProTx-II, a NaSpTx-3 family member is a potent $Na_V1.7$ inhibitor (IC_{50} 0.3 nM) yet also exhibits potent inhibition of off-target $Na_V1.4$ and $Na_V1.5$. Alanine and glutamine positional scanning of recombinantly expressed ProTx-II analogues in SAR studies against Na_V channels revealed an active face composed of hydrophobic and cationic residues [299], with residues Trp5, Lys26, Lys27, Arg13 and Arg22 critical for potency (10–125-fold). Additional residues Met6, Trp7, Met19, Val20, Trp24, Leu29 and Trp30 also result in a 10-fold loss of potency while an additional study identified some of these residues critical for $Na_V1.5$ activity.

JZTX-III is a NaSpTx 7 ICK isolated from the Chinese theraphosid *Chilobrachys jingzhao* and is a potent inhibitor of $K_V2.1$ channels and TTX-R currents in rat cardiac muscle. SAR investigations against $Na_V1.5$ using alanine substitutions revealed residues Asp1, Glu3, Trp8, Trp9, Trp28 and Trp30 as critical for activity while substituting arginine 13 for a glutamate residue enhanced $Na_V1.5$ inhibition 11-fold. JZTX-III does not inhibit the $Na_V1.7$ channel, which is attributed to an aspartate residue at position 816 in the $Na_V1.7$ channel. Substituting this aspartate for an arginine (D816R) within $Na_V1.7$ resulted in enhanced inhibitory properties of JZTX-III against $Na_V1.7$ [271].

Together with other similar studies on these peptides, optimised analogues and Na_V proteins, a consensus of prevalent features and key residues have been identified that are associated with activity. In the NaSpTx 1 family, amino acid residue positions 5 and 6 are typically occupied by the hydrophobic residue phenylalanine; similarly, in NaSpTx-3 these positions are occupied by tryptophan and methionine and are critical for activity. Increased potency can be achieved by reducing the negative charge at the N-terminus while modifications at the C-terminus by increasing positive charge or increased hydrophobicity enhance inhibition of Na_V channels.

Due to gene duplication events and subsequent natural amino acid mutations SAR tables can be assembled and analysed without the need for cloning or synthesis of a large library of systematic alanine substitutions. By utilising the natural mutations all the changes have a function as they have been subject to natural selection and thus the data is easily obtainable. Potassium channel toxins blocking Kv1.3 from bark scorpion (*Centruroides* spp.) venoms are just such a group containing intra and inter-specific mutations. Disulphide bonds have been conserved between Cys7-Cys29, Cys13-Cys34 and Cys17-Cys36 and thus the sizes of the loops between the cysteines are also maintained (as shown in Table 3) however other Kv1.3 toxins from other genera of scorpions do have different loop lengths. Linear regression analysis of the amino acid sequences along with potencies at Kv1.3 allows for the identification of key residues driving potency. This can also be performed with selectivity or pharmacokinetic data to optimised lead peptides.

Table 3 Structure-activity relationships of Kv1.3 toxins form bark scorpions (*Centruroides* spp) with amino acids indicated by single letter codes.

UniProt ID	Species	Toxins	1	2	3	4	5	6	7	8	9	10	11	12	13	14	15	16	17	18	19	20	21	22	23	24	25	26	27	28	29	30	31	32	33	34	35	36	37	38	39	Kv1.3 (nM)
P40755	*Centruroides margaritatus*	KTx 2.2	T	I	I	N	V	K	C	T	S	P	K	Q	C	L	P	P	C	K	A	Q	F	G	Q	S	A	G	A	K	C	M	N	G	K	C	K	C	Y	P	H	0.23
P0C162	*Centruroides elegans*	KTx 2.9	T	I	I	N	V	K	C	T	S	P	K	Q	C	L	K	P	C	K	D	L	Y	G	P	H	A	G	A	K	C	M	N	G	K	C	K	C	Y	N	N	0.25
P0C161	*Centruroides elegans*	KTx 2.8	T	V	I	N	V	K	C	T	S	P	K	Q	C	L	K	P	C	K	D	L	Y	G	P	H	A	G	A	K	C	M	N	G	K	C	K	C	Y	N	N	0.71
P0C164	*Centruroides elegans*	KTx 2.11	T	I	I	N	V	K	C	T	S	P	K	Q	C	L	L	P	C	K	E	I	Y	G	I	H	A	G	A	K	C	M	N	G	K	C	K	C	Y	K	I	0.98
P08815	*Centruroides noxius*	KTx 2.1	T	I	I	N	V	K	C	T	S	P	K	Q	C	S	K	P	C	K	E	L	Y	G	S	S	A	G	A	K	C	M	N	G	K	C	K	C	Y	N	N	1
P85529	*Centruroides suffusus*	KTx 2.13	I	F	I	N	V	K	C	S	S	P	Q	Q	C	L	K	P	C	K	A	A	F	G	I	S	A	G	G	K	C	I	N	G	K	C	K	C	Y	P	-	7.2
C0HJW6	*Centruroides tecomanus*	KTx 2.16	T	I	I	N	V	K	C	T	S	P	K	Q	C	L	K	P	C	K	D	L	Y	G	P	H	A	G	E	K	C	M	N	G	K	C	K	C	Y	K	I	26.2
P0C165	*Centruroides elegans*	KTx 2.12	T	I	I	N	V	K	C	T	S	P	K	Q	C	L	P	P	C	K	E	I	Y	G	R	H	A	G	A	K	C	M	N	G	K	C	H	C	S	K	I	69
P0C163	*Centruroides elegans*	KTx 2.10	I	F	I	N	V	K	C	S	L	P	Q	Q	C	L	R	P	C	K	D	R	F	G	Q	H	A	G	G	K	C	I	N	G	K	C	K	C	Y	P	-	366
C0HJW5	*Centruroides tecomanus*	KTx 2.17	T	I	I	N	V	K	C	T	S	P	K	Q	C	L	L	P	C	K	Q	I	Y	G	P	H	A	G	A	K	C	M	N	G	K	C	H	C	S	K	I	1000
P0DL70	*Centruroides limpidus*	KTx 2.18	T	T	I	N	V	K	C	T	S	P	K	Q	C	L	P	P	C	K	E	I	Y	G	R	H	A	G	A	K	C	I	N	G	K	C	H	C	S	K	I	1000

Linear regression analysis has identified five key residue positions driving potency with a correlation coefficient >0.6 (black) and four less important substitutions with a correlation coefficient >0.3 (grey).

Linear regression performed on data presented by Tanaka K-I, Sekino S, Ikegami M, Ikeda H, Kamei J. Antihyperalgesic effects of ProTx-II, a Nav1.7 antagonist, and A803467, a Nav1.8 antagonist, in diabetic mice. J Exp Pharmacol 2015;7:11–6.

The application of SAR to optimise Na_V channel modulators and help develop treatments for channelopathies and chronic pain has been demonstrated in rodent models, where the ProTX-II peptide and its optimised analogue (JNJ63955918) successfully reversed neuropathic and inflammatory pain when administered intrathecally or locally [89,94]. Similarly, intraperitoneal administration of the NaSpTx 1 peptides HwTx-IV and HNTX-IV reversed neuropathic and inflammatory pain [292].

10. Future perspective

To unleash the full potential of drugs and tools from venoms, further crossover is required between evolutionary biology, biochemistry, and drug discovery chemistry. Through the building of this interdisciplinary approach, further novel tools are likely to be found to meet the demands of drug discovery in a world dominated by tough targets, new therapeutic modalities, and emerging diseases. The intrinsic nature of venoms being stable, repurposed versions of many protein families, will continue to aid protein therapeutic development. Biotechnological entrepreneurs have already taken the stable spider venom ICK motif and recombinantly added it to the binding domain of humanised antibodies, thus delivering the binding potency and selectivity of the ICK peptide with the pharmacokinetic properties of antibodies. However, these still lack the desired oral bioavailability required for the patients' preferred dose route. Many venom peptides demonstrate exceptional stability in gut relevant pH ranges, protease stability and an ability to cross membranes. In addition as small cyclic peptides such as cyclosporin, have already shown oral bioavailability, the potential for oral bioavailability of more complex venom derived peptides is within in our grasp. Coupled with improvements in understanding of biotherapeutic delivery mechanisms, the ideal pharmacological properties of venom peptides could soon be making their way into the oral dose route. A greater use and understanding of venom peptides as therapeutics will open up the major class of difficult targets for small molecules—the protein-protein interactions. It is often cited that these targets are tough to modulate with small molecules due to the large interaction surface and surface chemistry, whereas nearly all venom peptides exert their actions through such interactions and thus will prove a great resource for this underrepresented target class.

Coupled with this greater understanding of peptide chemistry is the continued discovery of new venom systems, and new components and actions within known venoms. Charles Darwin said that "endless forms most beautiful and most wonderful have been, and are being, evolved in the natural world" and thus it is clear that we have only just scratched the surface of the potential substrates for new biotherapeutics from venoms.

References

[1] Herzig V, Cristofori-Armstrong B, Israel MR, Nixon SA, Vetter I, King GF. Animal toxins—nature's evolutionary-refined toolkit for basic research and drug discovery. Biochem Pharmacol 2020;181:114096.

[2] Zancolli G, Casewell NR. Venom systems as models for studying the origin and regulation of evolutionary novelties. Mol Biol Evol 2020;37(10):2777–90.

[3] Casewell NR. Evolution: gene co-option underpins venom protein evolution. Curr Biol 2017;27(13):R647–9.

[4] Martinson EO, Mrinalini, Kelkar YD, Chang CH, Werren JH. The evolution of venom by co-option of single-copy genes. Curr Biol 2017;27(13):2007–2013.e8.

[5] Hanf ZR, Chavez AS. A comprehensive multi-omic approach reveals a relatively simple venom in a diet generalist, the northern short-tailed shrew, *Blarina brevicauda*. Genome Biol Evol 2020;12(7):1148–66.

[6] Pineda SS, Chin YKY, Undheim EAB, Senff S, Mobli M, Dauly C, et al. Structural venomics reveals evolution of a complex venom by duplication and diversification of an ancient peptide-encoding gene. Proc Natl Acad Sci USA 2020;117(21): 11399–408.

[7] Casewell NR, Wüster W, Vonk FJ, Harrison RA, Fry BG. Complex cocktails: the evolutionary novelty of venoms. Trends Ecol Evol 2013;28(4):219–29.

[8] Casewell NR, Wagstaff SC, Wuster W, DAN C, FMS B, King SI, et al. Medically important differences in snake venom composition are dictated by distinct postgenomic mechanisms. Proc Natl Acad Sci USA 2014;111(25):9205–10.

[9] Tang X, Zhang Y, Hu W, Xu D, Tao H, Yang X, et al. Molecular diversification of peptide toxins from the tarantula *Haplopelma hainanum* (Ornithoctonus hainana) venom based on research articles. J Proteome Res 2010;9(5):2550–64.

[10] Gutiérrez JM, Calvete JJ, Habib AG, Harrison RA, Williams DJ, Warrell DA. Snakebite envenoming. Nat Rev Dis Primers 2017;3(1):17063.

[11] Rodrigo C, Gnanathasan A. Management of scorpion envenoming: a systematic review and meta-analysis of controlled clinical trials. Syst Rev 2017;6(1):74.

[12] Gelbart D, Donoughe JS. Spider bites. JAAPA 2019;32(3):49–50.

[13] Hardy MC, Daly NL, Mobli M, Morales RAV, King GF. Isolation of an orally active insecticidal toxin from the venom of an Australian tarantula. PLoS One 2013;8(9): e73136.

[14] Jin J, Agwa AJ, Szanto TG, Csóti A, Panyi G, Schroeder CI, et al. Weaponisation 'on the fly': convergent recruitment of knottin and defensin peptide scaffolds into the venom of predatory assassin flies. Insect Biochem Mol Biol 2020;118(October 2019):103310.

[15] Gorai B, Sivaraman T. Unfolding stabilities of two paralogous proteins from *Naja naja* naja (Indian cobra) as probed by molecular dynamics simulations. Toxicon 2013;72:11–22.

[16] Kim SM, Kim SY, Kim SH, Kim SZ. Dendroaspis natriuretic peptide is degraded by a metalloproteinase in the rat kidney. Mol Med Rep 2014;9(3):1037–43.

[17] Simonsen L, Holst JJ, Deacon CF. Exendin-4, but not glucagon-like peptide-1, is cleared exclusively by glomerular filtration in anaesthetised pigs. Diabetologia 2006;49(4):706–12.

[18] Mohamed Abd El-Aziz T, Garcia Soares A, Stockand JD. Snake venoms in drug discovery: valuable therapeutic tools for life saving. Toxins (Basel) 2019;11(10):564.
[19] Fry BG, Vidal N, Norman JA, Vonk FJ, Scheib H, Ramjan SFR, et al. Early evolution of the venom system in lizards and snakes. Nature 2006;439(7076):584–8.
[20] Fry BG, Casewell NR, Wüster W, Vidal N, Young B, Jackson TNW. The structural and functional diversification of the Toxicofera reptile venom system. Toxicon 2012;60(4):434–48.
[21] Hargreaves AD, Swain MT, Logan DW, Mulley JF. Testing the Toxicofera: comparative transcriptomics casts doubt on the single, early evolution of the reptile venom system. Toxicon 2014;92:140–56.
[22] Tasoulis T, Isbister GK. A review and database of snake venom proteomes. Toxins (Basel) 2017;9(9):290.
[23] Carregari VC, Rosa-Fernandes L, Baldasso P, Bydlowski SP, Marangoni S, Larsen MR, et al. Snake venom extracellular vesicles (SVEVs) reveal wide molecular and functional proteome diversity. Sci Rep 2018;8(1):12067.
[24] Senji Laxme RR, Suranse V, Sunagar K. Arthropod venoms: biochemistry, ecology and evolution. Toxicon 2019;158:84–103.
[25] Pineda SS, Undheim EAB, Rupasinghe DB, Ikonomopoulou MP, King GF. Spider venomics: implications for drug discovery. Future Med Chem 2014;6(15):1699–714.
[26] Bhere KV, Haney RA, Ayoub NA, Garb JE. Gene structure, regulatory control, and evolution of black widow venom latrotoxins. FEBS Lett 2014;588(21):3891–7.
[27] Zhang Y-Y, Huang Y, He Q-Z, Luo J, Zhu L, Lu S-S, et al. Structural and functional diversity of peptide toxins from tarantula *Haplopelma hainanum* (Ornithoctonus hainana) venom revealed by transcriptomic, peptidomic, and patch clamp approaches. J Biol Chem 2015;290(22):14192–207.
[28] Trim SA, Trim CM. Venom: the sharp end of pain therapeutics. Br J Pain 2013;7 (4):179–88.
[29] Meehan CJ, Olson EJ, Reudink MW, Kyser TK, Curry RL. Herbivory in a spider through exploitation of an ant-plant mutualism. Curr Biol 2009;19:R892–3.
[30] Bennie M, Loaring C, Trim S. Laboratory husbandry of arboreal tarantulas (Theraphosidae) and evaluation of environmental enrichment. Anim Technol J Inst Anim Tech 2011;10:163–9.
[31] Graudins A, Padula M, Broady K, Nicholson GM. Red-back spider (*Latrodectus hasselti*) antivenom prevents the toxicity of widow spider venoms. Ann Emerg Med 2001;37(2):154–60.
[32] Gaver-Wainwright MM, Zack RS, Foradori MJ, Lavine LC. Misdiagnosis of spider bites: bacterial associates, mechanical pathogen transfer, and hemolytic potential of venom from the hobo spider, *Tegenaria agrestis* (Araneae: Agelenidae). J Med Entomol 2011;48(2):382–8.
[33] Chaim OM, Trevisan-Silva D, Chaves-Moreira D, Wille ACM, Ferrer VP, Matsubara FH, et al. Brown spider (Loxosceles genus) venom toxins: tools for biological purposes. Toxins (Basel) 2011;3(3):309–44.
[34] Chaves-Moreira D, Matsubara FH, Schemczssen-Graeff Z, De Bona E, Heidemann VR, Guerra-Duarte C, et al. Brown spider (Loxosceles) venom toxins as potential biotools for the development of novel therapeutics. Toxins (Basel) 2019;11(6):355.
[35] Wang X, Connor M, Smith R, Maciejewski MW, Howden ME, Nicholson GM, et al. Discovery and characterization of a family of insecticidal neurotoxins with a rare vicinal disulfide bridge. Nat Struct Biol 2000;7(6):505–13.
[36] Xiao Y, Liang S. Inhibition of neuronal tetrodotoxin-sensitive Na+ channels by two spider toxins: hainantoxin-III and hainantoxin-IV. Eur J Pharmacol 2003;477(1):1–7.
[37] Chen G, Wu H, Wang N, Zhong S, Zhou Y, Liang B. A mitogenomic phylogeny of spiders and complete mitochondrial genome of Cyriopagopus hainanus (Araneae: Theraphosidae). Mitochondrial DNA B 2020;5(1):782–3.

[38] Evans ERJ, Northfield TD, Daly NL, Wilson DT. Venom costs and optimization in scorpions. Front Ecol Evol 2019;7:196.

[39] Krämer J, Pohl H, Predel R. Venom collection and analysis in the pseudoscorpion *Chelifer cancroides* (Pseudoscorpiones: Cheliferidae). Toxicon 2019;162(November 2018):15–23.

[40] Cabezas-Cruz A, Valdés JJ. Are ticks venomous animals? Front Zool 2014;11:47.

[41] Undheim EAB, Jenner RA, King GF. Centipede venoms as a source of drug leads. Expert Opin Drug Discovery 2016;11(12):1139–49.

[42] von Reumont BM, Blanke A, Richter S, Alvarez F, Bleidorn C, Jenner RA. The first venomous crustacean revealed by transcriptomics and functional morphology: remipede venom glands express a unique toxin cocktail dominated by enzymes and a neurotoxin. Mol Biol Evol 2014;31(1):48–58.

[43] von Reumont BM, Campbell LI, Richter S, Hering L, Sykes D, Hetmank J, et al. A Polychaete's powerful punch: venom gland transcriptomics of Glycera reveals a complex cocktail of toxin homologs. Genome Biol Evol 2014;6(9):2406–23.

[44] Walker AA, Robinson SD, Yeates DK, Jin J, Baumann K, Dobson J, et al. Entomo-venomics: the evolution, biology and biochemistry of insect venoms. Toxicon 2018;154:15–27.

[45] Carrijo-Carvalho LC, Chudzinski-Tavassi AM. The venom of the Lonomia caterpillar: an overview. Toxicon 2007;49(6):741–57.

[46] Nekaris KA-I, Moore RS, Rode EJ, Fry BG. Mad, bad and dangerous to know: the biochemistry, ecology and evolution of slow loris venom. J Venomous Anim Toxins Incl Trop Dis 2013;19(1):21.

[47] Arbuckle K. Evolutionary context of venom in animals. In: Evolution of venomous animals and their toxins; 2017. p. 3–31.

[48] Whittington CM, Papenfuss AT, Locke DP, Mardis ER, Wilson RK, Abubucker S, et al. Novel venom gene discovery in the platypus. Genome Biol 2010;11(9):R95.

[49] Casewell NR, Petras D, Card DC, Suranse V, Mychajliw AM, Richards D, et al. Solenodon genome reveals convergent evolution of venom in eulipotyphlan mammals. Proc Natl Acad Sci USA 2019;116(51):25745–55.

[50] Kowalski K, Marciniak P, Rosiński G, Rychlik L. Evaluation of the physiological activity of venom from the Eurasian water shrew *Neomys fodiens*. Front Zool 2017;14:46.

[51] Kita M, Nakamura Y, Okumura Y, Ohdachi SD, Oba Y, Yoshikuni M, et al. Blarina toxin, a mammalian lethal venom from the short-tailed shrew *Blarina brevicauda*: isolation and characterization. Proc Natl Acad Sci USA 2004;101(20):7542–7.

[52] Kita M, Okumura Y, Ohdachi SD, Oba Y, Yoshikuni M, Nakamura Y, et al. Purification and characterisation of blarinasin, a new tissue kallikrein-like protease from the short-tailed shrew *Blarina brevicauda*: comparative studies with blarina toxin. Biol Chem 2005;386(2):177–82.

[53] Prashanth JR, Dutertre S, Lewis RJ. Pharmacology of predatory and defensive venom peptides in cone snails. Mol Biosyst 2017;13(12):2453–65.

[54] Dutt M, Dutertre S, Jin A-H, Lavergne V, Alewood PF, Lewis RJ. Venomics reveals venom complexity of the piscivorous cone snail, Conus tulipa. Mar Drugs 2019;17(1):71.

[55] Neale V, Smout MJ, Seymour JE. Spine-bellied sea snake (Hydrophis curtus) venom shows greater skeletal myotoxicity compared with cardiac myotoxicity. Toxicon 2018;143:108–17.

[56] Rensch G, Murphy-Lavoie HM. Lionfish, scorpionfish, and stonefish toxicity. In: StatPearls [Internet]. Treasure Island (FL): StatPearls Publishing; 2021. https://pubmed.ncbi.nlm.nih.gov/29489204/.

[57] Harris RJ, Jenner RA. Evolutionary ecology of fish venom: adaptations and consequences of evolving a venom system. Toxins (Basel) 2019;11(2):60.
[58] Smith WL, Stern JH, Girard MG, Davis MP. Evolution of venomous cartilaginous and ray-finned fishes. Integr Comp Biol 2016;56(5):950–61.
[59] Gomes HL, Menezes TN, Malacarne PF, Roman-Campos D, Gondim AN, Cruz JS, et al. Cardiovascular effects of Sp-CTx, a cytolysin from the scorpionfish (*Scorpaena plumieri*) venom. Toxicon 2016;118:141–8.
[60] Mouchbahani-Constance S, Lesperance LS, Petitjean H, Davidova A, Macpherson A, Prescott SA, et al. Lionfish venom elicits pain predominantly through the activation of nonpeptidergic nociceptors. Pain 2018;159(11):2255–66.
[61] Khalil AM, Wahsha MA, Abu Khadra KM, Khalaf MA, Al-Najjar TH. Biochemical and histopathological effects of the stonefish (*Synanceia verrucosa*) venom in rats. Toxicon 2018;142:45–51.
[62] Monteiro Dos Santos J, Cardoso Dos Santos J, Marques EE, de Araújo GC, Seibert CS, Lopes-Ferreira M, et al. Stingray (Potamotrygon rex) maturity is associated with inflammatory capacity of the venom. Toxicon 2019;163:74–83.
[63] Hornbeak KB, Auerbach PS. Marine envenomation. Emerg Med Clin North Am 2017;35(2):321–37.
[64] Gorman LM, Judge SJ, Fezai M, Jemaà M, Harris JB, Caldwell GS. The venoms of the lesser (*Echiichthys vipera*) and greater (*Trachinus draco*) weever fish—a review. Toxicon X 2020;6:100025.
[65] Chhatwal I, Dreyer F. Isolation and characterization of dracotoxin from the venom of the greater weever fish *Trachinus draco*. Toxicon 1992;30(1):87–93.
[66] Fezai M, Slaymi C, Ben-Attia M, Lang F, Jemaà M. Purified lesser weever fish venom (*Trachinus vipera*) induces eryptosis, apoptosis and cell cycle arrest. Sci Rep 2016;6:39288.
[67] Robinson P, Trim S, Trim C. Non-invasive extraction of cnidarian venom through the use of autotomised tentacles. Anim Technol Welf 2019;18(3):167–73.
[68] Schmidt CA, Daly NL, Wilson DT. Coral venom toxins. Front Ecol Evol 2019;7:320.
[69] Jouiaei M, Casewell NR, Yanagihara AA, Nouwens A, Cribb BW, Whitehead D, et al. Firing the sting: chemically induced discharge of cnidae reveals novel proteins and peptides from box jellyfish (*Chironex fleckeri*) venom. Toxins (Basel) 2015;7(3):936–50.
[70] Madio B, King GF, Undheim EAB. Sea anemone toxins: a structural overview. Mar Drugs 2019;17(6):325.
[71] Lau M-T, Manion J, Littleboy JB, Oyston L, Khuong TM, Wang Q-P, et al. Molecular dissection of box jellyfish venom cytotoxicity highlights an effective venom antidote. Nat Commun 2019;10(1):1655.
[72] Bosmans F, Tytgat J. Sea anemone venom as a source of insecticidal peptides acting on voltage-gated Na+ channels. Toxicon 2007;49(4):550–60.
[73] Hifumi T, Fukuchi Y, Otani N. Marine envenomation. SN Compr Clin Med 2020;2:2288–92. https://doi.org/10.1007/s42399-020-00490.
[74] Haddad VJ, Virga R, Bechara A, da Silveira FL, Morandini AC. An outbreak of Portuguese man-of-war (*Physalia physalis*-Linnaeus, 1758) envenoming in Southeastern Brazil. Rev Soc Bras Med Trop 2013;46(5):641–4.
[75] Whitelaw BL, Cooke IR, Finn J, Zenger K, Strugnell JM. The evolution and origin of tetrodotoxin acquisition in the blue-ringed octopus (genus Hapalochlaena). Aquat Toxicol 2019;206:114–22.
[76] Berling I, Isbister G. Marine envenomations. Aust Fam Physician 2015;44(1–2):28–32.

[77] Lee C-C, Tsai W-S, Hsieh HJ, Hwang D-F. Hemolytic activity of venom from crown-of-thorns starfish *Acanthaster planci* spines. J Venomous Anim Toxins Incl Trop Dis 2013;19(1):22.
[78] McDermott A. News feature: venom back in vogue as a wellspring for drug candidates. Proc Natl Acad Sci USA 2020;117:10100–4.
[79] Pennington MW, Czerwinski A, Norton RS. Peptide therapeutics from venom: current status and potential. Bioorg Med Chem 2018;26(10):2738–58.
[80] Waheed H, Moin SF, Choudhary MI. Snake venom: from deadly toxins to life-saving therapeutics. Curr Med Chem 2017;24(17):1874–91.
[81] Camargo ACM, Ianzer D, Guerreiro JR, Serrano SMT. Bradykinin-potentiating peptides: beyond captopril. Toxicon 2012;59(4):516–23.
[82] Dinicolantonio JJ, Lavie CJ, O'Keefe JH. Not all angiotensin-converting enzyme inhibitors are equal: focus on ramipril and perindopril. Postgrad Med 2013;125 (4):154–68.
[83] Scarborough RM, Naughton MA, Teng W, Rose JW, Phillips DR, Nannizzi L, et al. Design of potent and specific integrin antagonists. Peptide antagonists with high specificity for glycoprotein IIb-IIIa. J Biol Chem 1993;268(2):1066–73.
[84] Levy DE, Trammel J, Wasiewski WW. Ancrod for acute ischemic stroke: a new dosing regimen derived from analysis of prior ancrod stroke studies. J Stroke Cerebrovasc Dis 2009;18(1):23–7.
[85] Sherman DG. Ancrod. Curr Med Res Opin 2002;18(Suppl. 2):s48–52.
[86] Sherman DG, Atkinson RP, Chippendale T, Levin KA, Ng K, Futrell N, et al. Intravenous ancrod for treatment of acute ischemic stroke: the STAT study: a randomized controlled trial. Stroke treatment with Ancrod trial. JAMA 2000;283 (18):2395–403.
[87] Hennerici MG, Kay R, Bogousslavsky J, Lenzi GL, Verstraete M, Orgogozo JM. Intravenous ancrod for acute ischaemic stroke in the European Stroke Treatment with Ancrod Trial: a randomised controlled trial. Lancet (London, England) 2006;368 (9550):1871–8.
[88] Ho S-J, Brighton TA. Ximelagatran: direct thrombin inhibitor. Vasc Health Risk Manag 2006;2(1):49–58.
[89] Schulman S, Wahlander K, Lundstrom T, Clason SB, Eriksson H, THRIVE III Investigators. Secondary prevention of venous thromboembolism with the oral direct thrombin inhibitor ximelagatran. N Engl J Med 2003;349(18):1713–21. https://doi.org/10.1056/NEJMoa030104.
[90] Huang DX, Gai LY, Wang SR, Li TD, Yang TS, Zhi G, et al. Defibrase, a purified fibrinolytic protease from snake venom in acute myocardial infarction. Acta Cardiol 1992;47(5):445–58.
[91] Cooperative Group for Reassessment of Defibrase. Reassessment of defibrase in treatment of acute cerebral infarction: a multicenter, randomized, double-blind, placebo-controlled trial. Chin Med Sci J 2005;20(3):151–8.
[92] Yao YT, Yuan X, Fang NX. Hemocoagulase reduces postoperative bleeding and blood transfusion in cardiac surgical patients: a PRISMA-compliant systematic review and meta-analysis. Medicine (Baltimore) 2019;98(52), e18534.
[93] Ding J, Zhou D, Hu Y, Elmadhoun O, Pan L, Ya J, et al. The efficacy and safety of Batroxobin in combination with anticoagulation on cerebral venous sinus thrombosis. J Thromb Thrombolysis 2018;46(3):371–8.
[94] Yao Y, Yuan X, Fang N. Hemocoagulase reduces postoperative bleeding and blood transfusion in cardiac surgical patients: a PRISMA-compliant systematic review and meta-analysis. Medicine (USA) 2019;98(52):e18534.
[95] Markland FS, Swenson S. Fibrolase: trials and tribulations. Toxins (Basel) 2010;2 (4):793–808.

[96] Warkentin TE. Bivalent direct thrombin inhibitors: hirudin and bivalirudin. Best Pract Res Clin Haematol 2004;17(1):105–25.
[97] Beiderlinden M, Werner P, Bahlmann A, Kemper J, Brezina T, Schäfer M, et al. Monitoring of argatroban and lepirudin anticoagulation in critically ill patients by conventional laboratory parameters and rotational thromboelastometry—a prospectively controlled randomized double-blind clinical trial. BMC Anesthesiol 2018;18(1):18.
[98] Miljanich G. Ziconotide: neuronal calcium channel blocker for treating severe chronic pain. Curr Med Chem 2004;11(23):3029–40.
[99] Kolosov A, Goodchild CS, Cooke I. CNSB004 (leconotide) causes antihyperalgesia without side effects when given intravenously: a comparison with ziconotide in a rat model of diabetic neuropathic pain. Pain Med 2010;11(2):262–73.
[100] Furman BL. The development of Byetta (exenatide) from the venom of the Gila monster as an anti-diabetic agent. Toxicon 2012;59(4):464–71.
[101] King GF, Gentz MC, Escoubas P, Nicholson GM. A rational nomenclature for naming peptide toxins from spiders and other venomous animals. Toxicon 2008;52 (2):264–76.
[102] Oliveira JS, Fuentes-Silva D, King GF. Development of a rational nomenclature for naming peptide and protein toxins from sea anemones. Toxicon 2012;60(4):539–50. Available from: https://doi.org/10.1016/j.toxicon.2012.05.020.
[103] Hargreaves AD, Mulley JF. A plea for standardized nomenclature of snake venom toxins. Toxicon 2014;90(August):351–3.
[104] Stocker K, Barlow GH. The coagulant enzyme from *Bothrops Atrox* venom (Batroxobin). Methods Enzymol 1976;45(C):214–23.
[105] Kamiguti AS, Matsunaga S, Spir M, Sano-Martins IS, Nahas L. Alterations of the blood coagulation system after accidental human inoculation by *Bothrops jararaca* venom. Braz J Med Biol Res 1986;19(2):199–204.
[106] Neumann C, Slagboom J, Somsen GW, Vonk F, Casewell NR, Cardoso CL, et al. Development of a generic high-throughput screening assay for profiling snake venom protease activity after high-resolution chromatographic fractionation. Toxicon 2020;178(February):61–8.
[107] Heo Y, Kwon YC, Bae SK, Hwang D, Yang HR, Choudhary I, et al. Cloning a chymotrypsin-like 1 (CTRL-1) protease cDNA from the jellyfish nemopilema nomurai. Toxins (Basel) 2016;8(7):1–10.
[108] Knittel PS, Long PF, Brammall L, Marques AC, Almeida MT, Padilla G, et al. Characterising the enzymatic profile of crude tentacle extracts from the South Atlantic jellyfish *Olindias sambaquiensis* (Cnidaria: Hydrozoa). Toxicon 2016; 119:1–7. Available from: https://doi.org/10.1016/j.toxicon.2016.04.048.
[109] Li R, Yu H, Xue W, Yue Y, Liu S, Xing R, et al. Jellyfish venomics and venom gland transcriptomics analysis of *Stomolophus meleagris* to reveal the toxins associated with sting. J Proteomics 2014;106:17–29.
[110] Rastogi A, Biswas S, Sarkar A, Chakrabarty D. Anticoagulant activity of moon jellyfish (*Aurelia aurita*) tentacle extract. Toxicon 2012;60(5):719–23.
[111] Barbaro KC, Lira MS, Malta MB, Soares SL, Garrone Neto D, Cardoso JLC, et al. Comparative study on extracts from the tissue covering the stingers of freshwater (Potamotrygon falkneri) and marine (*Dasyatis guttata*) stingrays. Toxicon 2007;50 (5):676–87.
[112] Ruder T, Sunagar K, Undheim EAB, Ali SA, Wai T-C, Low DHW, et al. Molecular phylogeny and evolution of the proteins encoded by coleoid (cuttlefish, octopus, and squid) posterior venom glands. J Mol Evol 2013;76(4):192–204.
[113] Teng ZW, Xiong SJ, Xu G, Gan SY, Chen X, Stanley D, et al. Protein discovery: combined transcriptomic and proteomic analyses of venom from the endoparasitoid cotesia chilonis (hymenoptera: Braconidae). Toxins (Basel) 2017;9(4):135.

[114] Baumann K, Dashevsky D, Sunagar K, Fry B. Scratching the surface of an itch: molecular evolution of aculeata venom allergens. J Mol Evol 2018;86(7):484–500.
[115] Bouzid W, Verdenaud M, Klopp C, Ducancel F, Noirot C, Vétillard A. De novo sequencing and transcriptome analysis for *Tetramorium bicarinatum*: a comprehensive venom gland transcriptome analysis from an ant species. BMC Genomics 2014; 15(1):1–18.
[116] Borges MH, Figueiredo SG, Leprevost FV, De Lima ME, Cordeiro MdN, Moresco J, et al. Venomous extract protein profile of Brazilian tarantula *Grammostola iheringi*: searching for potential biotechnological applications. J Proteom 2016;136:35–47. https://doi.org/10.1016/j.jprot.2016.01.013.
[117] Devaraja S, Girish KS, Gowtham YNJ, Kemparaju K. The hag-protease-II is a fibrin(ogen)ase from *Hippasa agelenoides* spider venom gland extract: purification, characterization and its role in hemostasis. Toxicon 2011;57(2):248–58.
[118] Brazón J, Guerrero B, D'Suze G, Sevcik C, Arocha-Piñango CL. Fibrin(ogen)olytic enzymes in scorpion (Tityus discrepans) venom. Comp Biochem Physiol B Biochem Mol Biol 2014;168(1):62–9.
[119] Boldrini-França J, Pinheiro-Junior EL, Peigneur S, Pucca MB, Cerni FA, Borges RJ, et al. Beyond hemostasis: a snake venom serine protease with potassium channel blocking and potential antitumor activities. Sci Rep 2020;10(1):1–11.
[120] Vulfius CA, Kasheverov IE, Kryukova EV, Spirova EN, Shelukhina IV, Starkov VG, et al. Pancreatic and snake venom presynaptically active phospholipases A2 inhibit nicotinic acetylcholine receptors. PLoS One 2017;12(10):1–15.
[121] Still KBM, Slagboom J, Kidwai S, Xie C, Zhao Y, Eisses B, et al. Development of high-throughput screening assays for profiling snake venom phospholipase A2 activity after chromatographic fractionation. Toxicon 2020;184(June):28–38.
[122] Rojas JM, Arán-Sekul T, Cortés E, Jaldín R, Ordenes K, Orrego PR, et al. Phospholipase d from *Loxosceles laeta* spider venom induces IL-6, IL-8, CXCL1/GRO-α, and CCL2/MCP-1 production in human skin fibroblasts and stimulates monocytes migration. Toxins (Basel) 2017;9(4):1–20.
[123] Amorim FG, Longhim HT, Cologna CT, Degueldre M, De Pauw E, Quinton L, et al. Proteome of fraction from Tityus serrulatus venom reveals new enzymes and toxins. J Venomous Anim Toxins Incl Trop Dis 2019;25, e148218.
[124] Memar B, Jamili S, Shahbazzadeh D, Bagheri KP. The first report on coagulation and phospholipase A2 activities of Persian Gulf lionfish, Pterois russelli, an Iranian venomous fish. Toxicon 2016;113:25–31.
[125] Möller C, Davis WC, Clark E, DeCaprio A, Marí F. Conodipine-P1-3, the first phospholipases A2 characterized from injected cone snail venom. Mol Cell Proteomics 2019;18(5):876–91.
[126] Fernandes LGR, Perez-Riverol A, Bazon ML, Abram DM, Brochetto-Braga MR, Zollner RL. Functional profile of antigen specific CD4 + T cells in the immune response to phospholipase A1 allergen from Polybia paulista venom. Toxins (Basel) 2020;12(6):379.
[127] Estrada-Gómez S, Vargas-Muñoz LJ, Saldarriaga-Córdoba M, Cifuentes Y, Perafan C. Identifying different transcribed proteins in the newly described Theraphosidae Pamphobeteus verdolaga. Toxicon 2017;129:81–8.
[128] Ferreira FRB, da Silva PM, Soares T, Gonçalves Machado L, de Araújo LCC, da Silva TG, et al. Evaluation of antimicrobial, cytotoxic, and hemolytic activities from venom of the spider Lasiodora sp. Toxicon 2016;122:119–26.
[129] Bickler PE. Amplification of snake venom toxicity by endogenous signaling pathways. Toxins (Basel) 2020;12(2):1–26.
[130] Gremski LH, Trevisan-Silva D, Ferrer VP, Matsubara FH, Meissner GO, Wille ACM, et al. Recent advances in the understanding of brown spider venoms: from the biology of spiders to the molecular mechanisms of toxins. Toxicon 2014;83:91–120.

[131] Xu N, Zhao HY, Yin Y, Shen SS, Shan LL, Chen CX, et al. Combined venomics, antivenomics and venom gland transcriptome analysis of the monocoled cobra (*Naja kaouthia*) from China. J Proteomics 2017;159:19–31. Internet. Available from: https://doi.org/10.1016/j.jprot.2017.02.018.
[132] Undheim EAB, Fry BG, King GF. Centipede venom: recent discoveries and current state of knowledge. Toxins (Basel) 2015;7(3):679–704.
[133] Kita M, Black DS, Ohno O, Yamada K, Kigoshi H, Uemura D. Duck-billed platypus venom peptides induce Ca^{2+} influx in neuroblastoma cells. J Am Chem Soc 2009;131(50):18038–9.
[134] Boolell M, Allen MJ, Ballard SA, Gepi-Attee S, Muirhead GJ, Naylor AM, et al. Sildenafil: an orally active type 5 cyclic GMP-specific phosphodiesterase inhibitor for the treatment of penile erectile dysfunction. Int J Impot Res 1996;8(2):47–52.
[135] Bjork W. Purification of phosphodiesterase from *Bothrops atrox* venom, with special consideration of the elimination of monophosphatases. J Biol Chem 1963;238(7):2487–90.
[136] Subbotina A, Ravna AW, Lysaa RA, Abagyan R, Bugno R, Sager G. Inhibition of PDE5A1 guanosine cyclic monophosphate (cGMP) hydrolysing activity by sildenafil analogues that inhibit cellular cGMP efflux. J Pharm Pharmacol 2017;69(6):675–83.
[137] Troncone LRP, Srougi M, Leite KRM, Andrade E, Magnoli FC, Ramos AT. Phoneutria nigriventer spider toxin Tx2-6 causes priapism and death: a histopathological investigation in mice. Toxicon 2012;60(5):797–801.
[138] Petricevich VL. Scorpion venom and the inflammatory response. Mediators Inflamm 2010;2010:903295.
[139] Senthilkumaran S, Williams HF, Patel K, Trim SA, Thirumalaikolundusubramanian P, Vaiyapuri S. Priapism following a juvenile Russell's viper bite: an unusual case report. PLoS Negl Trop Dis 2021. In press.
[140] Khoo HE, Yuen R, Poh CH, Tan CH. Biological activities of *Synanceja horrida* (stonefish) venom. Nat Toxins 1992;1(1):54–60.
[141] Sales PBV, Santoro ML. Nucleotidase and DNase activities in Brazilian snake venoms. Comp Biochem Physiol C Toxicol Pharmacol 2008;147(1):85–95.
[142] Lino-López GJ, Valdez-Velázquez LL, Corzo G, Romero-Gutiérrez MT, Jiménez-Vargas JM, Rodríguez-Vázquez A, et al. Venom gland transcriptome from *Heloderma horridum horridum* by high-throughput sequencing. Toxicon 2020;180:62–78.
[143] Walker AA, Dobson J, Jin J, Robinson SD, Herzig V, Vetter I, et al. Buzz kill: function and proteomic composition of venom from the giant assassin fly Dolopus genitalis (Diptera: Asilidae). Toxins (Basel) 2018;10(11):456.
[144] Layfield HJ, Williams HF, Ravishankar D, Mehmi A, Sonavane M, Salim A, et al. Repurposing cancer drugs batimastat and marimastat to inhibit the activity of a group I metalloprotease from the venom of the western diamondback rattlesnake, *Crotalus atrox*. Toxins 2020;12:309. 2020 May 9;12(5):309.
[145] Xie C, Slagboom J, Albulescu LO, Somsen GW, Vonk FJ, Casewell NR, et al. Neutralising effects of small molecule toxin inhibitors on nanofractionated coagulopathic Crotalinae snake venoms. Acta Pharm Sin B 2020;10:1835–45.
[146] Trevisan-Silva D, Bednaski AV, Gremski LH, Chaim OM, Veiga SS, Senff-Ribeiro A. Differential metalloprotease content and activity of three Loxosceles spider venoms revealed using two-dimensional electrophoresis approaches. Toxicon 2013;76:11–22.
[147] Ellsworth SA, Nystrom GS, Ward MJ, Freitas de Sousa LA, Hogan MP, Rokyta DR. Convergent recruitment of adamalysin-like metalloproteases in the venom of the red bark centipede (Scolopocryptops sexspinosus). Toxicon 2019;168:1–15.
[148] Andrikopoulos PC, Liu Y, Picchiotti A, Lenngren N, Kloz M, Chaudhari AS, et al. Femtosecond-to-nanosecond dynamics of flavin mononucleotide monitored by stimulated Raman spectroscopy and simulations. Phys Chem Chem Phys 2020;22(12):6538–52.

[149] Tempone AG, Andrade HF, Spencer PJ, Lourenço CO, Rogero JR, Nascimento N. Bothrops moojeni venom kills Leishmania spp. with hydrogen peroxide generated by its L-amino acid oxidase. Biochem Biophys Res Commun 2001;280(3):620–4.
[150] Bansal PS, Torres AM, Crossett B, Wong KKY, Koh JMS, Geraghty DP, et al. Substrate specificity of platypus venom L-to-D-peptide isomerase. J Biol Chem 2008;283(14):8969–75.
[151] Shikata Y, Watanabe T, Teramoto T, Inoue A, Kawakami Y, Nishizawa Y, et al. Isolation and characterization of a peptide isomerase from funnel web spider venom. J Biol Chem 1995;270(28):16719–23.
[152] Jacobsen R, Jimenez EC, Grilley M, Watkins M, Hillyard D, Cruz LJ, et al. The contryphans, a D-tryptophan-containing family of Conus peptides: interconversion between conformers. J Pept Res 1998;51(3):173–9.
[153] Kuwada M, Teramoto T, Kumagaye KY, Nakajima K, Watanabe T, Kawai T, et al. Omega-agatoxin-TK containing D-serine at position 46, but not synthetic omega-[L-Ser46]agatoxin-TK, exerts blockade of P-type calcium channels in cerebellar Purkinje neurons. Mol Pharmacol 1994;46(4):587–93.
[154] Koludarov I, Jackson TNW, op den Brouw B, Dobson J, Dashevsky D, Arbuckle K, et al. Enter the dragon: the dynamic and multifunctional evolution of Anguimorpha lizard venoms. Toxins (Basel) 2017;9(8):242. https://doi.org/10.3390/toxins9080242.
[155] Yamazaki Y, Morita T. Structure and function of snake venom cysteine-rich secretory proteins. Toxicon 2004;44(3):227–31.
[156] Brown RL, Lynch LL, Haley TL, Arsanjani R. Pseudechetoxin binds to the pore turret of cyclic nucleotide-gated ion channels. J Gen Physiol 2003;122(6):749–60.
[157] Hew BE, Pangburn MK, Vogel CW, Fritzinger DC. Identification of intermolecular bonds between human factor B and cobra venom factor important for C3 convertase stability. Toxicon 2020;184(March):68–77.
[158] Huang XD, Zhao W, Zhang LX, Yang XJ, Wang LH, Chen YS, et al. The role of complement activation in rhabdomyolysis-induced acute kidney injury. PLoS One 2018;13(2):1–20.
[159] Araujo-Voces M, Quesada V. Frequent birth-and-death events throughout perforin-1 evolution. BMC Evol Biol 2020;20(1):135. Internet. Available from: https://www.ncbi.nlm.nih.gov/pmc/articles/PMC7574235/.
[160] Ellisdon AM, Reboul CF, Panjikar S, Huynh K, Oellig CA, Winter KL, et al. Stonefish toxin defines an ancient branch of the perforin-like superfamily. Proc Natl Acad Sci USA 2015;112(50):15360–5.
[161] Calvete JJ, Marcinkiewicz C, Sanz L. Snake venomics of *Bitis gabonica gabonica*. Protein family composition, subunit organization of venom toxins, and characterization of dimeric disintegrins bitisgabonin-1 and bitisgabonin-2. J Proteome Res 2007;6(1):326–36.
[162] Urra FA, Araya-Maturana R. Targeting metastasis with snake toxins: molecular mechanisms. Toxins (Basel) 2017;9(12):390.
[163] Reynaud S, Ciolek J, Degueldre M, Saez NJ, Sequeira AF, Duhoo Y, et al. A venomics approach coupled to high-throughput toxin production strategies identifies the first venom-derived melanocortin receptor agonists. J Med Chem 2020;63(15):8250–64.
[164] Xiao L, Gurrola GB, Zhang J, Valdivia CR, SanMartin M, Zamudio FZ, et al. Structure-function relationships of peptides forming the calcin family of ryanodine receptor ligands. J Gen Physiol 2016;147(5):375–94.
[165] Chow CY, Chin YKY, Ma L, Undheim EAB, Herzig V, King GF. A selective NaV1.1 activator with potential for treatment of Dravet syndrome epilepsy. Biochem Pharmacol 2020;181:113991.

[166] Destriau R, Mccullough D, Rowlinson R, Abbott WM, Trim SA. Spider vs Snake—discovery of novel snake venom protease inhibitors from spider venom. In: Rowlinson, Abbott WM, Trim SA, editors. European Laboratory Research Group, Drug Discovery. Liverpool; 2019. p. 1.
[167] Santibáñez-López CE, Ontano AZ, Harvey MS, Sharma PP. Transcriptomic analysis of pseudoscorpion venom reveals a unique cocktail dominated by enzymes and protease inhibitors. Toxins (Basel) 2018;10(5):207.
[168] Earl STH, Richards R, Johnson LA, Flight S, Anderson S, Liao A, et al. Identification and characterisation of Kunitz-type plasma kallikrein inhibitors unique to Oxyuranus sp. snake venoms. Biochimie 2012;94(2):365–73.
[169] Ianzer D, Konno K, Marques-Porto R, Vieira Portaro FC, Stöcklin R, Martins De Camargo AC, et al. Identification of five new bradykinin potentiating peptides (BPPs) from *Bothrops jararaca* crude venom by using electrospray ionization tandem mass spectrometry after a two-step liquid chromatography. Peptides 2004;25(7):1085–92.
[170] Murayama N, Hayashi MAF, Ohi H, Ferreira LAF, Hermann VV, Saito H, et al. Cloning and sequence analysis of a *Bothrops jararaca* cDNA encoding a precursor of seven bradykinin-potentiating peptides and a C-type natriuretic peptide. Proc Natl Acad Sci USA 1997;94(4):1189–93.
[171] Roy A, Qingxiang S, Alex C, Rajagopalan N, Jobichen C, Sivaraman J, et al. Identification of a α-helical molten globule intermediate and structural characterization of β-cardiotoxin, an all β-sheet protein isolated from the venom of *Ophiophagus hannah* (king cobra). Protein Sci 2019;28(5):952–63.
[172] Näreoja K, Kukkonen JP, Rondinelli S, Toivola DM, Meriluoto J, Näsman J. Adrenoceptor activity of muscarinic toxins identified from mamba venoms. Br J Pharmacol 2011;164(2 B):538–50.
[173] Ciolek J, Reinfrank H, Quinton L, Viengchareun S, Stura EA, Vera L, et al. Green mamba peptide targets type-2 vasopressin receptor against polycystic kidney disease. Proc Natl Acad Sci USA 2017;114(27):7154–9.
[174] Vaiyapuri S, Hutchinson EG, Ali MS, Dannoura A, Stanley RG, Harrison RA, et al. Rhinocetin, a venom-derived integrin-specific antagonist inhibits collagen-induced platelet and endothelial cell functions. J Biol Chem 2012;287(31):26235–44.
[175] Cudic M, Condie BA, Weiner DJ, Lysenko ES, Xiang ZQ, Insug O, et al. Development of novel antibacterial peptides that kill resistant isolates. Peptides 2002;23(12):2071–83.
[176] Peigneur S, Orts DJB, Prieto da Silva AR, Oguiura N, Boni-Mitake M, de Oliveira EB, et al. Crotamine pharmacology revisited: novel insights based on the inhibition of KV channels. Mol Pharmacol 2012;82(1):90–6.
[177] Poillot C, Dridi K, Bichraoui H, Pêcher J, Alphonse S, Douzi B, et al. D-maurocalcine, a pharmacologically inert efficient cell-penetrating peptide analogue. J Biol Chem 2010;285(44):34168–80.
[178] Terwilliger TC, Eisenberg D. The structure of melittin. II. Interpretation of the structure. J Biol Chem 1982;257(11):6016–22.
[179] Lyu C, Fang F, Li B. Anti-tumor effects of melittin and its potential applications in clinic. Curr Protein Pept Sci 2019;20(3):240–50.
[180] Memariani H, Memariani M, Moravvej H, Shahidi-Dadras M. Melittin: a venom-derived peptide with promising anti-viral properties. Eur J Clin Microbiol Infect Dis 2020;39(1):5–17.
[181] Kostiza T, Meier J. Nerve growth factors from snake venoms: chemical properties, mode of action and biological significance. Toxicon 1996;34(7):787–806.
[182] Yamazaki Y, Matsunaga Y, Tokunaga Y, Obayashi S, Saito M, Morita T. Snake venom vascular endothelial growth factors (VEGF-Fs) exclusively vary their structures and functions among species. J Biol Chem 2009;284(15):9885–91.

[183] Ching ATC, Paes Leme AF, Zelanis A, Rocha MMT, Furtado MdF, Silva DA, et al. Venomics profiling of *Thamnodynastes strigatus* unveils matrix metalloproteinases and other novel proteins recruited to the toxin arsenal of rear-fanged snakes. J Proteome Res 2012;11(2):1152–62.
[184] Robinson SD, Mueller A, Clayton D, Starobova H, Hamilton BR, Payne RJ, et al. A comprehensive portrait of the venom of the giant red bull ant, *Myrmecia gulosa*, reveals a hyperdiverse hymenopteran toxin gene family. Sci Adv 2018;4(9):eaau4640.
[185] Shiomi K, Honma T, Ide M, Nagashima Y, Ishida M, Chino M. An epidermal growth factor-like toxin and two sodium channel toxins from the sea anemone *Stichodactyla gigantea*. Toxicon 2003;41(2):229–36.
[186] McCullough D, Atofanei C, Knight E, Trim SA, Trim CM. Kinome scale profiling of venom effects on cancer cells reveals potential new venom activities. Toxicon 2020;185:129–46.
[187] Al-Asmari AK, Kunnathodi F, Al Saadon K, Idris MM. Elemental analysis of scorpion venoms. J Venom Res 2016;7:16–20.
[188] Inceoglu B, Lango J, Jing J, Chen L, Doymaz F, Pessah IN, et al. One scorpion, two venoms: prevenom of Parabuthus transvaalicus acts as an alternative type of venom with distinct mechanism of action. Proc Natl Acad Sci USA 2003;100(3):922–7.
[189] Forster YM, Reusser S, Forster F, Bienz S, Bigler L. VenoMS—a website for the low molecular mass compounds in spider venoms. Metabolites 2020;10(8):1–16.
[190] Esmaeilishirazifard E, Usher L, Trim C, Denise H, Sangal V, Tyson GH, et al. Microbial adaptation to venom is common in snakes and spiders. bioRxiv; 2018 Jun 16. p. 348433. cited 2019 Aug 15. Internet. Available from: https://www.biorxiv.org/content/10.1101/348433v1.
[191] Ul-Hasan S, Rodríguez-Román E, Reitzel AM, Adams RMM, Herzig V, Nobile CJ, et al. The emerging field of venom-microbiomics for exploring venom as a microenvironment, and the corresponding initiative for venom associated microbes and parasites (iVAMP). Toxicon X 2019;4:100016.
[192] Rádis-Baptista G, Dodou HV, Prieto-da-Silva ÁRB, Zaharenko AJ, Kazuma K, Nihei K, et al. Comprehensive analysis of peptides and low molecular weight components of the giant ant *Dinoponera quadriceps* venom. Biol Chem 2020;401:945–54.
[193] Pallaghy PK, Nielsen KJ, Craik DJ, Norton RS. A common structural motif incorporating a cystine knot and a triple-stranded beta-sheet in toxic and inhibitory polypeptides. Protein Sci 1994;3(10):1833–9.
[194] Cardoso FC, Dekan Z, Rosengren KJ, Erickson A, Vetter I, Deuis JR, et al. Identification and characterization of ProTx-III [μ-TRTX-Tp1a], a new voltage-gated sodium channel inhibitor from venom of the tarantula *Thrixopelma pruriens*. Mol Pharmacol 2015;88(2):291–303.
[195] Bohlen CJ, Priel A, Zhou S, King D, Siemens J, Julius D. A bivalent tarantula toxin activates the capsaicin receptor, TRPV1, by targeting the outer pore domain. Cell 2010;141(5):834–45.
[196] Chassagnon IR, McCarthy CA, Chin YK-Y, Pineda SS, Keramidas A, Mobli M, et al. Potent neuroprotection after stroke afforded by a double-knot spider-venom peptide that inhibits acid-sensing ion channel 1a. Proc Natl Acad Sci USA 2017;114(14):3750–5.
[197] Sunagar K, Undheim EAB, Chan AHC, Koludarov I, Muñoz-Gómez SA, Antunes A, et al. Evolution stings: the origin and diversification of scorpion toxin peptide scaffolds. Toxins (Basel) 2013;5(12):2456–87.
[198] Rong M, Liu J, Zhang M, Wang G, Zhao G, Wang G, et al. A sodium channel inhibitor ISTX-I with a novel structure provides a new hint at the evolutionary link between two toxin folds. Sci Rep 2016;6:29691.

[199] Slagboom J, Kool J, Harrison RA, Casewell NR. Haemotoxic snake venoms: their functional activity, impact on snakebite victims and pharmaceutical promise. Br J Haematol 2017;177(6):947–59.
[200] Munawar A, Ali SA, Akrem A, Betzel C. Snake venom peptides: tools of biodiscovery. Toxins (Basel) 2018;10(11):474.
[201] Barber CM, Isbister GK, Hodgson WC. Alpha neurotoxins. Toxicon 2013;66:47–58.
[202] Osipov AV, Utkin YN. Snake venom toxins targeted at the nervous system. In: Gopalakrishnakone P, Inagaki H, Mukherjee AK, Rahmy TR, Vogel C-W, editors. Snake venoms. Dordrecht: Springer Netherlands; 2015. p. 1–21.
[203] Kessler P, Marchot P, Silva M, Servent D. The three-finger toxin fold: a multifunctional structural scaffold able to modulate cholinergic functions. J Neurochem 2017;142 (Suppl):7–18.
[204] Szolajska E, Poznanski J, Ferber ML, Michalik J, Gout E, Fender P, et al. Poneratoxin, a neurotoxin from ant venom. Structure and expression in insect cells and construction of a bio-insecticide. Eur J Biochem 2004;271(11):2127–36.
[205] Chen J, Guan S-M, Sun W, Fu H. Melittin, the major pain-producing substance of bee venom. Neurosci Bull 2016;32(3):265–72.
[206] Harris JB, Scott-Davey T. Secreted phospholipases A2 of snake venoms: effects on the peripheral neuromuscular system with comments on the role of phospholipases A2 in disorders of the CNS and their uses in industry. Toxins (Basel) 2013;5(12):2533–71.
[207] Saidemberg DM, Ferreira MAB, Takahashi TN, Gomes PC, Cesar-Tognoli LMM, da Silva-Filho LC, et al. Monoamine oxidase inhibitory activities of indolylalkaloid toxins from the venom of the colonial spider *Parawixia bistriata*: functional characterization of PwTX-I. Toxicon 2009;54(6):717–24.
[208] Walker AA, Weirauch C, Fry BG, King GF. Venoms of heteropteran insects: a treasure trove of diverse pharmacological toolkits. Toxins (Basel) 2016;8(2):43.
[209] Xiong X-F, Poulsen MH, Hussein RA, Nørager NG, Strømgaard K. Structure-activity relationship study of spider polyamine toxins as inhibitors of ionotropic glutamate receptors. ChemMedChem 2014;9(12):2661–70.
[210] Itagaki Y, Nakajima T. Acylpolyamines: mass spectrometric analytical methods for araneidae spider acylpolyamines. J Toxicol Toxin Rev 2000;19(1):23–52.
[211] Usherwood PN, Blagbrough IS. Spider toxins affecting glutamate receptors: polyamines in therapeutic neurochemistry. Pharmacol Ther 1991;52(2):245–68.
[212] McCormick KD, Meinwald J. Neurotoxic acylpolyamines from spider venoms. J Chem Ecol 1993;19(10):2411–51.
[213] Chiba T, Akizawa T, Mastukawa M, Nishi M, Kawai N, Yoshioka M. Total syntheses of spidamine and joramine, polyamine toxins from the Joro Spider, *Nephila clavata*. Chem Pharm Bull 1996;44(5):972–5.
[214] McCormick J, Li Y, McCormick K, Duynstee HI, van Engen AK, van der Marel GA, et al. Structure and total synthesis of HF-7, a neuroactive glyconucleoside disulfate from the funnel-web spider *Hololena curta*. J Am Chem Soc 1999;121(24):5661–5.
[215] Silva RCMC, Fox EGP, Gomes FM, Feijó DF, Ramos I, Koeller CM, et al. Venom alkaloids against chagas disease parasite: search for effective therapies. Sci Rep 2020;10 (1):10642.
[216] Gentilucci L, Cardillo G, Spampinato S, Tolomelli A, Squassabia F, De Marco R, et al. Antiangiogenic effect of dual/selective α5β1/αvβ3 integrin antagonists designed on partially modified retro-Inverso cyclotetrapeptide mimetics. J Med Chem 2010;53 (1):106–18.
[217] Noto PB, Abbadessa G, Cassone M, Mateo GD, Agelan A, Wade JD, et al. Alternative stabilities of a proline-rich antibacterial peptide in vitro and in vivo. Protein Sci 2008;17(7):1249–55.

[218] Herzig V, King GF. The cystine knot is responsible for the exceptional stability of the insecticidal spider toxin ω-Hexatoxin-Hv1a. Toxins (Basel) 2015;7(10):4366–80.
[219] Li ZJ, Cho CH. Peptides as targeting probes against tumor vasculature for diagnosis and drug delivery. J Transl Med 2012;10 Suppl 1(Suppl. 1):S1.
[220] Teesalu T, Sugahara KN, Kotamraju VR, Ruoslahti E. C-end rule peptides mediate neuropilin-1-dependent cell, vascular, and tissue penetration. Proc Natl Acad Sci USA 2009;106(38):16157–62.
[221] Albert R, Marbach P, Bauer W, Briner U, Fricker G, Bruns C, et al. SDZ CO 611: a highly potent glycated analog of somatostatin with improved oral activity. Life Sci 1993;53(6):517–25.
[222] Kihlberg J, Ahman J, Walse B, Drakenberg T, Nilsson A, Söderberg-ahlm C, et al. Glycosylated peptide hormones: pharmacological properties and conformational studies of analogues of [1-desamino,8-D-arginine]vasopressin. J Med Chem 1995;38 (1):161–9.
[223] Egleton RD, Davis TP. Development of neuropeptide drugs that cross the blood-brain barrier. NeuroRx 2005;2(1):44–53.
[224] Varamini P, Mansfeld FM, Blanchfield JT, Wyse BD, Smith MT, Toth I. Synthesis and biological evaluation of an orally active glycosylated endomorphin-1. J Med Chem 2012;55(12):5859–67.
[225] Lax R, Meenan C. Challenges for therapeutic peptides part 1: on the inside, looking out. Innov Pharm Technol 2012;42:54–6.
[226] Yu R, Seymour VAL, Berecki G, Jia X, Akcan M, Adams DJ, et al. Less is more: design of a highly stable disulfide-deleted mutant of analgesic cyclic α-conotoxin Vc1.1. Sci Rep 2015;5(May):1–11.
[227] Stepensky D. Pharmacokinetics of toxin-derived peptide drugs. Toxins (Basel) 2018;10(11):483.
[228] Duchin KL, McKinstry DN, Cohen AI, Migdalof BH. Pharmacokinetics of captopril in healthy subjects and in patients with cardiovascular diseases. Clin Pharmacokinet 1988;14(4):241–59.
[229] Smith BS, Gandhi PJ. Pharmacokinetics and pharmacodynamics of low-molecular-weight heparins and glycoprotein IIb/IIIa receptor antagonists in renal failure. J Thromb Thrombolysis 2001;11(1):39–48.
[230] Pope JE, Deer TR. Ziconotide: a clinical update and pharmacologic review. Expert Opin Pharmacother 2013;14(7):957–66.
[231] McGivern JG. Ziconotide: a review of its pharmacology and use in the treatment of pain. Neuropsychiatr Dis Treat 2007;3(1):69–85.
[232] Gurrola GB, Capes EM, Zamudio FZ, Possani LD, Valdivia HH. Imperatoxin a, a cell-penetrating peptide from scorpion venom, as a probe of Ca2+-release channels/ryanodine receptors. Pharmaceuticals 2010;3(4):1093–107.
[233] Williamson H, Bennett N, Walsh J. Activating challenging GPCR targets with venom peptides. In: Presented at the European Laboratory Research Group Drug Discovery conference; 2017. p. 132.
[234] Younis S, Rashid S. Alpha conotoxin-BuIA globular isomer is a competitive antagonist for oleoyl-L-alpha-lysophosphatidic acid binding to LPAR6; a molecular dynamics study. PLoS One 2017;12(12):1–18.
[235] Quinton L, Girard E, Maiga A, Rekik M, Lluel P, Masuyer G, et al. Isolation and pharmacological characterization of AdTx1, a natural peptide displaying specific insurmountable antagonism of the α 1A- adrenoceptor. Br J Pharmacol 2010;159 (2):316–25.
[236] Maïga A, Merlin J, Marcon E, Rouget C, Larregola M, Gilquin B, et al. Orthosteric binding of ρ-Da1a, a natural peptide of snake venom interacting selectively with the α1A-adrenoceptor. PLoS One 2013;8(7):1–11.

[237] Liang JS, Carsi-Gabrenas J, Krajewski JL, McCafferty JM, Purkerson SL, Santiago MP, et al. Anti-muscarinic toxins from *Dendroaspis angusticeps*. Toxicon 1996;34(11 – 12): 1257–67.
[238] Petrel C, Hocking HG, Reynaud M, Upert G, Favreau P, Biass D, et al. Identification, structural and pharmacological characterization of τ-CnVA, a conopeptide that selectively interacts with somatostatin sst3 receptor. Biochem Pharmacol 2013;85(11): 1663–71.
[239] Bourinet E, Zamponi GW. Block of voltage-gated calcium channels by peptide toxins. Neuropharmacology 2017;127:109–15.
[240] Klint JK, Smith JJ, Vetter I, Rupasinghe DB, Er SY, Senff S, et al. Seven novel modulators of the analgesic target NaV1.7 uncovered using a high-throughput venom-based discovery approach. Br J Pharmacol 2015;172(10):2445–58.
[241] Servent D, Winckler-Dietrich V, Hu HY, Kessler P, Drevet P, Bertrand D, et al. Only snake curaremimetic toxins with a fifth disulfide bond have high affinity for the neuronal alpha7 nicotinic receptor. J Biol Chem 1997;272(39):24279–86.
[242] Kennedy AC, Belgi A, Husselbee BW, Spanswick D, Norton RS, Robinson AJ. α-Conotoxin peptidomimetics: probing the minimal binding motif for effective analgesia. Toxins (Basel) 2020;12(8):1–25.
[243] Rosso JP, Schwarz JR, Diaz-Bustamante M, Céard B, Gutiérrez JM, Kneussel M, et al. MmTX1 and MmTX2 from coral snake venom potently modulate receptor activity modulate GABAa receptor activity. Proc Natl Acad Sci USA 2015;112(8):E891–900.
[244] Moradi M, Najafi R, Amini R, Solgi R, Tanzadehpanah H, Esfahani AM, et al. Remarkable apoptotic pathway of Hemiscorpius lepturus scorpion venom on CT26 cell line. Cell Biol Toxicol 2019;35(4):373–85.
[245] Amirgholami N, Karampour NS, Ghadiri A, Tagavi moghadam A, Ghasemi dehcheshmeh M, Pipelzadeh MH. A. crassicauda, M. eupeus and *H. lepturus* scorpion venoms initiate a strong in vivo anticancer immune response in CT26-tumor mice model. Toxicon 2020;180:31–8.
[246] Das Gupta S, Halder B, Gomes A, Gomes A. Bengalin initiates autophagic cell death through ERK-MAPK pathway following suppression of apoptosis in human leukemic U937 cells. Life Sci 2013;93(7):271–6.
[247] Swenson S, Costa F, Ernst W, Fujii G, Markland FS. Contortrostatin, a snake venom disintegrin with anti-angiogenic and anti-tumor activity. Pathophysiol Haemost Thromb 2005;34(4–5):169–76.
[248] Minea R, Helchowski C, Rubino B, Brodmann K, Swenson S, Markland FJ. Development of a chimeric recombinant disintegrin as a cost-effective anti-cancer agent with promising translational potential. Toxicon 2012;59(4):472–86.
[249] Schönthal AH, Swenson SD, Chen TC, Markland FS. Preclinical studies of a novel snake venom-derived recombinant disintegrin with antitumor activity: a review. Biochem Pharmacol 2020;181:114149.
[250] Calderon LA, Sobrinho JC, Zaqueo KD, de Moura AA, Grabner AN, Mazzi MV, et al. Antitumoral activity of snake venom proteins: new trends in cancer therapy. Biomed Res Int 2014;2014:203639.
[251] Chaisakul J, Hodgson WC, Kuruppu S, Prasongsook N. Effects of animal venoms and toxins on hallmarks of cancer. J Cancer 2016;7(11):1571–8.
[252] Abdelfatah S, Lu X, Schmeda-Hirschmann G, Efferth T. Cytotoxicity and antimitotic activity of *Rhinella schneideri* and *Rhinella marina* venoms. J Ethnopharmacol 2019;242:112049.
[253] Badr G, Garraud O, Daghestani M, Al-Khalifa MS, Richard Y. Human breast carcinoma cells are induced to apoptosis by samsum ant venom through an IGF-1-dependant pathway, PI3K/AKT and ERK signaling. Cell Immunol 2012;273 (1):10–6.

[254] Al-Tamimi J, Semlali A, Hassan I, Ebaid H, Alhazza IM, Mehdi SH, et al. Samsum ant venom exerts anticancer activity through immunomodulation in vitro and in vivo. Cancer Biother Radiopharm 2018;33(2):65–73.

[255] Su J-C, Lin K-L, Chien C-M, Chuang P-W, Chang L-S, Lin S-R. Concomitant inactivation of the epidermal growth factor receptor, phosphatidylinositol 3-kinase/Akt and Janus tyrosine kinase 2/signal transducer and activator of transcription 3 signalling pathways in cardiotoxin III-treated A549 cells. Clin Exp Pharmacol Physiol 2010;37 (8):833–40.

[256] Lee H, Bae SK, Kim M, Pyo MJ, Kim M, Yang S, et al. Anticancer effect of Nemopilema nomurai jellyfish venom on HepG2 cells and a tumor xenograft animal model. Evid Based Complement Alternat Med 2017;2017:2752716.

[257] Merquiol L, Romano G, Ianora A, D'Ambra I. Biotechnological applications of scyphomedusae. Mar Drugs 2019;17(11):604.

[258] Rady I, Siddiqui IA, Rady M, Mukhtar H. Melittin, a major peptide component of bee venom, and its conjugates in cancer therapy. Cancer Lett 2017;402:16–31.

[259] Fezai M, Slaymi C, Ben-Attia M, Kroemer G, Lang F, Jemaà M. Inhibition of Colon carcinoma cell migration following treatment with purified venom from lesser weever fish (*Trachinus Vipera*). Cell Physiol Biochem 2017;41(6):2279–88.

[260] Patil CG, Walker DG, Miller DM, Butte P, Morrison B, Kittle DS, et al. Phase 1 safety, pharmacokinetics, and fluorescence imaging study of Tozuleristide (BLZ-100) in adults with newly diagnosed or recurrent gliomas. Neurosurgery 2019;85 (4):E641–9.

[261] Dodds DR. Antibiotic resistance: a current epilogue. Biochem Pharmacol 2017;134:139–46.

[262] Perumal Samy R, Stiles BG, Franco OL, Sethi G, Lim LHK. Animal venoms as antimicrobial agents. Biochem Pharmacol 2017;134:127–38.

[263] Azim S, McDowell D, Cartagena A, Rodriguez R, Laughlin TF, Ahmad Z. Venom peptides cathelicidin and lycotoxin cause strong inhibition of *Escherichia coli* ATP synthase. Int J Biol Macromol 2016;87:246–51.

[264] Al-Tamimi J, Alhazza IM, Al-Khalifa M, Metwalli A, Rady A, Ebaid H. Potential effects of samsum ant, Brachyponera sennaarensis, venom on TNF-α/NF-κB mediated inflammation in CCL4-toxicity in vivo. Lipids Health Dis 2016;15(1):198.

[265] Li Q, Zhao Z, Zhou D, Chen Y, Hong W, Cao L, et al. Virucidal activity of a scorpion venom peptide variant mucroporin-M1 against measles, SARS-CoV and influenza H5N1 viruses. Peptides 2011;32(7):1518–25.

[266] Memariani H, Memariani M, Shahidi-Dadras M, Nasiri S, Akhavan MM, Moravvej H. Melittin: from honeybees to superbugs. Appl Microbiol Biotechnol 2019;103 (8):3265–76.

[267] Zhang S-K, Ma Q, Li S-B, Gao H-W, Tan Y-X, Gong F, et al. RV-23, a melittin-related peptide with cell-selective antibacterial activity and high hemocompatibility. J Microbiol Biotechnol 2016;26(6):1046–56.

[268] Kuleshina ON, Kruykova EV, Cheremnykh EG, Kozlov LV, Andreeva TV, Starkov VG, et al. Screening snake venoms for toxicity to *Tetrahymena pyriformis* revealed anti-protozoan activity of cobra cytotoxins. Toxins (Basel) 2020;12(5):325.

[269] Bhattacharya S, Ghosh P, De T, Gomes A, Gomes A, Dungdung SR. In vivo and in vitro antileishmanial activity of *Bungarus caeruleus* snake venom through alteration of immunomodulatory activity. Exp Parasitol 2013;135(1):126–33.

[270] Hou S, Liu Y, Tang Y, Wu M, Guan J, Li X, et al. Anti-toxoplasma gondii effect of two spider venoms in vitro and in vivo. Toxicon 2019;166:9–14.

[271] Cardoso FC, Lewis RJ. Structure-function and therapeutic potential of spider venom-derived cysteine knot peptides targeting sodium channels. Front Pharmacol 2019;10:366.

[272] Robinson SD, Safavi-Hemami H. Venom peptides as pharmacological tools and therapeutics for diabetes. Neuropharmacology 2017;127:79–86.
[273] Suntravat M, Nuchprayoon I, Pérez JC. Comparative study of anticoagulant and procoagulant properties of 28 snake venoms from families Elapidae, Viperidae, and purified Russell's viper venom-factor X activator (RVV-X). Toxicon 2010;56 (4):544–53.
[274] Estrada-Gomez S, Cardoso FC, Vargas-Muñoz LJ, Quintana-Castillo JC, Arenas Gómez CM, Pineda SS, et al. Venomic, transcriptomic, and bioactivity analyses of pamphobeteus verdolaga venom reveal complex disulfide-rich peptides that modulate calcium channels. Toxins (Basel) 2019;11(9):496.
[275] Moran Y, Gordon D, Gurevitz M. Sea anemone toxins affecting voltage-gated sodium channels—molecular and evolutionary features. Toxicon 2009;54(8):1089–101.
[276] Tajti G, Wai DCC, Panyi G, Norton RS. The voltage-gated potassium channel K(V) 1.3 as a therapeutic target for venom-derived peptides. Biochem Pharmacol 2020;181:114146.
[277] Kalman K, Pennington MW, Lanigan MD, Nguyen A, Rauer H, Mahnir V, et al. ShK-Dap22, a potent Kv1.3-specific immunosuppressive polypeptide. J Biol Chem 1998;273(49):32697–707.
[278] Rashid MH, Heinzelmann G, Huq R, Tajhya RB, Chang SC, Chhabra S, et al. A potent and selective peptide blocker of the Kv1.3 channel: prediction from free-energy simulations and experimental confirmation. PLoS One 2013;8(11): e78712.
[279] Prentis PJ, Pavasovic A, Norton RS. Sea anemones: quiet achievers in the field of peptide toxins. Toxins (Basel) 2018;10(1):36.
[280] Tarcha EJ, Olsen CM, Probst P, Peckham D, Muñoz-Elías EJ, Kruger JG, et al. Safety and pharmacodynamics of dalazatide, a Kv1.3 channel inhibitor, in the treatment of plaque psoriasis: a randomized phase 1b trial. PLoS One 2017;12(7):e0180762.
[281] Sunanda P, Krishnarjuna B, Peigneur S, Mitchell ML, Estrada R, Villegas-Moreno J, et al. Identification, chemical synthesis, structure, and function of a new KV1 channel blocking peptide from Oulactis sp. Pept Sci 2018;110(4):e24073.
[282] Richards V. The toxic principles of snake-venoms. Ind Med Gaz 1887;22(12):353–4.
[283] Barnett BB. The haemostatic uses of snake venoms. Proc R Soc Med 1935;2:1469–72.
[284] Favaloro EJ. The Russell viper venom time (RVVT) test for investigation of lupus anticoagulant (LA). Am J Hematol 2019;94(11):1290–6.
[285] La Duca FM. Method for clotting heparized blood by the addition of protamine, thrombin and snake venom. USA; 5089415; 1992.
[286] Bibic L, Herzig V, King GF, Stokes L. Development of high-throughput fluorescent-based screens to accelerate discovery of P2X inhibitors from animal venoms. J Nat Prod 2019;82:2559–67.
[287] Mayr LM, Fuerst P. The future of high-throughput screening. J Biomol Screen 2008;13(6):443–8.
[288] von Reumont B, Campbell L, Jenner R, von Reumont BM, Campbell LI, Jenner RA. Quo vadis venomics? A roadmap to neglected venomous invertebrates. Toxins (Basel) 2014;6(12):3488–551. Internet. cited 2018 Dec 26. Available from: http://www.mdpi.com/2072-6651/6/12/3488.
[289] Vetter I, Davis JL, Rash LD, Anangi R, Mobli M, Alewood PF, et al. Venomics: a new paradigm for natural products-based drug discovery. Amino Acids 2011;40(1):15–28.
[290] Currier RB, Calvete JJ, Sanz L, Harrison RA, Rowley PD, Wagstaff SC. Unusual stability of messenger RNA in snake venom reveals gene expression dynamics of venom replenishment. PLoS One 2012;7(8):1–10.
[291] Jia Y, Olvera P, Rangel F, Mendez B, Reddy S. Rapid identification of phospholipase A 2 transcripts from snake venoms. Toxins (Basel) 2019;11(2):1–12.

[292] Gui J, Liu B, Cao G, Lipchik AM, Perez M, Dekan Z, et al. A tarantula-venom peptide antagonizes the TRPA1 nociceptor ion channel by binding to the S1-S4 gating domain. Curr Biol 2014;24(5):473–83. Available from: https://doi.org/10.1016/j.cub.2014.01.013.
[293] Aili SR, Touchard A, Escoubas P, Padula MP, Orivel J, Dejean A, et al. Diversity of peptide toxins from stinging ant venoms. Toxicon 2014;92:166–78.
[294] Jimenez EC, Watkins M, Juszczak LJ, Cruz LJ, Olivera BM. Contryphans from Conus textile venom ducts. Toxicon 2001;39(6):803–8. Internet. Available from: http://www.ncbi.nlm.nih.gov/pubmed/11137539.
[295] Gonçalves TC, Benoit E, Kurz M, Lucarain L, Fouconnier S, Combemale S, et al. From identification to functional characterization of cyriotoxin-1a, an antinociceptive toxin from the spider *Cyriopagopus schioedtei*. Br J Pharmacol 2019;176(9):1298–314.
[296] Tjernberg A, Markova N, Griffiths WJ, Hallén D. DMSO-related effects in protein characterization. J Biomol Screen 2006;11(2):131–7.
[297] Murray JK, Ligutti J, Liu D, Zou A, Poppe L, Li H, et al. Engineering potent and selective analogues of GpTx-1, a tarantula venom peptide antagonist of the Na(V)1.7 sodium channel. J Med Chem 2015;58(5):2299–314.
[298] Zhang Y, Yang Q, Zhang Q, Peng D, Chen M, Liang S, et al. Engineering gain-of-function analogues of the spider venom peptide HNTX-I, a potent blocker of the hNa(V)1.7 sodium channel. Toxins (Basel) 2018;10(9):358.
[299] Smith JJ, Cummins TR, Alphy S, Blumenthal KM. Molecular interactions of the gating modifier toxin ProTx-II with NaV 1.5: implied existence of a novel toxin binding site coupled to activation. J Biol Chem 2007;282(17):12687–97.

CHAPTER TWO

PROTACs, molecular glues and bifunctionals from bench to bedside: Unlocking the clinical potential of catalytic drugs

M. Maneiro, E. De Vita, D. Conole, C.S. Kounde, Q. Zhang, and E.W. Tate*

Department of Chemistry, Molecular Sciences Research Hub, Imperial College London, London, United Kingdom

*Corresponding author: e-mail address: e.tate@imperial.ac.uk

Contents

Progress in Medicinal Chemistry, Volume 60
ISSN 0079-6468
https://doi.org/10.1016/bs.pmch.2021.01.002

Keywords: PROTAC, Protein degradation, Ubiquitin-proteasome system, Heterobifunctional molecules, Pharmacokinetics, Pharmacodynamics

1. PROTACs: Heterobifunctional molecules hijacking the UPS

Dynamic modulation of protein levels is a fundamental process for the maintenance of cellular homeostasis. Alongside protein expression and regulatory mechanisms, protein degradation machineries have evolved as essential players in the recycling and regulation of proteins [4]. Endogenous protein degradation needs to be tightly regulated to favour the disposal of aberrant or abnormal proteins, without incurring pathological or uncontrolled depletion of functional proteins and other cellular components. Different protein degradation pathways have been identified in cells, the most important being the ubiquitin-proteasome system (UPS) and lysosomal degradation [5]. PROTACs are compounds designed to hijack the UPS system with the intent of inducing the degradation of specific POIs.

The UPS is a suite of highly regulated intracellular protein degradation machinery that recycles most proteins in eukaryotes (Fig. 1A) [6]. Substrate proteins destined for degradation are covalently tagged with a highly conserved 76-amino acid protein termed ubiquitin (Ub). Specifically, this process is regulated stepwise by the activity of three classes of ubiquitinating enzymes: E1 (ubiquitin-activating enzyme), E2 (ubiquitin-conjugating enzyme) and E3 (ubiquitin-protein ligase). These enzymes sequentially activate, conjugate, and transfer an ubiquitin molecule to a lysine or the N-terminal of protein substrates in an ATP-dependent manner (Fig. 1A). Varied ligases sequences are associated with differing substrates. This process can be antagonised by another class of enzymes, the deubiquitinases (DUBs), further contributing to tight regulation of programmed protein degradation. DUBs are cysteine or metallo-proteases that catalyse the removal of mono- or poly-Ub chains from proteins (Fig. 1A), which can ultimately affect their localisation and activity, PPIs, and thus regulate degradation [7].

For protein degradation, the ubiquitination cycle is repeated at least four times to form a polyubiquitin chain on the protein substrate, which is thus

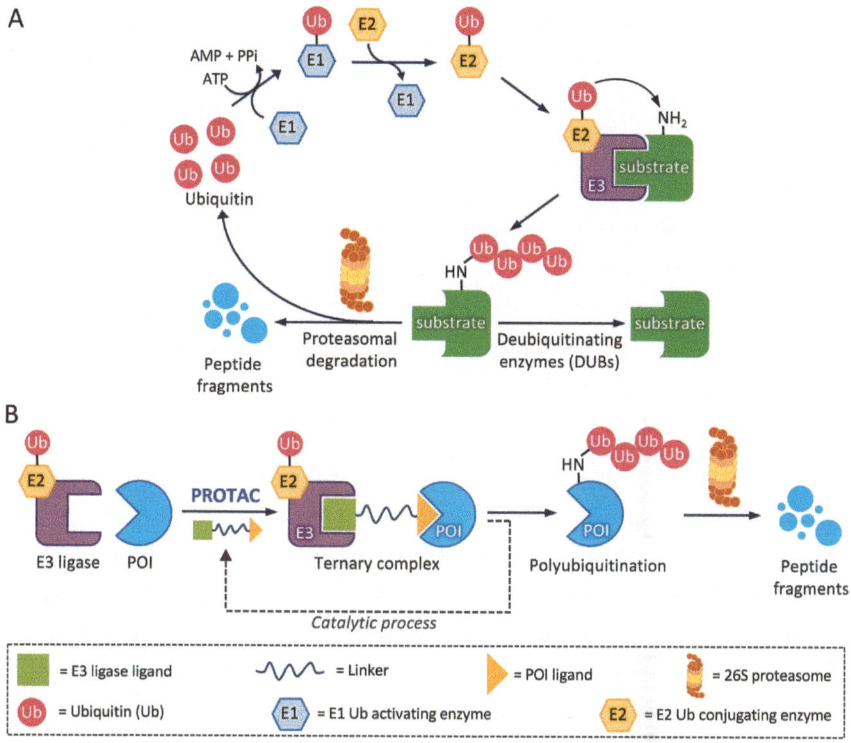

Fig. 1 (A) The ubiquitin-proteasome system (UPS) is one of the main regulatory mechanisms for protein homeostasis in eukaryotes. (B) The catalytic MOA of PROTACs involves the formation of a key ternary complex with an E3 ligase and a POI that results in polyubiquitination of the POI and subsequent proteasomal degradation.

tagged for recognition by the 26S proteasome. The proteasome is a macromolecular protein complex of cylindrical structure composed of a 20S proteolytic core, where threonine proteases with different substrate specificity catalyse the degradation of proteins, and a 19S regulatory cap that recognises ubiquitin-tagged proteins and directs them to the core for degradation. Proteolysis catalysed by the proteasome converts the protein into peptide fragments and ubiquitin molecules. These are released into the cytosol, where the fragments are recycled for anabolism [8–10].

On the basis of emerging understanding of the proteasomal degradation pathway, PROTACs have been designed as heterobifunctional molecules consisting of three basic components: a ligand that is able to recruit an E3 ligase, a specific warhead that binds to the POI, and a suitable linker to connect these units (Fig. 1B). This chimeric structure enables simultaneous recruitment of both the POI and the E3 ligase, forming a ternary complex

(E3-PROTAC-POI). Thus, the PROTAC is able to induce sufficient proximity between the E3 ligase and the POI to allow for the transfer of the first ubiquitin molecule from the E2 enzyme to an accessible lysine residue on the POI. This is followed by rapid transfer of additional ubiquitin units leading to polyubiquitin-tagged POI, which is then recognised by the 26S proteasome and degraded similarly to native UPS substrates.

1.1 Historical timeline and milestones of PROTACs

The concept of chimeric degraders was first described in the patent literature in 1999 by a biotech company, Proteinix, that reported bifunctional compounds consisting of an ubiquitination recognition element and a POI binding element to control protein levels and activity (Fig. 2) [11]. However, the name PROTACs was first introduced in 2001, in a pioneering work by the Crews group, who designed a phosphopeptide to recruit the $SCF^{\beta\text{-}TRCP}$ E3 ligase for the degradation of methionine aminopeptidase-2 (MetAP-2) [12]. In 2003, the same group reported PROTACs that were able to degrade the nuclear hormone receptors for androgen (AR) and oestrogen (ER), demonstrating broader applicability of this approach [13]. In 2004, a short hydroxyproline-containing peptide sequence derived from the von Hippel–Lindau tumour suppressor (VHL) natural substrate, the transcription factor hypoxia-inducible factor 1α (HIF1α), was used in the design of the first PROTACs to recruit the VHL E3 ligase [14]. In the presence of physiological oxygen levels, HIF1α gets enzymatically hydroxylated on a proline residue and is subsequently recognised by VHL for degradation, thus maintaining homeostasis by preventing expression of hypoxic-response genes [15]. Exploiting this specific PPI, these chimeric compounds were successful degraders of GFP-fused FKBP12 protein and AR. However, this first generation of PROTAC molecules was peptide-based and suffered from limited cell permeability, significantly compromising cellular activity. To improve drug-like properties, small molecule recruiters were pursued for new E3 ligases, culminating in the first report of a small molecule PROTAC in 2008 [16]. This compound (Fig. 2, PROTAC 14) recruited the Mouse Double Minute 2 homologue (MDM2) [17] E3 ligase and AR resulting in targeted degradation at 10 μM. In 2010, bestatin methyl esters were used to develop inhibitor-of-apoptosis-protein (IAP)-based PROTACs targeting cellular retinol- and retinoic acid-binding proteins (CRABP-I and II) [18]. These types of degraders were named 'specific and non-genetic inhibitor of apoptosis protein (IAP)-dependent protein erasers' (SNIPERs, Fig. 2),

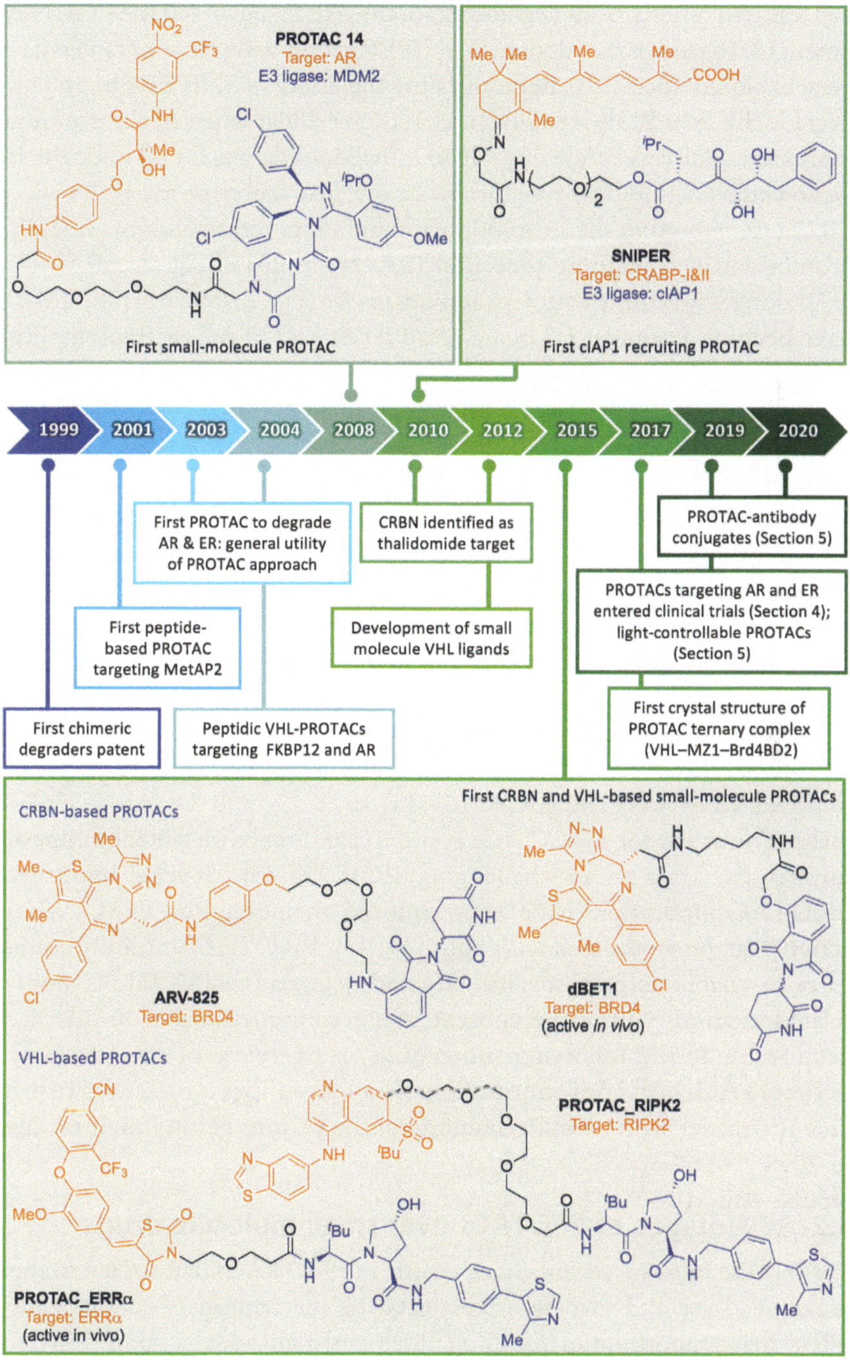

Fig. 2 Historical timeline of PROTAC development, showing the most significant milestones and some representative PROTAC structures.

and featured simultaneous degradation of the IAP E3 ligase and the POI. New potent IAP ligands were adopted in SNIPERs and a wide range of targets have been explored since 2010, demonstrating the utility of SNIPERs to potently degrade different POIs [18–24]. In 2010, the cellular target of the notorious immunomodulatory imide drug (IMiD) thalidomide was found to be the E3 ligase cereblon (CRBN) [25,26]. Since the first report of a CRBN-based PROTAC targeting the bromodomain and extra terminal domain (BET) bromodomain containing protein 4 (BRD4) in 2015 (Fig. 2, ARV-825) [27], along with confirmation of activity *in vivo* (Fig. 2, dBET1) [28], IMiDs have become a popular E3 ligase recruiter in PROTAC technology [29].

In 2015, VHL E3 ligase was first reported to be recruited by small molecule PROTACs [30,31]. The Ciulli group reported a VHL PROTAC targeting BRD4 (MZ1) [30], while the Crews group hijacked VHL to degrade the protein kinase RIPK2 in cells (Fig. 2, PROTAC_RIPK2) and the ERRα in mice (Fig. 2, PROTAC_ERRα) [31]. These proof-of-concept studies triggered subsequent successful PROTAC work and applications, leading to increased research interest from academia as well as investment from industry, which culminated in 2019 with two orally bioavailable PROTACs, targeting AR (ARV-110) and ER (ARV-471), progressing to Phase I clinical trials.

Several new E3 ligases have been successfully recruited and employed in PROTAC-mediated degradation, including KEAP1 [32] in 2018, RNF4 [33], RNF114 [34] and DCAF16 [35] in 2019, and DCAF15 [36] in 2020. The scope for PROTACs is still expanding, with potential opportunities to target more challenging POIs [37,38]. Several innovative PROTAC applications have been explored, including PROTACs whose activity can be modulated with light [39–43], PROTAC-antibody conjugates to enable antigen-specific targeted degradation [44,45], as well as related methods within the context of heterobifunctional molecules (see Section 5). In the following subsection, an overview of the advantages of PROTACs over conventional inhibition modalities is presented to provide the reader with an understanding of the reasons behind their success.

1.2 Advantages of PROTACs over small molecule drugs

Possibly the most successful characteristic of PROTACs that has led to their widespread use and success is their catalytic mechanism of action, which offers new opportunities for drug discovery that address some inherent limitations of traditional small molecule drugs. This MOA is described as an 'event-driven' process, where the formation of the ternary complex is the event that triggers ubiquitination and subsequent degradation of the POI [46]. This is in contrast to competitive, occupancy-driven traditional small molecules, for which the efficacy is determined by sustained POI binding, which usually requires high doses to achieve effective target engagement, often leading to off-target toxicity [47,48]. Non-covalent PROTACs can dissociate from the ternary complex once the POI is ubiquitinated, and are thus recycled for further ternary complex formation. Consequently, only sub-stoichiometric doses are required, mitigating the risk of off-target effects [49,50]. For example, VHL or CRBN-based PROTACs targeting BET family proteins (dBET1 [28], ARV-711 [51], ARV-825 [27] and MZ1 [30]), could induce more than 90% degradation of the target at nanomolar concentrations or even at a picomolar range [52]. This means that PROTACs can in principle deliver fast and quantitative degradation of the POI at low dosages, thus leading to potent and sustained phenotypic effects compared with small molecule drugs. Several examples in the literature highlight that PROTACs are more potent in phenotypic assays than the parent compound [53–55]. On the contrary, PROTAC concentration in extreme excess compared to the binding partners can favour binary interactions (i.e. PROTAC-POI and PROTAC-E3 ligase) over ternary complex formation thus inhibiting degradation, a phenomenon also known as the 'hook effect' (see Section 1.1.3) [56].

Another valuable aspect of PROTACs is the long lasting effects of protein degradation compared to modulatory binding of small molecules. Prolonged perturbance of protein levels has been observed *in cellulo* and also *in vivo*, where resynthesis of the target protein becomes the rate-determining step for duration of action, as opposed to dissociation and systemic clearance of the drug [55,57–59]. For example, an experiment in mice with a PROTAC against FKBP12 showed that, following a single day treatment with two 30 mg/kg injections, FKBP12 degradation could be observed for over a week [59].

An interesting phenomenon that has emerged thus far in the study of PROTACs is that a high-affinity POI warhead does not necessarily generate a more potent PROTAC [60]. Hence, it has been postulated that weak binding might be sufficient for effective PROTACs, in principle enabling previously unthinkable POIs to be targeted, as long as a ligand of weak affinity is known [38]. From a drug discovery perspective, some 75%–80% of the human proteome is currently considered pharmaceutically inaccessible to small molecule drugs and protein therapeutics [2]. This is mostly due to protein targets lacking well-defined binding pockets for small molecules, or to the inaccessibility of intracellular POIs for protein therapeutics, which can only target extracellular antigens [61,62]. Furthermore, many proteins have been shown to possess non-catalytic functions, such as scaffolding, that are essential in the regulation of multiple biological processes and cannot be always modulated by small molecule binding. These targets are deemed 'undruggable', or 'challenging to drug', and include transcription factors, scaffolding proteins, and non-enzymatic proteins [63,64]. Hence, PROTACs present a unique opportunity to degrade such POIs and modulate their function. A prominent example is the protein kinase class, which are ideal POIs for PROTACs, since many small molecule inhibitors are available [65]. For instance, while inhibitors of Fak kinase can only modulate kinase activity, the scaffolding functions of this protein remain intact, promoting cell migration and invasion. PROTACs against Fak have been shown to moderate kinase-independent scaffold signalling [66].

The ternary, proximity-dependent nature of PROTAC-mediated degradation can also lead to altered selectivity profiles compared to parent compounds, a phenomenon demonstrated in several proteome-wide studies of target depletion [28,31,55,67,68]. For example, a PROTAC based on the pan-CDK inhibitor SNS-032, shows selective degradation of CDK9 over other known CDK targets of SNS-032 [55]. In another report from the Crews group, PROTACs based on the promiscuous kinase inhibitor foretinib, which binds 133 kinases at 10 μM, showed binding to only 50 of those kinases, and effective degradation of an even smaller pool, which also varied depending on the E3 ligase recruited [38].

Another benefit of the event-driven MOA of PROTACs is the postulated potential to overcome conventional drug resistance mechanisms arising from active site mutations or compensatory upregulation [69]. Although mutation in a target protein may lead to weaker binary binding, PROTAC efficacy may remain, and further offers the opportunity to degrade multiple mutants simultaneously [70]. Indeed, several reported PROTACs outperform their parent inhibitor in treatment by degrading resistance-causing mutants of the POI [71–74]. Inhibition of BTK kinase with covalent inhibitor ibrutinib is effective against chronic lymphocytic leukaemia and other B-cell malignancies, but patients often develop drug resistance due to mutation of the target cysteine to serine in the kinase domain, which impairs the irreversible binding of the acrylamide covalent warhead of ibrutinib. PROTAC MT-802 with a reversible ibrutinib derivative warhead can degrade both wild-type and C481S BTK at equivalent potency [71]. Following PROTACs for BTK showed improved properties and were able to induce the degradation of six other BTK single-point mutants at the key 481 position with half-maximal effective concentrations around 30 nM [73].

Specific mechanisms of resistance against PROTACs have also been recently identified, caused by targeted E3 ligase under-regulation or loss-of-function mutations affecting the UPS system components [75,76]. Proteins involved in this functional axis play essential roles for cellular homeostasis, offering in principle an overall lower risk, albeit resistance mechanisms have been observed upon prolonged treatment with PROTACs [75]. While compensatory upregulation of the POI could be easily overcome by PROTACs, another theorised escape mechanism would involve upregulation of DUBs to promote POI deubiquitation over PROTAC-mediated ubiquitination. Nonetheless, different classes of DUBs inhibitors have been developed recently and co-treatment with PROTACs could help prevent resistance against TPD, at least in the deubiquitinating axis [77].

Last but not least, PROTACs can also serve as chemical biology probes and a useful alternative to genetic modification strategies such as RNA interference or clustered regularly interspaced short palindromic repeats (CRISPR)–Cas9 for protein knockdown or knockout. These techniques can be invaluable tools to study protein function, signalling pathways, cellular activities, and target validation in drug discovery, but they can require laborious and time-consuming manipulation of mRNA or genomic DNA, have limited use for essential genes, and do not provide information regarding post-translational activities of a POI in a complex biological system [78]. Hence, PROTACs can provide a convenient, rapid and selective alternative for protein knockdown at the post-translational level, with the potential for temporal and even spatial control [79].

1.3 Key parameters to evaluate PROTAC activity

PROTACs represent a new paradigm to externally trigger a cellular event, and thus differ significantly from conventional small molecule drugs, be it inhibitory or activatory binders of proteins. According to traditional classifications, PROTACs could be considered as agonists of the UPS system.

Hence, the way these compounds are evaluated both *in vitro* and *in vivo* requires different approaches to assess the occurrence of a 'cellular event', i.e., ternary complex formation, and at the same time quantify potency in terms of concentration and maximal efficacy. For *in vitro* assays for binary

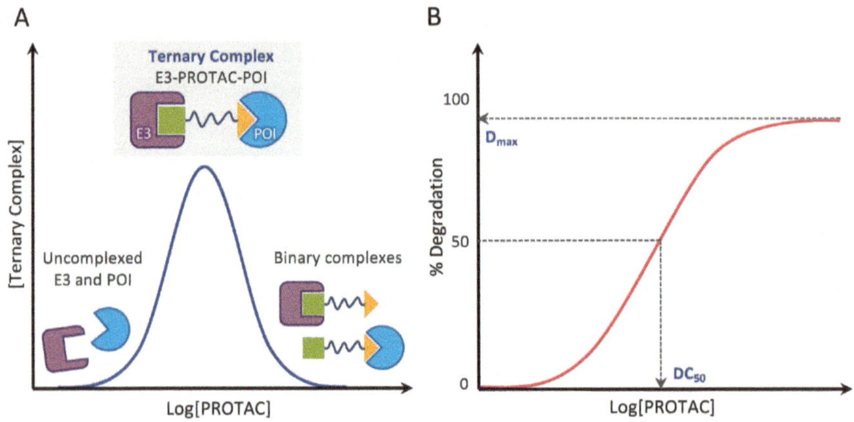

Fig. 3 (A) The concentration-response curve for PROTAC-mediated ternary complex formation is bell-shaped. Maximal ternary complex is obtained at an optimal PROTAC concentration. At higher PROTAC concentrations, the presence of binary complexes is predominant, and this phenomenon is known as 'hook effect'. (B) Dose-response curve for POI degradation with PROTACs. DC_{50} and D_{max} values are obtained by interpolation.

interaction with POI or E3 ligase, K_d values can be used to compare binding affinity to the parent compounds. Ternary complex formation should be evaluated next over a range of concentrations that allows determination of the optimal PROTAC concentration relative to POI and E3 ligase. In this context, the concentration-response curve may be bell-shaped rather than the conventional sigmoidal, highlighting the delicate equilibrium required to favour ternary over binary interactions (Fig. 3A) [68]. This is largely due to the hook effect [46,80], which describes a phenomenon whereby POI degradation decreases beyond an optimal PROTAC concentration as a result of the PROTAC driving binary complex formation at the expense of ternary interactions. Increasing concentrations of a PROTAC will lead to an increase in ternary complex formation until a maximum is reached, following which binary complexes will prevail due to the PROTAC excess. Common assays to quantify ternary complex formation are presented in Section 2.4.

Quantification of successful 'events' as opposed to simple target engagement is key when testing PROTACs *in cellulo*. For small molecules, dose-response curves are employed to determine EC_{50}, i.e., the concentration that gives 50% of the effect upon cell treatment. Similarly the 'effect' of PROTACs can be identified as POI degradation, which is quantifiable; thus DC_{50}, i.e., concentration that gives 50% degradation, can be used to rank

PROTACs (Fig. 3B). As discussed, POI degradation is a complex event, which is not only regulated by polyubiquitination, but also by POI expression and further compensatory mechanisms. Hence, a PROTAC cannot always induce complete POI degradation, and D_{max}, or the maximal degradation, is another fundamental parameter to consider in a PROTAC optimisation campaign (Fig. 3B). Since excess concentration may be deleterious for activity optimisation of ternary complex cooperativity and kinetics should be sought to improve DC_{50} and D_{max} values, and cooperativity should be balanced to avoid reduced catalytic rates of PROTACs with extremely high ternary complex affinities [56,81,82].

1.4 General considerations for PROTAC design

As the underlying principles governing PROTAC efficacy differ significantly from conventional small molecules, novel design strategies are needed to achieve effective POI degradation. The formation of a stable ternary complex is favourable for ubiquitin transfer, and ultimately determines the occurrence and efficiency of target degradation [31,68]. A potential breakthrough in understanding PROTAC ternary complexes was provided by X-ray crystallographic structure studies (Fig. 4A) [68,83,84]. Furthermore, assays and technologies to study ternary complex formation under more dynamic conditions have been developed (see Section 2). It has become clear that multiple factors can affect the structure and stability of ternary complexes, including the type of E3 ligase, design of the PROTAC linker, the choice of POI ligand and binding site exit vector, and the nature of the PPI interface between the E3 ligase and the POI. Modular design is often employed to explore PROTAC structure-activity relationships (SAR), with different combinations of PROTAC components often resulting in varied and unpredictable degradation profiles. While optimisation of ternary complex stability is the key to designing effective and potent PROTACs, predicting the effect of specific PROTAC design still represents the major challenge in the field [37]. For instance, study of PROTAC linkers alone, which has been named 'linkerology', shows these influence the degradation, permeability, efficacy and selectivity of the designed PROTAC through minimal variations in the chemical nature, length, and geometry of the linker (see Section 2) [83,85,86].

Over 600 E3 ligases have been identified in the human proteome but only a handful have been successfully recruited to interact with small-molecule PROTACs [87,88]. The most used for PROTAC design include

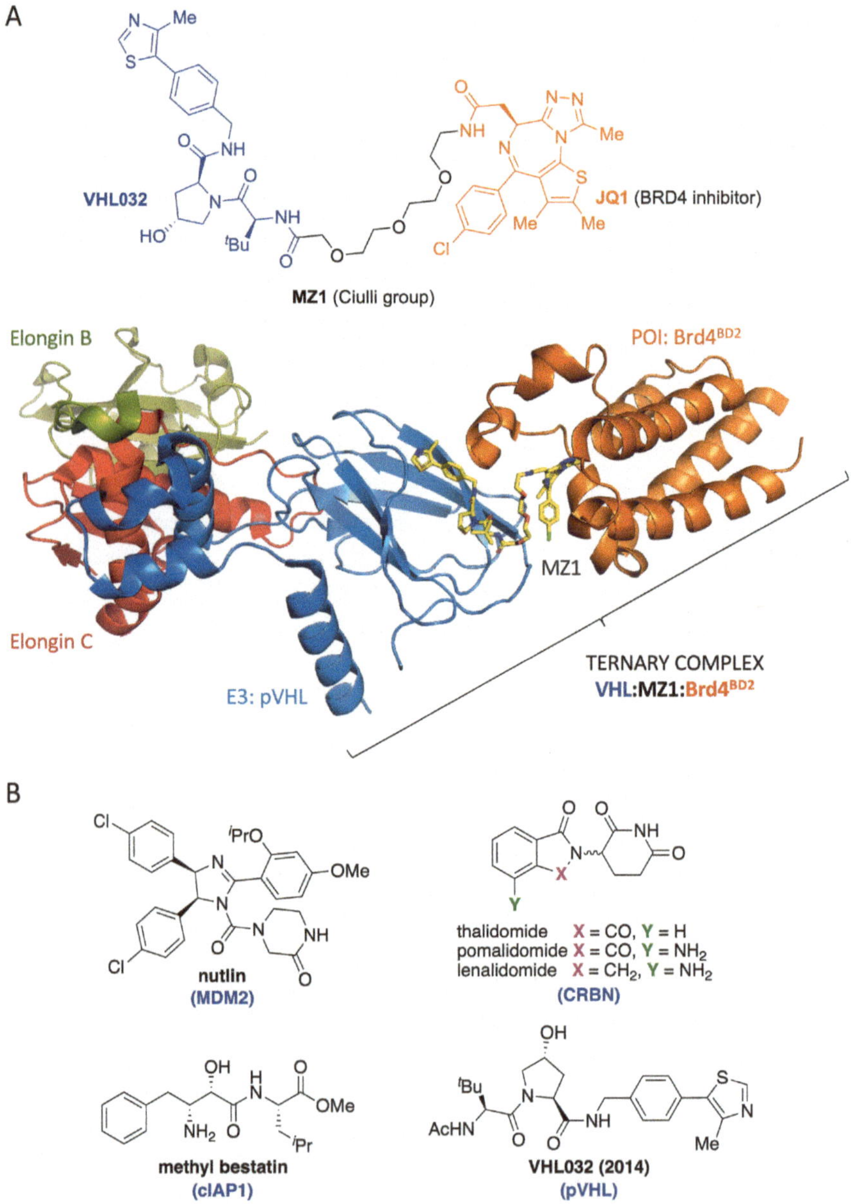

Fig. 4 Structural considerations of PROTACs. (A) The first determined crystallographic structure of a PROTAC ternary complex: BRD4 degrader MZ1 fold in place to create a cooperative binding between human VHL and BRD4 bromodomain (Brd4^{BD2}) (PDB: 5T35, 2.70 Å). (B) Representative E3 ligase ligands used for PROTAC design.

CRBN [27,28], VHL [14,30], IAPs [18,89], and MDM2 [16] (Fig. 4B) [16]. Among these E3 ligases, VHL and CRBN are the most commonly recruited, targeted mechanistically in over 80% of the reported PROTACs to date since 2015 [90]. In recent times, the scope of available E3 ligase ligands has expanded, with an expanded set of small-molecule ligands now successfully employed for PROTAC mediated targeted degradation (see Section 3). It is anticipated that an increase in the accessibility and understanding of various E3 ligases will allow the fine tuning of PROTACs for particular proteins and disease contexts [38,91].

PROTAC technology has the potential to be applied to any POI for which a ligand has been optimised, and to repurpose low-affinity ligands or binders which themselves have no effect on protein activity. To date, PROTACs have been used to degrade a broad range of protein classes, including epigenetic targets, kinases, nuclear hormone receptors and transcription factors. However, almost infinite possibilities for modular design are now at the hands of researchers, with the fast expansion of liganded E3 ligases, increased understanding of linkerology effects and modest requirements for warhead affinity. With these opportunities will come substantial challenges of activity-guided PROTAC design, cryptic side-effects of liganding E3 ligases, and the potential for promiscuous degradation due to potent degradation of low-affinity off-targets, requiring careful deconvolution through proteomic analysis.

1.5 PROTACs and 'molecular glues': Two sub-modalities for targeted degradation emerge

PROTACs has been described as inherently heterobifunctional molecules consisting of two different ligands and a linker, which are designed to bring together a POI and an E3 ligase to form a non-native ternary complex that causes a programmed biological degradation event. However, compact 'monofunctional' molecules lacking a linker which can trigger a similar ternary complex have also been reported, and these have been classified as 'molecular glues' (Fig. 5A and B). The plant hormone auxin is an example of molecular glue in nature, which degrades transcriptional regulators Aux/IAAs by promoting their interactions with the SCF^{TIR1} ubiquitin ligase complex [92]. Rather than inducing POI and E3 ligase proximity by linking binary binding interactions, molecular glues enhance or reprogram protein–protein interactions between the two partners resulting in a tighter, cooperative ternary complex [93–95]. As such, the binding affinity

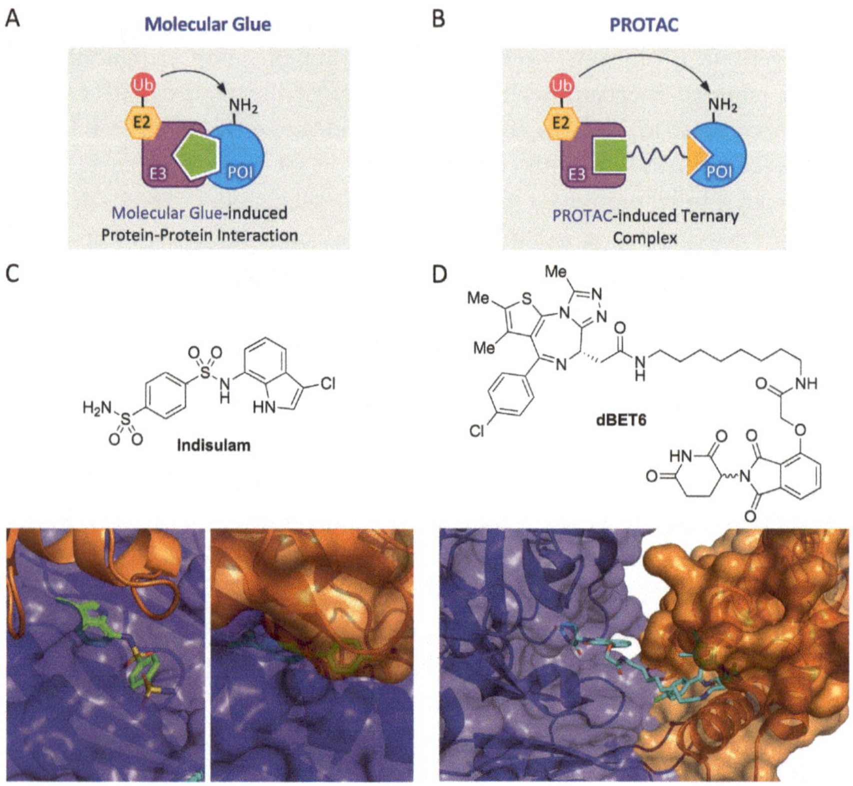

Fig. 5 Schematic representation for the comparison of molecular glue (A) and PROTAC (B) approaches. (C) Snapshots of the crystal structure of the human DDB1-DDA1-DCAF15 E3 ubiquitin ligase (blue) in complex with RBM39 (orange) and Indisulam (green) (PDB: 6SJ7, 3.54 Å) Indisulam induces significant PPI between DCAF15 and RBM39. (D) Crystal structure of the ternary complex between DDB1-CRBN E3 ligase (blue), BRD4(BD1) (orange) and PROTAC dBET6 (cyan) (PDB: 6BOY, 3.33 Å). A non-cooperative ternary complex was induced by dBET6 binding with no significantly induced PPI.

of a molecular glue for each binding partner can be surprisingly low since the binding affinity is derived mainly from the protein–protein interface [92,96–98].

In the context of drug discovery, molecular glues afford more favourable pharmaceutical properties than PROTACs (see Section 4) [99,100]. However, due to their smaller size and serendipitous PPI induction profiles, design of novel and potent molecular glues is challenging, and their identification has arisen primarily from retrospective or empirical studies [101,102]. In contrast, the PROTAC heterodimeric structure allows for modular design, which, as discussed in Section 1.3, often entails mix and match of the best binders and linkers to find an optimal combination.

A canonical example of a molecular glue is the anticancer sulfonamide indisulam (Fig. 5C), which was reported to induce degradation of RBM39 splicing factor by recruitment of the DCAF15 E3 ligase [103–106]. Indisulam promotes significant PPIs and cooperative binding between the E3 ligase and the substrate protein, despite possessing negligible *in vitro* binding affinity to RBM39 [107]. This is in contrast to PROTAC dBET6 (Fig. 5D), which does not require cooperativity for ternary complex stability [83]. Further studies of IMiDs (Fig. 4A) have shown that these compounds can recruit cancer-related neosubstrates, such as transcription factors lkaros (IKZF1) and Aiolos (IKZF3) to CRBN, thus acting as molecular glues, contributing to the teratogenic effects of this class of compounds [25,108–110]. Subtle changes in structure can drastically change neosubstrate recruitment profiles of IMiDs, with GSPT1 [111,112], SALL4 [113], and ARID2 [114], all recently identified as IMiD-CRBN neosubstrates. Though molecular glues are not the principal subject of this chapter, their correlation with PROTAC advancement is evident and thus the state of the art for molecular glues, including discovery approaches, will also be discussed in Section 3.

2. From concept to practice: Challenges and opportunities in PROTAC development

Following conceptualisation and first successes of PROTACs, many scientists have contributed to the field, which has grown exponentially over a relatively short time. In this section, we will present standard approaches to PROTAC optimisation that have been delineated over the last two decades.

2.1 Considerations and methodologies for POI and E3 ligase selection

While a POI may constitute a well-validated therapeutic target by gene knockout or knockdown strategies, whether this particular protein is amenable to a pharmacologically driven, TPD therapeutic strategy constitutes an entirely new query. In order for a POI to be degradable, a series of endogenous target protein features need to be considered.

First of all, protein expression and localisation should be evaluated. PROTAC mediated targeted protein degradation relies on ligand-induced proximity, therefore spatial and temporal co-expression of the E3 ubiquitin ligase with the POI are critical [3]. Hence, POI expression profiles across different tissues and cellular compartments, as well as co-localisation with the recruited E3 ligase need to be carefully evaluated when choosing a POI [115]. A second factor to consider in the target choice is represented by protein synthesis and clearance rates. As discussed in Section 1.2, POI resynthesis and adaptive mechanisms will influence the efficacy of a degrader, and therefore its translational potential [3,116]. Resynthesis rates vary widely for different proteins, and the degrader must overcome POI homeostatic regulation to achieve sustained suppression of protein levels. This will be more challenging for proteins with naturally elevated levels of resynthesis.

Even if the POI presents favourable pre-requisites for TPD, it is essential to ensure compatibility with an active E3 ligase complex for the successful development of a PROTAC. An appropriate geometry and orientation of the ternary complex is crucial to achieve effective POI polyubiquitination. Structural and molecular dynamics simulation studies with the cullin-RING E3 ligase subclass suggest that the geometry of the substrate-E3 interaction and its compatibility with E2 recruitment are critical determinants of efficient ubiquitin transfer and subsequent POI degradation [117]. Even if potent binding is observed for each of the binary interactions (i.e. ligand-POI and ligand-E3 ligase), and a ternary complex forms *in vitro*, this may still not translate into efficient ubiquitination in cells in the absence of the correct complex geometry. This will also be ultimately affected by the hijacked E3 ligase. The selection of a particular E3 ligase to trigger the degradation of a given POI may in turn compromise E3 ligase activity, by either disrupting endogenous degradation profiles, or inducing the unnatural degradation of neosubstrates. This phenomenon has been well documented in the case of CRBN, for which both IMiD molecular glues and IMiD-based PROTACs were found to reprogram the 'degradome', leading to off-target effects [113,118–120]. Similarly, hijacking of cIAP for neosubstrates has been shown to lead to undesired auto-ubiquitination of this E3 ligase [18], thereby limiting its use as a proteasome recruiting element in PROTAC design [121].

Taking all these considerations into account, a variety of molecular biology methods have been developed to evaluate whether a POI of interest is suitable for TPD in cells before embarking on development of a POI ligand or PROTAC. The main genetic and macromolecular methodologies will be presented in the following subsections, while the reader is directed to a review by the Fischer group for a more extensive overview [122].

2.1.1 Genetic fusion technologies

Genetic fusion technologies have been widely employed to assess POI degradability for PROTAC development. Broadly, these methods work by appending the sequence for a degradation tag to the POI gene, and offer orthogonal, fast and precise proof-of-concept tools to assess POI degradability via the use of established ligands. While numerous variations of this technology have been reported, including IKZF3 peptidic [123] or Auxin inducible [124] degrons, destabilising domains [125], and the SMASh system [126], the most commonly used genetic fusion techniques remain the dTAG [127] and HaloTag® [128,129] systems.

HaloTag® technology [130] was implemented for TPD (termed HaloPROTACs) by the Crews lab in 2015 [129], and has been the most widely used for this purpose. The HaloTag® consists of a modified haloalkane dehalogenase, which reacts irreversibly with a chloroalkane ligand (the HaloTag® ligand, Fig. 6A). The dehalogenase enzyme originates from bacteria, so it represents a bio-orthogonal tagging system in mammalian cells

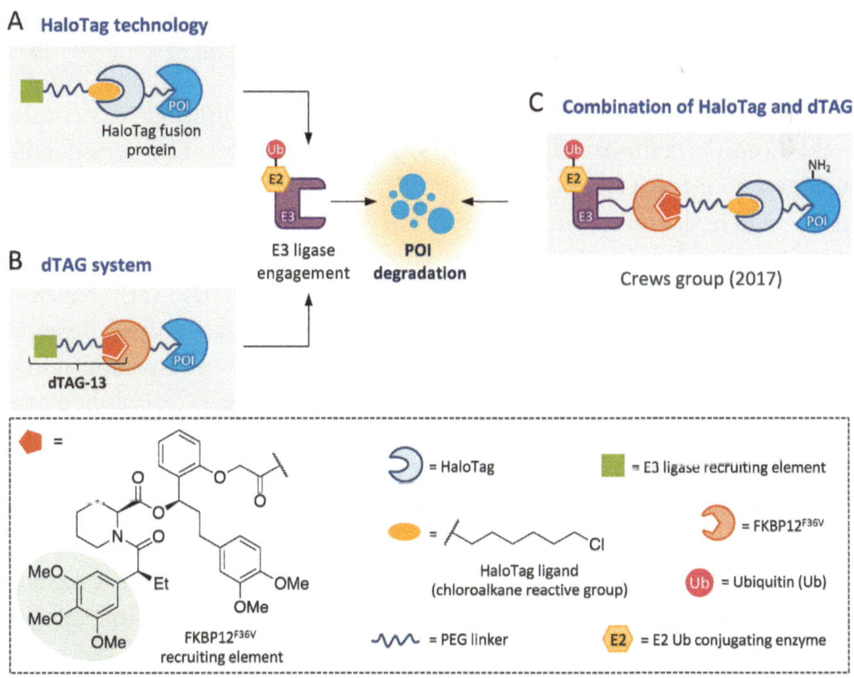

Fig. 6 Genetic fusion strategies to assess POI degradability. HaloTag® (A) and dTAG (B) technologies. (C) A combination of the two genetic fusion strategies to probe new E3 ligases for their ability to degrade POIs [131].

[132]. HaloTag® fusion POIs can thus be labelled with a heterobifunctional probe featuring a chloroalkane moiety linked to a proteasome recruiting element. Initial work in this area employed an adamantyl group that triggers degradation via the unfolded protein response [128], whereby this highly hydrophobic group is recognised by the UPS as an unfolded protein and thus degraded. Later efforts focused on attaching the chloroalkane reactive moiety to small molecule ligands for VHL or cIAP E3 ligases, generating the first HaloPROTACs [129,132].

Established by Bradner, Gray and co-workers [127], the dTAG system involves a POI fused to a mutant cytosolic signalling protein (FKBP12^{F36V}), which can be recruited by a specifically-designed dTAG small molecule degrader (dTAG13) that is selective for FKBP12^{F36V} over FKBP12^{WT} (Fig. 6B). This selectivity was achieved by use of a 'bump-and-hole' strategy, whereby the FKBP12^{F36V} accommodates a new engineered pocket (hole) for a bump-containing dTAG degrader that, due to its size, does not fit in the pocket of FKBP12^{WT} [133,134]. The dTAG13 degrader also carries an IMiD-based ligand to recruit CRBN and induce POI degradation. The degradation kinetics of the dTAG system are rapid due to the high availability of ubiquitination sites on FKBP12, which is valuable for temporal probing of protein function.

One clear advantage of the dTAG system over HaloPROTACs consists in the fusion protein size (12 kDa vs 33 kDa, respectively). The considerably smaller size of the dTAG can reduce PPI perturbation in cellular compartments with high protein concentration [135]. On the other hand, a wide variety of HaloTag® fusion constructs are commercially available, thus increasing accessibility of this technology with respect to dTAG. HaloTag® has been used extensively in chemical biology as a tool to investigate protein degradation and endogenous protein function as well as in live-cell imaging studies [124]. Interestingly, before Bradner and co-workers published the generalisable dTAG system, the Crews lab published an approach that combines the use of the fusion mutant FKBP12 ^{F36V} protein and 'bump' ligand with HaloTag®-fused E3 ligases to show that five out of six previously unexplored E3 ligases could be hijacked to degrade the fusion FKBP12^{F36V} protein (Fig. 6C) [131]. While this study used highly engineered systems, it clearly highlights the hidden potential of the available pool of unexploited E3 ligases for TPD.

More recently, these systems have been used in combination with CRISPR/Cas9 genome-editing technology to allow for more physiologically relevant expression levels to be used. CRISPR-inserted HaloTags® have been used to degrade endogenous levels of phospholipase PNPLA3 in complex *in vivo* models of hepatic steatosis [136]. Coupled with CRISPR-mediated knock-in, dTAG technology has been applied to understand the contextual relevance of the kinase MELK to basal-like breast cancer [123], and to validate the discovery of a transcription factor (ENL-YEATS) for a well-known oncogene in mixed-lineage leukaemia [137].

A third gene fusion-based method to evaluate TPD is AdPROM, a relatively new technology that exploits nanobodies (i.e. antibody fragments of reduced size) against GFP to induce degradation of GFP-fusion POI constructs [138]. This strategy achieves rapid and selective degradation by expressing the GFP nanobody tethered to an E3 ligase (such as VHL) in cells that have a GFP-tag knocked into the POI using CRISPR-Cas9 gene editing. The anti-GFP nanobody construct can either be expressed constitutively, or with a tetracycline-inducible expression system. More recently, in a collaboration between Sapkota and Ciulli [124] a 'ligand-inducible' version of this system was developed, termed L-AdPROM, which uses HaloTag® technology to switch on or off the AdPROM degradation system. In this work, the authors employ an antigen-stabilised anti-GFP nanobody construct, which is only stable when bound to the GFP-fused POI. The GFP tag still needs to be knocked-in to the POI, but this system is easier to manipulate, and more accurately reflects an endogenous system and a degradation mechanism that is more akin to small molecule PROTACs.

Each of the presented methods has its own drawbacks, such as exogenous expression requirements, potential off-target effects, the need for constant ligand treatment or slow degradation kinetics [122]. Nonetheless, the main caveat associated with these methods is that, depending on the size of the genetically encoded fusion tag, it may significantly modify POI structure and possibly its PPI and post-translational modification (PTM) profiles. In addition, these tools often utilise overexpression systems, which may further affect physiological relevance. Hence, the degradation observed with these engineered systems may not be representative of that induced by a small molecule PROTAC on the endogenous POI, and therefore must be used with caution in drug target validation studies. Nonetheless, genetic tagging strategies are important chemical biology tools that can help validate TPD for a POI in a particular disease state, and often uncover new insights about protein function.

2.1.2 Macromolecular strategies

Due to the swift development of gene editing technologies, efforts have been historically focused on genetically engineered proteins, but more recently, whole proteins and/or protein complexes have been employed to achieve targeted POI degradation. One such example is the 'Trim-away' strategy [139], an approach that leverages the tight binding between ubiquitin ligase TRIM21 and the Fc domain of a specific antibody, which is used to recruit the POI. Main issues associated with this technique include the need for electroporation to deliver the cell-impermeable antibody, off-target degradation of POI binding partners, and TRIM21 auto-ubiquitination, which results in rapid POI level recovery. Nevertheless, the Trim-away system is able to induce targeted degradation of unmodified proteins even in the absence of a known ligand. Other macromolecular techniques have been designed to hijack the lysosomal degradation pathway, such as LYTACs [140] (see Section 5.2.2 for more detail) and peptides [141], to engineer targeted POI degradation. However, these methods are not as generalisable as HaloTag®, dTAG or Trim-away, with LYTACs being designed specifically to target challenging cell surface and extracellular proteins, and peptide directed methods relying on structural biology data on key PPIs of POI sub-domains.

2.1.3 Protein modification strategies to expand the E3 ligase toolbox

While it is clear that degradability of a certain POI mostly depends on intrinsic properties, it has also been shown that E3 ligase selection can lead to drastically different PROTAC degradation profiles. There are approximately 600 different E3 ligases in the cell, and unsuccessful POI degradation with PROTACs could be due to the targeted E3 ligase not presenting appropriate substrate-E3 geometry at the multi-protein ligase interface for a particular POI. To evaluate the E3 ligase influence on the effectiveness of a TPD approach, scientists at Novartis [142] devised COFFEE (Covalent Functionalisation Followed by E3 Electroporation into live cells), a novel strategy that, provided a potent POI ligand is available, can assist in determining the optimal E3 ligase for POI degradation. As suggested by the name, the method consists of introducing a covalent warhead on the POI ligand, which can be used to attach different recombinant E3 ligases via available cysteine residues. These chimeras are then transported into the cell via electroporation. While this approach is highly artificial in that the concentrations of electroporated E3 ligases will be much higher than physiological levels, it may provide useful information concerning the most suitable E3 ligase for TPD studies of a POI. Furthermore, it allows novel E3 ligases to be evaluated quickly for TPD even in the absence of optimised ligands.

2.2 Optimisation of PROTAC properties and the importance of linkerology

Due to their heterobifunctional nature, PROTACs are inherently larger in size and more lipophilic than standard small molecule drug candidates, resulting in compounds with reduced cell membrane permeability, potential metabolic and chemical lability, and low aqueous solubility. These inherent issues of PROTACs are currently being addressed either through medicinal chemistry efforts, or via entirely new assay screening strategies tailored specifically for PROTACs. While the approaches implemented in a pharmaceutical industry context will be discussed in Section 4, cutting edge assays recently developed to assess and rank PROTAC physicochemical properties will be presented here. A particular focus will be given to the emerging concept of linkerology, a word coined to indicate the unpredictable effect of linker properties on the efficacy of PROTACs.

2.2.1 Assays to evaluate physicochemical properties of PROTACs in vitro

The main shortfall of PROTACs is their often limited cellular permeability, which is mostly due to their size and to the presence of numerous polar groups necessary to guarantee noncovalent interactions with both POI and E3 ligase targets. Hence, PROTAC-specific permeability assays are currently under development. To cite one, the Fischer group [83,143] established a competitive CRBN engagement assay to quantitatively rank the cell permeability of their PROTACs. In this assay, CRBN engagement is assessed by competition of the PROTAC against any POI with an optimised CRBN-BRD4 PROTAC in a cell system expressing GFP-tagged BRD4 protein. The loss of GFP-BRD4 green fluorescent signal is thus inversely correlated with the permeability of the novel PROTAC. One could argue that CRBN target engagement potency of each PROTAC still significantly influences the assay signal, and furthermore that the assay limit is the intrinsic permeability of the CRBN-BRD4 PROTAC, but nonetheless, the simultaneous assessment of different PROTAC parameters is a useful proof of concept, with potential for translation to other E3 systems.

More recently, a HaloTag®-based cell penetration assay was developed by the Frye lab as a quantitative method to evaluate PROTAC permeability [144]. The method involves tagging the prospective PROTAC with the HaloTag® ligand. If the tagged PROTAC is cell permeable, it will be covalently attached to a HaloTag®-fused GFP protein anchored to the outer mitochondrial membrane. Upon PROTAC treatment and subsequent washout, the unlabelled HaloTag®-GFP fusion proteins are allowed to react with a fluorescent HaloTag® ligand. The resulting fluorescence intensity, which is inversely proportional to the permeability of the tagged-PROTAC can be then quantified by flow cytometry. While this method is quantitative, sensitive and adequately isolates the permeability parameter for measurement, the use of the HaloTag® ligand, with its non-negligible mass and hydrophobicity, may either be synthetically challenging, or significantly change the permeability profile of the original molecule. Thus, the results, albeit indicative within a PROTAC series, should be carefully evaluated in orthogonal and more direct assays.

2.2.2 PROTAC linkerology

Since ligands for the POI and E3 ligase are usually well optimised for their respective binary interactions, the linker portion of the molecule is the primary opportunity for structure based PROTAC degradation optimisation. The linker design must achieve a delicate balance in order to trigger formation of a stable ternary complex and efficient ubiquitination, while facilitating cellular uptake and PROTAC solubility.

To the bemusement of medicinal chemists, the optimisation of PROTAC linkers has remained largely empirical and case-dependent [145,146]. Linker design and its influence on the formation of a productive ternary complex has been the subject of great controversy over the last few years.

Emerging evidence suggests that optimal linkers may be those that promote a complex network of non-cognate PPIs between the POI and E3 ligase, with some or many of these interactions taking place distal to the PROTAC binding site [68,147]. This paradigm would suggest that PROTACs promoting cooperative binding interactions between the E3 ligase and POI, like molecular glues, are optimal. In contrast, a team from Pfizer recently reported that linkers promoting thermodynamic cooperativity between POI and E3 ligase proteins are not necessary for efficient degradation of BTK [81]. In fact, this study suggests that linkers that are large enough to avoid steric interactions between the two proteins facilitate ternary complex formation. Moreover, the Fischer group showed that efficient degradation of BRD4 can be induced even with negative cooperativity (i.e. weaker ternary over binary complex formation in the presence of PROTAC) [83]. Consequently, in the absence of biophysical and structural data for the ternary complex, and potentially even when this is available, rational structure-based design of PROTAC linkers remains challenging.

There is some recent evidence that computational approaches, which have been successfully employed for structure-based design in drug discovery, may represent an useful tool to support rational design of PROTACs, although these methods are still in their infancy [83,148,149]. The main limitation again is the lack of structural data for POIs and E3 ligases.

Attempts to establish general rules based on rational design remain sparse and are often limited to a specific PROTAC series. At a fairly simple level, it has been suggested that incorporation of multiple amide motifs in PROTAC linker structures should be avoided to preserve compound permeability [66]. A study from the Ciulli group [84] demonstrated using a Caco-2 permeability assay that replacement of a polyethylene glycol (PEG) based linker with a 1,4-disubstituted phenyl ring markedly improved PROTAC cellular permeability, while maintaining the appropriate linker geometry for ternary complex formation, but further validation would be necessary to assess whether this trend is broadly applicable for PROTAC design. Provided there is a favourable or minimally negative effect on target binding, conventional methods such as inserting basic nitrogen into aromatic rings or alkyl linker regions have proved fruitful for improving PROTAC solubility [143,150]. Flexible linkers containing groups such as PEG are commonly employed due to their hydrophilicity and high aqueous solubility, but a tight balance of hydrophilic and hydrophobic properties should be sought to achieve simultaneous aqueous solubility and cell permeability.

Altogether, the growing experimental data available in the PROTAC field allows a series of guidelines to be compiled for PROTAC linkerology, i.e., for the *de novo* design of linkers.

It has been postulated that a 'sweet spot' for linker length can be achieved for any new PROTAC [85]. If the linker is too short, steric clashes may affect one or both of the binding events, while if it is too long, PPIs that are potentially important for ternary complex formation, or POI ubiquitination, may not occur [151]. Due to the use of modular synthetic approaches in the preparation of PROTACs, this parameter is generally explored first by systematically exploring heterogeneous PEG linkers. Linker chemical composition is also a crucial factor, and while PEG linkers have been extensively employed [151], piperidine linkers have recently emerged as a highly soluble alternative [150], with various commercially available building blocks containing these motifs. Similarly, intermediates amenable to Cu(I)-catalysed azide–alkyne cycloaddition chemistry are widely available, providing a fast and robust way to connect POI and E3 ligands and explore PROTAC structure-activity relationships [152].

Crystal structures of the binary complexes between either the POI or E3 ligase with their respective ligands have facilitated the identification of solvent exposed areas, which are best suited to introduce PROTAC linkers [151]. Furthermore, the exit vector for linker attachment may play a crucial role in degradation. In fact, it has been shown that different degradation selectivity profiles for related protein isoforms of p38 MAPK can be achieved by varying the linker attachment point on the E3 ligase ligand (VHL) [86].

2.3 Assessment of targeted protein degradation in cells

A key step in PROTAC development is biological evaluation of degradation properties in an endogenous and physiologically relevant system, and several methods have been established to acquire qualitative and quantitative insight into the endogenous degradation potential of novel PROTACs. These range from conventional immunoassay methods and proteomics approaches to cutting-edge techniques involving genetic modification of targets, as described in the following section.

2.3.1 Immunoassays

Traditionally, western blotting (WB) [118,119] has been the most widely used method to analyse endogenous protein abundance [28,33,35,153,154]. This technique mainly relies on the availability of a suitable primary antibody against the POI. The main disadvantages of this method are in the low throughput, labour intensive protocols, and only semi-quantitative measurement. Consequently, it is advised that data derived from WB should be complemented or supplanted by other, high-throughput and more quantitative methods where possible.

Improved assay timeframes are offered by capillary electrophoresis immunoassays [155], which minimise the electrophoresis process, allowing lower sample consumption and shorter analysis time. This method has been successfully used to characterise PROTAC degradation for pirin [156] and BTK [81] proteins, but remains relatively low throughput.

To increase assay throughput, plate based-techniques such as enzyme-linked immunosorbent assays (ELISA) are more effective. Recently, the Tang lab reported an in-cell version of ELISA, whereby cells are fixed in the plate and treated with antibodies for protein quantification. This is a sensitive, robust, and quantitative method to measure degradation, as shown for HDAC6 deacetylase in MM1S cells [157]. The high throughput nature of this method enabled competition assays with a model PROTAC to profile the affinity of novel E3 ligase ligands, leading to identification of hits that were derivatised into new and efficient PROTACs [158].

Meso Scale Discovery (MSD) Multi-Array Technology employs a similar principle to ELISA; however, the detection system measures electrochemical luminescence, rather than standard chemiluminescence [159]. Each well contains a working electrode that, after a workflow similar to ELISA is performed, can apply a voltage to trigger an electrochemiluminescent reaction with an MSD SULFO-TAGGED secondary antibody, which is proportional to the POI levels in the sample. Due to its spot-based coating and orthogonal detection systems, this method offers conservation of reagents, higher sensitivity, and greater versatility. In addition, once an MSD assay is optimised, different plate-types are available for multiplexing, allowing for the investigation of more than one POI simultaneously. Despite it being available since 2014, only scientists at Amgen have published work using this technology in the PROTAC field [152]. The reason for the limited implementation of this technology may be due to its requirement for a bespoke voltage-based imaging system, and higher consumables cost [159].

2.3.2 Reporter assays

Tagged mutants of the POI have been used extensively to measure protein abundance in response to PROTAC treatment in a time-resolved manner. These methods generally offer high sensitivity and superior selectivity over the immunoassay methods. As with all genetic fusion strategies, these approaches present the caveat that the tagged-POI will be expressed in an artificial manner with respect to its native context, and therefore it may not be subject to endogenous regulators of its expression. This could be prevented by the use of CRISPR-Cas9 knock-in strategies, albeit clone selection might lead to labour intensive and extended timelines. In the case of PROTACs, it is important to consider, particularly when developing PROTACs for proteins where little is known about their homeostatic regulation, that a reporter-based system may miss key factors, which ultimately contribute to resistance in the clinic.

2.3.2.1 Fluorescence-based reporter assays

Recent work from the Tate [40] and Deiters group [160] showed that degradation of GFP-tagged BRD4 induced by a transient transfection system in HEK293 cells and degraded by photocaged PROTACs could be measured using live-cell fluorescence microscopy [40]. The Zhong and Pan labs also applied this technology to monitor GFP-tagged BRD4 degradation in transfected zebrafish embryos [43]. Around the same time, the group of Westover et al. demonstrated that degradation of a GFP-tagged $KRAS^{G12C}$ GTPase mutant, created by stable transfection of Flp-In™ HEK293 cells using a Flp-*FRT* recombination strategy [143], could also be used to measure PROTAC induced degradation, albeit with modest D_{max} values.

2.3.2.2 Luminescence-based reporter assays

Recently, the Promega Corporation has developed a highly sophisticated luminescence-based endogenous tagging assay system to assess PROTAC degradation, termed NanoBiT (Fig. 7) [161]. In this method, the N-terminus of a POI is tagged with a 'HiBiT' peptide (consisting of 11 amino acids) using CRISPR/Cas9 technology in HEK293 cells stably expressing a 'LgBiT' protein (18 kDa), randomly inserted into the genome. The HiBiT peptide possesses picomolar affinity for the 'LgBiT' protein, and when these two components bind, fully functional NanoLuc enzyme is generated, which emits a luminescent signal in the presence of a luciferase substrate. The HiBiT peptide is small compared to most POIs, and is therefore assumed to cause minimal perturbation of POI activity. When a PROTAC degrades the HiBiT-tagged POI, this can be measured quantitatively and in high throughput as a loss in bioluminescent signal. The endogenous tagging systems employed in this technology can be used to quantify degradation kinetics in live cells, and it more closely reflects native POI expression levels and transcriptional regulation than previous systems. The kinetic nature of this assay is exemplified by its ability to differentiate the degradation of BET BRD2, 3 and 4 isoforms by MZ1 (Fig. 4B) in a concentration and time-resolved manner [161]. This protocol has also been applied to measure in-cell degradation kinetics of BRD7 and BRD9 [162]. Disadvantages

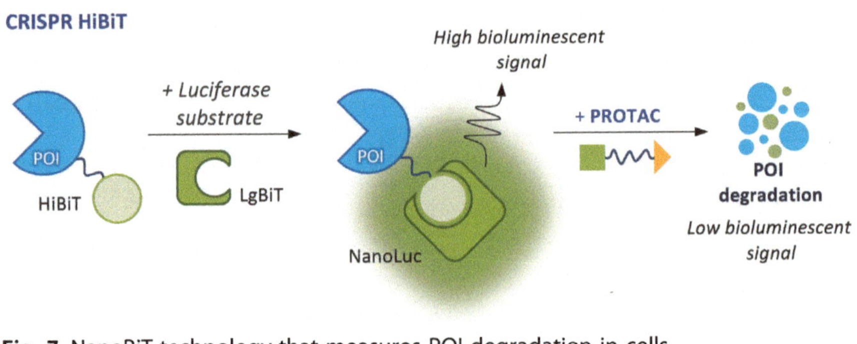

Fig. 7 NanoBiT technology that measures POI degradation in cells.

include relatively high consumables costs, and the need for endogenous tagging in a cell line of interest, which can be very time consuming to implement. While these can be circumvented by overexpression of all components, this leads to challenging multiparameter assay optimisation (e.g. to optimise signal-to-noise) and removes a key advantage of the approach.

2.3.3 Mass spectrometry

Owing to its high sensitivity and label-free nature, mass spectrometry represents an important method in the toolbox to measure TPD. Bottom-up proteomic workflows enable detection of changes in the abundance of thousands of proteins in response to PROTAC treatment, allowing assessment of PROTAC selectivity and identification of off-targets [163]. In addition, metabolic labelling strategies such as Stable Isotope Labelling by/with Amino acids in Cell culture (SILAC) [164] and 'spike-in' variants [165], as well as isobaric labelling methods (e.g. Tandem Mass Tagging (TMT) [166] or isobaric Tags for Relative and Absolute Quantitation (iTRAQ) [167]) render this approach highly quantitative, and allow evaluation of multiple conditions simultaneously. A sophisticated application of this technology for PROTACs is multiplexed Proteome Dynamics Profiling, developed by Cellzome, which uses a combination of SILAC and TMT isotopic labelling to distinguish mature and nascent proteins [168]. This allows measurement of PROTAC effects on protein resynthesis, which is an important parameter for PROTAC efficacy. Due to the complexity of the MS component, workflows, and cost of the consumables, these assays are typically not high throughput, nor accessible to all labs. In light of this, these experiments have recently been frequently adopted as an orthogonal technique to those previously described to measure degradation, and to evaluate PROTAC selectivity, which is an essentially inaccessible parameter without proteomics [38,58,81,84,169–171].

2.3.4 Ubiquitin-proteasome system dependency assays

In addition to quantifying degradation in an endogenous system, it is important to confirm that the novel PROTAC hijacks the UPS to induce POI degradation. Various researchers have shown that it is possible to hijack endogenous degradation mechanisms that are independent of the UPS [139,141]. Therefore, it is possible that, despite careful design, candidate molecules induce degradation via pathways other than the UPS. In order to validate the UPS-dependent MOA of the designed PROTAC, it is necessary to incorporate control experiments into the methods mentioned above. This assessment is usually performed by the evaluation of the degradation properties of the PROTAC in the presence of pharmacological agents that block some aspect of the ubiquitin-proteasome degradation pathway. For example, degradation dependency on E3 ligase recruitment can also be evaluated either using competition experiments with E3 ligase ligands or by genetic knockout/knockdown of the targeted E3 protein. Other pharmacological controls frequently employed include, but are not limited to: 26S proteasome inhibitors (bortezomib, epoxomicin, carfilzomib, ixazomib and MG132) [172]; neddylation/NAE1 inhibitors (MLN4924) [173]; and inactive PROTAC analogue controls for established E3 ligases such as VHL [31,174,175] and CRBN [28,38].

2.4 Quantification of key intracellular events for targeted protein degradation

Dissection of key cellular events upon PROTAC treatment is crucial to validate the expected MOA and assess PROTAC efficiency. Hence, specific assays have been designed to detect and measure ternary complex formation and POI ubiquitination, both *in vitro* and *in cellulo*.

2.4.1 Ternary complex formation

Effective formation and stability of the ternary complex are often critical for proximity-dependent POI ubiquitination and efficient TPD [68]. Traditional small-molecule ligands typically involve binary interactions that can be characterised by straightforward Langmuir-Hill or Michaelis-Menten models; however, due to the ternary binding interactions of PROTACs (see Section 1.3), predicting and optimising ternary complex formation and dissociation is more challenging and requires more sophisticated models [176]. Experiments to isolate and quantify these ternary interactions, whether biochemically or in cells, are required to understand and develop improved PROTACs. A number of techniques have been used to examine ternary complex formation, including but not limited to X-ray crystallography [83,84], size exclusion chromatography [177,178], co-immunoprecipitation [179] and mass spectrometry [81,180], each with varying degrees of advantages and disadvantages, and while the reader is encouraged to review these works, it is not within the scope of this sub-chapter to discuss all of these in detail. Within this section, the focus will be on technologies that have demonstrated broad utility for measuring ternary complex formation and which are most amenable to drug discovery effects and high-throughput screening.

A number of well-established, *in vitro* binding methods are available to characterise ternary complex kinetics. While these experiments do not replicate the physiological environment of the cell, they can be designed to isolate the phenomenon of ternary binding, which can be helpful to gain insights into SAR.

If a fluorescently labelled ligand of either the POI or E3 ligase binds tightly to its cognate partner, then it can be used in standard Fluorescence Polarisation (FP) competitive binding studies to measure ternary complex equilibrium [181]. Concentration-response curves can be constructed in the absence or presence of the protein that does not bind the fluorescently labelled ligand to qualitatively determine PROTAC binding cooperativity. This has been applied to both CRBN [83] and VHL E3 ligases [162] in the context of BET protein degradation. One caveat of this approach is that it generally uses large quantities of each protein to saturate probe binding, which limits its use in high-throughput screening.

Alternatively, Time-Resolved Fluorescence Energy Transfer (TR-FRET) can also be used to quantify binding by fluorescence. However, TR-FRET uses two fluorescently labelled proteins that will undergo a measurable energy transfer event when brought into proximity by a PROTAC resulting in a change in fluorescence lifetime in the donor. As TR-FRET is measured ratiometrically it is less susceptible to assay variability and is highly sensitive. This technique has been primarily applied to measure ternary complex formation for CRBN-based PROTACs, but should be widely applicable [81,83].

A biophysical technique commonly used to evaluate PPIs and that can be adapted to measure ternary complex formation is Surface Plasmon Resonance (SPR), which involves measuring refractive index changes induced by the binding of an analyte to an immobilised ligand. A distinct advantage of this approach over FP is that, provided the interactions are equilibrium-based, it allows delineation of on-and-off rates by control of PROTAC binary complex flow over a chip immobilised with either the POI or the E3 ligase, requiring only a very small amount of the target that is immobilised to the chip. The Ciulli group used this technique to propose structural differences in BRD2, 3 and 4 isoforms as the reason for differential ternary complex formation [82]. They also demonstrated that differences in SPR-derived ternary complex dissociation rates of various BRD isoforms with MZ1 correlate with their initial rates of intracellular degradation, as measured using Promega's NanoBRET™ technology [161]. In a similar manner, the Crews lab used SPR to discover that construction of a promiscuous kinase targeted PROTAC with different vectors off the VHL E3 ligase ligand significantly affected its ternary complex formation and dissociation of p38δ, and offered this as a rationale for observed differences in degradation selectivity in cells [86].

Another highly successful technique for measuring ternary complex formation is the Amplified Luminescent Proximity Homogenous Assay Screen (ALPHAScreen). This method is a highly sensitive bead-based photosensitiser assay that emits a chemiluminescent signal resulting from the transfer of singlet oxygen only when an analyte (such as a PROTAC) induces proximity between streptavidin donor beads immobilising the POI and biotinylated acceptor beads immobilising the E3 ligase [182]. This method has been applied to complexes of BRD isoforms [68,162], BAF chromatin ATPase subunits (SMARCA2 and SMARCA4) [84], BCL-X_L [171], and p38α [86] with VHL, and BRD4 with CRBN [28,181]. The assay can be adapted to 96- and 384-well microplates, and thus is quite scalable for high-throughput screening; however, it can be subject to many assay artefacts, such as compounds mimicking the effect of biotin, or those that quench singlet oxygen or scatter light [182]. As a result, the company that markets the technology offers a TruHit kit that offers expedient identification of false positives [183]. This approach also requires the use of tagged proteins, which may not be readily available.

Isothermal Titration Calorimetry (ITC), which measures the heat released in binding events, can also be used to measure cooperativity of a ternary complex. Usefully, ITC is a label free-method; however, it requires a large amount of protein, and one sample measurement takes approximately 40 min so it is overall a low-throughput assay. Therefore, it has been commonly employed as secondary assay to investigate PROTAC cooperativity, for example, between VHL and BRD isoforms [68,162] or SMARCA2 [84].

Due to the fact that ternary complex formation may depend on cellular context, or POIs and E3 ligases may be difficult to isolate and purify, Promega have developed an approach to measure this phenomenon *in cellulo* (Fig. 8A), i.e., NanoBioluminescence Resonance Energy Transfer (NanoBRET™) [161]. In this technology, BRET drives detection of ternary complex formation. One binding partner (e.g. the POI) is fused to Nano luciferase (NanoLuc) enzyme, while the other binding partner (e.g. the E3 ligase) is fused to a HaloTag® (see Section 2.1.1). A luciferase substrate is added to cells to excite the NanoLuc POI, and a chloroalkane–fluorophore conjugate added separately to label the Halo-tagged E3 ligase. If a PROTAC

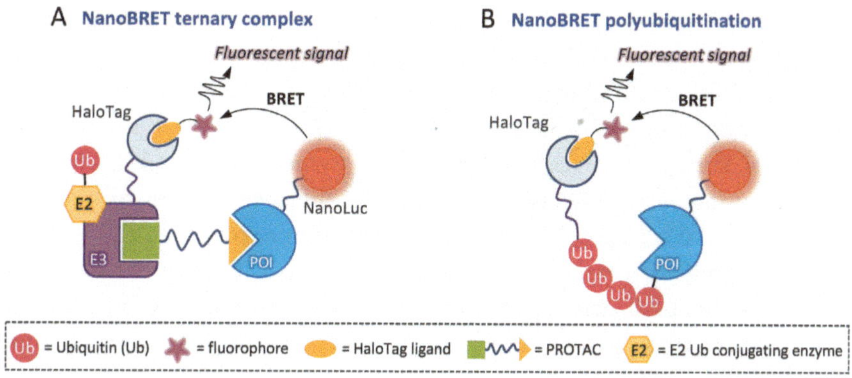

Fig. 8 NanoBRET™ technology to quantify key intracellular PROTAC events, such as ternary complex formation (A) and POI ubiquitination (B).

brings these fusion proteins together, BRET occurs and this is detected as a change in the fluorescent signal of the HaloTag® acceptor. Contrary to *in vitro* AlphaScreen® results, this approach has been employed to demonstrate that a VHL-based PROTAC (DT2216) could only form a ternary complex with BCL-X_L and not BCL-2, exemplifying the importance of measuring this phenomenon in cells [171]. Apart from this example and Promega's 2018 publication [161], there are no further published examples to date of the NanoBRET™ technology being used for this purpose; however, the Wang group recently used it to characterise the target engagement of their PROTACs separately with each POI (BTK) and E3 ligase (CRBN) partner [169]. At the time of writing, this technology is relatively new but is becoming widely adopted in industry, and it is possible that there will be many more publications in this space.

2.4.2 Target ubiquitination

As discussed in Section 2.1 an appropriate geometrical orientation of the ternary complex is essential in order to promote efficient poly-ubiquitin transfer from the E1, E2, E3 enzymatic cascade to lysine residues on the target protein [184]. This step is critical to trigger POI recognition by the 26S proteasome complex that performs protein degradation. The formation of a stable ternary complex does not always result in expedient ubiquitination, as recently documented for a covalent PROTAC targeting the clinically important $KRAS^{G12C}$ mutation [143,185]. Therefore, methods are needed that can isolate, characterise and quantify this important intracellular event in TPD. While various cell-free systems have been utilised, including FP [184],

AlphaScreen® [184], TR-FRET [184], and nanopore technology [186], these systems require addition of multiple co-factors (ATP) and enzymes (E1, E2 and E3), and the complexity involved in reconstituting these systems *in vitro* brings into question the physiological relevance of these assays. Therefore, in this section cell-based methods to measure POI ubiquitination and the available approaches for elucidating the specific sites of ubiquitination will be discussed. An important point to note for all methods is that a ubiquitinated POI needs to accumulate to detect ubiquitination; for an effective PROTAC this species typically undergoes rapid degradation, and therefore these assays often require co-treatment with a proteasome inhibitor to force accumulation (see Section 2.3.4).

A straightforward, qualitative technique to measuring ubiquitination in cells is to perform an immunoprecipitation (IP) for the POI from cell lysates treated with the PROTAC, followed by western blot analysis with antibodies against ubiquitin [184]. Tagged POIs are also amenable to this approach; with the ubiquitination of FLAG-tagged p38α measured using this technique [86]. This method can also offer insight on the extent of poly-ubiquitination that occurs, as evidenced by the band profile in the WB membrane. A more quantitative, ELISA-based version of immunoblotting is the Dissociation Enhanced Lanthanide Fluoroimmunoassay (DELFIA), which has been used to measure PROTAC-mediated ubiquitination of epidermal growth factor receptor (EGFR) [184]. The detection method relies on a lanthanide (Europium)-conjugated antibody that produces a highly sensitive fluorescent signal.

Promega's previously described proximity-based NanoBRET™ system (see Section 2.4.1) can also be adapted to measure target protein ubiquitination (Fig. 8B) [161]. Rather than employing a Halo-Tagged E3 ligase protein, as with the system that measures live-cell ternary complex formation, the HaloTag® acceptor is attached to ubiquitin. Therefore, upon accumulation of HaloTagged® ubiquitin monomers to the NanoLuc®-tagged POI, a proximity-dependent increase in fluorescence (from BRET) is observed. Despite the apparent utility of this assay, its uptake appears limited to date. A variant of this assay has, however, recently been employed to measure the kinetics of BRD7 and 9 ubiquitination induced by a VHL-based PROTAC (VZ185) [162].

The process of poly-ubiquitin chain elongation within the cell is complex. A total of eight poly-ubiquitin linkage types are known; however, lysine-48 (K48) linked chains are the most common in the context of proteasomal degradation [186]. These diverse linkage types result in different poly-ubiquitin chain topologies that ultimately dictate the cellular fate or role of the ubiquitinated POI. Understanding specifically where a POI is endogenously poly-ubiquitinated can aid design of PROTACs that not only form a stable ternary complex but also accommodate efficient POI ubiquitination. In addition to quantification of POI degradation, mass spectrometry is a powerful tool to quantify protein ubiquitination in response to PROTAC treatment. In combination with tandem ubiquitin binding entities or anti-Gly-Gly antibody enrichment techniques, mass spectrometry also allows identification of specific ubiquitination types [187] and sites [188], respectively. In the PROTAC field, mass spectrometry has been employed to identify multiple ubiquitination sites on BRD4 [68], and a single ubiquitination site on BCL-X_L [171], both induced by VHL-based PROTACs.

2.5 PROTAC development workflow

De novo development of heterobifunctional protein degraders is not a trivial task, but an arduous process which ideally, and often after an extended optimisation exercise, culminates in a fine-tuned design with rapid and efficient degradation at low PROTAC concentrations. Since the MOA of PROTACs is complex and successful degradation results from the orchestration of multiple factors, it is logical that the development of an efficient degrader involves a variety of queries ranging from chemical design to biological function. As delineated in the previous sections, the standard PROTAC drug discovery workflow is a multi-step process in which the information obtained at each stage could feed back into PROTAC design and optimisation (Fig. 9). The starting point when embarking in a

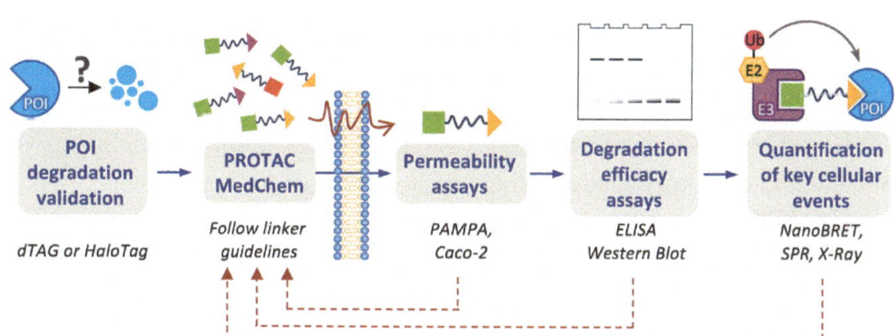

Fig. 9 A general PROTAC drug discovery workflow. Solid blue arrows indicate the initial order of assays, whereas dashed red arrows represent potential feedback workflow.

PROTAC discovery campaign involves consideration on the degradability potential of the POI of interest. This can be addressed using genetic fusion technologies (e.g. HaloTag® or dTAG), which, among other information, allow the identification of the better suited E3 ligases to degrade a given POI. The next step consists in the chemical synthesis of a PROTAC library, leveraging different linker designs attending to the principles outlined in Section 2.2.2 (e.g. linker length and composition). A critical factor in PROTAC activity lies in compound intrinsic cell permeability, which can be assessed with routine permeability assays (e.g. Caco-2, PAMPA). Subsequent evaluation of PROTAC degradation properties *in cellulo*, as well as their dependence on the UPS, can be carried out with primary (e.g. in-cell ELISA) and secondary (e.g. WB) screening assays. Finally, for those identified as successful degraders, key intracellular events for POI degradation (i.e. ternary complex formation and ubiquitination) can be characterised using live-cell methods (e.g. Nano-BRET™). Alternatively, ternary complex formation and structural features could also be addressed using biophysical methods (e.g. SPR, X-ray crystallography). (Fig. 9).

Two decades after the discovery of TPD, the study of key factors governing efficient TPD remains a highly dynamic field, with new technologies and essential discoveries established at a blistering pace. In addition to the above in-depth discussion, we provide a quick reference guide to the utility of the described approaches (Table 1).

3. Scope of targeted protein degradation

Among the more than 20,000 protein coding genes that constitute the human genome, almost 4500 are considered to be directly linked to disease-causing proteins, and constitute the pool of potential protein targets for drug discovery [189]. In their publication 'A comprehensive map of molecular drug targets', the Overington group reviewed 1578 US FDA approved drugs and identified close to 700 human protein targets [190]. Alongside enzymes, protein families such as GPCRs, ion channels, kinases and nuclear receptors accounted for 44% of drugged targets. Over the past decade, TPD has leveraged several small molecule binders developed against those validated targets, and a multitude of well-characterised and potent ligands have been successfully converted into PROTACs [191]. Within this section, an exhaustive review of the proteins degraded so far via TPD is initially presented, providing an insight into the breadth of PROTAC-mediated protein knockdown as a novel drug modality. The opportunity

Table 1 A summary of the utilities of various PROTAC assays and technologies.

	PROTAC design elements					Desired drug discovery characteristics					
				Important intracellular events							
Assay system	POI degradation POC	Cellular perme-ability	Cellular POI degradation	III° complex	POI Ub	Throughput?	Quantitative?	Physiological relevance?	Validation for PROTACs in lit.?	Distinct advantages?	Potential drawbacks?
HaloTag®	✓	✓				Low	Qual.	Medium	High	Commercially available plasmids	
dTAG	✓					Low	Qual.	Medium	High	Rapid degradation	
PAMPA		✓				High	Yes	Low	Medium		Artificial membrane
Caco-2		✓				High	Yes	Medium	Medium	Cellular system	
Western blotting			✓		✓	Low	Semi	High	High	Endogenous system	Labour intensive—error prone
Capillary Immunoassay			✓		✓	Low	Semi	High	High		
ELISA			✓			High	Yes	High	High	Endogenous system	
Reporter assays (e.g. NanoBRET™)			✓	✓	✓	High	Yes	Medium	High	Can quantify in live cells	
MSD			✓			High	Yes		Medium	Highly sensitive	
Mass spec			✓	✓	✓	Low	Yes	High	High	Site of labelling analysis	

Continued

Table 1 A summary of the utilities of various PROTAC assays and technologies.—cont'd

	PROTAC design elements					Desired drug discovery characteristics					
				Important intracellular events							
Assay system	POI degradation POC	Cellular perme-ability	Cellular POI degradation	III° complex	POI Ub	Throughput?	Quantitative?	Physiological relevance?	Validation for PROTACs in lit.?	Distinct advantages?	Potential drawbacks?
FP				✓		High	Yes	Low	High		Requires labels
SPR				✓		Medium	Yes	Low	High	Label-free Can measure real-time kinetics	
AlphaScreen®				✓		High	Yes	Low	High		Susceptible to false positives
TR-FRET				✓		High	Yes	Low	High		
ITC				✓		Very Low	Yes	Low	High	Informs on thermodynamics	
X-ray crystallography				✓		Very Low	Qual.	Low	High	Offers unique structural insight	
IP +immunoblotting					✓	Low	Qual.	Medium	High	Endogenous system	
DELFIA					✓	High	Yes	High	Med		

✓ denotes an assay system that is useful for a specific PROTAC design element. The level of throughput for an assay is based on the following criteria. *Very Low*: Non-plate-based and slow readout, *Low*: Non-plate-based, *Medium*: Plate-based but slow readout, *High*: Plate-based, fast readout. The categories for whether an assay is quantitative or not are *Yes*: Quantitative, *Semi*: partially quantitative, and *Qual.*: qualitative but not quantitative. Physiological relevance was classified on this basis *Low*: cell-free systems, *Medium*: Cell-based but genetically engineered, *High*: cell-based and endogenous. Literature validation assignment is based on the following criteria: *High*: Used successfully by two independent research groups; *Medium*: Used successfully by one research group.

offered by the large number of existing yet untapped E3 ligases and emerging methods for their discovery are discussed. Finally, notable examples supporting the use of PROTACs to tackle challenging targets are provided.

3.1 Proteins degraded: An overview

While an estimated 16% of disease-relevant human proteins have been targeted by a drug (small molecule or biologic) [190], the data we have collected for PROTACs indicates that 2% of the same pool of proteins have been successfully knocked down via TPD (Fig. 10). At the time this manuscript was prepared, a total of 83 human and 2 viral proteins had been degraded by small molecule and peptide-based PROTACs altogether (Table 2).

Considering the relatively recent emergence of TPD as a protein knockdown strategy, the percentage of degraded targets recorded to date is in fact extremely promising. With multiple research projects both in academia and industry exploiting available inhibitors, but also binding ligands which do not occupy protein functional sites, it is anticipated that the number of targets amenable for TPD will grow rapidly in the years to come [46].

A closer examination of the proteins listed in Table 2 reveals that oncology targets have been at the forefront of TPD, and kinases (including RTKs) encompass 44% of proteins degraded to date (Fig. 11A and C). The substantial variety of scaffolds investigated for kinases in past drug discovery projects accounts in part for the success of this target class in TPD, often giving structure-guided and facile access to solvent exposed exit vectors.

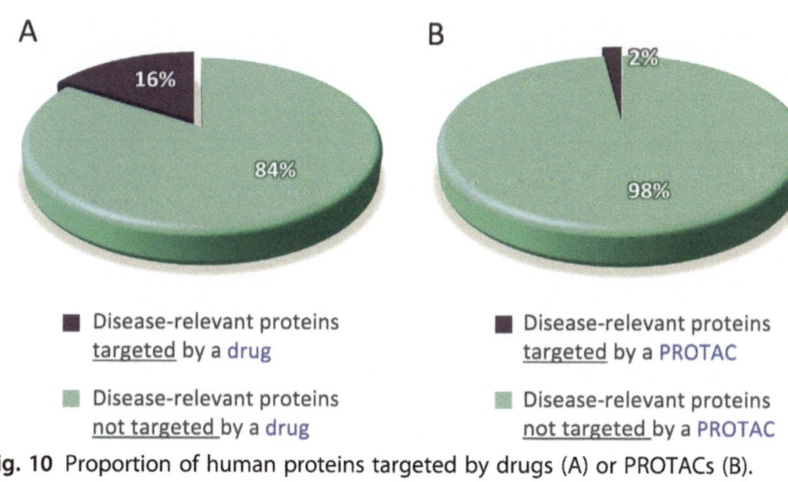

Fig. 10 Proportion of human proteins targeted by drugs (A) or PROTACs (B).

Table 2 Summary of proteins degraded by small molecule and peptide-based PROTACs.

Target	Target class	Therapeutic area[a]	References
ADRA1A	GPCR	ONC	[192]
AHR	Transcription factor	ONC	[193]
AKT1/2/3	Kinase	ONC	[194]
ALK	RTK	ONC	[195–197]
AR	NHR	ONC	[23,74,150,198]
AURKA/B	Kinase	ONC	[199]
BCL2	Antiapoptotic protein	ONC	[200]
BCL6	Transcription factor	ONC	[201]
Bcl-xL	Antiapoptotic protein	ONC	[171]
BCR-ABL	Kinase	ONC	[91,202,203]
BLK	Kinase	ONC	[204]
BRAF	Kinase	ONC	[205]
BRAF (V600E)	Kinase	ONC	[206]
BRD2/3/4	Epigenetic target	ONC	[27,28,30,44,207]
BRD7	Epigenetic target	ONC	[162]
BRD9	Epigenetic target	ONC	[67,162]
BTK	Kinase	ONC	[71,208]
BTK (C481S)	Kinase	ONC	[71]
c-ABL	Kinase	ONC	[91]
CDC20	Kinase	ONC	[209]
CDK2	Kinase	ONC	[210,211]
CDK4	Kinase	ONC	[212]
CDK5	Kinase	ONC	[211]
CDK6	Kinase	ONC	[212]
CDK8	Kinase	ONC	[213]
CDK9	Kinase	ONC	[210,214]
c-IAP (BIRC2)	E3 ligase	ONC	[21]

Table 2 Summary of proteins degraded by small molecule and peptide-based PROTACs.—cont'd

Target	Target class	Therapeutic area[a]	References
CK2	Kinase	ONC	[215]
CRABPI/II	Carrier protein	ONC	[21,216]
CRBN	E3 ligase	ONC/IA	[217–219]
CREPT (RPRD1B)	Transcription modulator	ONC	[220]
EED (PRC2 complex)	Transcription factor	ONC	[221,222]
EGFR	RTK	ONC	[54]
EGFR (L858R/ T790M)	RTK	ONC	[223,224]
ER	NHR	ONC	[24,225,226]
ERK1/2 (MAPK3/1)	Kinase	ONC	[227]
ERRα	NHR	ONC	[31,228]
FAK (PTK2)	Kinase	ONC	[66,229]
FAK2 (PTK2B)	Kinase	ONC	[37]
FKBP12	Immunophilin	CNS/IA	[28,35,59]
FLT3	RTK	ONC	[230]
FRS2	RTK effector	ONC	[231]
GSPT1	Translation factor	ONC	[111,118]
HDAC1/2/3	Epigenetic target	ONC	[232]
HDAC6	Epigenetic target	ONC	[157,233]
HER2	RTK	ONC	[54]
HGFR (c-Met)	RTK	ONC	[54]
HMGCR	Oxidoreductase	CVD	[234]
HTT	Scaffolding protein	CNS	[235]
IRAK4	Kinase	IA	[236]
JAK1/2	Kinase	ONC	[237]
KRAS (G12C)	Hydrolase	ONC	[238]

Continued

Table 2 Summary of proteins degraded by small molecule and peptide-based PROTACs.—cont'd

Target	Target class	Therapeutic area[a]	References
LRRK2	Kinase	ONC/CNS	[239]
MAPK11 (p38β)	Kinase	ONC	[240]
MAPK13 (p38δ)	Kinase	ONC/IA	[86]
MAPK14 (p38α)	Kinase	ONC/IA	[86,240]
MCL1	Antiapoptotic protein	ONC	[200]
MDM2	E3 ligase	ONC	[241,242]
MetAP2	Peptidase	ONC	[12]
NS3/4A	HCV protease	ID	[243]
PARP1	Glycosyltransferase	ONC	[244,245]
PCAF/GCN5 (KAT2B)	Epigenetic target	ONC/IA	[246]
PDE4	Hydrolase	IA	[24]
PI3K	Kinase	ONC	[247]
PIR	Transcription coregulator	ONC	[156]
PLK1	Kinase	ONC	[248]
RAR	Transcription factor	ONC	[89]
RIPK2	Kinase	IA	[31]
Rpn13 (ADRM1)	Ubiquitin receptor	ONC	[249]
SGK3	Kinase	ONC	[250]
SHP2 (PTPN11)	Phosphatase	ONC	[251]
SIRT2	Epigenetic target	ONC	[252]
SLC91A	Carrier protein	ONC	[253]
SMAD3	Transcription modulator	ONC/FBR	[254]
SMARCA2/4	Transcription factor	ONC	[84]
SNCA (α-Synuclein)	Synapse control	CNS	[255]
STAT3	Transcription factor	ONC	[256]
TACC3	Motor spindle protein	ONC	[257]

Table 2 Summary of proteins degraded by small molecule and peptide-based PROTACs.—cont'd

Target	Target class	Therapeutic area[a]	References
Tau	Microtubule stabilisation	CNS	[258]
TBK1	Kinase	ONC	[259]
TRIM24	Transcription factor	ONC	[260]
TrkC	RTK	ONC	[261]
VHL	E3 ligase	BLD	[177]
WEE1	Kinase	ONC	[262]
X-Protein	HBV viral protein	ID	[263]

[a]BLD: Blood disorders; CNS: Central nervous system diseases; CVD: cardiovascular diseases; FBR: Fibrotic diseases; IA: Inflammation and autoimmune diseases; ID: Infectious diseases; ONC: Oncology.

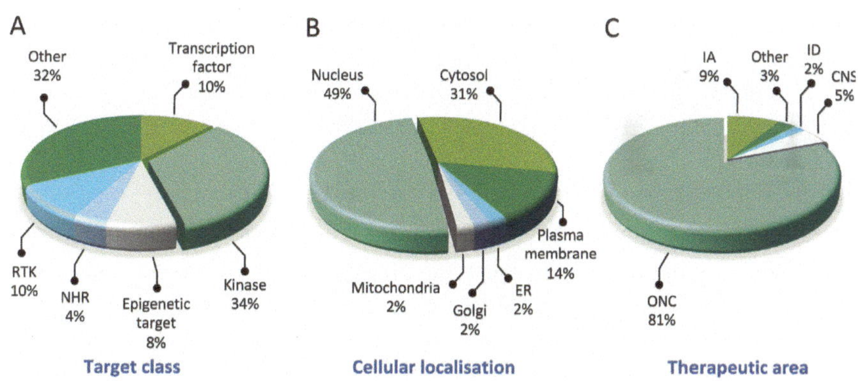

Fig. 11 Distribution of proteins degraded by PROTACs according to (A) target class (B) cellular localisation and (C) therapeutic area. NHR, Nuclear hormone receptor; RTK, Receptor-tyrosine kinase; ER, Endoplasmic reticulum; CNS, Central nervous system diseases; IA, Inflammation and autoimmune diseases; ID, Infectious diseases; ONC, Oncology.

Transcription factors, epigenetic targets and nuclear hormone receptors are also prominent in the spectrum of cancer-related proteins pursued for degradation via PROTACs. Although GPCRs and ion channels have been explored widely with small molecule drugs, the number of reports describing the degradation of these target classes remains scarce. The ability of E3 ligases to form a productive ternary complex with multi-pass *trans*-membrane protein may represent a challenge; a recent report suggest that such proteins may equally be depleted by PROTACs, although care should be taken to distinguish TPD from standard endocytotic degradation of plasma membrane proteins which may also occur in response to a ligand [253]. A wide variety of enzymes other than kinase (added in the group 'other' in Fig. 11A) have been tackled via TPD and include small molecule hydrolases (e.g. PDE4) [24], transferases (e.g. PARP1) [244,245], and peptidases (e.g. MetAP2) [12].

With regard to cellular localisation, we note that a large proportion of degraded proteins is located in the nucleus (49%) and/or the cytosol (32%) while membrane-bound targets, although less represented in the dataset, might be tagged for degradation via their intracellular domain (e.g. RTKs) (Fig. 11B). Ubiquitous expression of E3 ligases such as CRBN and VHL in both the cytosol and the nucleus of various cell lines allows for the efficient degradation of intracellular proteins. Finally, besides cancer-associated targets, several research groups have been successful in knocking down proteins related to the central nervous system (e.g. Tau, Huntingtin) [235,258], inflammation and autoimmunity (e.g. IRAK4, RIPK2) [31,236], and infectious diseases (e.g. HBV protein X, NS3/4A HCV protease) [243,263] (Fig. 11C).

3.2 Oncology targets at the forefront of TPD

The field of oncology has been the primary therapeutic area to benefit from PROTACs, with degraders showing pharmacodynamic superiority both *in vitro* and *in vivo* over modulation of activity of the same POI with small-molecule ligands [54]. A summary of the most relevant PROTACs developed against proteins with a recognised role in various malignancies is given below (Tables 3–7). The data collected provides information about the recruited E3 ligase, the degradation capacity of the PROTAC as a DC_{50} value, as well as the cell line used in the study. For an exhaustive set of data combining all bifunctional degraders reported to date, we refer the reader to a recently launched database that offers detailed information about chemical structures, biological activities, and physicochemical properties for individual compounds [265].

Overall, a significant number of potent degraders have been developed against a wide range of well-studied and validated oncological targets. The DC_{50} values reported for these degraders are often below 100 nM (58% of the collected data) and this favourable activity range frequently translates

Table 3 PROTACs targeting kinases.

Target	E3 ligase	PROTAC	DC_{50} (nM)	Cell line	References
AKT1/2/3	CRBN	INY-03-041	10–100	MDA-MD-468	[194]
AURKA	CRBN	PROTAC-D	<250	U2OS/HeLa	[199]
BCR-ABL	IAP	DAS-IAP	3–10	K-562	[202]
	CRBN	DAS-CRBN	30–100	K-562	[202]
	VHL	GMB-475	340	K-562	[203]
BLK	VHL	PROTAC 7	220	Ramos	[204]
BRAF	CRBN	Compound 2	5000	MCF-7	[205]
BRAF (V600E)	CRBN	Compound 12	500	A375	[206]
BTK (C481S)	CRBN	MT-802	14.9	XLA	[71]
BTK (WT)	CRBN	MT-802	9.1	Namalwa	[71]
			14.6	XLA	[71]
	IAP	PROTAC 3	200	THP-1	[208]
c-ABL	VHL	DAS-VHL	<1000	K562	[91]
CDK2	CRBN	A9	<1000	PC-3	[210]
		F3	62	PC-3	[210]
		TMX-2172	<250	Jurkat	[211]
CDK4	CRBN	BSJ-04-132	<500	Jurkat	[212]
CDK5	CRBN	TMX-2172	<250	Jurkat	[211]
CDK6	CRBN	BSJ-03-123	<500	Jurkat	[212]
CDK8	CRBN	JH-XI-10-02	<1000	Jurkat	[213]
CDK9	CRBN	F3	33	PC-3	[210]
CK2	CRBN	Compound 2	5000	MDA-MB-231	[215]
ERK1/2	CRBN	ERK-CLIPTAC6	N/A	A375	[227]
FAK (PTK2)	CRBN	BI-3663	27	A549	[229]
	VHL	PROTAC-3	3	PC-3	[66]

Continued

Table 3 PROTACs targeting kinases.—cont'd

Target	E3 ligase	PROTAC	DC_{50} (nM)	Cell line	References
IRAK4	VHL	Compound 9	151	PBMC	[236]
JAK1/2	IAP	JP-6	2500	THP-1	[237]
MAPK11 (p38β)	CRBN	NR-6a	19	MDA-MB-231	[240]
MAPK13 (p38δ)	VHL	SJFδ	46	MDA-MB-231	[86]
MAPK14 (p38α)	VHL	SJFα	7	MDA-MB-231	[86]
PI3K	CRBN	PROTAC-D	10,000	HepG2	[247]
PLK1	CRBN	HBL-4	20	MV4-11	[248]
RIPK2	VHL	PROTAC_RIPK2	1.4	THP-1	[31]
SGK3	VHL	SGK3-PROTAC1	300	HEK293	[250]
TBK1	VHL	Compound 3i	12	A549/H23	[259]
WEE1	CRBN	ZNL-02-096	10–100	MOLT4	[262]

Table 4 PROTACs targeting transcription factors.

Target	E3 ligase	PROTAC	DC_{50} (nM)	Cell line	References
BCL6	CRBN	AZ15	<100	SU-DHL-4	[201]
EED (PRC2 complex)	VHL	PROTAC 2	<120	Karpas 422	[221]
		UNC6852	790	HeLa	[222]
PIR	CRBN	CCT367766	0.5–1	SK-OV-3	[156]
RAR	IAP	Compound 9	1000–10,000	HT1080	[89]
SMAD3	VHL	Smad3 PROTAC	N/A	ACHN	[254]
SMARCA2	VHL	ACBI1	6	MV-4-11	[84]
SMARCA4	VHL	ACBI1	11	MV-4-11	[84]
STAT3	CRBN	SD-36	60	Molm-16	[256]
TRIM24	VHL	Dtrim24	2500–5000	293FT	[260]

Table 5 PROTACs targeting receptor tyrosine kinases (RTKs).

Target	E3 ligase	PROTAC	DC_{50} (nM)	Cell line	References
ALK	CRBN	MS4078	11	SU-DHL-1	[195]
		TL13-12	10	NCI-H3122	[196]
	VHL	TD-004	<300	NCI-H3122	[197]
EGFR	VHL	GB-RTK1	39.2	OVCAR8	[54]
EGFR (L858R/T790M)	VHL	Compound 14o	5.9	H1975	[223]
EGFR (L858R/T790M)	CRBN	DDC-01-163	<100	Ba/F3	[224]
FLT3	VHL	FLT-3 PROTAC	<5	MV-4-11	[230]
HER2	VHL	GB-RTK1	<100	OVCAR8	[54]
HGFR (c-Met)	VHL	GB-RTK7	<500	MDA-MB-231	[54]
TrkC	CRBN	IY-IY-Pom 4	100–1000	Hs578t	[261]

Table 6 PROTACs targeting epigenetic targets.

Target	E3 ligase	PROTAC	DC_{50} (nM)	Cell line	References
BRD4	CRBN	dBET-1	430	SUM149	[28]
	MDM2	A1874	32	HCT-116	[207]
	VHL	MZ1	<100	HeLa	[30]
		ARV-825	1	Namalwa	[27]
		GNE987	0.03	EOL-1	[44]
BRD7	VHL	VZ185	4.5	Ri-1	[162]
BRD9	CRBN	dBRD9	<50	Molm-13	[67]
	VHL	VZ185	1.8	Ri-1	[162]
HDAC1/2/3	VHL	PROTAC 4	1000	HCT-116	[232]
HDAC6	VHL	Compound 3j	7.1	MM1S	[233]
	CRBN	Compound 12d	1.6	MM1S	[157]
PCAF/GCN5 (KAT2B)	CRBN	GSK983	1.5	THP-1	[246]
SIRT2	CRBN	Compound 12	200–1000	HeLa	[252]

Table 7 PROTACs targeting nuclear hormone receptors.

Target	E3 ligase	PROTAC	DC_{50} (nM)	Cell line	References
AR	IAP	SNIPER (AR)-51	1000–3000	22Rv1	[23]
	VHL	ARCC-4	5	VCaP	[74]
		ARD-69	0.86	LNCaP	[150]
ER	VHL	ERD-308	0.17	MCF-7	[225]
		AM-A3	1.1	MCF-7	[226]
	IAP	Compound 5	10,000	MCF-7	[264]
	XIAP	SNIPER (ER)-87	1–3	MCF-7	[24]
ERRα	VHL	PROTAC_ERRα	100	MCF-7	[31]
	VHL	Compound 6c	3–10	MDA-MB-231	[228]

into profound inhibition of cell growth in various cell lines. CRBN and VHL E3 ligase ligands are by far the most used in degrader design (92% of the reported PROTACs), while IAP-recruiting ligands are occasionally included as a third choice when assessing degradation. Additionally, PROTACs targeting the mutant forms of cancer-driving proteins deliver a means to address selectivity and resistance issues [71,223]. Thus, we anticipate that assessment of the growing arsenal of cancer-targeting degraders in animal models followed by their translation into the clinic will make a significant impact on cancer treatment [266].

3.3 The search for novel E3 ligase ligands: Opportunities in TPD

While the number of successfully degraded target proteins with PROTACs has been growing rapidly, the choice of suitable E3 ligase ligands has remained limited. Novel E3 ligase ligands will not only expand the number of POI-E3 ligase combinations available to a PROTAC discovery campaign, but additionally, variation in E3 ligase expression levels and in their tissue localisation may be exploited to achieve degradation with higher selectivity [171]. Among the putative 600 E3 ligases of the human proteome, optimised binders suitable for PROTAC design have been identified for only few of these enzymes and therefore, the large pool of available E3 ligases represents untapped potential for TPD. Attempts to discover new E3 ligase binders via innovative approaches are highlighted below.

3.3.1 *Chemoproteomics as a tool for the identification of E3 ligase ligands*

Recent studies have brought to light ingenious strategies to recruit previously inaccessible E3 ligases to broaden the scope of TPD with PROTACs. The Cravatt group at the Scripps Research Institute has used a chemoproteomic approach based on the design of bifunctional compounds incorporating cysteine-reactive electrophilic fragments (referred to as scout fragments/probes) to identify potential ligands for novel E3 ligases [35]. Hit fragments were linked to reversible ligands specific for their studied proteins (BRD4 and prolyl *cis-trans* isomerase FKBP12), and the designed chemical tools were found to promote degradation of BRD4 and FKBP12 via the recruitment of the E3 ligase substrate receptor DCAF16. The cysteine-reactive chemoproteomic platform underlying the discovery of DCAF16 is known as isoTOP-ABPP (isotopic tandem orthogonal proteolysis–activity-based protein profiling) and uses reactive chemical probes, such as scout fragment KB02, to map ligandable hotspots in whole proteomes (Figs. 12 and 13) [267]. The workflow for this approach involves the treatment of intact cells or lysates with a reactive electrophilic fragment or vehicle. Subsequently, samples are incubated with a broad-spectrum cysteine-reactive probe (e.g. iodoacetamide (IA)-alkyne), then chemically conjugated to isotopically labelled azide-biotin tags via click chemistry. After enrichment with streptavidin beads, protease digestion provides fragment-labelled isotopic peptide pairs, which are submitted for mass-spectrometry analysis and quantification.

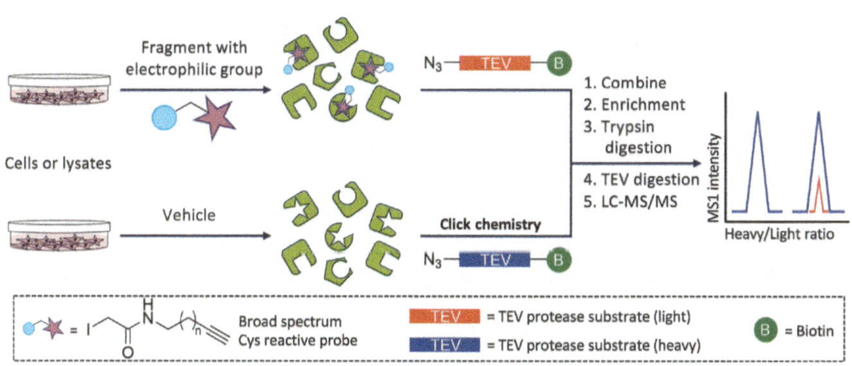

Fig. 12 Quantitative chemoproteomic approach developed by the Cravatt lab to identify covalent ligands in native biological systems. isoTOP-ABPP consists of several sequential steps: proteome labelling, click-chemistry for the incorporation of isotopically labelled cleavable tags, and proteolytic digestion to provide fragment-labelled peptides for mass spectrometry analysis. TEV, Tobacco Etch Virus nuclear-inclusion-a endopeptidase.

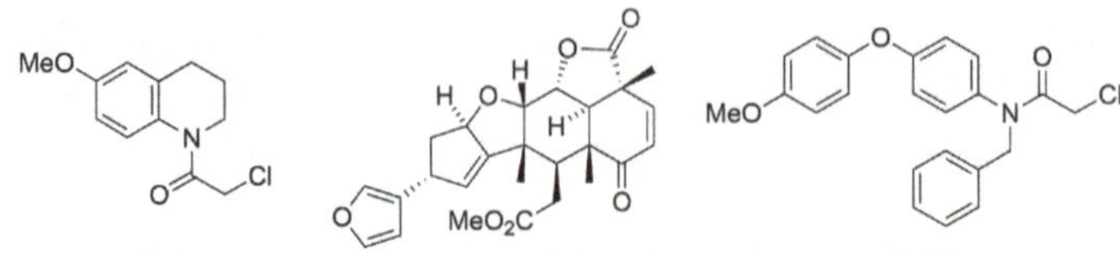

Compound or covalent fragment	KB02	Nimbolide	CCW16
E3 ligase recruited	DCAF16	RNF114	RNF4
Protein degraded	BRD4 FKBP12	BRD4 BCR-ABL	BRD4

Fig. 13 Covalent molecules used in the ABPP-chemoproteomic approaches to recruit novel E3 ligases.

Cravatt's work provides an elegant strategy to uncover novel E3 ligases by covalent recruitment in a native biological system. Besides, only 10%–40% covalent modification of DCAF16 was sufficient to achieve degradation without perturbing the endogenous function of the E3 ligase. A spinout biotech company leveraging Cravatt's findings, Vividion Therapeutics, was launched in 2017, and focuses on the discovery of novel protein–ligand interactions, especially those involving E3 ligases.

Following a similar approach, Nomura's lab used isoTOP-ABPP methodology to unveil the mechanism of action of the anticancer natural product nimbolide (Fig. 13) [34]. It was found that disruption of the interaction between the E3 ligase RNF114 and its substrate, the tumour suppressor p21, through covalent modification of a key cysteine present in RNF114, contributes to its antiproliferative effect. By tethering nimbolide to JQ1 and more recently to the tyrosine kinase inhibitor dasatinib [268], RNF114-recruiting PROTACs were generated to degrade BRD4 and BCR-ABL, respectively.

In a search for ligands that could bind the RNF4 E3 ligase which targets SUMOylated proteins for degradation, the Nomura group used an ABPP strategy to identify chloroacetamide ligand CCW16 as a RNF4 covalent binder [33]. A PROTAC featuring this covalent warhead linked to JQ1 (CCW28-3) efficiently degraded BRD4 via a RNF4-dependent mechanism, further highlighting the strength of chemoproteomic-based platforms for the discovery of new chemical tool to recruit E3 ligases.

3.3.2 Rational identification of molecular glues

As discussed in Section 1.5, molecular glues (such as IMiDs and indisulam) induce non-physiological PPIs with E3 ligases to achieve targeted degradation of neosubstrates. Although the discovery of such compounds has often been serendipitous and remains a difficult endeavour, molecular glues are extremely attractive chemical tools due to their small size often imparting good physicochemical properties, their a priori low affinity for neosubstrates and their potential for incorporation into PROTACs. Recent attempts at rational discovery of molecular glues provide a strong basis for further investigation. A transatlantic consortium led by the Broad institute of MIT and Harvard conducted a data mining exercise in which cytotoxicity data for several preclinical and clinical compounds was correlated to the intracellular level of expression of several E3 ligase components [269]. The authors discovered that pan-CDK inhibitor CR8 (Fig. 14) induced degradation of cyclin K (a CDK activator) by promoting key intermolecular interactions between the substrate adaptor DDB1 of Cullin-RING ligase Cul4B and CDK12 via a molecular glue-type mechanism. Of note, CDK12 is not a constitutive E3 ligase component in this context but is reprogrammed to become a substrate receptor by CR8, highlighting how a binding event between an inhibitor and a protein can unravel new gain-of-function properties such as glue-induced degradation.

Following this methodology, several chemical entities dependent on E3 ligase components to exert a toxic effect were identified in the hit list and are currently under investigation. Nonetheless, due to the nature of the screen the initial dataset could not capture effective molecular glues for which downstream effects would not impair cell viability.

The Winter lab tackled the same challenge of rationally discovering molecular glues via a different and original approach. Comparative drug profiling of 2000 cytostatic and cytotoxic compounds was conducted in cells lacking activated E3 ligases due to abrogation of neddylation, which is essential to activate E3 ligase activity through attachment of the ubiquitin-like

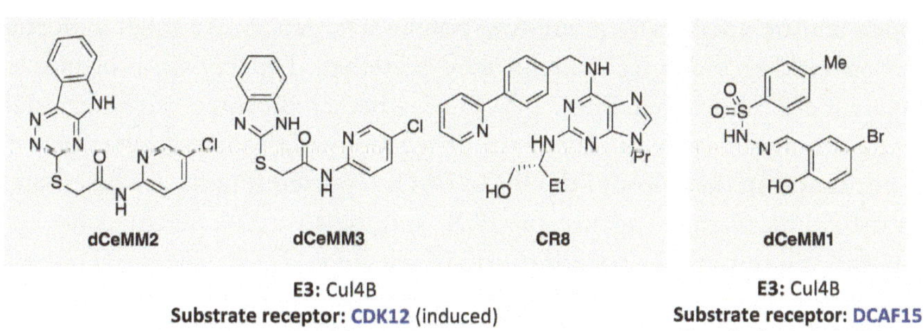

Fig. 14 Structures of rationally discovered molecular glues. The E3 ligase complex, substrate receptor and target degraded are indicated.

protein, NEDD8 [102]. Chemical series which depend on neddylation levels to induce cytotoxicity were identified (Fig. 14). First, a sulfonamide scaffold, dCeMM1, was found to redirect the activity of CUL4-DCAF15 to degrade RBM39 in a mechanism similar to indisulam. Second, two other compounds, dCeMM2 and dCeMM3, promoted the degradation of Cyclin K through a glue-like mechanism, recruiting CDK12 as a substrate receptor to mediate the ubiquitination of Cyclin K by CUL4B, in the same manner as CR8. Although the search for new ligands to recruit alternative E3 ligases to the commonly used CRBN and VHL still presents substantive challenges (less than 1% of known E3 ligases are currently being exploited), the examples described in this section represent a starting point for the development of ingenious methods and platforms that will change the landscape of E3 ligases accessibility and ultimately broaden the scope of TPD. In particular, we anticipate that systematic biochemical profiling of hundreds of members of the E3 ligase family using encoded library technologies will lead to disclosure of many novel ligands over the next several years and is likely to be particularly valuable for PROTACs for which relatively low affinity, pharmacologically inactive binders are desired.

3.4 Degrading 'challenging' targets

The emergence of PROTACs as a modality to efficiently knockdown a wide range of target proteins has brought with it the tantalising promise that proteins once labelled undruggable may now be accessible to drug discovery [64]. The term undruggable often refers to proteins lacking a well-defined ligand binding or enzyme active site and for which identification of a ligand remains extremely arduous. Non-enzymatic proteins in this category include transcription factors, regulatory and scaffolding proteins, many of which are therapeutically interesting potential targets. Since small molecule binders lacking biological activity may be tethered to E3 ligase ligands to generate efficacious degraders, PROTAC-mediated proteolysis is regarded as an exciting modality to address the druggability of those difficult targets. It is worth noting that most of the PROTACs reported thus far are based on a tight-binding warhead against the POI, and the so-called undruggable targets will still require the identification of selective, if not necessarily potent, binders in order to achieve on-target compound-induced interaction with an E3 ligase. Nonetheless, TPD may still lower the requirements needed to develop a successful tool compound or drug for these challenging proteins. As an example, the mutated $KRAS^{G12C}$ form of the notorious proto-oncogene

GTPase KRAS was recently degraded using a PROTAC (LC-2) made by the Crews lab from a VHL E3 ligase recruiting motif and a covalent warhead targeting specifically Cys12 in mutant $KRAS^{G12C}$ [185,238]. Although LC-2 did not improve the antiproliferative properties of the parent warhead MRTX-849, it is the first reported chemical tool to achieve sustained knockdown of endogenous mutated KRAS with an effect on its downstream signalling pathway. In addition to mutant KRAS, other disease-relevant and challenging targets have been tackled via a PROTAC-based strategy. Herein, two selected examples are discussed.

3.4.1 Signal transducer and activator of transcription 3

Signal transducer and activator of transcription 3 (STAT3) is a well-studied transcription factor for which deregulation via persistent activation is associated with cancer cell survival and proliferation [270]. STAT3 relays extracellular signals through binding from its Src homology 2 (SH2) domain to membrane-bound receptors. The subsequent phosphorylation of its Tyr705 residue results in dimerisation and translocation to the nucleus. Inhibition of STAT3 dimerisation via blockade of its SH2 domain has been extensively investigated in the past 20 years. Nonetheless, the inhibitors developed suffered from either poor physicochemical properties (peptidomimetics) or weak binding affinities. Additionally, high sequence homology between proteins of the STAT family has hindered the discovery of STAT3-selective compounds. Using a scaffold hopping approach, Wang's lab identified an inhibitor (SI-109) of the STAT3 SH2 domain [256]. After thorough linker length optimisation, a potent and selective STAT3 degrader (SD-36, $DC_{50} = 60\,nM$) was obtained from SI-109 (Fig. 15). SD-36 achieved efficient and sustained degradation of STAT3 resulting in the inhibition of its transcriptional activity and displayed antitumour activity superior to the parent inhibitor SI-109 in a xenograft mouse model, highlighting the benefit of targeting protein degradation over inhibition for this challenging target.

3.4.2 Tau

Tau is a microtubule-associated protein highly abundant in neurons and the central nervous system. Pathological aggregation of hyperphosphorylated Tau (also known as neurofibrillary tangles) is linked to neurological disorders such as Alzheimer's and Parkinson's diseases [271]. Targeting Tau as a therapeutic approach to treat associated CNS diseases remains challenging due to its lack of well-defined active sites and ordered tertiary structure. The use of antisense oligonucleotides to decrease Tau expression levels has shown

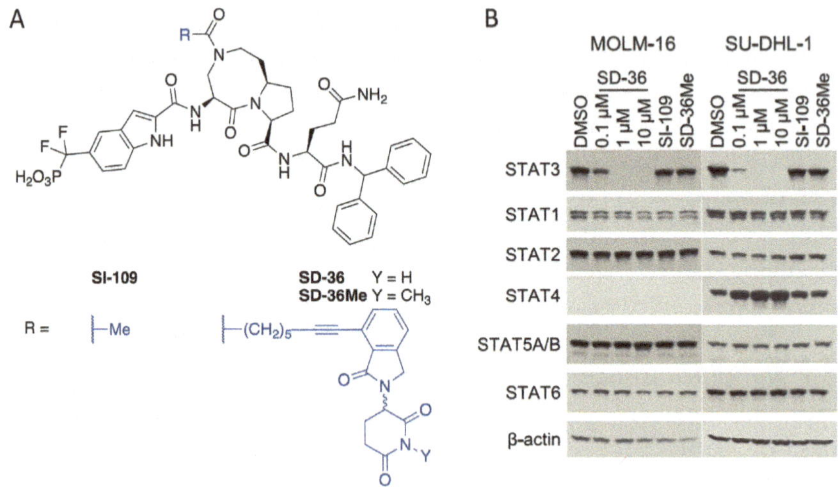

Fig. 15 Selective degradation of STAT3. (A) Structures of parent compound SI-109 and PROTACs SD-36 and SD-36Me derived from SI-109. (B) STAT protein western blot profiles. Cells were treated with increasing concentrations of SD-36, SI-109 (10 μM) or SD-36Me (10 μM) for 3 h (MOLM-16) or 24 h (SU-DHL-1) before harvesting. *Image adapted with permission from Zhou H, Bai L, Xu R, Zhao Y, Chen J, McEachern D, et al. Structure-based discovery of SD-36 as a potent, selective, and efficacious PROTAC degrader of STAT3 protein. J Med Chem 2019;62:11280–300. https://doi.org/10.1021/acs.jmedchem.9b01530, Copyright American Chemical Society.*

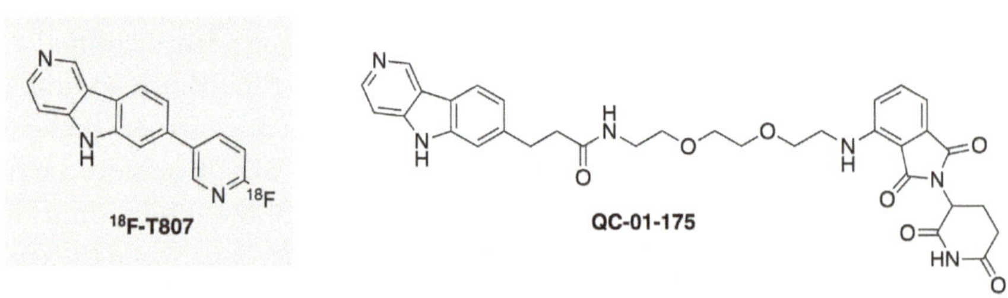

Fig. 16 Structure of PET probe ^{18}F-T807 and Tau degrader QC-01-175.

positive results but was limited by intracellular delivery and effective crossing of the blood-brain barrier. Ingeniously, a multidisciplinary team from Harvard medical school turned a reported Tau PET tracer, ^{18}F-T807, into a bifunctional degrader, QC-01-175 (Fig. 16) [258]. The efficacious Tau-targeting PROTAC emerged as a moderately potent degrader ($DC_{50} < 1$ μM) in patient-derived cellular models, among a library of 25 heterobifunctional molecules. These results set the precedent for the future application of this approach to other proteinopathies, or for targets for which high quality PET tracers are available as a basis for novel degraders.

4. Current progress in PROTAC translational research

Significant advancement has been made since the first conceptualisation of PROTACs in 2001 [12], which led to the first clinical trial of a PROTAC in 2019 (Fig. 2). The timeframe for development of drugs with standard mechanisms often exceeds 10 years [272], and it is remarkable that a new therapeutic modality has been optimised and progressed this far into the drug discovery pipeline within two decades.

One of the main reasons for widespread enthusiasm for PROTACs resides in the mode of action of this therapeutic strategy [1]. The PROTAC catalytic mechanism offers unusual pharmacodynamic (PD) properties, including exquisite potency and extremely low efficacious concentration, and protracted duration of action, which in most cases depends on protein resynthesis rate. Moreover, degradation rather than inhibition of undesired proteins can result in a superior PD response spectrum [54]. These PD properties have been considered highly promising for therapeutic applications. A recent example from scientists at GlaxoSmithKline (GSK), who developed PROTACs against RIPK2, demonstrated that *in vivo* PD properties of degraders can persist beyond the pharmacokinetic (PK) lifetime of the administered compound [57]. In fact, following a single 0.5 mg/kg subcutaneous administration of an optimised PROTAC in mice, protracted target degradation was observed over 4 days despite compound blood levels having already dropped significantly at 24 h. The observed effect, which could be extended with repeated dosages, was in line with the experimentally determined half-life of RIPK2 ($t_{1/2} = 50$ h), thus making this strategy highly efficacious to target slowly resynthesised proteins [54].

Moreover, it has been demonstrated that PROTACs are more successful against pathologies that are susceptible to the occurrence of resistance mechanisms, such as cancer [273], while further obvious applications would include pathologies caused by misfolded protein accumulation, such as synucleinopathies [255]. Cancer has been by far the most studied field for PROTACs due to the abundance of extremely challenging protein targets (Section 3) which have become more accessible with this new therapeutic modality, and to the ability of PROTACs to overcome upregulation resistance mechanisms [49].

Altogether, PROTAC-based therapies might deliver not only more efficacious treatment for challenging diseases but also improved patient compliance due to reduced dosing regimens and consequent potential for lower side effects. Moreover, the duration of action might be suited for options such as intravenous (IV) or intramuscular delivery, thus significantly reducing the need to overcome the hurdles of oral bioavailability. Similarly, advanced drug delivery opportunities may be explored, such as antibody-PROTAC conjugates which should have extended half-life and tissue-selective distribution (see Section 5.1). However, given that orally administered drugs remain the preferred option, considering both development costs and therapeutic compliance, current efforts have focused on the optimisation of PROTAC therapies for oral delivery. Unfavourable PK properties are often associated with the chemical nature of PROTACs, whose properties lie beyond the canonical 'rule of 5' dogma (see Section 4.1) and thus appear poorly suited to oral administration. Nevertheless, as will be discussed in the following section, pioneering work in the pharmaceutical industry has shown that PROTACs have the potential to be optimised into oral drugs [274,275].

4.1 Transitioning PROTACs from the bench to the clinic

4.1.1 The hurdles of pharmacokinetic optimisation

Lipinski's rule-of-5 (Ro5), a simple yet successful set of physicochemical properties designed particularly to support development of orally bioavailable drugs, has contributed to a reduction in the number of candidate drugs failing at clinical stages due to poor PK properties [276]. Ro5 parameters can be predicted *in silico*, thus allowing for efficient filtering processes at early stages in the drug discovery pipeline. However, recent reports have highlighted how reliance on Ro5 might also have resulted in lost opportunities [277]. Retrospective analyses have shown that many drugs on the market and in clinical trials actually contain active principles that break two or more Ro5 criteria, indicating that it is possible to develop less conventional molecules for therapy.

Even following several rounds of optimisation, PROTACs fall by necessity beyond Ro5 (bRo5) space, thus raising questions of tractability as orally administered drugs [278]. For example, it seems unlikely that VHL-ligand (VH032) based PROTACs can reach sufficient oral bioavailability according to data collected at AstraZeneca [275]. Compared to parenteral administration, oral drugs need to be stable and soluble under both acidic and near-neutral pHs (in the stomach and intestine, respectively), permeate cell membranes (absorption) and survive first-pass hepatic metabolism before reaching bioavailability in plasma. Despite these additional hurdles for optimisation, oral administration is the preferred route not only for patient compliance but also for advantageous drug storage and safer administration. While still sparse, recent reports by scientists at Roche [274] and

AstraZeneca [275] support the feasibility of orally administered PROTAC formulations. In a recent work from the Yang group, a prodrug strategy was successfully implemented to improve the PK of IMiD-based degraders enabling their oral administration. Protection of the glutarimide of lenalidomide with a pivaloyl group increased PROTAC lipophilicity, resulting in an increase in oral bioavailability in rats [279]. Nevertheless, the current roadblock at the industry level for the development of PROTAC drugs is the need to develop specific techniques for PK evaluation, as most routine assays have been found incompatible with the physicochemical properties of PROTACs. While the Ro5 is by definition a 'rule-of-thumb', several requirements need to be met for a drug to advance through the PK testing pipeline. Standard properties assessed include PSA (polar surface area), p*K*a, LogD (where D is the partition coefficient between lipid and aqueous phase at a defined pH), plasma protein binding, cell permeability, solubility, clearance and elimination. While some of these properties can be predicted *in silico* for small molecules (e.g. PSA), the size and chemical properties of PROTACs include many more variables and rotatable bonds, making automated predictions more difficult. For example, chameleonic behaviour, i.e., spatial molecular rearrangement in response to the environment, and 3D shape can have a huge impact on PSA and solubility, sometimes allowing unexpectedly favourable PK properties for PROTACs. Testing these big and unpredictable compounds has proven challenging with standard assays, and alternatives are currently under optimisation for high throughput assay platforms.

For instance, scientists at Roche have devised *in silico* p*K*a determination of degrader fragments rather than the entire molecule under the assumption that the actual values will not vary significantly once in the whole structure. The reliability of these predictions remains uncertain and needs to be assessed on a case-by-case basis, but this method seems to yield better predictions than the results from standard spectral gradient analysis, potentiometric titration or capillary electrophoresis [274].

Both Roche and AstraZeneca scientists agree on the use of chromatographic LogD (in lieu of LogD values, albeit not directly comparable), due to generally low solubility of PROTACs in standard assay buffers and the tendency of these compounds to bind to labware non-specifically [274,275]. For the same reason, plasma protein binding, usually measured via RED (rapid equilibrium dialysis) [280] assay, requires optimisation via a protocol that uses pre-saturation of labware surfaces. Similarly, compound permeability assessed with conventional PAMPA assay is often misleading.

With some optimisation (e.g. the use of 10% fetal bovine serum to promote solubility), cell-based permeability assays offer a better correlation to the expected impact of PROTAC size on cellular permeability, and moreover give the possibility to evaluate efflux [274]. However, these *in cellulo* data do not always correlate to *in vivo* scenarios, which represent the preferred option for testing PROTACs permeability at low throughput [275].

Overall, solubility and permeability were found to be the principal parameters influencing PROTAC oral absorption. Solubility is conventionally tested in high throughput, by determining kinetic solubility following a precipitation step. However, while this measurement correlates to thermodynamic solubility for small molecules, it has been proven to overestimate the solubility of highly flexible compounds [274]. Thermodynamic solubility in more biorelevant systems, e.g., in medium rather than aqueous buffers, may therefore be preferable. Also in this case, higher throughput methodologies need to be developed to implement efficient screening early in the PK optimisation pipeline.

Finally, clearance and elimination rates need to be carefully balanced to reach an optimal equilibrium of sufficient drug levels in plasma and target tissues, as well as to avoid toxic drug overaccumulation. Very little is known about how the physicochemical properties of PROTACs affect their clearance pathways. For example, metabolic fragments that are still able to bind either the E3 ligase or the target protein could hinder PROTAC action by outcompeting ternary complex formation. Preliminary studies conducted by Roche on first generation degraders showed that the main metabolism took place in the liver and affected the linker portion of the molecule, while ligand elimination was low [274]. Hence, PROTACs with optimised linkers could lead to significantly improved half-lives and adequate distribution.

In contrast to PROTACs, emerging molecular glue compounds (see Section 1.5) offer improved PK properties due to their physicochemical properties, often adhering to the Ro5 in a manner similar to small molecule drugs. However, rational molecular glue design is more challenging than for PROTACs and tractability might be limited to a subset of targets in cells, as ternary complex formation is driven by specific elements of the protein-protein interface. Meanwhile, as discussed above, the interest of big pharma in PROTACs is providing solutions to include these novel chemical entities in routine PK assays, and this new therapeutic modality may soon be well-integrated in drug discovery pipelines.

4.1.2 PROTACs degrade target proteins in vivo

While the most prominent scientific literature on PROTACs has been based on *in vitro* and *in cellulo* proof-of-concept studies, therapeutic applicability demands *in vivo* studies to assess the druggability of specific targets in systemic models. In addition, pharmacological protein knockdown in animal models presents a powerful tool that is orthogonal to genetic manipulation for the study of protein function *in vivo* (see Section 1.2). Pioneering and comprehensive work was conducted by the Rao group [59], who designed a PROTAC against FKBP12 that successfully depleted this highly conserved protein across different species including mice, rats, Bama pigs and rhesus monkeys. Interestingly, both intraperitoneal (IP) and oral administration were effective systemically in mice, with the exception of the nervous system compartment, producing long lasting FKBP12 degradation after a single day of treatment. On the contrary, intracerebroventricular administration in mice resulted in brain-confined degradation of FKBP12, thus highlighting the poor penetration of the compound through the blood-brain barrier, which could be convenient for the treatment of proteins with aberrant behaviour limited to the nervous system. While the use of primates for studies *in vivo* may present particular challenges of ethics and accessibility, this study further supports the use of this protein control modality across different mammalian species.

A significant challenge in the study of targeted protein degradation *in vivo* consists in dissecting the relation between protein levels and biological effect (partial protein degradation may be sufficient, or even preferable), as well as understanding the pharmacology of drug distribution versus duration of action [281]. Most of the literature describing *in vivo* studies has focused on degradation of BET proteins via recruitment of VHL, with compound ARV-771 being the most used for mouse experiments, demonstrating higher anti-tumour efficacy upon degradation compared to BRD4 inhibition [53]. Further protein targets successfully degraded in mice include RIPK2 [57], BTK [282], ALK [197], ER [31] and MDM2 [241]; while the ligases used to date *in vivo* are VHL, CRBN and IAP3 [281]. PROTAC administration in these studies usually involves IP or subcutaneous daily injections in mouse xenograft models, followed by drug and protein level monitoring in both plasma and tissues at different time points. Albeit promising, these results offer only a partial view on the efficacy of PROTACs for treatment, with dosages that are often not in line with expected compound potency. Overall, our understanding of this new therapeutic modality remains limited *in vivo*, as the lack of robust PK/PD models for targeted protein degradation testifies [281].

As previously mentioned, the endogenous rate of protein resynthesis is an essential variable affecting PROTAC activity and should always be determined prior to *in vivo* studies. Furthermore, it is not uncommon for significant differences in phenotype to be detected upon treatment with closely related PROTACs, which can show unexpected variations in selectivity in the cellular environment. These observations are relatively recent and have been linked to the characteristics of ternary complex formation, yet another PD variable that is still extremely challenging to predict and design in advance, thus limiting PROTAC development to a largely empirical approach (see Sections 1.4 and 2.2) [86]. Hence, in parallel to basic research *in vitro*, there is an urgent need to explore a wider range of targets and phenotypes of targeted degradation *in vivo* to facilitate future advancement in therapies.

4.2 PROTACs in clinical trials: The race for the first approval

Since 2019 two PROTAC-based drugs have undergone Phase I clinical trials. ARV-110 is an AR androgen receptor degrader indicated for patients with metastatic castration resistant prostate cancer [283] and ARV-471 is an ER degrader indicated for patients with ER+/HER2− (ER positive and human epidermal growth factor receptor negative) locally advanced or metastatic breast cancer [283]. Both compounds are formulated for oral administration and sponsored by Arvinas, a leading PROTAC biopharmaceutical company, founded by Craig M. Crews in 2013.

While Arvinas might appear to be the most advanced competitor in the race for the first approval of a PROTAC-based medication, the field has seen development worldwide. In the years following the discovery of ubiquitin-mediated protein degradation (2004 Chemistry Nobel Prize to Aaron Ciechanover, Avram Hershko and Irwin Rose) and the advancement of PROTAC research, start-ups and biotech companies have flourished in the field of targeted degradation, via PROTACs and beyond (see Section 5), and most major pharmaceutical companies have significantly contributed to understanding and optimisation of these novel therapeutics (see Table 8). Major pharmaceutical companies such as AbbVie, AstraZeneca, Genentech and Novartis have established in-house platforms for PROTAC discovery, and collaboration between industry and academia has also proven successful, as evidenced by joint publications [40,84]. It is expected that a considerable

Table 8 Main current biotech companies investing in targeted protein degradation research.

Name	Year[a]	Area of interest	Prominent partners/investors
Amphista Therapeutics	2017	Human Therapeutics	Advent Life Sciences, Scottish Investment Bank, BioMotiv
Arvinas	2013	Human Therapeutics	Genentech, Pfizer, Bayer
BeyondSpring	2010	Cancer	Stock
BioTheryx	2007	Human Therapeutics	MSD Partners
C4 Therapeutics	2016	Human Therapeutics	Roche, Calico, Biogen
Captor Therapeutics	2015	Autoimmune Diseases, Cancer	European Funds, National Centre of Research and Development, European Union
Creative Biolabs$_{PROTAC}$	–	Services for PROTAC discovery	Creative Biolabs
Cullgen	2018	Human Therapeutics	GNI group, Sequoia Capital China, Highlight Capital
Kymera Therapeutics	2016	Human Therapeutics	GSK, Leukaemia and Lymphoma society, Sanofi, Vertex
Lycia therapeutic	2019	Human Therapeutics	Versant Ventures
Monte Rosa Therapeutics	–	Human Therapeutics	Versant Ventures
Nurix Therapeutics	2009	Human Therapeutics	Sanofi, Gilead Sciences
OerthBio	2019	Plant pest and disease control	Bayer, Arvinas
PolyProx Therapeutics	2018	Human Therapeutics	Cambridge Innovation Capital, RT Capital, Cambridge Enterprise, LifeArc
Progenra	2002	Human Therapeutics	Ono Pharmaceuticals
Ranok Therapeutics	2018	Cancer	Stock
Vividion Therapeutics	2017	Human Therapeutics	Celgene, Roche

[a]Only shown for biotech/collaborations founded on UPS targeting and targeted protein degradation.

number of novel chemical entities with a targeted protein degradation or related mechanism of action will enter clinical trials over the next few years [284].

While initial clinical results seem promising in terms of safety [285], and there has been common interest among pharma companies, as testified by the rising number of players in the field, the efficacy of PROTACs as human therapeutics has yet to be confirmed. Limitations of PROTACs have been identified, and include the insurgence of resistance mechanisms that have been already shown for this new therapeutic modality (see Section 1.2) [75]. Nevertheless, the future of these powerful molecules is on the rise as further barriers are constantly being challenged. Of particular note are current promising efforts, again led by Arvinas, to develop blood-brain barrier penetrant PROTACs for the treatment of incurable neurodegenerative diseases such as Alzheimer's and Parkinson's [286].

5. Emerging approaches in targeted protein degradation

During the last few years, the scientific community has witnessed the blooming of new paradigms arising from the TPD concept, including controllable PROTACs and degraders hijacking cellular systems other than the UPS. This research effort recently culminated in repurposing of the underlying concept of 'heterobifunctional molecules' beyond protein degradation.

5.1 Targeted delivery and conditional activation of PROTACs

In spite of the huge arsenal of degraders developed to date, most PROTACs rely on E3 ligases with broad expression profiles, hence not discriminating against different cell types. Moreover, the physicochemical characteristics of certain degraders account for poor oral bioavailability, solubility or *in vivo* PK properties, hindering *in vivo* application (see Section 4.1.1) [287]. On the other hand, the catalytic nature of PROTACs along with the lack of external control over their function, could potentially lead to uncontrolled protein degradation, and systemic toxicity issues limiting clinical development. During the last couple of years, several strategies have emerged directed towards selective delivery of PROTACs and spatiotemporal control of degradation.

5.1.1 Tissue-selective protein degradation: Antibody-PROTAC conjugates

Researchers have recently applied the concept of Antibody-Drug Conjugates (ADCs) [288,289] to PROTACs, pursuing degraders with tissue selectivity as well as providing an alternative delivery option for these degraders *in vivo*. Traditional ADCs rely on the specificity of monoclonal antibodies to selectively deliver a highly cytotoxic payload to targeted cells, ideally reducing off-target effects. However, one of the major liabilities of ADCs is the low amount of drug typically delivered into cells, thus often requiring extremely cytotoxic agents that may be dose-limiting even at very low levels of non-specific release [290]. Hence, PROTACs might be excellent ADC payloads since they act catalytically, and generally benefit from rapid and sustained protein degradation at low doses. By analogy with ADCs, Antibody-PROTAC conjugates are designed to bind to over-expressed antigens on the surface of target cells, thus allowing their endosomal internalisation and subsequent lysosomal release of the active degrader inside the cell. In contrast, cells lacking antigen overexpression display none or reduced conjugate uptake, thus precluding protein degradation on antigen-negative phenotypes (Fig. 17A). In a proof-of-concept study, the Tate group [45] reported Ab-PROTAC **3**, (Fig. 17B) a trifunctional chimera conjugating trastuzumab with a BRD4 degrader. Ab-PROTAC **3** triggered BRD4 degradation exclusively in HER2+ cells while sparing HER2− phenotypes. Moreover, studies of internalisation kinetics by live-cell confocal microscopy using an AlexaFluor™488-tagged fluorescent conjugate (AF488-Ab-PROTAC **3**), confirmed that this conjugate was endocytosed and released intracellularly through the endosomal-lysosomal pathway, in line with the mechanism of action of internalising ADCs.

As in the case of ADCs, a key feature for the development of an Ab-PROTAC conjugate consists in the design of an appropriate linker to connect the PROTAC and the antibody, since the linker has a high impact on conjugate stability and release [291]. The linker needs to be stable enough to maintain conjugate integrity under assay conditions, while allowing for intracellular release of the active PROTAC. Thus, careful selection of chemical composition and point of attachment to the PROTAC is necessary to prevent poor solubility or unsuitable stability [292]. The majority of reported Ab-PROTACs to date are derived from VHL-based degraders and take advantage of the hydroxyproline moiety of the VHL ligand to attach the linker. Following this strategy, different types of connections to the hydroxyproline were studied (Fig. 17B). Tate and co-workers connected

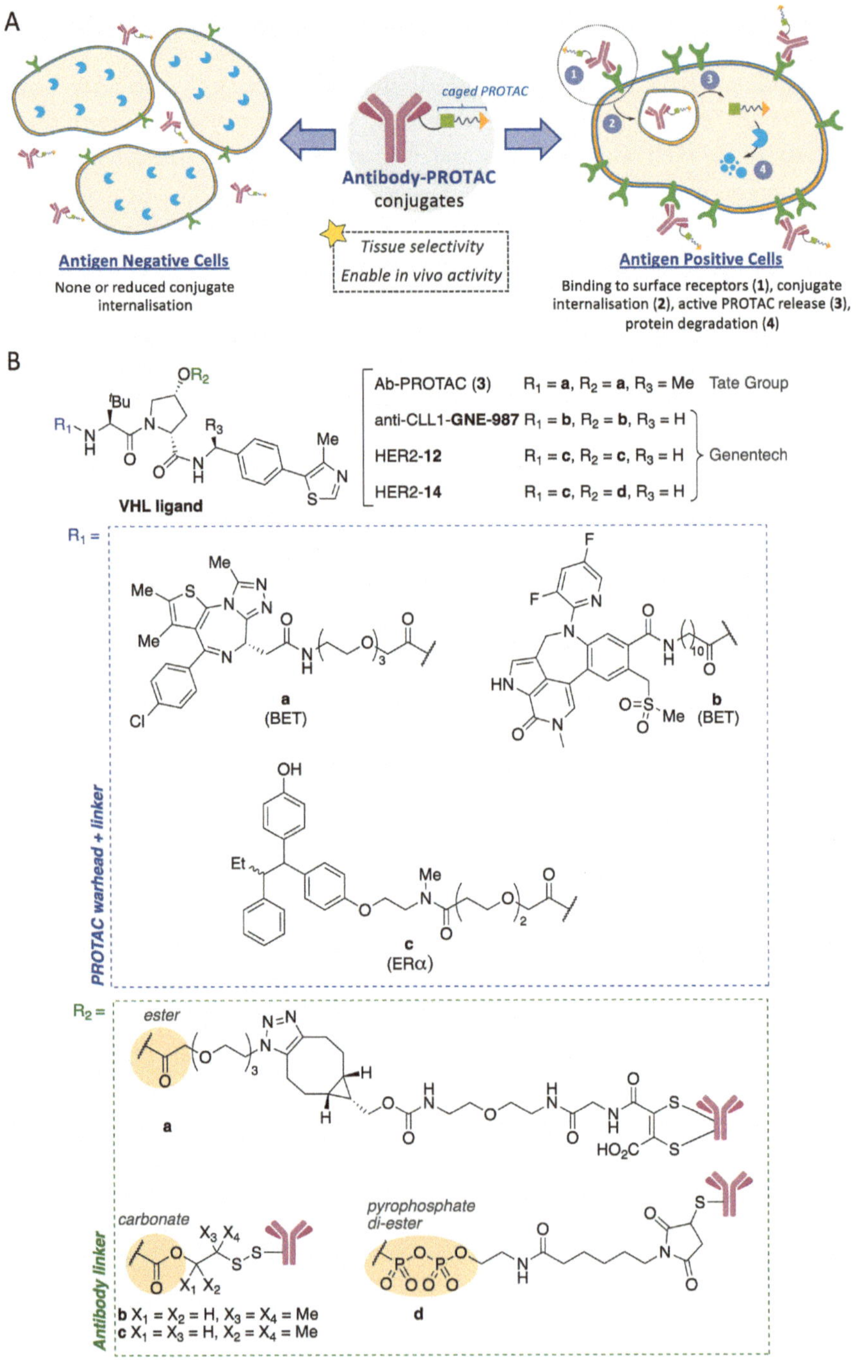

Fig. 17 (A) Mechanism of action of Antibody-PROTAC conjugates. (B) Chemical structures of selected Antibody-PROTAC conjugates.

the antibody linker to the E3 ligase through a labile ester moiety to afford Ab-PROTAC **3** [45]. Alternatively, reducible disulphide linkers have been successfully employed in Ab-PROTAC development. The resulting conjugates, although partially unstable in cell culture conditions, exhibited good *in vivo* stability properties, as previously observed for ADCs bearing this type of linker [293]. Researchers from Genentech employed a self-immolative disulphide linker attached to the VHL ligand via a carbonate to conjugate the BRD4 degrader GNE-987 to a CLL1 antibody (anti-CLL1–GNE-987, Fig. 17B) [44]. This same strategy was followed for the conjugation of an ERα degrader (conjugate HER-**12**, Fig. 17B) [292]. Furthermore, this latter degrader was also conjugated through a pyrophosphate di-ester group, which allows intracellular release of the active degrader via a phosphodiesterase cleavage mechanism (HER-**14**, Fig. 17B). Nonetheless, these ERα degrader-conjugates were moderately stable in mice, keeping 80% of the drug-antibody ratio after 72h in circulation [292].

In addition to selective delivery, conjugation of a PROTAC to an antibody proved to be a good strategy towards improving the PK properties of a degrader, hence enabling *in vivo* activity. The anti-CLL1–GNE-987 conjugate reported by Genentech showed enhanced PK properties compared to parent PROTAC GNE-987, a potent BRD4 degrader [44]. Whereas GNE-987 showed no *in vivo* efficacy following IV or oral administration to mice, a single IV dose of the anti-CLL1–GNE-987 conjugate caused dose-dependent tumour growth inhibition in mice harbouring HL-60 or EOL-1 acute myeloid leukaemia xenografts. Four days following administration, substantial amounts of GNE-987 persisted in EOL-1 tumours, accounting for the observed sustained BRD4 depletion. This is in line with the *in vivo* stability profile of this conjugate (>70% of the originally attached linker-drugs remaining after 7 days of circulation in mice) as well as with its exquisite PK behaviour showing long half-lives as along with low clearance. Although this work is an important advancement in the field of Ab-PROTACs, further studies are required to assess whether comparable *in vivo* outcomes could be achieved with different combinations of chimeric degraders and antibodies.

5.1.2 Spatiotemporal control of protein degradation: Conditional activation of PROTACs with light

Modulation of PROTAC activity may contribute to the development of targeted therapies with improved therapeutic index and reduced off-target effects, and could be achieved with degraders that only become activated under specific conditions, or 'conditional PROTACs'. The first example of PROTAC conditional activation was reported in 2013 by the Crews lab, with the development of phosphoPROTACs [231]. These molecules need to be phosphorylated by an activated RTK in order to recruit phosphotyrosine-binding (PTB) and SH2 domain-containing proteins for subsequent degradation. However, the peptide-based design of phosphoPROTACs limits their applications. In the search for alternative conditional activation approaches, several research groups leveraged the attributes of light for achieving spatiotemporal control over PROTAC activity [121,294]. The application of photopharmacology in drug development has been the subject of great interest during the last decade [295,296], since light is a non-invasive and easily modulatable precision tool that enables regulation of drug activity to reduce undesired side effects [297,298]. Moreover, photodynamic therapy is quite widely used in oncology, providing a paradigm example of the potential applications of light in medicine [299–302]. The modulation of protein degradation with light has been explored following two different approaches, which entailed development of photocaged or photoswitchable PROTACs [303]. Photocaged PROTACs rely on the functionalisation of PROTACs with photolabile moieties, which are used to cage functional groups essential for POI/E3 ligase engagement or ternary complex formation. Hence, protein degradation activity is prevented until the photoprotecting group is irreversibly removed by irradiation with UV light (Fig. 18A). A second photopharmacological method to control PROTAC activity relies on the use of photoswitches, which have been extensively employed in chemical biology to endow molecules with and on/off functionality, thus allowing the external modulation of bioactivity [298,304]. The introduction of a photoswitch on the PROTAC scaffold enables reversible photoisomerisation of the drug leading to an active/inactive pair of isomers (*E* and *Z*) with different degradation profiles triggered by the external stimulus of light irradiation at specific wavelengths. The rationale behind photoswitchable PROTACs derives from the hypothesis that different conformations might account for favourable or unfavourable formation of the key ternary complex that would lead to efficient polyubiquitination and eventual degradation (Fig. 18B).

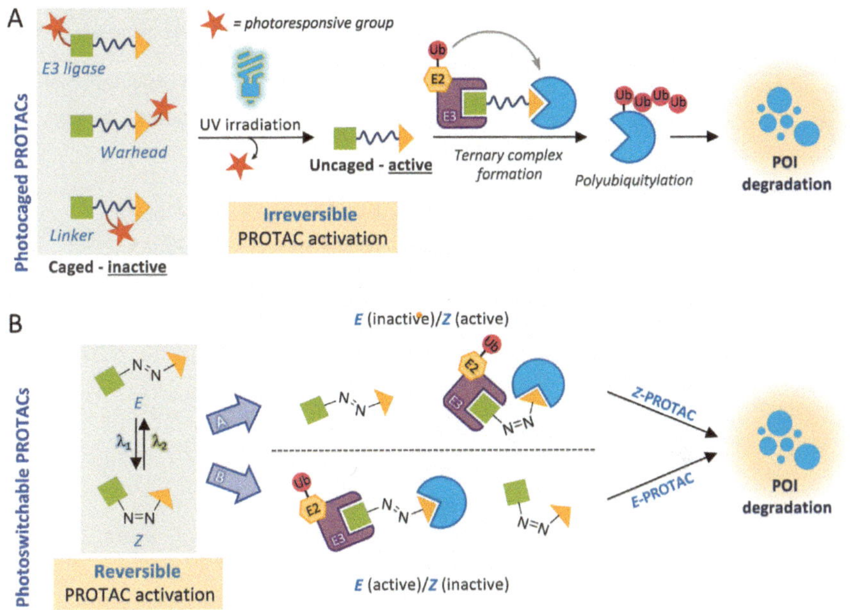

Fig. 18 Mode of action of light-activatable PROTACs, which allow for the *irreversible*—photocaged PROTACS (A)—and *reversible*—photoswitchable PROTACs (B) spatiotemporal control of protein degradation.

(A) Photocaged PROTACs

During 2019, a number of research groups developed photocaged PROTACs adopting different strategies [40,43,160,305]. Caging of the E3 ligase ligand and consequent impairment of E3 ligase engagement has been the preferred option, with various examples effectively caging the VHL ligand through the hydroxyproline moiety, essential for binding to pVHL (Fig. 19) [174,175]. Caging of thalidomide or pomalidomide based degraders has also proved successful via the imide moiety of the glutarimide ring, thus impairing binding to CRBN (Fig. 19) [97]. This IMiD caging strategy was also validated by Liu, Chen and co-workers with the development of an 'opto-pomalidomide' that was unable to bind CRBN and therefore to degrade IKZF1 until uncaged [305]. Given the prevalence of VHL and CRBN E3 ligase-based degraders in the PROTAC field, the reported caging designs are versatile for the degradation of different POI (Fig. 19). Nevertheless, other alternatives have also been considered, such

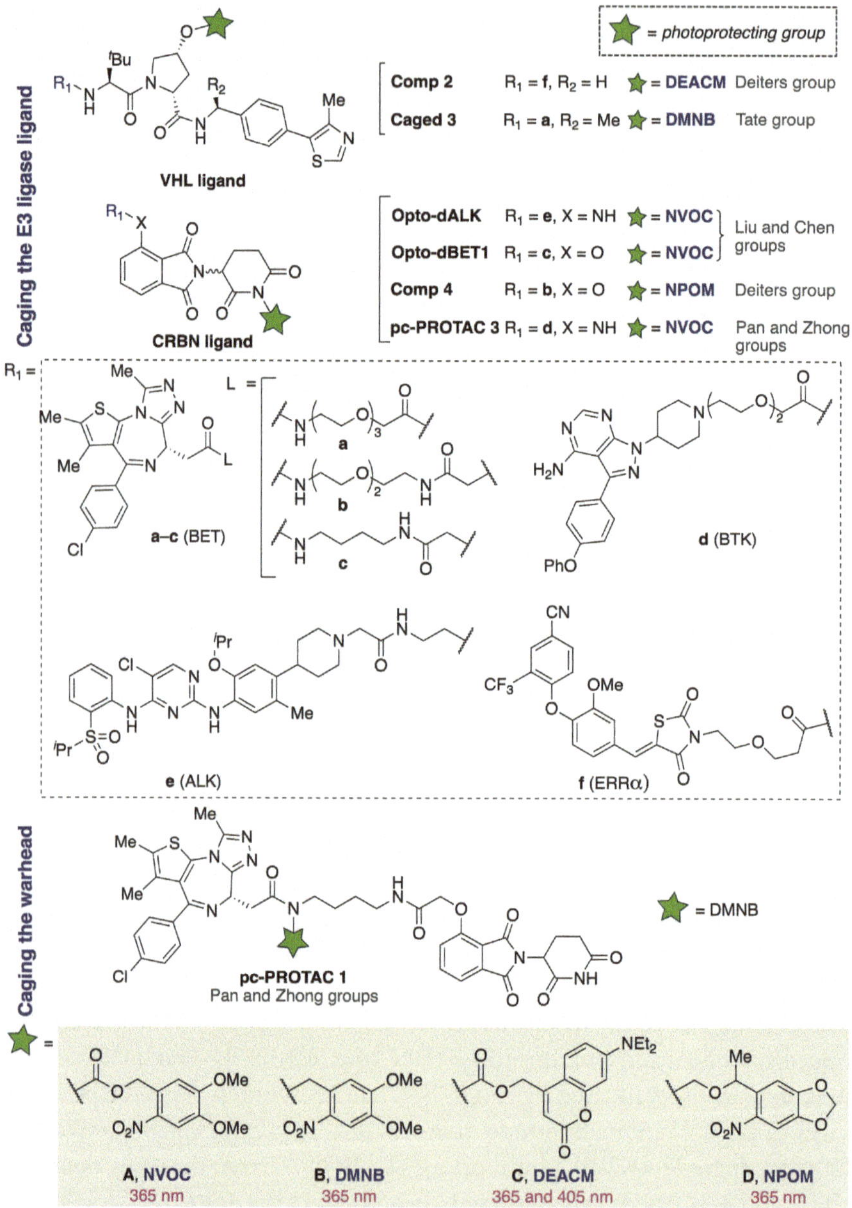

Fig. 19 Structures of the reported photocaged PROTACs. The different caging strategies and the photoresponsive groups employed are highlighted. The wavelengths used for uncaging of each group are highlighted in pink.

as introducing the caging group on the warhead (Fig. 19, pc-PROTAC 1) [43]. From a therapeutic point of view, this strategy would present the additional advantage of attenuating any pharmacological effect resulting from POI engagement before active PROTAC release. However, E3 ligase caging would be optimal for a PROTAC using a warhead lacking pharmacology in its own right (i.e. that binds at a non-functional site on the POI), since promiscuous degradation due to ligase binding presents an off-target pharmacology risk for a caged PROTAC. Alternatively, the E3 ligase ligand and warhead could be caged by a tether that prevents either from binding, although this design has not been implemented to date.

Successful PROTAC caging was achieved with different photoresponsive groups including coumarins and different nitroaromatic compounds, which can be easily deprotected under UV light irradiation at 365 or 405 nm depending on the structure (Fig. 19). Effective degradation upon conditional activation was achieved for BRD4 (Caged 3, Opto-dBET1, Comp 4 and pc-PROTAC 1), BTK (pc-PROTAC 3), ERRα (Comp 2) and ALK (Opto-dALK), demonstrating broad applicability across the degradable proteome (structures of the photocaged PROTACs are depicted in Fig. 19). The caged degraders proved stable and therefore inactive in the dark, displaying dose and UV-dependent protein degradation activity in cells only following photolysis, acting through a bona fide PROTAC mechanism. Moreover, degradation kinetics after uncaging were comparable to those of the parent uncaged PROTACs. Cell viability assays demonstrated that, as expected from their UV-dependent activity, caged degraders prevented the therapeutic response elicited from the effect of protein depletion [40,43]. Nonetheless, since these caged degraders are still cell permeable, it is not surprising that in some cases, a cytostatic effect derived from POI inhibition could be observed in cell proliferation studies [40]. Furthermore, applicability of photocaged PROTACs *in vivo* was demonstrated by Pan and Zhong groups, who evaluated pc-PROTAC 1 (Fig. 19) in zebrafish models [43]. Following irradiation of zebrafish embryos treated with pc-PROTAC 1 clear phenotypic changes correlated with a decrease in BRD4 levels, suggesting that light-mediated uncaging of pc-PROTAC 1 affected zebrafish embryo development and setting a precedent for the implementation of this technology in other *in vivo* models.

(B) Photoswitchable PROTACs

Several examples of photoswitchable degraders have been reported, namely, PhotoPROTACs, PHOTACS and Azo-PROTACs (Fig. 20), aimed at conditional degradation of varied targets including BET bromodomain proteins, FKBP12, BCR-ABL fusion protein and ABL [39,41,42]. In the design

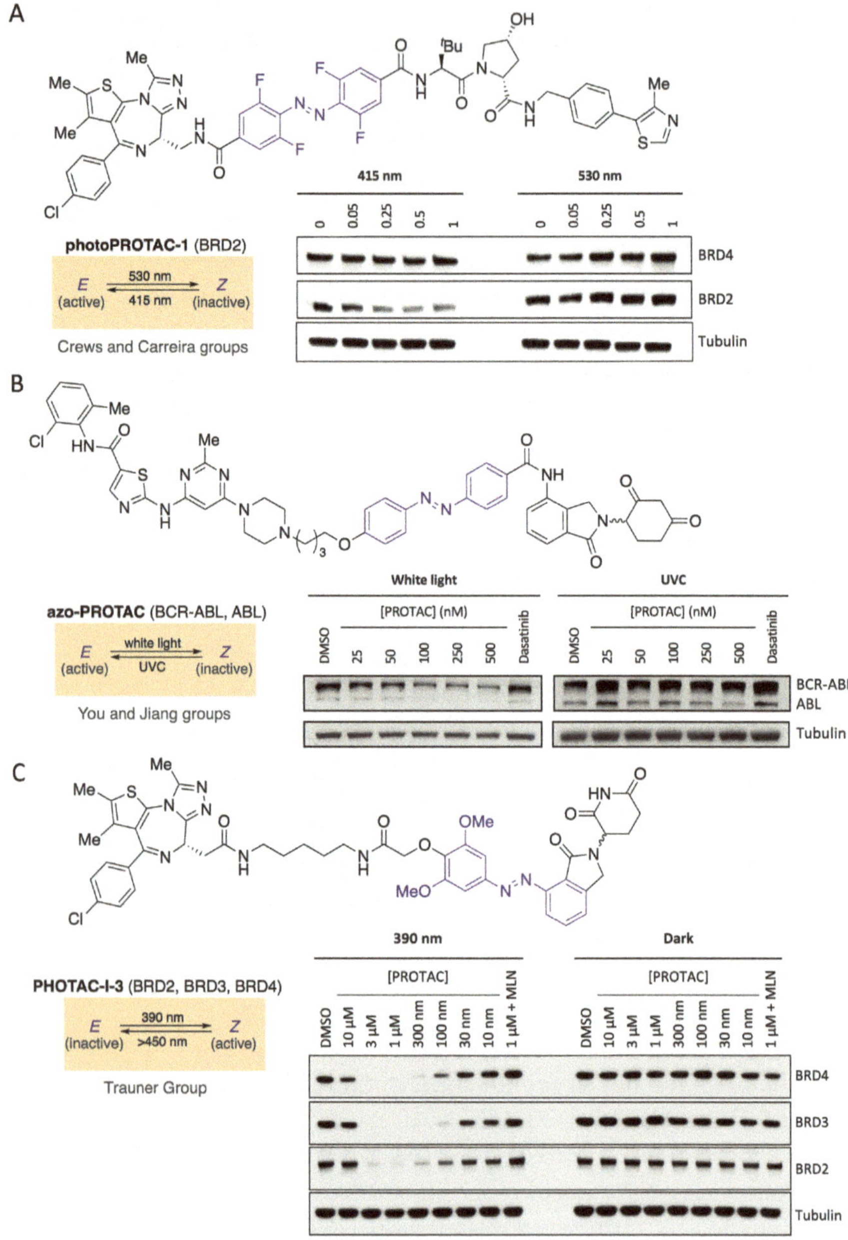

Fig. 20 Structures of photoswitchable PROTACs including (A) photoPROTACs, (B) azo-PROTACs and (C) PHOTACs. Western blot analysis of protein abundance upon PROTAC treatment shows the different degradation profiles among *E* and *Z* conformations. *Images adapted with permission from Pfaff P, Samarasinghe KTG, Crews CM, Carreira EM. Reversible Spatiotemporal Control of Induced Protein Degradation by Bistable PhotoPROTACs. ACS Cent Sci 2019;5:1682–90. https://doi.org/10.1021/acscentsci.9b00713; Jin Y-H, Lu M-C, Wang Y, Shan W-X, Wang X-Y, You Q-D, et al. Azo-PROTAC: Novel Light-Controlled Small-Molecule Tool for Protein Knockdown. J Med Chem 2020;63: 4644–54. https://doi.org/10.1021/acs.jmedchem.9b02058; Reynders M, Matsuura BS, Bérouti M, Simoneschi D, Marzio A, Pagano M, et al. PHOTACs enable optical control of protein degradation. Sci Adv 2020;6:eaay5064. https://doi.org/10.1126/sciadv.aay5064, Copyright American Chemical Society and American Association for the Advancement of Science.*

of these switchable degraders, the photoswitchable moiety can be included as part of the linker, changing its length and orientation (PhotoPROTACs and Azo-PROTACs) [39,41] (Fig. 20A and B) or it can be part of the E3 ligase ligand or the warhead ligand, primarily modulating the affinity for the E3 ligase or the POI, respectively (PHOTACs) (Fig. 20C) [42]. Concerning chemical structure, the reported photoswitchable PROTACs bear the widely studied azobenzene moiety with differential substitutions on the benzene rings, offering versatility and synthetic tractability [304]. Azobenzenes show decisive and predictable geometrical changes between *Z* and *E* states as well as photophysical properties that can be tuned by varying the substitution pattern in the aromatic rings [306,307]. The photoPROTACs designed by the Crews and Carreira groups [39] incorporate an *ortho*-F_4-azobenzene moiety with prolonged half-life resulting in bistable *Z* (inactive) and *E* (active) isomers that can be switched by a single period of irradiation at 415 or 530 nm, respectively. Interestingly, the active *E* isomer achieves degradation only of BRD2, showing no activity against BRD4 (Fig. 20A). On the other hand, PHOTACs reported by the Trauner group [42] require repeated pulsed irradiation at 390 nm or >450 nm in order to maintain high proportion of either *E* (inactive) or *Z* (active) conformations. In the case of azo-PROTACs, *E* (active) and *Z* (inactive) conformations can be interchanged by UVC or visible light irradiation. The *Z* conformation of these degraders undergoes thermal relaxation, isomerising to the thermodynamically more stable *E* isomer within 8–10 h for PHOTACs and azo-PROTACs, whereas photoPROTACs proved stable in the dark for several days. Furthermore, each of these photoswitchable degraders displayed dynamic degradation profiles in cells as a result of their reversible nature, meaning that the degradation properties of these compounds can be effectively switched on and off *in situ* by irradiation at appropriate wavelengths. Thermal reversion enables an interesting potential *in vivo* application by imparting automatic 'switch off' behaviour, since a PROTAC initially administered as an active *Z* isomer will switch off bioactivity when reverting to the *E* isomer over a timeframe programmed by the chemistry of the switch itself. However, whether this proves useful in practise in the face of the already very complex PK/PD profile of PROTACs remains an open question, and drugs with two stable pharmacoforms present interesting challenges for drug safety and approval.

These first insights into conditional PROTAC activation with light settled a precedent in the protein degradation field and have provided useful tools for chemical biology. Nonetheless, implementation in a clinical setting requires further studies to resolve the caveats associated with these methods. Despite the diverse uses of UV light in medicine, it is well-known that UV light exposure causes DNA damage [308]. Furthermore, therapeutic application of these approaches in oncology is restricted to a few types of cancer such as blood or skin tumours, due to the poor tissue penetration of UV light. In line with this, the use of other photoresponsive groups with absorptions in the IR region would be desirable to achieve enhanced tissue penetration [309,310].

5.2 Alternative approaches for targeted protein degradation

The degradation of proteins within the cell is controlled via two principal regulatory mechanisms which are known as the UPS, discussed in Section 1, and the lysosomal pathway. The delivery of intracellular proteins for degradation to lysosomes, takes place through a process called autophagy [311]. The lysosomal pathway is also involved in the degradation of extracellular material via the so-called endosome-lysosome pathway [312]. In the endosome-lysosome pathway the endocytosed material is sequentially delivered via fusion with several membrane-bound intracellular compartments prior to lysosome fusion. In contrast, selective autophagy relies on ubiquitination as a degradation tag, by analogy with proteasomal degradation [313]. These protein degradation pathways have been exploited to artificially degrade several disease-relevant proteins, with PROTAC technology being the most widely studied approach to hijack the UPS. Nonetheless, alternative approaches for targeted protein degradation have recently been disclosed [314], including strategies that take advantage either of the proteasomal or the lysosomal pathway, as discussed below.

5.2.1 Proteasomal degradation: BioPROTACs

Proteasomal degradation of targets induced by PROTACs is an efficient strategy for targeting proteins for which small-molecule binders are known. Nonetheless, there are some targets of great biological relevance that are deemed undruggable due to the lack of apparent binding sites, which hampers the discovery of small-molecule binders. As a complementary approach to PROTACs and aiming at targeting the degradation of undruggable scaffolds, the Partridge group developed bioPROTACs [315]. BioPROTACs are fusion proteins in which the E3 ligase has been reengineered by replacing its natural substrate binding domain with a peptide or miniprotein that binds a POI. Although the concept underlying bioPROTACs was already known, and previously employed for the degradation of challenging proteins such as KRAS [316] or c-Myc [317], the Partridge group devised a new and systematic approach for the development of bioPROTACs. Taking advantage of GFP-tagged proteins as model substrates, the authors demonstrated that bioPROTAC design is extensively flexible allowing for the use of a variety of POI binders (monobodies, nanobodies, DARPin and αRep) and the recruitment of a wide spectrum of E3 ligases for target degradation. It is important to highlight that since bioPROTACs are genetically encoded, they can leverage the entire repertoire of available E3 ligases (>600), in contrast with PROTACs that hijack the limited number of E3 ligases for which small molecule binders have been optimised.

Combining the in-house identified $SPOP_{167-374}$ E3 ligase with a proliferating cell nuclear antigen (PCNA) binding peptide (Con1), Partridge and co-workers developed a bioPROTAC that triggers rapid and potent degradation of PCNA, which results in pronounced growth inhibition in HEK293 cells. Although bioPROTACs benefit from great degradation potency, their dependence on genetic encoding limits their therapeutic applications. Nonetheless, this biological tool could be instrumental for the study of new small-molecule degraders, allowing assessment of target degradability, study of cellular outcomes derived from target depletion, and guiding of E3 ligase selection prior to lengthy chemical design and synthesis of a PROTAC.

5.2.2 Lysosomal degradation: LYTACs, AUTACs and ATTECs

5.2.2.1 LYTACs: Protein degradation through the endosome/lysosome pathway

The LYsosome TArgeting Chimera (LYTAC) technology recently developed by the Bertozzi group [140] enables targeted degradation of extracellular and membrane-associated proteins via the endosome/lysosome pathway. These chimeric molecules can bind simultaneously to a lysosome-targeting receptor (LTR) and a POI, thus inducing endosomal internalisation and subsequent lysosomal progression and degradation of the POI (Fig. 21A). In their pioneering work, Bertozzi and colleagues developed first generation LYTACs by conjugating a small molecule or an antibody for a POI with a chemically synthesised glycopeptide agonist of the cation independent mannose 6 phosphate receptor (CI-M6PR), a prototypical LTR which is involved in the lysosomal degradation of glycosylated proteins (Fig. 21B) [318]. In a proof-of-concept experiment, the authors demonstrated that a biotinylated poly(M6P) or poly(M6Pn) was able to uptake NeutrAvidin protein to endosomes and lysosomes. The scope of LYTAC technology was demonstrated through the successful LYTAC-mediated degradation of different therapeutically relevant targets like EGFR, CD71 and PDL-1. However, similar to PROTACs, the success of a particular LYTAC relies on a combination of factors that need to be taken into account

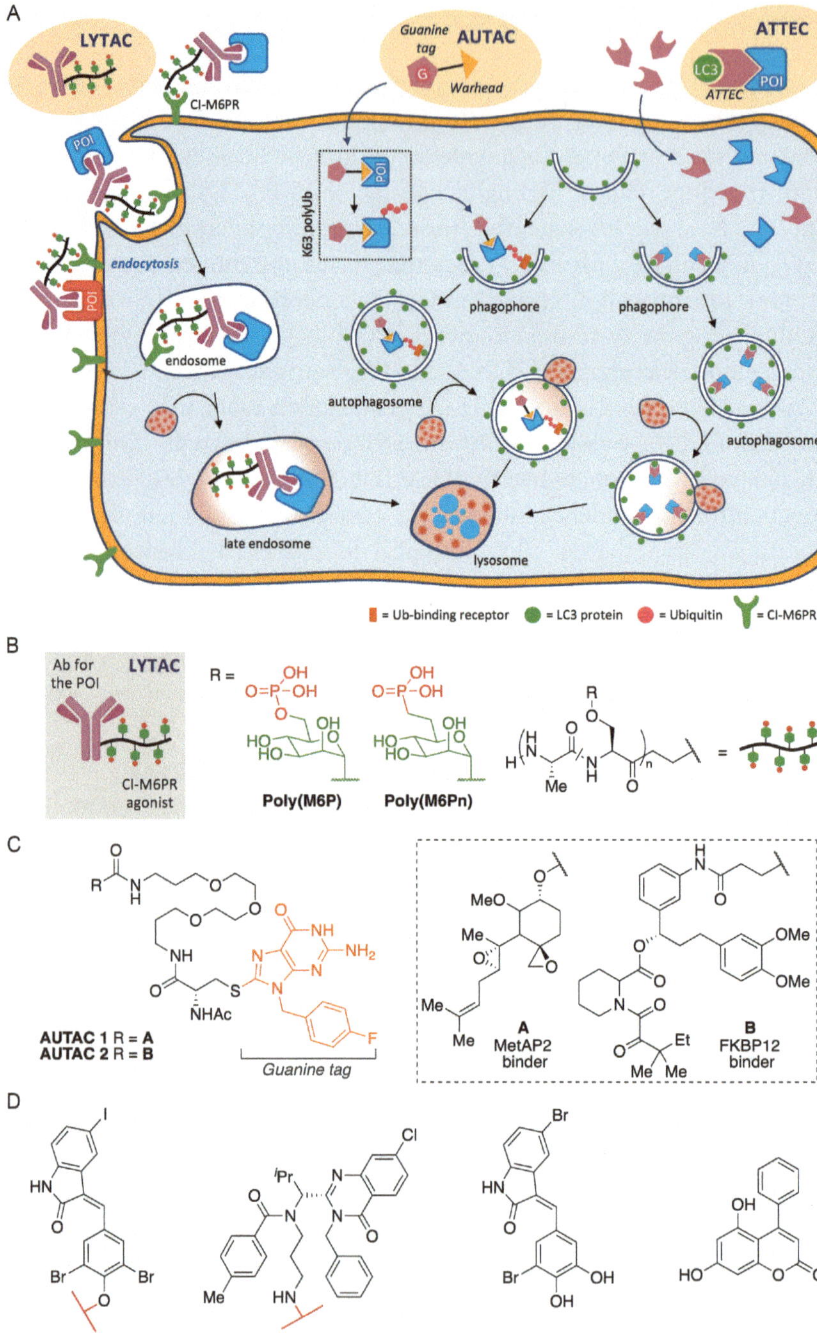

Fig. 21 (A) Mechanisms of action of degraders harnessing the lysosomal pathway. LYTACs mediate the endosomal internalisation and lysosomal trafficking of extracellular or membrane-associated proteins. AUTACs and ATTEC hijack the autophagy pathway for the degradation of intracellular targets. (B) General structure of LYTACs, containing an antibody to bind the POI and a CI-M6PR binding motif. (C) Structures of representative AUTACs targeting MetAP2 and FBPK12. The guanine tag is highlighted in orange. (D) Structures of four ATTEC molecules identified by high throughput screening.

at the time of design, which include internalisation kinetics, expression levels of surface receptors or susceptibility to lysosomal transportation through the endocytic pathway. From a pharmacological point of view, the major limitation of LYTACs is their large size, lacking the desirable properties of small molecule drugs. Nevertheless, the development of LYTACs as biochemical tools for studying receptor trafficking and extracellular protein degradation holds great potential. LYTAC technology also presents a novel and promising therapeutic alternative to conventional degraders, raising interest within the pharmaceutical industry. Lycia Therapeutics recently emerged with $50 million investment from Versant Ventures (see Table 8, Section 4) to develop next generation LYTACs, aiming to exploit further receptors to tackle extracellular proteins that remain intractable when employing other technologies [319].

5.2.2.2 AUTACs and ATTEC: Protein degradation through the autophagy pathway

Autophagy is a cell regulatory mechanism that clears redundant cytoplasmic material or dysfunctional organelles by controlling degradation and recycling of damaged cellular components [311]. This pathway becomes naturally activated under stress conditions such as viral infection or the unfolded protein response. Three different autophagic routes are known in the delivery pathway of cytoplasmic material to the lysosomal lumen: (a) macroautophagy, (b) microautophagy, and (c) chaperone-mediated autophagy. Among these, macroautophagy is considered the main autophagy mechanism, and involves a multi-step process in which the cytoplasmic material to be degraded is recognised and engulfed into autophagosomes, double-membrane vesicles that sequester the cargo molecules. Next, autophagosomes fuse with lysosomes leading to autolysosomes, which catalyse the degradation and recycling of cargo molecules [320]. While autophagosomes can target cytosolic material in a non-specific fashion, there is clear evidence for selective autophagy of protein aggregates, mitochondria, and microbes [321,322]. Although the mechanism of selective autophagy is not fully understood, it is known that Ub plays a key role in this process, and K63-linked Ub chains have been shown to be implicated in the proteolytic degradation of misfolded and aggregated proteins by the autophagic route [323]. Autophagy-mediated clearance of protein aggregates requires Ub-binding receptors (p62 and NBR1), which simultaneously bind to Ub and the autophagosome-associated Ub-like proteins (LC3/GABARAP proteins). These interactions mediate tethering of K63-polyubiquitinated protein aggregates to the autophagosome, thus enabling their subsequent transportation to the lysosome for degradation [321]. Based on this principle, the Arimoto group developed AUtophagy TArgeting Chimera (AUTAC) technology [324,325]. In a similar manner to PROTACs, AUTACs are chimeric molecules that function by promoting ubiquitination and subsequent degradation of the POI. However, AUTACs prompt K63 ubiquitination, which, as previously mentioned, is selectively recognised by the autophagy pathway (Fig. 21A) [313,323]. AUTAC molecules consist of a ligand to bind a POI connected to a degradation tag that mimics *S*-guanylation (guanine tag, Fig. 21A and C) [326]. In their work, Arimoto and colleagues [324,325] suggest that *S*-guanylation is a standalone tag that marks the POI for selective autophagy by inducing K63 polyubiquitination. This technology proved efficient at degrading different cytosolic proteins (MetAp2 and FKBP12) as well as fragmented organelles (mitochondria), which highlights the potential of AUTACs towards the degradation of more complex targets. However, possibly due to the cytoplasmic localisation of autophagy, AUTACs were less efficient at degrading the nuclear protein BRD4. Moreover, this technology is relatively recent and has not been deeply studied so far, therefore a significant number of unanswered questions remain around the AUTAC concept. These molecules function through a complex multi-step mechanism that is poorly understood and, specifically, the molecular mechanism mediating *S*-guanylation-triggered K63 ubiquitination of cargoes needs to be elucidated and further validated. Furthermore, in order to evaluate the phenotypic response to harnessing of the autophagy pathway, efficiency and potential off-target effects of AUTACs need to be considered.

The AuTophagosome-TEthering Compounds (ATTECs) developed by Ding, Fei, Lu and co-workers [327,328] represent a ubiquitin-independent alternative to hijack the autophagy machinery for protein degradation. In particular, ATTECs take advantage of the LC3 protein, which is considered a specific marker of autophagosomes [329]. ATTEC molecules act as linkers between a POI and the LC3 protein, tethering the POI to the autophagosome for its subsequent degradation (Fig. 21A). The authors explored ATTEC-mediated autophagic degradation of the mutant huntingtin protein (mHTT), which has an expanded polyglutamine (polyQ) stretch with at least 36 glutamine residues and is the causative agent of Huntington's disease (HD) [330]. In this report, the authors described the development of a high-throughput screening strategy which led to the identification of four mHTT-degraders (Fig. 21D). These compounds interacted with both mHTT and LC3, selectively targeting mHTT to phagophores for autophagic degradation while sparing wild-type HTT. ATTECs proved active *in cellulo* and *in vivo* (*drosophila* and mouse models). *In vitro* experiments showed that these compounds interact with the unique structural features of the expanded polyQ stretch of mHTT. Since the occurrence of proteins with expanded polyQ tracts is a common feature of many neurodegenerative disorders, ATTECs may be useful for other disease-relevant proteins bearing expanded polyQ motifs such as mutant ATXN3, which causes spinocerebellar ataxia III. Moreover, some of these molecules are active at low concentrations (100 nM) and penetrate the blood-brain barrier, providing an encouraging starting point for anti-neurodegenerative drug discovery. The discovery of ATTECs brings the opportunity to elucidate the core chemical structure that interacts with LC3 without affecting its function, which could be conjugated to specific POI binders to develop novel heterobifunctional degraders for degradation of specific targets.

In summary, recently developed compounds that mediate lysosomal degradation of targets (LYTACs, AUTACs and ATTECs) could open further opportunities for targeted protein degradation [314]. The emergence of these new strategies has paved the way for novel therapeutic concepts, targeting cellular material other than soluble pathogenic proteins. The potential of these technologies to target protein aggregates, nucleic acids or even dysfunctional organelles is encouraging for the treatment of many neurodegenerative disorders, which are frequently associated with the formation of misfolded protein aggregates (e.g. Alzheimer's, Parkinson's, and Huntington's diseases) [331]. Nevertheless, a key challenge is selectively targeting these dysfunctional proteins without interfering with their wild type counterparts, using molecules able to penetrate the blood-brain barrier [332]. Hence, the design of degraders with lower molecular weight and higher lipophilicity would be desirable from a pharmacological perspective. Moreover, these autophagic degraders could provide a novel approach for the clearance of bacterial infections via xenophagy. Overall, although they are promising, there are still unanswered questions around these new technologies. Among them, whether these compounds act catalytically needs to be addressed. A catalytic MOA seems plausible for AUTACs and ATTECs, since these small-molecule compounds may be recycled after lysosomal degradation of the target proteins. On the contrary, the macromolecular size of LYTACs might result in their destruction along with the target protein. Furthermore,.these modalities interface with large-scale and highly regulated cellular functions and are probably subject to complex dynamics and off-target effects; future proteomics studies are required to elucidate specificity and the potential off-targets effects of these degraders.

5.3 From chimeric degraders to heterobifunctionals: Expanding the proximity-induction paradigm

The majority of natural processes occurring within the cells are modulated by scaffolding proteins that enable co-localisation of enzymes and substrates in an appropriate orientation, thus causing an increase in reaction rates [333]. Hence, nature benefits from multimeric binders to tightly regulate cell homeostasis. During the last few decades, a branch of scientific research has focused on the study of bifunctional molecules as a means to interrogate biological systems, and a powerful approach to tackle the undruggable proteome [334]. Such protein-proximity inducers have the potential to artificially trigger cellular processes that would not otherwise spontaneously occur [335]. The bioactivity of these proximity modulators relies on their ability to form a multimeric complex with the different partners arranged in a particular conformation, stabilised by a network of intermolecular interactions. Among proximity-inducers, PROTACs, widely discussed throughout this chapter, have emerged as the most prominent examples in the heterobifunctional small-molecule superfamily, although the concept was already well-established by the mid-1990s as a chemical biology tool in the form of genetically encoded inducible dimerisers [336]. Recent success in the PROTAC development pipeline together with the versatility of proximity-induction has encouraged the emergence of further generations of heterobifunctional molecules beyond PROTACs, which are discussed below.

5.3.1 *RIBOTACs*

RNA cellular functions extend far beyond protein biosynthesis, and RNA molecules that are not involved in the biogenesis of proteins, known as non-coding RNAs, represent around 70% of total cellular RNA. These non-coding RNAs have been shown to play a myriad of biochemical roles, ranging from catalytic to regulatory processes [337]. Hence, targeting these non-coding RNAs is considered an attractive therapeutic approach to intervene in specific disease scenarios. Among the huge set of non-coding RNAs, micro RNAs (miRNAs) are known to play a key role in gene regulation, with implied functions in cancer and other malignancies [338]. Traditionally, RNA has been targeted with oligonucleotides or small molecule binders, which suffer from considerable disadvantages such as poor cellular uptake, or low tissue-specific delivery [339]. Recently, the Disney

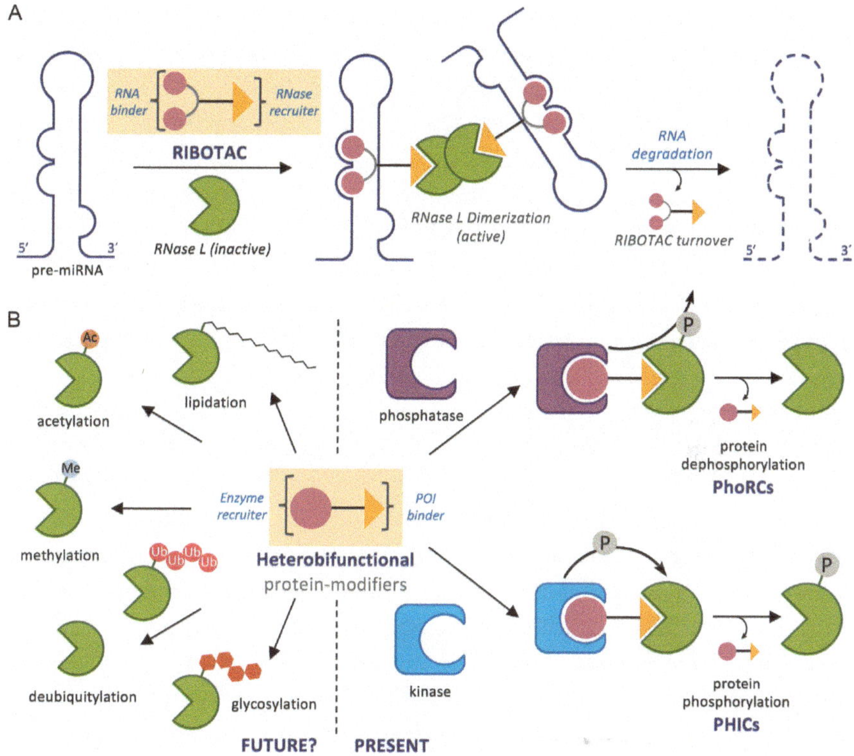

Fig. 22 (A) Mechanism of action of RIBOTACs. (B) Heterobifunctional protein proximity inducers are able to mediate protein dephosphorylation (PhoRCs) and phosphorylation (PHICS). Future development of these molecules could afford new heterobifunctional molecules triggering different PTMs.

group developed the RIBOnuclease TArgeting Chimera (RIBOTAC) technology, which enables targeting miRNAs for degradation by recruiting them to a ribonuclease enzyme (Fig. 22A) [340]. One of the biggest challenges for RIBOTAC development is the design of suitable RNA binders with marked specificity towards a particular RNA. To address this, the Disney group developed Inforna [341], a computational platform that enables sequence-based design of small molecule RNA binders. The first generation of RIBOTAC molecules consisted of an RNA binding small molecule linked to an oligoadenylate moiety [2′-5′-poly(A)] that binds and activates RNase L, a ubiquitously expressed RNase that activates via oligoadenylate-promoted dimerisation [342]. In a proof-of-concept study, RIBOTACs based on Targaprimir-96 [343], an in-house designed ligand that binds to the

Drosha endonuclease processing site of miRNA-96, proved to mediate selective oncogenic pri-miR-96 cleavage by RNase L *in vitro* and *in cellulo* in a catalytic fashion [344]. The authors showed that miR-96 decay triggers apoptosis in a breast cancer cell model due to the derepression of proapoptotic transcription factor FOXO1. In a second example, the Disney group described a RIBOTAC that mediates selective degradation of pre-miR-210, a precursor of miR-210 that is essential for cancer survival in hypoxic niches [345]. This compound engages pre-miR-210 and RNase L leading to a decrease in miR-210 levels along with concomitant apoptosis in hypoxic cancer cells. A second generation of RIBOTACs was next devised by substituting the 2′-5′ poly(A) motif with a heterocyclic module to recruit latent RNase L [346]. This new design was similarly efficacious for the degradation of miR-21 and miR-210 in cells. In particular, studies carried out with a RIBOTAC targeting miR-21 demonstrated that this compound triggers fast and sustained degradation of miR-21, in a catalytic and RNase-dependent manner, showing a 20-fold enhanced activity with respect to its parent miRNA binder. Since miR-21 stimulates an invasive phenotype, miR-21 RIBOTAC showed an inhibitory effect on cell invasiveness in cancer cell lines expressing miR-21. The applicability of RIBOTACs *in vivo* was also demonstrated, with miR-21 RIBOTAC able to inhibit metastasis in mice, highlighting the translational potential of this technology. Studies carried out with RIBOTACs showed clear advantages over the use of oligonucleotides to tackle RNA, including a substoichometric MOA and improved PK properties. One of the hallmarks of RIBOTACs is improved selectivity compared with their RNA-binding counterparts, probably due to the chimeric nature of these molecules. However, RIBOTAC design is challenging and there remain concerns around RNA selectivity and limited cellular uptake, so extensive further studies are required in order to translate RIBOTACs into medicines.

5.3.2 Heterobifunctional molecules: Expanding the post-translational modifiers toolbox

PTM of proteins involves the covalent appendage or cleavage of functional groups of proteins following their biosynthesis. This enzyme-mediated process increases the functional diversity of the proteome by several orders of magnitude and involves protein transformations of at least 1000 different types at over 2 million sites in the proteome. These range from peptide cleavage to covalent addition of biomolecules such as lipids, carbohydrates or proteins, to name just a few. PTMs are crucial in regulating cell signalling processes, and they can potentially influence protein function, localisation or stability, thus playing a key role in cell biology regulation and pathogenesis [347,348]. As discussed in Section 1, protein polyubiquitination is a natural regulatory process occurring at the post-translational level, which marks proteins for eventual degradation by the proteasome [349]. In this context, PROTACs have been developed to hijack proteasomal degradation for therapeutic purposes [1], and the proximity-inducing properties of PROTACs could be translated in principle to new heterobifunctional molecules recruiting different PTM machineries to a POI. During 2019, the first examples of heterobifunctional 'protein modifiers' emerged (Fig. 22B) [350]. In a proof-of-concept work, researchers from Genentech explored modulation of kinase dephosphorylation by recruiting a phosphatase enzyme [351]. Specifically, by using a HaloTag® approach, the authors were able to induce effective dephosphorylation of two oncogenic kinases (AKT and EGFR) by protein-phosphatase 1 (PP1) in cells. This concept was next translated into heterobifunctional molecules, namely, Phosphatase-Recruiting Chimeras (PhoRCs), which combine a ligand to engage the kinase with a phosphatase-binding motif. In particular, PhoRC **4a**, which uses an AKT allosteric inhibitor and the shortest PP1-activator sequence reported (RVSP tetrapeptide) [352] to facilitate drug uptake (Fig. 23A), proved to decrease $pAKT^{T308}$ and $pAKT^{S473}$ levels in LNCaP cells at 10 μM after 8 h treatment, and in a phosphatase-dependent manner. Since kinase activity is tightly affected by phosphorylation state, these heterobifunctional molecules promoting kinase dephosphorylation offer a promising opportunity to regulate kinase activity, thus allowing modulation or creation of altered signalling pathways. Moreover, this work provided evidence that PTMs other than ubiquitination can be effectively modulated using small molecules. Nonetheless, some of the molecules reported suffer from poor cellular permeability, mainly due to the peptidic nature of the PP1 activator, and development of non-peptidic phosphatase binders would be desirable to improve the cellular performance of these heterobifunctionals. Following the precedent set by Genentech, the Choudhary group recently reported PHosphorylation-Inducing Chimeric Small molecules (PHICS) (Fig. 22B) [353]. These heterobifunctional molecules are designed to rewire kinase activity, hijacking a kinase enzyme to enable phosphorylation of target proteins that would otherwise not be substrates for that particular kinase. PHICS features a small molecule binder of a kinase

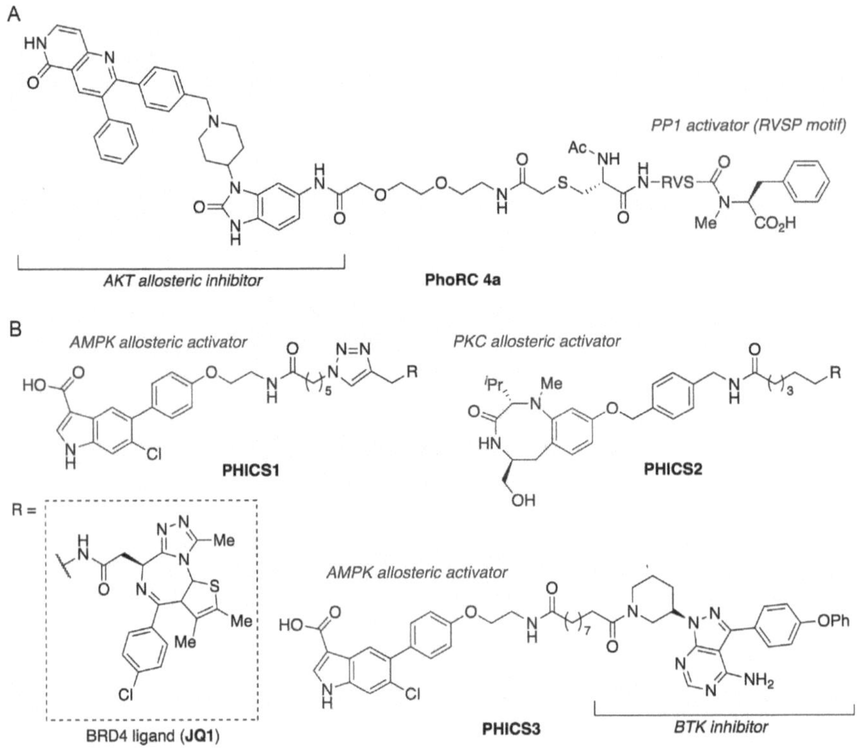

Fig. 23 (A) Chemical structure of PhoRC **4a**. (B) Chemical structure of PHICS1–3.

(allosteric activator) with a warhead to recruit the POI, through an appropriately designed linker. In this work, PHICS were employed to study modulation of substrate specificity for AMP-activated protein kinase (AMPK) and protein kinase C (PKC). Appropriately designed PHICS molecules induced native and neophosphorylation of BRD4 by AMPK (PHICS1) and PKC (PHICS2) *in vitro* (Fig. 23B). Moreover, PHICS3 (Fig. 23B) successfully induced dose-dependent AMPK-mediated phosphorylation of Bruton's Tyrosine Kinase (BTK) at Ser180 in HEK293 cells transfected with a tagged version of BTK. Site-directed mutagenesis and competition studies demonstrated that BTK phosphorylation by PHICS3 requires ternary complex formation, thus validating the hypothesis of proximity-induced phosphorylation. In analogy with PROTACs, PHICS share the classic hallmarks of bifunctional molecules including hook effect, isoform specificity, dose and time dependent phosphorylation, dependence on proximity, and catalytic MOA.

6. Summary

PROTACs are a powerful new therapeutic modality that could cause a paradigm shift in the way we treat diseases (Section 1). In this chapter, we aimed to illuminate the PROTAC state of the art, including assays to measure efficacy and mechanisms (Section 2), a comprehensive review of PROTACs and targets under development (Section 3), the challenges associated with translating this modality to the clinic (Section 4), and the pioneering and future directions of the field (Section 5).

The emerging picture shows that PROTAC-mediated proteolysis has already been successfully applied to a significant number of proteins and the current research efforts around this modality will undoubtedly widen the pool of disease-relevant targets amenable for degradation. We note that since TPD relies on the UPS machinery, it is limited to knockdown of intracellularly located proteins only. Nevertheless, emerging complementary degradation tools, harnessing other biological degradation machineries, have enabled expansion of TPD to extracellular and membrane-associated proteins [314], as well as other cellular components (e.g. RNA and organelles) and will bring additional weight to what has already been achieved with PROTACs.

In summary, the emergence of these promising therapeutic strategies, from PROTACs to heterobifunctionals in general, has opened a new era in the drug discovery landscape, paving the way for the future development of novel therapeutic approaches to finely modulate a wide variety of disease-relevant PTMs, such as ubiquitination, lipidation, deubiquitylation, glycosylation, etc. Altogether, these new paradigms would allow a better understanding of cell biology and could provide an excellent opportunity to interfere with biological processes for therapeutic purposes.

Appendix

A List of abbreviations

Abbreviation	Definition
ADC	Antibody-Drug Conjugate
AdPROM	Affinity-directed PROtein Missile system
ALPHA	Amplified Luminescent Proximity Homogeneous Assay

Continued

—cont'd

Abbreviation	Definition
ATP	Adenosine triphosphate
ATTEC	AuTophagosome-TEthering Compounds
AUTAC	AUtophagy TArgeting Chimera
αRep	alpha Repeat proteins
BET	Bromodomain and ExtraTerminal
BLD	Blood disorders
BRET	Bioluminescence Resonance Energy Transfer
bRo5	beyond Rule of five
CNS	Central Nervous System
COFFEE	Covalent Functionalisation Followed by E3 Electroporation into live cells
CRISPR/Cas9	Clustered Regularly Interspaced Short Palindromic Repeats/ CRISPR associated protein 9
CRL	Cullin Ring Ligase
CVD	Cardiovascular diseases
DARPin	Designed ankyrin repeat proteins
DNA	Deoxyribose Nucleic Acid
FBR	Fibrotic diseases
Fc	Fragment crystallisable
FDA	Food and Drug Administration
Flp-FRT	Flippase - Flippase Recognition Target
FP	Fluorescence Polarisation
GPCR	G-Protein Coupled Receptor
GSK	GlaxoSmithKline
GTPase	Guanosine triphosphatase
HCV	Hepatitis C Virus
IA	Inflammation and Autoimmune diseases
ID	Infectious Diseases
IM	Intramuscular
IMiD	Immmuno-Modulatory Drugs

—cont'd

Abbreviation	Definition
IP	Intraperitoneal
IR	Infrared
isoTOP-ABPP	isotopic Tandem Orthogonal Proteolysis- Activity-Based Protein Profiling
ITC	Isothermal Titration Calorimetry
iTRAQ	isobaric Tags for Relative and Absolute Quantitation
IV	Intravenous
L-AdPROM	"ligand-inducible" Affinity-directed protein missile system
LCMS	Liquid Chromatography-Mass Spectrometry
LTR	Lysosome-Targeting Receptor
LYTACs	LYsosome TArgeting Chimeras
miRNA	micro Ribose Nucleic Acid
MOA	Mechanism Of Action
poly(M6P/Pn)	poly(mannose-6-phosphate/phosphonate)
NanoBiT	NanoLuc® Binary Technology
NanoLuc	NanoLuciferase
NHR	Nuclear Hormone Receptor
ONC	Oncology
PAMPA	Parallel Artificial Membrane Permeability Assay
PD	Pharmacodynamic
PDT	Photodynamic therapy
PEG	polyethylene glycol
PET	Positron Emission Tomography
PHICS	PHosphorylation-Inducing Chimeric Small molecule
PhoRC	Phosphatase-Recruiting Chimera
PK	Pharmacokinetic
PKC	Protein Kinase C
POI	Protein Of Interest
polyQ	polyglutamine

Continued

—cont'd

Abbreviation	Definition
PPI	Protein-Protein Interaction
PROTAC	PROteolysis Targeting Chimera
PSA	Polar Surface Area
PTB	Phosphotyrosine-binding
PTM	Post-Translational Modification
RIBOTAC	RIBOnuclease TArgeting Chimeras
RNAi	RNA interference
RNase	Ribonuclease
Ro5	Rule of five
RTK	Receptor Tyrosine Kinase
SAR	Structure-Activity Relationships
SCF	Skp, Cullin, F-box
SILAC	Stable Isotope Labelling by/with Amino acids in Cell culture
SMASh	Small Molecule Assistant Shutoff
SPR	Surface Plasmon Resonance
SUMO	Small Ubiquitin-like MOdifier
TMT	Tandem Mass Tagging
TPD	Targeted Protein Degradation
TR-FRET	Time-Resolved Fluorescence Resonance Energy Transfer
Ub	Ubiquitin
UPS	Ubiquitin Proteasome System
WB	Western Blotting

B List of protein ID

Entry	Entry name	Protein names	Gene names	Organism	Length
A0A024R7N6	A0A024R7N6_HUMAN	Phosphoinositide-3-kinase, regulatory subunit 2 (P85 beta), isoform CRA_a	PIK3R2 hCG_1777142	*Homo sapiens* (Human)	728
A0A2X0SFG1	A0A2X0SFG1_HUMAN	Phosphatidylinositol 3-kinase 85 kDa regulatory subunit alpha (Phosphatidylinositol 3-kinase regulatory subunit alpha) (Fragment)	PIK3R1	Homo sapiens (Human)	724
P00519	ABL1_HUMAN	Tyrosine-protein kinase ABL1 (EC 2.7.10.2) (Abelson murine leukaemia viral oncogene homologue 1) (Abelson tyrosine-protein kinase 1) (Proto-oncogene c-Abl) (p150)	ABL1 ABL JTK7	Homo sapiens (Human)	1130
P35348	ADA1A_HUMAN	Alpha-1A adrenergic receptor (Alpha-1A adrenoreceptor) (Alpha-1A adrenoceptor) (Alpha-1C adrenergic receptor) (Alpha-adrenergic receptor 1c)	ADRA1A ADRA1C	Homo sapiens (Human)	466
Q16186	ADRM1_HUMAN	Proteasomal ubiquitin receptor ADRM1 (110 kDa cell membrane glycoprotein) (Gp110) (Adhesion-regulating molecule 1) (ARM-1) (Proteasome regulatory particle non-ATPase 13) (hRpn13) (Rpn13 homologue)	ADRM1 GP110	Homo sapiens (Human)	407
P35869	AHR_HUMAN	Aryl hydrocarbon receptor (Ah receptor) (AhR) (Class E basic helix-loop-helix protein 76) (bHLHe76)	AHR BHLHE76	Homo sapiens (Human)	848
P31749	AKT1_HUMAN	RAC-alpha serine/threonine-protein kinase (EC 2.7.11.1) (Protein kinase B) (PKB) (Protein kinase B alpha) (PKB alpha) (Proto-oncogene c-Akt) (RAC-PK-alpha)	AKT1 PKB RAC	Homo sapiens (Human)	480

Continued

—cont'd

Entry	Entry name	Protein names	Gene names	Organism	Length
P31751	AKT2_HUMAN	RAC-beta serine/threonine-protein kinase (EC 2.7.11.1) (Protein kinase Akt-2) (Protein kinase B beta) (PKB beta) (RAC-PK-beta)	AKT2	Homo sapiens (Human)	481
Q9Y243	AKT3_HUMAN	RAC-gamma serine/threonine-protein kinase (EC 2.7.11.1) (Protein kinase Akt-3) (Protein kinase B gamma) (PKB gamma) (RAC-PK-gamma) (STK-2)	AKT3 PKBG	Homo sapiens (Human)	479
Q9UM73	ALK_HUMAN	ALK tyrosine kinase receptor (EC 2.7.10.1) (Anaplastic lymphoma kinase) (CD antigen CD246)	ALK	Homo sapiens (Human)	1620
P10275	ANDR_HUMAN	Androgen receptor (Dihydrotestosterone receptor) (Nuclear receptor subfamily 3 group C member 4)	AR DHTR NR3C4	Homo sapiens (Human)	920
Q68CP9	ARID2_HUMAN	AT-rich interactive domain-containing protein 2 (ARID domain-containing protein 2) (BRG1-associated factor 200) (BAF200) (Zinc finger protein with activation potential) (Zipzap/p200)	ARID2 BAF200 KIAA1557	Homo sapiens (Human)	1835
P54252	ATX3_HUMAN	Ataxin-3 (EC 3.4.19.12) (Machado-Joseph disease protein 1) (Spinocerebellar ataxia type 3 protein)	ATXN3 ATX3 MJD MJD1 SCA3	Homo sapiens (Human)	361
O14965	AURKA_HUMAN	Aurora kinase A (EC 2.7.11.1) (Aurora 2) (Aurora/IPL1-related kinase 1) (ARK-1) (Aurora-related kinase 1) (hARK1) (Breast tumour-amplified kinase) (Serine/threonine-protein kinase 15) (Serine/threonine-protein kinase 6) (Serine/threonine-protein kinase aurora-A)	AURKA AIK AIRK1 ARK1 AURA AYK1 BTAK IAK1 STK15 STK6	Homo sapiens (Human)	403

Q96GD4	AURKB_HUMAN	Aurora kinase B (EC 2.7.11.1) (Aurora 1) (Aurora- and IPL1-like midbody-associated protein 1) (AIM-1) (Aurora/IPL1-related kinase 2) (ARK-2) (Aurora-related kinase 2) (STK-1) (Serine/threonine-protein kinase 12) (Serine/threonine-protein kinase 5) (Serine/threonine-protein kinase aurora-B)	AURKB AIK2 AIM1 AIRK2 ARK2 STK1 STK12 STK5	Homo sapiens (Human)	344
B0B3E3	B0B3E3_9HEPC	Genome polyprotein (EC 3.6.1.15) (EC 3.6.4.13) (Serine protease NS3) (Fragment)	NS3/4A	Hepacivirus C	685
Q07817	B2CL1_HUMAN	Bcl-2-like protein 1 (Bcl2-L-1) (Apoptosis regulator Bcl-X)	BCL2L1 BCL2L BCLX	Homo sapiens (Human)	233
Q92934	BAD_HUMAN	Bcl2-associated agonist of cell death (BAD) (Bcl-2-binding component 6) (Bcl-2-like protein 8) (Bcl2-L-8) (Bcl-xL/Bcl-2-associated death promoter) (Bcl2 antagonist of cell death)	BAD BBC6 BCL2L8	Homo sapiens (Human)	168
P41182	BCL6_HUMAN	B-cell lymphoma 6 protein (BCL-6) (B-cell lymphoma 5 protein) (BCL-5) (Protein LAZ-3) (Zinc finger and BTB domain-containing protein 27) (Zinc finger protein 51)	BCL6 BCL5 LAZ3 ZBTB27 ZNF51	Homo sapiens (Human)	706
P11274	BCR_HUMAN	Breakpoint cluster region protein (EC 2.7.11.1) (Renal carcinoma antigen NY-REN-26)	BCR BCR1 D22S11	Homo sapiens (Human)	1271
Q13490	BIRC2_HUMAN	Baculoviral IAP repeat-containing protein 2 (EC 2.3.2.27) (Cellular inhibitor of apoptosis 1) (C-IAP1) (IAP homologue B) (Inhibitor of apoptosis protein 2) (hIAP-2) (hIAP2) (RING finger protein 48) (RING-type E3 ubiquitin transferase BIRC2) (TNFR2-TRAF-signalling complex protein 2)	BIRC2 API1 MIHB RNF48	Homo sapiens (Human)	618

Continued

—cont'd

Entry	Entry name	Protein names	Gene names	Organism	Length
P51451	BLK_HUMAN	Tyrosine-protein kinase Blk (EC 2.7.10.2) (B lymphocyte kinase) (p55-Blk)	BLK	Homo sapiens (Human)	505
P15056	BRAF_HUMAN	Serine/threonine-protein kinase B-raf (EC 2.7.11.1) (Proto-oncogene B-Raf) (p94) (v-Raf murine sarcoma viral oncogene homologue B1)	BRAF BRAF1 RAFB1	Homo sapiens (Human)	766
P25440	BRD2_HUMAN	Bromodomain-containing protein 2 (O27.1.1) (Really interesting new gene 3 protein)	BRD2 KIAA9001 RING3	Homo sapiens (Human)	801
Q8K2F0	BRD3_MOUSE	Bromodomain-containing protein 3 (Bromodomain-containing FSH-like protein FSRG2)	Brd3 Fsrg2	*Mus musculus* (Mouse)	726
O60885	BRD4_HUMAN	Bromodomain-containing protein 4 (Protein HUNK1)	BRD4 HUNK1	Homo sapiens (Human)	1362
Q9NPI1	BRD7_HUMAN	Bromodomain-containing protein 7 (75 kDa bromodomain protein) (Protein CELTIX-1)	BRD7 BP75 CELTIX1	Homo sapiens (Human)	651
Q9H8M2	BRD9_HUMAN	Bromodomain-containing protein 9 (Rhabdomyosarcoma antigen MU-RMS-40.8)	BRD9 UNQ3040/ PRO9856	Homo sapiens (Human)	597
Q06187	BTK_HUMAN	Tyrosine-protein kinase BTK (EC 2.7.10.2) (Agammaglobulinemia tyrosine kinase) (ATK) (B-cell progenitor kinase) (BPK) (Bruton tyrosine kinase)	BTK AGMX1 ATK BPK	Homo sapiens (Human)	659
P22362	CCL1_HUMAN	C-C motif chemokine 1 (Small-inducible cytokine A1) (T lymphocyte-secreted protein I-309)	CCL1 SCYA1	Homo sapiens (Human)	96

Q9NYV4	CDK12_HUMAN	Cyclin-dependent kinase 12 (EC 2.7.11.22) (EC 2.7.11.23) (Cdc2-related kinase, arginine/serine-rich) (CrkRS) (Cell division cycle 2-related protein kinase 7) (CDC2-related protein kinase 7) (Cell division protein kinase 12) (hCDK12)	CDK12 CRK7 CRKRS KIAA0904	Homo sapiens (Human)	1490
P24941	CDK2_HUMAN	Cyclin-dependent kinase 2 (EC 2.7.11.22) (Cell division protein kinase 2) (p33 protein kinase)	CDK2 CDKN2	Homo sapiens (Human)	298
Q8IZL9	CDK20_HUMAN	Cyclin-dependent kinase 20 (EC 2.7.11.22) (CDK-activating kinase p42) (CAK-kinase p42) (Cell cycle-related kinase) (Cell division protein kinase 20) (Cyclin-dependent protein kinase H) (Cyclin-kinase-activating kinase p42)	CDK20 CCRK CDCH	Homo sapiens (Human)	346
P11802	CDK4_HUMAN	Cyclin-dependent kinase 4 (EC 2.7.11.22) (Cell division protein kinase 4) (PSK-J3)	CDK4	Homo sapiens (Human)	303
Q00535	CDK5_HUMAN	Cyclin-dependent-like kinase 5 (EC 2.7.11.1) (Cell division protein kinase 5) (Serine/threonine-protein kinase PSSALRE) (Tau protein kinase II catalytic subunit) (TPKII catalytic subunit)	CDK5 CDKN5	Homo sapiens (Human)	292
Q00534	CDK6_HUMAN	Cyclin-dependent kinase 6 (EC 2.7.11.22) (Cell division protein kinase 6) (Serine/threonine-protein kinase PLSTIRE)	CDK6 CDKN6	Homo sapiens (Human)	326
P50750	CDK9_HUMAN	Cyclin-dependent kinase 9 (EC 2.7.11.22) (EC 2.7.11.23) (C-2K) (Cell division cycle 2-like protein kinase 4) (Cell division protein kinase 9) (Serine/threonine-protein kinase PITALRE) (Tat-associated kinase complex catalytic subunit)	CDK9 CDC2L4 TAK	Homo sapiens (Human)	372

Continued

—cont'd

Entry	Entry name	Protein names	Gene names	Organism	Length
Q96SW2	CRBN_HUMAN	Protein cereblon	CRBN AD-006	Homo sapiens (Human)	442
P68400	CSK21_HUMAN	Casein kinase II subunit alpha (CK II alpha) (EC 2.7.11.1)	CSNK2A1 CK2A1	Homo sapiens (Human)	391
P19784	CSK22_HUMAN	Casein kinase II subunit alpha' (CK II alpha') (EC 2.7.11.1)	CSNK2A2 CK2A2	Homo sapiens (Human)	350
Q8NEV1	CSK23_HUMAN	Casein kinase II subunit alpha 3 (CK II alpha 3) (EC 2.7.11.1) (Casein kinase II alpha 1 polypeptide pseudogene)	CSNK2A3 CSNK2A1P	Homo sapiens (Human)	391
P67870	CSK2B_HUMAN	Casein kinase II subunit beta (CK II beta) (Phosvitin) (Protein G5a)	CSNK2B CK2N G5A	Homo sapiens (Human)	215
Q13620	CUL4B_HUMAN	Cullin-4B (CUL-4B)	CUL4B KIAA0695	Homo sapiens (Human)	913
Q66K64	DCA15_HUMAN	DDB1- and CUL4-associated factor 15	DCAF15 C19orf72	Homo sapiens (Human)	600
Q9NXF7	DCA16_HUMAN	DDB1- and CUL4-associated factor 16	DCAF16 C4orf30	Homo sapiens (Human)	216
O75530	EED_HUMAN	Polycomb protein EED (hEED) (Embryonic ectoderm development protein) (WD protein associating with integrin cytoplasmic tails 1) (WAIT-1)	EED	Homo sapiens (Human)	441

P00533	EGFR_HUMAN	Epidermal growth factor receptor (EC 2.7.10.1) (Proto-oncogene c-ErbB-1) (Receptor tyrosine-protein kinase erbB-1)	EGFR ERBB ERBB1 HER1	Homo sapiens (Human)	1210
Q03111	ENL_HUMAN	Protein ENL (YEATS domain-containing protein 1)	MLLT1 ENL LTG19 YEATS1	Homo sapiens (Human)	559
P04626	ERBB2_HUMAN	Receptor tyrosine-protein kinase erbB-2 (EC 2.7.10.1) (Metastatic lymph node gene 19 protein) (MLN 19) (Proto-oncogene Neu) (Proto-oncogene c-ErbB-2) (Tyrosine kinase-type cell surface receptor HER2) (p185erbB2) (CD antigen CD340)	ERBB2 HER2 MLN19 NEU NGL	Homo sapiens (Human)	1255
P15170	ERF3A_HUMAN	Eukaryotic peptide chain release factor GTP-binding subunit ERF3A (Eukaryotic peptide chain release factor subunit 3a) (eRF3a) (G1 to S phase transition protein 1 homologue)	GSPT1 ERF3A	Homo sapiens (Human)	499
P11474	ERR1_HUMAN	Steroid hormone receptor ERR1 (Oestrogen receptor-like 1) (Oestrogen-related receptor alpha) (ERR-alpha) (Nuclear receptor subfamily 3 group B member 1)	ESRRA ERR1 ESRL1 NR3B1	Homo sapiens (Human)	423
P03372	ESR1_HUMAN	Oestrogen receptor (ER) (ER-alpha) (Estradiol receptor) (Nuclear receptor subfamily 3 group A member 1)	ESR1 ESR NR3A1	Homo sapiens (Human)	595
Q92731	ESR2_HUMAN	Oestrogen receptor beta (ER-beta) (Nuclear receptor subfamily 3 group A member 2)	ESR2 ESTRB NR3A2	Homo sapiens (Human)	530

Continued

—cont'd

Entry	Entry name	Protein names	Gene names	Organism	Length
Q05397	FAK1_HUMAN	Focal adhesion kinase 1 (FADK 1) (EC 2.7.10.2) (Focal adhesion kinase-related nonkinase) (FRNK) (Protein phosphatase 1 regulatory subunit 71) (PPP1R71) (Protein-tyrosine kinase 2) (p125FAK) (pp125FAK)	PTK2 FAK FAK1	Homo sapiens (Human)	1052
Q14289	FAK2_HUMAN	Protein-tyrosine kinase 2-beta (EC 2.7.10.2) (Calcium-dependent tyrosine kinase) (CADTK) (Calcium-regulated non-receptor proline-rich tyrosine kinase) (Cell adhesion kinase beta) (CAK-beta) (CAKB) (Focal adhesion kinase 2) (FADK 2) (Proline-rich tyrosine kinase 2) (Related adhesion focal tyrosine kinase) (RAFTK)	PTK2B FAK2 PYK2 RAFTK	Homo sapiens (Human)	1009
Q9Y297	FBW1A_HUMAN	F-box/WD repeat-containing protein 1A (E3RSIkappaB) (Epididymis tissue protein Li 2a) (F-box and WD repeats protein beta-TrCP) (pIkappaBalpha-E3 receptor subunit)	BTRC BTRCP FBW1A FBXW1A	Homo sapiens (Human)	605
P68106	FKB1B_HUMAN	Peptidyl-prolyl cis-trans isomerase FKBP1B (PPIase FKBP1B) (EC 5.2.1.8) (12.6 kDa FK506-binding protein) (12.6 kDa FKBP) (FKBP-12.6) (FK506-binding protein 1B) (FKBP-1B) (Immunophilin FKBP12.6) (Rotamase) (h-FKBP-12)	FKBP1B FKBP12.6 FKBP1L FKBP9 OTK4	Homo sapiens (Human)	108
P36888	FLT3_HUMAN	Receptor-type tyrosine-protein kinase FLT3 (EC 2.7.10.1) (FL cytokine receptor) (Fetal liver kinase-2) (FLK-2) (Fms-like tyrosine kinase 3) (FLT-3) (Stem cell tyrosine kinase 1) (STK-1) (CD antigen CD135)	FLT3 CD135 FLK2 STK1	Homo sapiens (Human)	993

Q8WU20	FRS2_HUMAN	Fibroblast growth factor receptor substrate 2 (FGFR substrate 2) (FGFR-signalling adaptor SNT) (Suc1-associated neurotrophic factor target 1) (SNT-1)	FRS2	Homo sapiens (Human)	508
O95166	GBRAP_HUMAN	Gamma-aminobutyric acid receptor-associated protein (GABA(A) receptor-associated protein) (MM46)	GABARAP FLC3B HT004	Homo sapiens (Human)	117
P42212	GFP_AEQVI	Green fluorescent protein	GFP	*Aequorea victoria* (Jellyfish)	238
P42858	HD_HUMAN	Huntingtin (Huntington disease protein) (HD protein) [Cleaved into: Huntingtin, myristoylated N-terminal fragment]	HTT HD IT15	Homo sapiens (Human)	3142
Q13547	HDAC1_HUMAN	Histone deacetylase 1 (HD1) (EC 3.5.1.98)	HDAC1 RPD3L1	Homo sapiens (Human)	482
Q92769	HDAC2_HUMAN	Histone deacetylase 2 (HD2) (EC 3.5.1.98)	HDAC2	Homo sapiens (Human)	488
O15379	HDAC3_HUMAN	Histone deacetylase 3 (HD3) (EC 3.5.1.98) (RPD3–2) (SMAP45)	HDAC3	Homo sapiens (Human)	428
Q9UBN7	HDAC6_HUMAN	Histone deacetylase 6 (HD6) (EC 3.5.1.98) (Tubulin-lysine deacetylase HDAC6) (EC 3.5.1)	HDAC6 KIAA0901 JM21	Homo sapiens (Human)	1215
Q16665	HIF1A_HUMAN	Hypoxia-inducible factor 1-alpha (HIF-1-alpha) (HIF1-alpha) (ARNT-interacting protein) (Basic-helix-loop-helix-PAS protein MOP1) (Class E basic helix-loop-helix protein 78) (bHLHe78) (Member of PAS protein 1) (PAS domain-containing protein 8)	HIF1A BHLHE78 MOP1 PASD8	Homo sapiens (Human)	826

Continued

—cont'd

Entry	Entry name	Protein names	Gene names	Organism	Length
P04035	HMDH_HUMAN	3-Hydroxy-3-methylglutaryl-coenzyme A reductase (HMG-CoA reductase) (EC 1.1.1.34)	HMGCR	Homo sapiens (Human)	888
Q9UKT9	IKZF3_HUMAN	Zinc finger protein Aiolos (Ikaros family zinc finger protein 3)	IKZF3 ZNFN1A3	Homo sapiens (Human)	509
Q9NWZ3	IRAK4_HUMAN	Interleukin-1 receptor-associated kinase 4 (IRAK-4) (EC 2.7.11.1) (Renal carcinoma antigen NY-REN-64)	IRAK4	Homo sapiens (Human)	460
P23458	JAK1_HUMAN	Tyrosine-protein kinase JAK1 (EC 2.7.10.2) (Janus kinase 1) (JAK-1)	JAK1 JAK1A JAK1B	Homo sapiens (Human)	1154
O60674	JAK2_HUMAN	Tyrosine-protein kinase JAK2 (EC 2.7.10.2) (Janus kinase 2) (JAK-2)	JAK2	Homo sapiens (Human)	1132
Q92831	KAT2B_HUMAN	Histone acetyltransferase KAT2B (EC 2.3.1.48) (Histone acetyltransferase PCAF) (Histone acetylase PCAF) (Lysine acetyltransferase 2B) (P300/CBP-associated factor) (P/CAF) (Spermidine acetyltransferase KAT2B) (EC 2.3.1.57)	KAT2B PCAF	Homo sapiens (Human)	832
Q14145	KEAP1_HUMAN	Kelch-like ECH-associated protein 1 (Cytosolic inhibitor of Nrf2) (INrf2) (Kelch-like protein 19)	KEAP1 INRF2 KIAA0132 KLHL19	Homo sapiens (Human)	624
Q5S007	LRRK2_HUMAN	Leucine-rich repeat serine/threonine-protein kinase 2 (EC 2.7.11.1) (EC 3.6.5) (Dardarin)	LRRK2 PARK8	Homo sapiens (Human)	2527
P50579	MAP2_HUMAN	Methionine aminopeptidase 2 (MAP 2) (MetAP 2) (EC 3.4.11.18) (Initiation factor 2-associated 67 kDa glycoprotein) (p67) (p67eIF2) (Peptidase M)	METAP2 MNPEP P67EIF2	Homo sapiens (Human)	478

Q07820	MCL1_HUMAN	Induced myeloid leukaemia cell differentiation protein Mcl-1 (Bcl-2-like protein 3) (Bcl2-L-3) (Bcl-2-related protein EAT/mcl1) (mcl1/EAT)	MCL1 BCL2L3	Homo sapiens (Human)	350
Q00987	MDM2_HUMAN	E3 ubiquitin-protein ligase Mdm2 (EC 2.3.2.27) (Double minute 2 protein) (Hdm2) (Oncoprotein Mdm2) (RING-type E3 ubiquitin transferase Mdm2) (p53-binding protein Mdm2)	MDM2	Homo sapiens (Human)	491
Q14680	MELK_HUMAN	Maternal embryonic leucine zipper kinase (hMELK) (EC 2.7.11.1) (Protein kinase Eg3) (pEg3 kinase) (Protein kinase PK38) (hPK38) (Tyrosine-protein kinase MELK) (EC 2.7.10.2)	MELK KIAA0175	Homo sapiens (Human)	651
Q15759	MK11_HUMAN	Mitogen-activated protein kinase 11 (MAP kinase 11) (MAPK 11) (EC 2.7.11.24) (Mitogen-activated protein kinase p38 beta) (MAP kinase p38 beta) (p38b) (Stress-activated protein kinase 2b) (SAPK2b) (p38–2)	MAPK11 PRKM11 SAPK2 SAPK2B	Homo sapiens (Human)	364
O15264	MK13_HUMAN	Mitogen-activated protein kinase 13 (MAP kinase 13) (MAPK 13) (EC 2.7.11.24) (Mitogen-activated protein kinase p38 delta) (MAP kinase p38 delta) (Stress-activated protein kinase 4)	MAPK13 PRKM13 SAPK4	Homo sapiens (Human)	365
O15264	MK13_HUMAN	Mitogen-activated protein kinase 13 (MAP kinase 13) (MAPK 13) (EC 2.7.11.24) (Mitogen-activated protein kinase p38 delta) (MAP kinase p38 delta) (Stress-activated protein kinase 4)	MAPK13 PRKM13 SAPK4	Homo sapiens (Human)	365

Continued

—cont'd

Entry	Entry name	Protein names	Gene names	Organism	Length
Q16539	MK14_HUMAN	Mitogen-activated protein kinase 14 (MAP kinase 14) (MAPK 14) (EC 2.7.11.24) (Cytokine suppressive anti-inflammatory drug-binding protein) (CSAID-binding protein) (CSBP) (MAP kinase MXI2) (MAX-interacting protein 2) (Mitogen-activated protein kinase p38 alpha) (MAP kinase p38 alpha) (Stress-activated protein kinase 2a) (SAPK2a)	MAPK14 CSBP CSBP1 CSBP2 CSPB1 MXI2 SAPK2A	Homo sapiens (Human)	360
P20645	MPRD_HUMAN	Cation-dependent mannose-6-phosphate receptor (CD Man-6-P receptor) (CD-MPR) (46 kDa mannose 6-phosphate receptor) (MPR 46)	M6PR MPR46 MPRD	Homo sapiens (Human)	277
P01106	MYC_HUMAN	Myc proto-oncogene protein (Class E basic helix-loop-helix protein 39) (bHLHe39) (Proto-oncogene c-Myc) (Transcription factor p64)	MYC BHLHE39	Homo sapiens (Human)	439
Q14596	NBR1_HUMAN	Next to BRCA1 gene 1 protein (Cell migration-inducing gene 19 protein) (Membrane component chromosome 17 surface marker 2) (Neighbour of BRCA1 gene 1 protein) (Protein 1A1-3B)	NBR1 1A13B KIAA0049 M17S2 MIG19	Homo sapiens (Human)	966
Q15843	NEDD8_HUMAN	NEDD8 (Neddylin) (Neural precursor cell expressed developmentally down-regulated protein 8) (NEDD-8) (Ubiquitin-like protein Nedd8)	NEDD8	Homo sapiens (Human)	81
Q16288	NTRK3_HUMAN	NT-3 growth factor receptor (EC 2.7.10.1) (GP145-TrkC) (Trk-C) (Neurotrophic tyrosine kinase receptor type 3) (TrkC tyrosine kinase)	NTRK3 TRKC	Homo sapiens (Human)	839

O42033	O42033_HBV	Protein X (HBx) (Peptide X) (pX)	X	Hepatitis B virus (HBV)	154
Q92569	P55G_HUMAN	Phosphatidylinositol 3-kinase regulatory subunit gamma (PI3-kinase regulatory subunit gamma) (PI3K regulatory subunit gamma) (PtdIns-3-kinase regulatory subunit gamma) (Phosphatidylinositol 3-kinase 55 kDa regulatory subunit gamma) (PI3-kinase subunit p55-gamma) (PtdIns-3-kinase regulatory subunit p55-gamma) (p55PIK)	PIK3R3	Homo sapiens (Human)	461
P27986	P85A_HUMAN	Phosphatidylinositol 3-kinase regulatory subunit alpha (PI3-kinase regulatory subunit alpha) (PI3K regulatory subunit alpha) (PtdIns-3-kinase regulatory subunit alpha) (Phosphatidylinositol 3-kinase 85 kDa regulatory subunit alpha) (PI3-kinase subunit p85-alpha) (PtdIns-3-kinase regulatory subunit p85-alpha)	PIK3R1 GRB1	Homo sapiens (Human)	724
O00459	P85B_HUMAN	Phosphatidylinositol 3-kinase regulatory subunit beta (PI3-kinase regulatory subunit beta) (PI3K regulatory subunit beta) (PtdIns-3-kinase regulatory subunit beta) (Phosphatidylinositol 3-kinase 85 kDa regulatory subunit beta) (PI3-kinase subunit p85-beta) (PtdIns-3-kinase regulatory subunit p85-beta)	PIK3R2	Homo sapiens (Human)	728
P09874	PARP1_HUMAN	Poly [ADP-ribose] polymerase 1 (PARP-1) (EC 2.4.2.30) (ADP-ribosyltransferase diphtheria toxin-like 1) (ARTD1) (DNA ADP-ribosyltransferase PARP1) (EC 2.4.2) (NAD(+) ADP-ribosyltransferase 1) (ADPRT 1) (Poly[ADP-ribose] synthase 1) (Protein poly-ADP-ribosyltransferase PARP1) (EC 2.4.2)	PARP1 ADPRT PPOL	Homo sapiens (Human)	1014

Continued

—cont'd

Entry	Entry name	Protein names	Gene names	Organism	Length
P12004	PCNA_HUMAN	Proliferating cell nuclear antigen (PCNA) (Cyclin)	PCNA	Homo sapiens (Human)	261
Q9NZQ7	PD1L1_HUMAN	Programmed cell death 1 ligand 1 (PD-L1) (PDCD1 ligand 1) (Programmed death ligand 1) (hPD-L1) (B7 homologue 1) (B7-H1) (CD antigen CD274)	CD274 B7H1 PDCD1L1 PDCD1LG1 PDL1	Homo sapiens (Human)	290
Q86H13	PDE4_DICDI	cAMP-specific 3′,5′-cAMP phosphodiesterase 4 (EC 3.1.4.53) (Phosphodiesterase 4) (DdPDE4)	Pde4 DDB_G0289121	Dictyostelium discoideum (Slime mould)	1039
O00625	PIR_HUMAN	Pirin (EC 1.13.11.24) (Probable quercetin 2,3-dioxygenase PIR) (Probable quercetinase)	PIR	Homo sapiens (Human)	290
P53350	PLK1_HUMAN	Serine/threonine-protein kinase PLK1 (EC 2.7.11.21) (Polo-like kinase 1) (PLK-1) (Serine/threonine-protein kinase 13) (STPK13)	PLK1 PLK	Homo sapiens (Human)	603
Q06124	PTN11_HUMAN	Tyrosine-protein phosphatase non-receptor type 11 (EC 3.1.3.48) (Protein-tyrosine phosphatase 1D) (PTP-1D) (Protein-tyrosine phosphatase 2C) (PTP-2C) (SH-PTP2) (SHP-2) (Shp2) (SH-PTP3)	PTPN11 PTP2C SHPTP2	Homo sapiens (Human)	593
Q8N381	Q8N381_HUMAN	Phosphoinositide-3-kinase, regulatory subunit 3 (Gamma)	PIK3R3	Homo sapiens (Human)	461
Q96CK7	Q96CK7_HUMAN	PIK3R2 protein (Fragment)	PIK3R2	Homo sapiens (Human)	324

Q96EV4	Q96EV4_HUMAN	PIK3R2 protein (Fragment)	PIK3R2	Homo sapiens (Human)	388
P29762	RABP1_HUMAN	Cellular retinoic acid-binding protein 1 (Cellular retinoic acid-binding protein I) (CRABP-I)	CRABP1 RBP5	Homo sapiens (Human)	137
P29373	RABP2_HUMAN	Cellular retinoic acid-binding protein 2 (Cellular retinoic acid-binding protein II) (CRABP-II)	CRABP2	Homo sapiens (Human)	138
P10276	RARA_HUMAN	Retinoic acid receptor alpha (RAR-alpha) (Nuclear receptor subfamily 1 group B member 1)	RARA NR1B1	Homo sapiens (Human)	462
P10826	RARB_HUMAN	Retinoic acid receptor beta (RAR-beta) (HBV-activated protein) (Nuclear receptor subfamily 1 group B member 2) (RAR-epsilon)	RARB HAP NR1B2	Homo sapiens (Human)	455
P13631	RARG_HUMAN	Retinoic acid receptor gamma (RAR-gamma) (Nuclear receptor subfamily 1 group B member 3)	RARG NR1B3	Homo sapiens (Human)	454
P01116	RASK_HUMAN	GTPase KRas (EC 3.6.5.2) (K-Ras 2) (Ki-Ras) (c-K-ras) (c-Ki-ras) [Cleaved into: GTPase KRas, N-terminally processed]	KRAS KRAS2 RASK2	Homo sapiens (Human)	189
Q14498	RBM39_HUMAN	RNA-binding protein 39 (CAPER alpha) (CAPERalpha) (Hepatocellular carcinoma protein 1) (RNA-binding motif protein 39) (RNA-binding region-containing protein 2) (Splicing factor HCC1)	RBM39 HCC1 RNPC2	Homo sapiens (Human)	530
P62877	RBX1_HUMAN	E3 ubiquitin-protein ligase RBX1 (EC 2.3.2.27) (EC 2.3.2.32) (E3 ubiquitin-protein transferase RBX1) (Protein ZYP) (RING finger protein 75) (RING-box protein 1) (Rbx1) (Regulator of cullins 1) (ROC1) [Cleaved into: E3 ubiquitin-protein ligase RBX1, N-terminally processed (E3 ubiquitin-protein transferase RBX1, N-terminally processed)]	RBX1 RNF75 ROC1	Homo sapiens (Human)	108

Continued

—cont'd

Entry	Entry name	Protein names	Gene names	Organism	Length
O43353	RIPK2_HUMAN	Receptor-interacting serine/threonine-protein kinase 2 (EC 2.7.11.1) (CARD-containing interleukin-1 beta-converting enzyme-associated kinase) (CARD-containing IL-1 beta ICE-kinase) (RIP-like-interacting CLARP kinase) (Receptor-interacting protein 2) (RIP-2) (Tyrosine-protein kinase RIPK2) (EC 2.7.10.2)	RIPK2 CARDIAK RICK RIP2 UNQ277/PRO314/PRO34092	Homo sapiens (Human)	540
Q9Y508	RN114_HUMAN	E3 ubiquitin-protein ligase RNF114 (EC 2.3.2.27) (RING finger protein 114) (RING-type E3 ubiquitin transferase RNF114) (Zinc finger protein 228) (Zinc finger protein 313)	RNF114 ZNF228 ZNF313	Homo sapiens (Human)	228
P78317	RNF4_HUMAN	E3 ubiquitin-protein ligase RNF4 (EC 2.3.2.27) (RING finger protein 4) (RING-type E3 ubiquitin transferase RNF4) (Small nuclear ring finger protein) (Protein SNURF)	RNF4 SNURF RES4–26	Homo sapiens (Human)	190
P19474	RO52_HUMAN	E3 ubiquitin-protein ligase TRIM21 (EC 2.3.2.27) (52 kDa Ro protein) (52 kDa ribonucleoprotein autoantigen Ro/SS-A) (RING finger protein 81) (RING-type E3 ubiquitin transferase TRIM21) (Ro(SS-A)) (Sjoegren syndrome type A antigen) (SS-A) (Tripartite motif-containing protein 21)	TRIM21 RNF81 RO52 SSA1	Homo sapiens (Human)	475
Q9NQG5	RPR1B_HUMAN	Regulation of nuclear pre-mRNA domain-containing protein 1B (Cell cycle-related and expression-elevated protein in tumour)	RPRD1B C20orf77 CREPT	Homo sapiens (Human)	326

Q9UJQ4	SALL4_HUMAN	Sal-like protein 4 (Zinc finger protein 797) (Zinc finger protein SALL4)	SALL4 ZNF797	Homo sapiens (Human)	1053
Q96BR1	SGK3_HUMAN	Serine/threonine-protein kinase Sgk3 (EC 2.7.11.1) (Cytokine-independent survival kinase) (Serum/glucocorticoid-regulated kinase 3) (Serum/glucocorticoid-regulated kinase-like)	SGK3 CISK SGKL	Homo sapiens (Human)	496
Q8IXJ6	SIR2_HUMAN	NAD-dependent protein deacetylase sirtuin-2 (EC 2.3.1.286) (Regulatory protein SIR2 homologue 2) (SIR2-like protein 2)	SIRT2 SIR2L SIR2L2	Homo sapiens (Human)	389
P19634	SL9A1_HUMAN	Sodium/hydrogen exchanger 1 (APNH) (Na(+)/H(+) antiporter, amiloride-sensitive) (Na(+)/H(+) exchanger 1) (NHE-1) (Solute carrier family 9 member 1)	SLC9A1 APNH1 NHE1	Homo sapiens (Human)	815
P84022	SMAD3_HUMAN	Mothers against decapentaplegic homologue 3 (MAD homologue 3) (Mad3) (Mothers against DPP homologue 3) (hMAD-3) (JV15–2) (SMAD family member 3) (SMAD 3) (Smad3) (hSMAD3)	SMAD3 MADH3	Homo sapiens (Human)	425
P51531	SMCA2_HUMAN	Probable global transcription activator SNF2L2 (EC 3.6.4) (ATP-dependent helicase SMARCA2) (BRG1-associated factor 190B) (BAF190B) (Protein brahma homologue) (hBRM) (SNF2-alpha) (SWI/SNF-related matrix-associated actin-dependent regulator of chromatin subfamily A member 2)	SMARCA2 BAF190B BRM SNF2A SNF2L2	Homo sapiens (Human)	1590

Continued

—cont'd

Entry	Entry name	Protein names	Gene names	Organism	Length
P51532	SMCA4_HUMAN	Transcription activator BRG1 (EC 3.6.4) (ATP-dependent helicase SMARCA4) (BRG1-associated factor 190A) (BAF190A) (Mitotic growth and transcription activator) (Protein BRG-1) (Protein brahma homologue 1) (SNF2-beta) (SWI/SNF-related matrix-associated actin-dependent regulator of chromatin subfamily A member 4)	SMARCA4 BAF190A BRG1 SNF2B SNF2L4	Homo sapiens (Human)	1647
O43791	SPOP_HUMAN	Speckle-type POZ protein (HIB homologue 1) (Roadkill homologue 1)	SPOP	Homo sapiens (Human)	374
Q13501	SQSTM_HUMAN	Sequestosome-1 (EBI3-associated protein of 60 kDa) (EBIAP) (p60) (Phosphotyrosine-independent ligand for the Lck SH2 domain of 62 kDa) (Ubiquitin-binding protein p62)	SQSTM1 ORCA OSIL	Homo sapiens (Human)	440
P40763	STAT3_HUMAN	Signal transducer and activator of transcription 3 (Acute-phase response factor)	STAT3 APRF	Homo sapiens (Human)	770
P37840	SYUA_HUMAN	Alpha-synuclein (Non-A beta component of AD amyloid) (Non-A4 component of amyloid precursor) (NACP)	SNCA NACP PARK1	Homo sapiens (Human)	140
Q9Y6A5	TACC3_HUMAN	Transforming acidic coiled-coil-containing protein 3 (ERIC-1)	TACC3 ERIC1	Homo sapiens (Human)	838
P10636	TAU_HUMAN	Microtubule-associated protein tau (Neurofibrillary tangle protein) (Paired helical filament-tau) (PHF-tau)	MAPT MAPTL MTBT1 TAU	Homo sapiens (Human)	758

Q9UHD2	TBK1_HUMAN	Serine/threonine-protein kinase TBK1 (EC 2.7.11.1) (NF-kappa-B-activating kinase) (T2K) (TANK-binding kinase 1)	TBK1 NAK	Homo sapiens (Human)	729
P02786	TFR1_HUMAN	Transferrin receptor protein 1 (TR) (TfR) (TfR1) (Trfr) (T9) (p90) (CD antigen CD71) [Cleaved into: Transferrin receptor protein 1, serum form (sTfR)]	TFRC	Homo sapiens (Human)	760
O15164	TIF1A_HUMAN	Transcription intermediary factor 1-alpha (TIF1-alpha) (EC 2.3.2.27) (E3 ubiquitin-protein ligase TRIM24) (RING finger protein 82) (RING-type E3 ubiquitin transferase TIF1-alpha) (Tripartite motif-containing protein 24)	TRIM24 RNF82 TIF1 TIF1A	Homo sapiens (Human)	1050
Q570C0	TIR1_ARATH	Protein TRANSPORT INHIBITOR RESPONSE 1 (Weak ethylene-insensitive protein 1)	TIR1 FBL1 WEI1 At3g62980 T20O10.80	*Arabidopsis thaliana* (Mouse-ear cress)	594
Q13564	ULA1_HUMAN	NEDD8-activating enzyme E1 regulatory subunit (Amyloid beta precursor protein-binding protein 1, 59 kDa) (APP-BP1) (Amyloid protein-binding protein 1) (Proto-oncogene protein 1)	NAE1 APPBP1 HPP1	Homo sapiens (Human)	534
P40337	VHL_HUMAN	von Hippel-Lindau disease tumour suppressor (Protein G7) (pVHL)	VHL	Homo sapiens (Human)	213
P30291	WEE1_HUMAN	Wee1-like protein kinase (WEE1hu) (EC 2.7.10.2) (Wee1A kinase)	WEE1	Homo sapiens (Human)	646

References

[1] Churcher I. Protac-induced protein degradation in drug discovery: breaking the rules or just making new ones? J Med Chem 2018;61:444–52. https://doi.org/10.1021/acs.jmedchem.7b01272.

[2] Dang CV, Reddy EP, Shokat KM, Soucek L. Drugging the "undruggable" cancer targets. Nat Rev Cancer 2017;17:502–8. https://doi.org/10.1038/nrc.2017.36.

[3] Chamberlain PP, Hamann LG. Development of targeted protein degradation therapeutics. Nat Chem Biol 2019;15:937–44. https://doi.org/10.1038/s41589-019-0362-y.

[4] Pickart CM. Back to the future with ubiquitin. Cell 2004;116:181–90. https://doi.org/10.1016/S0092-8674(03)01074-2.

[5] Cooper GM. Protein degradation. In: Cell A molecular approach. 2nd ed. Sinauer Associates; 2000.

[6] Schwartz AL, Ciechanover A. Targeting proteins for destruction by the ubiquitin system: implications for human pathobiology. Annu Rev Pharmacol Toxicol 2008;49:73–96. https://doi.org/10.1146/annurev.pharmtox.051208.165340.

[7] Clague MJ, Urbé S, Komander D. Breaking the chains: deubiquitylating enzyme specificity begets function. Nat Rev Mol Cell Biol 2019;20:338–52. https://doi.org/10.1038/s41580-019-0099-1.

[8] Nalepa G, Rolfe M, Harper JW. Drug discovery in the ubiquitin—proteasome system. Nat Rev Drug Discov 2006;5:596–613. https://doi.org/10.1038/nrd2056.

[9] Kerscher O, Felberbaum R, Hochstrasser M. Modification of proteins by ubiquitin and ubiquitin-like proteins. Annu Rev Cell Dev Biol 2006;22:159–80. https://doi.org/10.1146/annurev.cellbio.22.010605.093503.

[10] Huang X, Dixit VM. Drugging the undruggables: exploring the ubiquitin system for drug development. Cell Res 2016;26:484–98. https://doi.org/10.1038/cr.2016.31.

[11] Kenten JH, Roberts SF. Controlling protein levels in eucaryotic organisms; 1999. US patent USOO6306663B1.

[12] Sakamoto KM, Kim KB, Kumagai A, Mercurio F, Crews CM, Deshaies RJ. Protacs: chimeric molecules that target proteins to the Skp1-Cullin-F box complex for ubiquitination and degradation. Proc Natl Acad Sci U S A 2001;98:8554–9. https://doi.org/10.1073/pnas.141230798.

[13] Sakamoto KM, Kim KB, Verma R, Ransick A, Stein B, Crews CM, et al. Development of Protacs to target cancer-promoting proteins for ubiquitination and degradation. Mol Cell Proteomics 2003;2:1350–8. https://doi.org/10.1074/mcp.T300009-MCP200.

[14] Schneekloth JS, Fonseca FN, Koldobskiy M, Mandal A, Deshaies R, Sakamoto K, et al. Chemical genetic control of protein levels: selective in vivo targeted degradation. J Am Chem Soc 2004;126:3748–54. https://doi.org/10.1021/ja039025z.

[15] Metzen E, Ratcliffe PJ. HIF hydroxylation and cellular oxygen sensing. Biol Chem 2004;385:223–30. https://doi.org/10.1515/BC.2004.016.

[16] Schneekloth AR, Pucheault M, Tae HS, Crews CM. Targeted intracellular protein degradation induced by a small molecule: en route to chemical proteomics. Bioorg Med Chem Lett 2008;18:5904–8. https://doi.org/10.1016/j.bmcl.2008.07.114.

[17] Vassilev LT, Vu BT, Graves B, Carvajal D, Podlaski F, Filipovic Z, et al. In vivo activation of the p53 pathway by small-molecule antagonists of MDM2. Science 2004;303:844–8. https://doi.org/10.1126/science.1092472.

[18] Itoh Y, Ishikawa M, Naito M, Hashimoto Y. Protein knockdown using methyl bestatin-ligand hybrid molecules: design and synthesis of inducers of ubiquitination-mediated degradation of cellular retinoic acid-binding proteins. J Am Chem Soc 2010;132:5820–6. https://doi.org/10.1021/ja100691p.

[19] Okuhira K, Ohoka N, Sai K, Nishimaki-Mogami T, Itoh Y, Ishikawa M, et al. Specific degradation of CRABP-II via cIAP1-mediated ubiquitylation induced by hybrid molecules that crosslink cIAP1 and the target protein. FEBS Lett 2011;585:1147–52. https://doi.org/10.1016/j.febslet.2011.03.019.
[20] Sekine K, Takubo K, Kikuchi R, Nishimoto M, Kitagawa M, Abe F, et al. Small molecules destabilize cIAP1 by activating auto-ubiquitylation. J Biol Chem 2008;283:8961–8. https://doi.org/10.1074/jbc.M709525200.
[21] Itoh Y, Ishikawa M, Kitaguchi R, Okuhira K, Naito M, Hashimoto Y. Double protein knockdown of cIAP1 and CRABP-II using a hybrid molecule consisting of ATRA and IAPs antagonist. Bioorg Med Chem Lett 2012;22:4453–7. https://doi.org/10.1016/j.bmcl.2012.04.134.
[22] Okuhira K, Demizu Y, Hattori T, Ohoka N, Shibata N, Nishimaki-Mogami T, et al. Development of hybrid small molecules that induce degradation of estrogen receptor-alpha and necrotic cell death in breast cancer cells. Cancer Sci 2013;104: 1492–8. https://doi.org/10.1111/cas.12272.
[23] Shibata N, Nagai K, Morita Y, Ujikawa O, Ohoka N, Hattori T, et al. Development of protein degradation inducers of androgen receptor by conjugation of androgen receptor ligands and inhibitor of apoptosis protein ligands. J Med Chem 2018;61: 543–75. https://doi.org/10.1021/acs.jmedchem.7b00168.
[24] Ohoka N, Okuhira K, Ito M, Nagai K, Shibata N, Hattori T, et al. In vivo knockdown of pathogenic proteins via specific and nongenetic inhibitor of apoptosis protein (IAP)-dependent protein erasers (SNIPERs). J Biol Chem 2017;292:4556–70. https://doi.org/10.1074/jbc.M116.768853.
[25] Ito T, Ando H, Suzuki T, Ogura T, Hotta K, Imamura Y, et al. Identification of a primary target of thalidomide teratogenicity. Science 2010;327:1345–50. https://doi.org/10.1126/science.1177319.
[26] Lopez-Girona A, Mendy D, Ito T, Miller K, Gandhi AK, Kang J, et al. Cereblon is a direct protein target for immunomodulatory and antiproliferative activities of lenalidomide and pomalidomide. Leukemia 2012;26:2326–35. https://doi.org/10.1038/leu.2012.119.
[27] Lu J, Qian Y, Altieri M, Dong H, Wang J, Raina K, et al. Hijacking the E3 ubiquitin ligase cereblon to efficiently target BRD4. Chem Biol 2015;22:755–63. https://doi.org/10.1016/j.chembiol.2015.05.009.
[28] Winter GE, Buckley DL, Paulk J, Roberts JM, Souza A, Dhe-Paganon S, et al. Phthalimide conjugation as a strategy for in vivo target protein degradation. Science 2015;348:1376–81. https://doi.org/10.1126/science.aab1433.
[29] Gao S, Wang S, Song Y. Novel immunomodulatory drugs and neo-substrates. Biomark Res 2020;8:2. https://doi.org/10.1186/s40364-020-0182-y.
[30] Zengerle M, Chan KH, Ciulli A. Selective small molecule induced degradation of the BET bromodomain protein BRD4. ACS Chem Biol 2015;10:1770–7. https://doi.org/10.1021/acschembio.5b00216.
[31] Bondeson DP, Mares A, Smith IED, Ko E, Campos S, Miah AH, et al. Catalytic in vivo protein knockdown by small-molecule PROTACs. Nat Chem Biol 2015;11:611–7. https://doi.org/10.1038/nchembio.1858.
[32] Lu M, Liu T, Jiao Q, Ji J, Tao M, Liu Y, et al. Discovery of a Keap1-dependent peptide PROTAC to knockdown Tau by ubiquitination-proteasome degradation pathway. Eur J Med Chem 2018;146:251–9. https://doi.org/10.1016/j.ejmech.2018.01.063.
[33] Ward CC, Kleinman JI, Brittain SM, Lee PS, Chung CYS, Kim K, et al. Covalent ligand screening uncovers a RNF4 E3 ligase recruiter for targeted protein degradation applications. ACS Chem Biol 2019;14:2430–40. https://doi.org/10.1021/acschembio.8b01083.

[34] Spradlin JN, Hu X, Ward CC, Brittain SM, Jones MD, Ou L, et al. Harnessing the anti-cancer natural product nimbolide for targeted protein degradation. Nat Chem Biol 2019;15:747–55. https://doi.org/10.1038/s41589-019-0304-8.

[35] Zhang X, Crowley VM, Wucherpfennig TG, Dix MM, Cravatt BF. Electrophilic PROTACs that degrade nuclear proteins by engaging DCAF16. Nat Chem Biol 2019;15:737–46. https://doi.org/10.1038/s41589-019-0279-5.

[36] Li L, Mi D, Pei H, Duan Q, Wang X, Zhou W, et al. In vivo target protein degradation induced by PROTACs based on E3 ligase DCAF15. Signal Transduct Target Ther 2020;5:4–6. https://doi.org/10.1038/s41392-020-00245-0.

[37] Huang HT, Dobrovolsky D, Paulk J, Yang G, Weisberg EL, Doctor ZM, et al. A chemoproteomic approach to query the degradable kinome using a multi-kinase degrader. Cell Chem Biol 2018;25:88–99.e6. https://doi.org/10.1016/j.chembiol.2017.10.005.

[38] Bondeson DP, Smith BE, Burslem GM, Buhimschi AD, Hines J, Jaime-Figueroa S, et al. Lessons in PROTAC design from selective degradation with a promiscuous warhead. Cell Chem Biol 2018;25:78–87.e5. https://doi.org/10.1016/j.chembiol.2017.09.010.

[39] Pfaff P, Samarasinghe KTG, Crews CM, Carreira EM. Reversible spatiotemporal control of induced protein degradation by bistable PhotoPROTACs. ACS Cent Sci 2019;5:1682–90. https://doi.org/10.1021/acscentsci.9b00713.

[40] Kounde CS, Shchepinova MM, Saunders CN, Muelbaier M, Rackham MD, Harling JD, et al. A caged E3 ligase ligand for PROTAC-mediated protein degradation with light. Chem Commun 2020;56:5532–5. https://doi.org/10.1039/d0cc00523a.

[41] Jin Y-H, Lu M-C, Wang Y, Shan W-X, Wang X-Y, You Q-D, et al. Azo-PROTAC: novel light-controlled small-molecule tool for protein knockdown. J Med Chem 2020;63:4644–54. https://doi.org/10.1021/acs.jmedchem.9b02058.

[42] Reynders M, Matsuura BS, Bérouti M, Simoneschi D, Marzio A, Pagano M, et al. PHOTACs enable optical control of protein degradation. Sci Adv 2020;6, eaay5064. https://doi.org/10.1126/sciadv.aay5064.

[43] Xue G, Wang K, Zhou D, Zhong H, Pan Z. Light-induced protein degradation with photocaged PROTACs. J Am Chem Soc 2019;141:18370–4. https://doi.org/10.1021/jacs.9b06422.

[44] Pillow TH, Adhikari P, Blake RA, Chen J, Del Rosario G, Deshmukh G, et al. Antibody conjugation of a chimeric BET degrader enables in vivo activity. ChemMedChem 2020;15:17–25. https://doi.org/10.1002/cmdc.201900497.

[45] Maneiro M, Forte N, Shchepinova MM, Kounde CS, Chudasama V, Baker JR, et al. Antibody-PROTAC conjugates enable HER2-dependent targeted protein degradation of BRD4. ACS Chem Biol 2020;15:1306–12. https://doi.org/10.1021/acschembio.0c00285.

[46] Pettersson M, Crews CM. PROteolysis TArgeting chimeras (PROTACs)—past, present and future. Drug Discov Today Technol 2019;31:15–27. https://doi.org/10.1016/j.ddtec.2019.01.002.

[47] Sun X, Gao H, Yang Y, He M, Wu Y, Song Y, et al. Protacs: great opportunities for academia and industry. Signal Transduct Target Ther 2019;4:64. https://doi.org/10.1038/s41392-019-0101-6.

[48] An S, Fu L. Small-molecule PROTACs: an emerging and promising approach for the development of targeted therapy drugs. EBioMedicine 2018;36:553–62. https://doi.org/10.1016/j.ebiom.2018.09.005.

[49] Salami J, Crews CM. Waste disposal—an attractive strategy for cancer therapy. Science 2017;355:1163–7. https://doi.org/10.1126/science.aam7340.

[50] Moreau K, Coen M, Zhang AX, Pachl F, Castaldi MP, Dahl G, et al. Proteolysis-targeting chimeras in drug development: a safety perspective. Br J Pharmacol 2020;177:1709–18. https://doi.org/10.1111/bph.15014.

[51] Raina K, Lu J, Qian Y, Altieri M, Gordon D, Rossi AMK, et al. PROTAC-induced BET protein degradation as a therapy for castration-resistant prostate cancer. Proc Natl Acad Sci U S A 2016;113:7124–9. https://doi.org/10.1073/pnas.1521738113.
[52] Zhou B, Hu J, Xu F, Chen Z, Bai L, Fernandez-Salas E, et al. Discovery of a small-molecule degrader of bromodomain and extra-terminal (BET) proteins with picomolar cellular potencies and capable of achieving tumor regression. J Med Chem 2018;61:462–81. https://doi.org/10.1021/acs.jmedchem.6b01816.
[53] Sun B, Fiskus W, Qian Y, Rajapakshe K, Raina K, Coleman K, et al. BET protein proteolysis targeting chimera (PROTAC) exerts potent lethal activity against mantle cell lymphoma cells. Leukemia 2018;32:343–52. https://doi.org/10.1038/leu.2017.207.
[54] Burslem GM, Smith BE, Lai AC, Jaime-Figueroa S, McQuaid DC, Bondeson DP, et al. The advantages of targeted protein degradation over inhibition: an RTK case study. Cell Chem Biol 2018;25:67–77.e3. https://doi.org/10.1016/j.chembiol.2017.09.009.
[55] Olson CM, Jiang B, Erb MA, Liang Y, Doctor ZM, Zhang Z, et al. Pharmacological perturbation of CDK9 using selective CDK9 inhibition or degradation. Nat Chem Biol 2018;14:163–70. https://doi.org/10.1038/nchembio.2538.
[56] Roy RD, Rosenmund C, Stefan MI. Cooperative binding mitigates the high-dose hook effect. BMC Syst Biol 2017;11:1–10. https://doi.org/10.1186/s12918-017-0447-8.
[57] Mares A, Miah AH, Smith IED, Rackham M, Thawani AR, Cryan J, et al. Extended pharmacodynamic responses observed upon PROTAC-mediated degradation of RIPK2. Commun Biol 2020;3:1–13. https://doi.org/10.1038/s42003-020-0868-6.
[58] Bai L, Zhou H, Xu R, Zhao Y, Chinnaswamy K, McEachern D, et al. A potent and selective small-molecule degrader of STAT3 achieves complete tumor regression in vivo. Cancer Cell 2019;36:498–511.e17. https://doi.org/10.1016/j.ccell.2019.10.002.
[59] Sun X, Wang J, Yao X, Zheng W, Mao Y, Lan T, et al. A chemical approach for global protein knockdown from mice to non-human primates. Cell Discov 2019;5:1–13. https://doi.org/10.1038/s41421-018-0079-1.
[60] Chan K-H, Zengerle M, Testa A, Ciulli A. Impact of target warhead and linkage vector on inducing protein degradation: comparison of bromodomain and extra-terminal (BET) degraders derived from triazolodiazepine (JQ1) and tetrahydroquinoline (I-BET726) BET inhibitor scaffolds. J Med Chem 2018;61:504–13. https://doi.org/10.1021/acs.jmedchem.6b01912.
[61] Fisher SL, Phillips AJ. Targeted protein degradation and the enzymology of degraders. Curr Opin Chem Biol 2018;44:47–55. https://doi.org/10.1016/j.cbpa.2018.05.004.
[62] Kim J, Kim H, Park SB. Privileged structures: efficient chemical "navigators" toward unexplored biologically relevant chemical spaces. J Am Chem Soc 2014;136: 14629–38. https://doi.org/10.1021/ja508343a.
[63] Toure M, Crews CM. Small-molecule PROTACS: new approaches to protein degradation. Angew Chem Int Ed 2016;55:1966–73. https://doi.org/10.1002/anie.201507978.
[64] Crews CM. Targeting the undruggable proteome: the small molecules of my dreams. Chem Biol 2010;17:551–5. https://doi.org/10.1016/j.chembiol.2010.05.011.
[65] Rauch J, Volinsky N, Romano D, Kolch W. The secret life of kinases: functions beyond catalysis. Cell Commun Signal 2011;9:1–28. https://doi.org/10.1186/1478-811X-9-23.
[66] Cromm PM, Samarasinghe KTG, Hines J, Crews CM. Addressing kinase-independent functions of Fak via PROTAC-mediated degradation. J Am Chem Soc 2018;140:17019–26. https://doi.org/10.1021/jacs.8b08008.
[67] Remillard D, Buckley DL, Paulk J, Brien GL, Sonnett M, Seo H-S, et al. Degradation of the BAF complex factor BRD9 by heterobifunctional ligands HHS public access. Angew Chem Int Ed Eng 2017;56:5738–43. https://doi.org/10.1002/anie.201611281.

[68] Gadd MS, Testa A, Lucas X, Chan KH, Chen W, Lamont DJ, et al. Structural basis of PROTAC cooperative recognition for selective protein degradation. Nat Chem Biol 2017;13:514–21. https://doi.org/10.1038/nchembio.2329.

[69] Rubin BP, Duensing A. Mechanisms of resistance to small molecule kinase inhibition in the treatment of solid tumors. Lab Investig 2006;86:981–6. https://doi.org/10.1038/labinvest.3700466.

[70] Pisa R, Kapoor TM. Chemical strategies to overcome resistance against targeted anticancer therapeutics. Nat Chem Biol 2020;16:817–25. https://doi.org/10.1038/s41589-020-0596-8.

[71] Buhimschi AD, Armstrong HA, Toure M, Jaime-Figueroa S, Chen TL, Lehman AM, et al. Targeting the C481S Ibrutinib-resistance mutation in Bruton's tyrosine kinase using PROTAC-mediated degradation. Biochemistry 2018;57:3564–75. https://doi.org/10.1021/acs.biochem.8b00391.

[72] Sun Y, Zhao X, Ding N, Gao H, Wu Y, Yang Y, et al. PROTAC-induced BTK degradation as a novel therapy for mutated BTK C481S induced ibrutinib-resistant B-cell malignancies. Cell Res 2018;28:779–81. https://doi.org/10.1038/s41422.

[73] Sun Y, Ding N, Song Y, Yang Z, Liu W, Zhu J, et al. Degradation of Bruton's tyrosine kinase mutants by PROTACs for potential treatment of ibrutinib-resistant non-Hodgkin lymphomas. Leukemia 2019;33:2105–10. https://doi.org/10.1038/s41375-019-0440-x.

[74] Salami J, Alabi S, Willard RR, Vitale NJ, Wang J, Dong H, et al. Androgen receptor degradation by the proteolysis-targeting chimera ARCC-4 outperforms enzalutamide in cellular models of prostate cancer drug resistance. Commun Biol 2018;1:100. https://doi.org/10.1038/s42003-018-0105-8.

[75] Ottis P, Palladino C, Thienger P, Britschgi A, Heichinger C, Berrera M, et al. Cellular resistance mechanisms to targeted protein degradation converge toward impairment of the engaged ubiquitin transfer pathway. ACS Chem Biol 2019;14:2215–23. https://doi.org/10.1021/acschembio.9b00525.

[76] Zhang L, Riley-Gillis B, Vijay P, Shen Y. Acquired resistance to BET-ProTACS (proteolysis-targeting chimeras) caused by genomic alterations in core components of E3 ligase complexes. Mol Cancer Ther 2019;18:1302–11. https://doi.org/10.1158/1535-7163.MCT-18-1129.

[77] Conole D, Mondal M, Majmudar JD, Tate EW. Recent developments in cell permeable deubiquitinating enzyme activity-based probes. Front Chem 2019;7:876. https://doi.org/10.3389/fchem.2019.00876.

[78] Cong F, Cheung AK, Huang S-MA. Chemical genetics–based target identification in drug Discovery. Annu Rev Pharmacol Toxicol 2012;52:57–78. https://doi.org/10.1146/annurev-pharmtox-010611-134639.

[79] Lai AC, Crews CM. Induced protein degradation: an emerging drug discovery paradigm. Nat Rev Drug Discov 2017;16:101–14. https://doi.org/10.1038/nrd.2016.211.

[80] Rodbard D, Feldman Y, Jaffe ML, Miles LEM. Kinetics of two-site immunoradiometric ('sandwich') assays-II. Studies on the nature of the -high-dose hook effect. Mol Immunol 1978;15:77–82. https://doi.org/10.1016/0161-5890(78)90046-9.

[81] Zorba A, Nguyen C, Xu Y, Starr J, Borzilleri K, Smith J, et al. Delineating the role of cooperativity in the design of potent PROTACs for BTK. Proc Natl Acad Sci U S A 2018;115:E7285–92. https://doi.org/10.1073/pnas.1803662115.

[82] Roy MJ, Winkler S, Hughes SJ, Whitworth C, Galant M, Farnaby W, et al. SPR-measured dissociation kinetics of PROTAC ternary complexes influence target degradation rate. ACS Chem Biol 2019;14:361–8. https://doi.org/10.1021/acschembio.9b00092.

[83] Nowak RP, DeAngelo SL, Buckley D, He Z, Donovan KA, An J, et al. Plasticity in binding confers selectivity in ligand-induced protein degradation. Nat Chem Biol 2018;14:706–14. https://doi.org/10.1038/s41589-018-0055-y.
[84] Farnaby W, Koegl M, Roy MJ, Whitworth C, Diers E, Trainor N, et al. BAF complex vulnerabilities in cancer demonstrated via structure-based PROTAC design. Nat Chem Biol 2019;15:672–80. https://doi.org/10.1038/s41589-019-0294-6.
[85] Cyrus K, Wehenkel M, Choi EY, Han HJ, Lee H, Swanson H, et al. Impact of linker length on the activity of PROTACs. Mol BioSyst 2011;7:359–64. https://doi.org/10.1039/c0mb00074d.
[86] Smith BE, Wang SL, Jaime-Figueroa S, Harbin A, Wang J, Hamman BD, et al. Differential PROTAC substrate specificity dictated by orientation of recruited E3 ligase. Nat Commun 2019;10:1–13. https://doi.org/10.1038/s41467-018-08027-7.
[87] Schapira M, Calabrese MF, Bullock AN, Crews CM. Targeted protein degradation: expanding the toolbox. Nat Rev Drug Discov 2019;18:949–63. https://doi.org/10.1038/s41573-019-0047-y.
[88] Chen Y, Jin J. The application of ubiquitin ligases in the PROTAC drug design. Acta Biochim Biophys Sin Shanghai 2020;52:776–90. https://doi.org/10.1093/abbs/gmaa053.
[89] Itoh Y, Kitaguchi R, Ishikawa M, Naito M, Hashimoto Y. Design, synthesis and biological evaluation of nuclear receptor-degradation inducers. Bioorg Med Chem 2011;19:6768–78. https://doi.org/10.1016/j.bmc.2011.09.041.
[90] Maple HJ, Clayden N, Baron A, Stacey C, Felix R. Developing degraders: principles and perspectives on design and chemical space. MedChemComm 2019;10:1755–64. https://doi.org/10.1039/c9md00272c.
[91] Lai A, Momar T, Hellerschmieda D, Salamia J, Ko SJ-FE, Hinesa J, et al. Modular PROTAC design for the degradation of oncogenic BCR-ABL. Angew Chem Int Ed 2016;55:807–10. https://doi.org/10.1002/anie.201507634.
[92] Tan X, Irina Calderon-Villalobos LA, Sharon M, Zheng C, Robinson CV, Estelle M, et al. Mechanism of auxin perception by the TIR1 ubiquitin ligase. Nature 2007;446:640–5. https://doi.org/10.1038/nature05731.
[93] Thiel P, Kaiser M, Ottmann C. Small-molecule stabilization of protein-protein interactions: An underestimated concept in drug discovery? Angew Chem Int Ed 2012;51:2012–8. https://doi.org/10.1002/anie.201107616.
[94] Hanan EJ, Liang J, Wang X, Blake R, Blaquiere N, Staben ST, et al. Monomeric targeted protein degraders. J Med Chem 2020;63:11330–61.
[95] Burslem GM, Crews CM. Small-molecule modulation of protein homeostasis. Chem Rev 2017;117:11269–301. https://doi.org/10.1021/acs.chemrev.7b00077.
[96] Chamberlain PP, Cathers BE. Cereblon modulators: low molecular weight inducers of protein degradation. Drug Discov Today Technol 2019;31:29–34. https://doi.org/10.1016/j.ddtec.2019.02.004.
[97] Fischer ES, Böhm K, Lydeard JR, Yang H, Stadler MB, Cavadini S, et al. Structure of the DDB1-CRBN E3 ubiquitin ligase in complex with thalidomide. Nature 2014;512:49–53. https://doi.org/10.1038/nature13527.
[98] Chamberlain PP, Lopez-Girona A, Miller K, Carmel G, Pagarigan B, Chie-Leon B, et al. Structure of the human Cereblon-DDB1-lenalidomide complex reveals basis for responsiveness to thalidomide analogs. Nat Struct Mol Biol 2014;21:803–9. https://doi.org/10.1038/nsmb.2874.
[99] Lipinski CA, Lombardo F, Dominy BW, Feeney PJ. Experimental and computational approaches to estimate solubility and permeability in drug discovery and development settings. Adv Drug Deliv Rev 2012;46:3–26. https://doi.org/10.1016/j.addr.2012.09.019.

[100] Konstantinidou M, Li J, Zhang B, Wang Z, Shaabani S, Ter Brake F, et al. PROTACs-a game-changing technology. Expert Opin Drug Discovery 2019;14:1255–68. https://doi.org/10.1080/17460441.2019.1659242.
[101] den Besten W, Lipford JR. Prospecting for molecular glues. Nat Chem Biol 2020;16:1157–8. https://doi.org/10.1038/s41589-020-0620-z.
[102] Mayor-Ruiz C, Bauer S, Brand M, Kozicka Z, Siklos M, Imrichova H, et al. Rational discovery of molecular glue degraders via scalable chemical profiling. Nat Chem Biol 2020;16:1199–207. https://doi.org/10.1038/s41589-020-0594-x.
[103] Ozawa Y, Sugi NH, Nagasu T, Owa T, Watanabe T, Koyanagi N, et al. E7070, a novel sulphonamide agent with potent antitumour activity in vitro and in vivo. Eur J Cancer 2001;37:2275–82.
[104] Han T, Goralski M, Gaskill N, Capota E, Kim J, Ting TC, et al. Anticancer sulfonamides target splicing by inducing RBM39 degradation via recruitment to DCAF15. Science 2017;356:397. https://doi.org/10.1126/science.aal3755.
[105] Uehara T, Minoshima Y, Sagane K, Sugi NH, Mitsuhashi KO, Yamamoto N, et al. Selective degradation of splicing factor CAPERα by anticancer sulfonamides. Nat Chem Biol 2017;13:675–80. https://doi.org/10.1038/nchembio.2363.
[106] Bussiere DE, Xie L, Srinivas H, Shu W, Burke A, Be C, et al. Structural basis of indisulam-mediated RBM39 recruitment to DCAF15 E3 ligase complex. Nat Chem Biol 2020;16:15–23. https://doi.org/10.1038/s41589-019-0411-6.
[107] Faust TB, Yoon H, Nowak RP, Donovan KA, Li Z, Cai Q, et al. Structural complementarity facilitates E7820-mediated degradation of RBM39 by DCAF15. Nat Chem Biol 2020;16:7–14. https://doi.org/10.1038/s41589-019-0378-3.
[108] Krönke J, Udeshi ND, Narla A, Grauman P, Hurst SN, McConkey M, et al. Lenalidomide causes selective degradation of IKZF1 and IKZF3 in multiple myeloma cells. Science 2014;343:301–5. https://doi.org/10.1126/science.1244851.
[109] Lu G, Middleton RE, Sun H, Naniong MV, Ott CJ, Mitsiades CS, et al. The myeloma drug lenalidomide promotes the cereblon-dependent destruction of ikaros proteins. Science 2014;343:305–9. https://doi.org/10.1126/science.1244917.
[110] Krönke J, Fink EC, Hollenbach PW, MacBeth KJ, Hurst SN, Udeshi ND, et al. Lenalidomide induces ubiquitination and degradation of CK1α in del(5q) MDS. Nature 2015;523:183–8. https://doi.org/10.1038/nature14610.
[111] Matyskiela ME, Lu G, Ito T, Pagarigan B, Lu C-C, Miller K, et al. A novel cereblon modulator recruits GSPT1 to the CRL4 CRBN ubiquitin ligase. Nature 2016;535:252–7. https://doi.org/10.1038/nature18611.
[112] Powell CE, Du G, Che J, He Z, Donovan KA, Yue H, et al. Selective degradation of GSPT1 by Cereblon modulators identified via a focused combinatorial library. ACS Chem Biol 2020;15:2722–30. https://doi.org/10.1021/acschembio.0c00520.
[113] Donovan KA, An J, Nowak RP, Yuan JC, Fink EC, Berry BC, et al. Thalidomide promotes degradation of SALL4, a transcription factor implicated in Duane radial ray syndrome. elife 2018;7:1–25. https://doi.org/10.7554/eLife.38430.
[114] Yamamoto J, Suwa T, Murase Y, Tateno S, Mizutome H, Asatsuma-Okumura T, et al. ARID2 is a pomalidomide-dependent CRL4CRBN substrate in multiple myeloma cells. Nat Chem Biol 2020;16:1208–17. https://doi.org/10.1038/s41589-020-0645-3.
[115] He Y, Zhang X, Chang J, Kim H, Zhang P, Wang Y, et al. Using proteolysis-targeting chimera technology to reduce navitoclax platelet toxicity and improve its senolytic activity. Nat Commun 2020;11:1996. https://doi.org/10.1038/s41467-020-15838-0.
[116] Daurio NA, Zhou H, Chen Y, Sheth PR, Imbriglio JE, McLaren DG, et al. Examining targeted protein degradation from physiological and analytical perspectives: enabling translation between cells and subjects. ACS Chem Biol 2020; 15:2623–35. https://doi.org/10.1021/acschembio.0c00380.

[117] Liu J, Nussinov R. The mechanism of ubiquitination in the cullin-RING E3 ligase machinery: conformational control of substrate orientation. PLoS Comput Biol 2009;5, e1000527. https://doi.org/10.1371/journal.pcbi.1000527.

[118] Ishoey M, Chorn S, Singh N, Jaeger MG, Brand M, Paulk J, et al. Translation termination factor GSPT1 is a phenotypically relevant off-target of heterobifunctional phthalimide degraders. ACS Chem Biol 2018;13:553–60. https://doi.org/10.1021/acschembio.7b00969.

[119] An J, Ponthier CM, Sack R, Seebacher J, Stadler MB, Donovan KA, et al. PSILAC mass spectrometry reveals ZFP91 as IMiD-dependent substrate of the CRL4 CRBN ubiquitin ligase. Nat Commun 2017;8:1–11. https://doi.org/10.1038/ncomms15398.

[120] Matyskiela ME, Couto S, Zheng X, Lu G, Hui J, Stamp K, et al. SALL4 mediates teratogenicity as a thalidomide-dependent cereblon substrate. Nat Chem Biol 2018;14:981–7. https://doi.org/10.1038/s41589-018-0129-x.

[121] Nalawansha DA, Crews CM. PROTACs: an emerging therapeutic modality in precision medicine. Cell Chem Biol 2020;27:998–1014. https://doi.org/10.1016/j.chembiol.2020.07.020.

[122] Wu T, Yoon H, Xiong Y, Dixon-Clarke SE, Nowak aw P, Fischer ES. Targeted protein degradation as a powerful research tool in basic biology and drug target discovery. Nat Struct Mol Biol 2020;27:605–14. https://doi.org/10.1038/s41594-020-0438-0.

[123] Koduri V, McBrayer SK, Liberzon E, Wang AC, Briggs KJ, Cho H, et al. Peptidic degron for IMiD-induced degradation of heterologous proteins. elife 2017;116:2539–44. https://doi.org/10.1073/pnas.1818109116.

[124] Simpson LM, Macartney TJ, Nardin A, Ciulli A, Ganley IG, Sapkota Correspondence GP, et al. Inducible degradation of target proteins through a tractable affinity-directed protein missile system ll inducible degradation of target proteins through a tractable affinity-directed protein missile system. Cell Chem Biol 2020;27:1–17. https://doi.org/10.1016/j.chembiol.2020.06.013.

[125] Banaszynski LA, Chun CL, Maynard-Smith LA, Ooi AGL, Wandless TJ. A rapid, reversible, and tunable method to regulate protein function in living cells using synthetic small molecules. Cell 2006;126:995–1004. https://doi.org/10.1016/j.cell.2006.07.025.

[126] Chung HK, Jacobs CL, Huo Y, Yang J, Krumm SA, Plemper RK, et al. Tunable and reversible drug control of protein production via a self-excising degron. Nat Chem Biol 2015;11:713–20. https://doi.org/10.1038/nchembio.1869.

[127] Nabet B, Roberts JM, Buckley DL, Paulk J, Dastjerdi S, Yang A, et al. The dTAG system for immediate and target-specific protein degradation. Nat Chem Biol 2018;14:431–41. https://doi.org/10.1038/s41589-018-0021-8.

[128] Neklesa TK, Tae HS, Schneekloth AR, Stulberg MJ, Corson TW, Sundberg TB, et al. Small-molecule hydrophobic tagging-induced degradation of HaloTag fusion proteins. Nat Chem Biol 2011;7:538–43. https://doi.org/10.1038/nchembio.597.

[129] Buckley DL, Raina K, Darricarrere N, Hines J, Gustafson JL, Smith IE, et al. HaloPROTACS: use of small molecule PROTACs to induce degradation of HaloTag fusion proteins. ACS Chem Biol 2015;10:1831–7. https://doi.org/10.1021/acschembio.5b00442.

[130] Los GV, Encell LP, McDougall MG, Hartzell DD, Karassina N, Zimprich C, et al. HaloTag: a novel protein labeling technology for cell imaging and protein analysis. ACS Chem Biol 2008;3:373–82. https://doi.org/10.1021/cb800025k.

[131] Ottis P, Toure M, Cromm PM, Ko E, Gustafson JL, Crews CM. Assessing different E3 ligases for small molecule induced protein ubiquitination and degradation. ACS Chem Biol 2017;12:2570–8. https://doi.org/10.1021/acschembio.7b00485.

[132] Tomoshige S, Naito M, Hashimoto Y, Ishikawa M. Degradation of HaloTag-fused nuclear proteins using bestatin-HaloTag ligand hybrid molecules. Org Biomol Chem 2015;13:9746–50. https://doi.org/10.1039/c5ob01395j.

[133] Yang W, Rozamus LW, Narula S, Rollins CT, Yuan R, Andrade LJ, et al. Investigating protein-ligand interactions with a mutant FKBP possessing a designed specificity pocket. J Med Chem 2000;43:1135–42. https://doi.org/10.1021/jm9904396.
[134] Clackson T, Yang W, Rozamus LW, Hatada M, Amara JF, Rollins CT, et al. Redesigning an FKBP-ligand interface to generate chemical dimerizers with novel specificity. Proc Natl Acad Sci U S A 1998;95:10437–42. https://doi.org/10.1073/pnas.95.18.10437.
[135] Burslem GM, Crews CM. Proteolysis-targeting chimeras as therapeutics and tools for biological Discovery. Cell 2020;181:102–14. https://doi.org/10.1016/j.cell.2019.11.031.
[136] BasuRay S, Wang Y, Smagris E, Cohen JC, Hobbs HH. Accumulation of PNPLA3 on lipid droplets is the basis of associated hepatic steatosis. Proc Natl Acad Sci U S A 2019;116:9521–6. https://doi.org/10.1073/pnas.1901974116.
[137] Erb MA, Scott TG, Li BE, Xie H, Paulk J, Seo HS, et al. Transcription control by the ENL YEATS domain in acute leukaemia. Nature 2017;543:270–4. https://doi.org/10.1038/nature21688.
[138] Fulcher LJ, Macartney T, Bozatzi P, Hornberger A, Rojas-Fernandez A, Sapkota GP. An affinity-directed protein missile system for targeted proteolysis. Open Biol 2016;6:160255. https://doi.org/10.1098/rsob.160255.
[139] Clift D, McEwan WA, Labzin LI, Konieczny V, Mogessie B, James LC, et al. A method for the acute and rapid degradation of endogenous proteins. Cell 2017;171:1692–1706.e18. https://doi.org/10.1016/j.cell.2017.10.033.
[140] Banik SM, Pedram K, Wisnovsky S, Ahn G, Riley NM, Bertozzi CR. Lysosome-targeting chimaeras for degradation of extracellular proteins. Nature 2020;584:291–7. https://doi.org/10.1038/s41586-020-2545-9.
[141] Fan X, Jin WY, Lu J, Wang J, Wang YT. Rapid and reversible knockdown of endogenous proteins by peptide-directed lysosomal degradation. Nat Neurosci 2014;17:471–80. https://doi.org/10.1038/nn.3637.
[142] Pinch BJ, Buckley DL, Gleim S, Brittain SM, Tandeske L, Harvey PLDEP, et al. A strategy to assess the cellular activity of E3 ligases against neo-substrates using electrophilic probes. BioRxiv 2020. https://doi.org/10.1017/CBO9781107415324.004.
[143] Zeng M, Xiong Y, Safaee N, Nowak RP, Donovan KA, Yuan CJ, et al. Exploring targeted degradation strategy for oncogenic KRASG12C. Cell Chem Biol 2020;27:19–31.e6. https://doi.org/10.1016/j.chembiol.2019.12.006.
[144] Foley CA, Potjewyd F, Lamb KN, James LI, Frye SV. Assessing the cell permeability of bivalent chemical degraders using the chloroalkane penetration assay. ACS Chem Biol 2020;15:290–5. https://doi.org/10.1021/acschembio.9b00972.
[145] Kostic M, Jones LH. Critical assessment of targeted protein degradation as a research tool and pharmacological modality. Trends Pharmacol Sci 2020;41:305–17. https://doi.org/10.1016/j.tips.2020.02.006.
[146] Troup RI, Fallan C, Baud MGJ. Current strategies for the design of PROTAC linkers: a critical review. Explor Target Antitumor Ther 2020;1:273–312. https://doi.org/10.37349/etat.2020.00018.
[147] Chung CW, Dai H, Fernandez E, Tinworth CP, Churcher I, Cryan J, et al. Structural insights into PROTAC-mediated degradation of Bcl-xL. ACS Chem Biol 2020;15:2316–23. https://doi.org/10.1021/acschembio.0c00266.
[148] Bai N, Kirubakaran P, Karanicolas J. Rationalizing PROTAC-mediated ternary complex formation using Rosetta. Bioxriv 2020. https://doi.org/10.1101/2020.05.27.119347.
[149] Zaidman D, Prilusky J, London N. ProsetTac: rosetta based modeling of PROTAC mediated ternary complexes. J Chem Inf Model 2020;60:4894–903. https://doi.org/10.1021/acs.jcim.0c00589.

[150] Han X, Wang C, Qin C, Xiang W, Fernandez-Salas E, Yang CY, et al. Discovery of ARD-69 as a highly potent proteolysis targeting chimera (PROTAC) degrader of androgen receptor (AR) for the treatment of prostate Cancer. J Med Chem 2019;62:941–64. https://doi.org/10.1021/acs.jmedchem.8b01631.
[151] Wang Y, Jiang X, Feng F, Liu W, Sun H. Degradation of proteins by PROTACs and other strategies. Acta Pharm Sin B 2020;10:207–38. https://doi.org/10.1016/j.apsb.2019.08.001.
[152] Wurz RP, Dellamaggiore K, Dou H, Javier N, Lo MC, McCarter JD, et al. A "click chemistry platform" for the rapid synthesis of bispecific molecules for inducing protein degradation. J Med Chem 2018;61:453–61. https://doi.org/10.1021/acs.jmedchem.6b01781.
[153] Buckley DL, Van Molle I, Gareiss PC, Tae HS, Michel J, Noblin DJ, et al. Targeting the von Hippel-Lindau E3 ubiquitin ligase using small molecules to disrupt the VHL/HIF-1α interaction. J Am Chem Soc 2012;134:4465–8. https://doi.org/10.1021/ja209924v.
[154] Tong B, Luo M, Xie Y, Spradlin JN, Tallarico JA, McKenna JM, et al. Bardoxolone conjugation enables targeted protein degradation of BRD4. Sci Rep 2020;10:1–8. https://doi.org/10.1038/s41598-020-72491-9.
[155] Yeung WSB, Luo GA, Wang QG, Ou JP. Capillary electrophoresis-based immunoassay. J Chromatogr B Anal Technol Biomed Life Sci 2003;797:217–28. https://doi.org/10.1016/S1570-0232(03)00489-6.
[156] Chessum NEA, Sharp SY, Caldwell JJ, Pasqua AE, Wilding B, Colombano G, et al. Demonstrating in-cell target engagement using a Pirin protein degradation probe (CCT367766). J Med Chem 2018;61:918–33. https://doi.org/10.1021/acs.jmedchem.7b01406.
[157] Wu H, Yang K, Zhang Z, Leisten ED, Li Z, Xie H, et al. Development of multifunctional histone deacetylase 6 degraders with potent antimyeloma activity. J Med Chem 2019;62:7042–57. https://doi.org/10.1021/acs.jmedchem.9b00516.
[158] Yang K, Zhao Y, Nie X, Yang K, Zhao Y, Nie X, et al. A cell-based target engagement assay for the identification of Cereblon E3 ubiquitin ligase ligands and their application in HDAC6 degraders resource a cell-based target engagement assay for the identification of Cereblon E3 ubiquitin ligase ligands. Cell Chem Biol 2020;27(7):866–876.e8. https://doi.org/10.1016/j.chembiol.2020.04.008.
[159] Mesoscale Discovery. MESO SCALE DISCOVERY's MULTI-ARRAY® technology—product sheet. Mesoscale Discovery; 2014. p. 1–29.
[160] Naro Y, Darrah K, Deiters A. Optical control of small molecule-induced protein degradation. J Am Chem Soc 2020;142:2193–7. https://doi.org/10.1021/jacs.9b12718.
[161] Riching KM, Mahan S, Corona CR, McDougall M, Vasta JD, Robers MB, et al. Quantitative live-cell kinetic degradation and mechanistic profiling of PROTAC mode of action. ACS Chem Biol 2018;13:2758–70. https://doi.org/10.1021/acschembio.8b00692.
[162] Zoppi V, Hughes SJ, Maniaci C, Testa A, Gmaschitz T, Wieshofer C, et al. Iterative design and optimization of initially inactive proteolysis targeting chimeras (PROTACs) identify VZ185 as a potent, fast, and selective von Hippel-Lindau (VHL) based dual degrader probe of BRD9 and BRD7. J Med Chem 2019;62:699–726. https://doi.org/10.1021/acs.jmedchem.8b01413.
[163] Zhang Y, Fonslow BR, Shan B, Baek MC, Yates JR. Protein analysis by shotgun/bottom-up proteomics. Chem Rev 2013;113:2343–94. https://doi.org/10.1021/cr3003533.
[164] Ong SE, Blagoev B, Kratchmarova I, Kristensen DB, Steen H, Pandey A, et al. Stable isotope labeling by amino acids in cell culture, SILAC, as a simple and accurate approach to expression proteomics. Mol Cell Proteomics 2002;1:376–86. https://doi.org/10.1074/mcp.M200025-MCP200.

[165] Geiger T, Wisniewski JR, Cox J, Zanivan S, Kruger M, Ishihama Y, et al. Use of stable isotope labeling by amino acids in cell culture as a spike-in standard in quantitative proteomics. Nat Protoc 2011;6:147–57. https://doi.org/10.1038/nprot.2010.192.

[166] Thompson A, Schäfer J, Kuhn K, Kienle S, Schwarz J, Schmidt G, et al. Tandem mass tags: a novel quantification strategy for comparative analysis of complex protein mixtures by MS/MS. Anal Chem 2003;75:1895–904. https://doi.org/10.1021/ac0262560.

[167] Aggarwal K, Choe LH, Lee KH. Shotgun proteomics using the iTRAQ isobaric tags. Brief Funct Genomic Proteomic 2006;5:112–20. https://doi.org/10.1093/bfgp/ell018.

[168] Savitski MM, Zinn N, Faelth-Savitski M, Poeckel D, Gade S, Becher I, et al. Multiplexed proteome dynamics profiling reveals mechanisms controlling protein homeostasis. Cell 2018;173:260–274.e25. https://doi.org/10.1016/j.cell.2018.02.030.

[169] Guo WH, Qi X, Yu X, Liu Y, Chung CI, Bai F, et al. Enhancing intracellular accumulation and target engagement of PROTACs with reversible covalent chemistry. Nat Commun 2020;11:1–16. https://doi.org/10.1038/s41467-020-17997-6.

[170] Gabizon R, Shraga A, Gehrtz P, Livnah E, Shorer Y, Gurwicz N, et al. Efficient targeted degradation via reversible and irreversible covalent PROTACs. J Am Chem Soc 2020;142:11734–42. https://doi.org/10.1021/jacs.9b13907.

[171] Khan S, Zhang X, Lv D, Zhang Q, He Y, Zhang P, et al. A selective BCL-XL PROTAC degrader achieves safe and potent antitumor activity. Nat Med 2019;25:1938–47. https://doi.org/10.1038/s41591-019-0668-z.

[172] Nunes AT, Annunziata CM. Proteasome inhibitors: structure and function. Semin Oncol 2017;44:377–80. https://doi.org/10.1053/j.seminoncol.2018.01.004.

[173] Soucy TA, Smith PG, Milhollen MA, Berger AJ, Gavin JM, Adhikari S, et al. An inhibitor of NEDD8-activating enzyme as a new approach to treat cancer. Nature 2009;458:732–6. https://doi.org/10.1038/nature07884.

[174] Min JH, Yang H, Ivan M, Gertler F, Kaelin WG, Pavietich NP. Structure of an HIF-1α-pVHL complex: hydroxyproline recognition in signaling. Science 2002;296: 1886–9. https://doi.org/10.1126/science.1073440.

[175] Hon WC, Wilson MI, Harlos K, Claridge TDW, Schofield CJ, Pugh CW, et al. Structural basis for the recognition of hydroxyproline in HIF-1α by pVHL. Nature 2002;417:975–8. https://doi.org/10.1038/nature00767.

[176] Douglass EF, Miller CJ, Sparer G, Shapiro H, Spiegel DA. A comprehensive mathematical model for three-body binding equilibria. J Am Chem Soc 2013;135:6092–9. https://doi.org/10.1021/ja311795d.

[177] Maniaci C, Hughes SJ, Testa A, Chen W, Lamont DJ, Rocha S, et al. Homo-PROTACs: bivalent small-molecule dimerizers of the VHL E3 ubiquitin ligase to induce self-degradation. Nat Commun 2017;8:830. https://doi.org/10.1038/s41467-017-00954-1.

[178] Zhao Q, Ren C, Liu L, Chen J, Shao Y, Sun N, et al. Discovery of SIAIS178 as an effective BCR-ABL degrader by recruiting von Hippel-Lindau (VHL) E3 ubiquitin ligase. J Med Chem 2019;62:9281–98. https://doi.org/10.1021/acs.jmedchem.9b01264.

[179] Simonetta KR, Taygerly J, Boyle K, Basham SE, Padovani C, Lou Y, et al. Prospective discovery of small molecule enhancers of an E3 ligase-substrate interaction. Nat Commun 2019;10:1402. https://doi.org/10.1038/s41467-019-09358-9.

[180] Beveridge R, Kessler D, Rumpel K, Ettmayer P, Meinhart A, Clausen T. Native mass spectrometry can effectively predict PROTAC efficacy. ACS Cent Sci 2020;6: 1223–30. https://doi.org/10.1021/acscentsci.0c00049.

[181] Huang X. Fluorescence polarization competition assay: the range of resolvable inhibitor potency is limited by the affinity of the fluorescent ligand. J Biomol Screen 2003;8:34–8. https://doi.org/10.1177/1087057102239666.

[182] Yasgar A, Jadhav A, Simeonov A, Coussens NP. AlphaScreen-based assays: ultra-high-throughput screening for small-molecule inhibitors of challenging enzymes and protein-protein interactions. Methods Mol Biol 2016;1439:77–98. https://doi.org/10.1007/978-1-4939-3673-1.
[183] PerkinElmer. TruHit AlphaScreen Kit n.d. https://www.perkinelmer.com/it/product/alphascreen-truhits-kits-1000-pts-6760627d. Accessed 22 February 2021.
[184] Oh E, Akopian D, Rape M. Principles of ubiquitin-dependent signaling. Annu Rev Cell Dev Biol 2018;34:137–62. https://doi.org/10.1146/annurev-cellbio-100617-062802.
[185] De Vita E, Maneiro M, Tate EW. The missing link between (un)druggable and degradable KRAS. ACS Cent Sci 2020;6:1281–4. https://doi.org/10.1021/acscentsci.0c00920.
[186] Wloka C, Van Meervelt V, Van Gelder D, Danda N, Jager N, Williams CP, et al. Label-free and real-time detection of protein ubiquitination with a biological nanopore. ACS Nano 2017;11:4387–94. https://doi.org/10.1021/acsnano.6b07760.
[187] Hjerpe R, Aillet F, Lopitz-Otsoa F, Lang V, England P, Rodriguez MS. Efficient protection and isolation of ubiquitylated proteins using tandem ubiquitin-binding entities. EMBO Rep 2009;10:1250–8. https://doi.org/10.1038/embor.2009.192.
[188] Udeshi ND, Mertins P, Svinkina T, Carr SA. Large-scale identification of ubiquitination sites by mass spectrometry. Nat Protoc 2013;8:1950–60. https://doi.org/10.1038/nprot.2013.120.
[189] Finan C, Gaulton A, Kruger FA, Lumbers RT, Shah T, Engmann J, et al. The druggable genome and support for target identification and validation in drug development. Sci Transl Med 2017;9, eeag1166. https://doi.org/10.1126/scitranslmed.aag1166.
[190] Santos R, Ursu O, Gaulton A, Bento AP, Donadi RS, Bologa CG, et al. A comprehensive map of molecular drug targets. Nat Rev Drug Discov 2016;16:19–34. https://doi.org/10.1038/nrd.2016.230.
[191] Verma R, Mohl D, Deshaies RJ. Harnessing the power of proteolysis for targeted protein inactivation. Mol Cell 2020;77:446–60. https://doi.org/10.1016/j.molcel.2020.01.010.
[192] Li Z, Lin Y, Song H, Qin X, Yu Z, Zhang Z, et al. First small-molecule PROTACs for G protein-coupled receptors: inducing α1A-adrenergic receptor degradation. Acta Pharm Sin B 2020;10:1669–79. https://doi.org/10.1016/j.apsb.2020.01.014.
[193] Lee H, Puppala D, Choi EY, Swanson H, Kim KB. Targeted degradation of the aryl hydrocarbon receptor by the PROTAC approach: a useful chemical genetic tool. Chembiochem 2007;8:2058–62. https://doi.org/10.1002/cbic.200700438.
[194] You I, Erickson EC, Donovan KA, Eleuteri NA, Fischer ES, Gray NS, et al. Discovery of an AKT degrader with prolonged inhibition of downstream signaling. Cell Chem Biol 2020;27:66–73.e7. https://doi.org/10.1016/j.chembiol.2019.11.014.
[195] Zhang C, Han XR, Yang X, Jiang B, Liu J, Xiong Y, et al. Proteolysis targeting chimeras (PROTACs) of anaplastic lymphoma kinase (ALK). Eur J Med Chem 2018;151:304–14. https://doi.org/10.1016/j.ejmech.2018.03.071.
[196] Powell CE, Gao Y, Tan L, Donovan KA, Nowak RP, Loehr A, et al. Chemically induced degradation of anaplastic lymphoma kinase (ALK). J Med Chem 2018;61:4249–55. https://doi.org/10.1021/acs.jmedchem.7b01655.
[197] Kang CH, Lee DH, Lee CO, Du Ha J, Park CH, Hwang JY. Induced protein degradation of anaplastic lymphoma kinase (ALK) by proteolysis targeting chimera (PROTAC). Biochem Biophys Res Commun 2018;505:542–7. https://doi.org/10.1016/j.bbrc.2018.09.169.
[198] Han X, Zhao L, Xiang W, Qin C, Miao B, Xu T, et al. Discovery of highly potent and efficient PROTAC degraders of androgen receptor (AR) by employing weak binding affinity VHL E3 ligase ligands. J Med Chem 2019;62:11218–31. https://doi.org/10.1021/acs.jmedchem.9b01393.

[199] Wang R, Abdelbaki A, Ascanelli C, Fung A, Rasmusson T, Roberts K, et al. Degradation tool. BioRxiv 2020. 20200722215814 https://doi.org/10.1101/2020.07.22.215814.
[200] Wang Z, He N, Guo Z, Niu C, Song T, Guo Y, et al. Proteolysis targeting chimeras for the selective degradation of Mcl-1/Bcl-2 derived from nonselective target binding ligands. J Med Chem 2019;62:8152–63. https://doi.org/10.1021/acs.jmedchem.9b00919.
[201] McCoull W, Cheung T, Anderson E, Barton P, Burgess J, Byth K, et al. Development of a novel B-cell lymphoma 6 (BCL6) PROTAC to provide insight into small molecule targeting of BCL6. ACS Chem Biol 2018;13:3131–41. https://doi.org/10.1021/acschembio.8b00698.
[202] Shibata N, Shimokawa K, Nagai K, Ohoka N, Hattori T, Miyamoto N, et al. Pharmacological difference between degrader and inhibitor against oncogenic BCR-ABL kinase. Sci Rep 2018;8:13549. https://doi.org/10.1038/s41598-018-31913-5.
[203] Burslem GM, Schultz AR, Bondeson DP, Eide CA, Stevens SLS, Druker BJ, et al. Targeting BCR-ABL1 in chronic myeloid leukemia by PROTAC-mediated targeted protein degradation. Cancer Res 2019;79:4744–53. https://doi.org/10.1158/0008-5472.CAN-19-1236.
[204] Xue G, Chen J, Liu L, Zhou D, Zuo Y, Fu T, et al. Protein degradation through covalent inhibitor-based PROTACs. Chem Commun 2020;56:1521–4. https://doi.org/10.1039/c9cc08238g.
[205] Chen H, Chen F, Pei S, Gou S. Pomalidomide hybrids act as proteolysis targeting chimeras: synthesis, anticancer activity and B-Raf degradation. Bioorg Chem 2019;87:191–9. https://doi.org/10.1016/j.bioorg.2019.03.035.
[206] Han XR, Chen L, Wei Y, Yu W, Chen Y, Zhang C, et al. Discovery of selective small molecule degraders of BRAF-V600E. J Med Chem 2020;63:4069–80. https://doi.org/10.1021/acs.jmedchem.9b02083.
[207] Hines J, Lartigue S, Dong H, Qian Y, Crews CM. MDM2-recruiting PROTAC offers superior, synergistic antiproliferative activity via simultaneous degradation of BRD4 and stabilization of p53. Cancer Res 2019;79:251–62. https://doi.org/10.1158/0008-5472.CAN-18-2918.
[208] Tinworth CP, Lithgow H, Dittus L, Bassi ZI, Hughes SE, Muelbaier M, et al. PROTAC-mediated degradation of Bruton's tyrosine kinase is inhibited by covalent binding. ACS Chem Biol 2019;14:342–7. https://doi.org/10.1021/acschembio.8b01094.
[209] Chi J(J), Li H, Zhou Z, Izquierdo-Ferrer J, Xue Y, Wavelet CM, et al. A novel strategy to block mitotic progression for targeted therapy. EBioMedicine 2019;49:40–54. https://doi.org/10.1016/j.ebiom.2019.10.013.
[210] Zhou F, Chen L, Cao C, Yu J, Luo X, Zhou P, et al. Development of selective mono or dual PROTAC degrader probe of CDK isoforms. Eur J Med Chem 2020;187:111952. https://doi.org/10.1016/j.ejmech.2019.111952.
[211] Teng M, Jiang J, He Z, Kwiatkowski NP, Donovan KA, Mills CE, et al. Development of CDK2 and CDK5 dual degrader TMX-2172. Angew Chem Int Ed 2020;59:13865–70. https://doi.org/10.1002/anie.202004087.
[212] Jiang B, Wang ES, Donovan KA, Liang Y, Fischer ES, Zhang T, et al. Development of dual and selective degraders of cyclin-dependent kinases 4 and 6. Angew Chem Int Ed 2019;58:6321–6. https://doi.org/10.1002/anie.201901336.
[213] Hatcher JM, Wang ES, Johannessen L, Kwiatkowski N, Sim T, Gray NS. Development of highly potent and selective steroidal inhibitors and degraders of CDK8. ACS Med Chem Lett 2018;9:540–5. https://doi.org/10.1021/acsmedchemlett.8b00011.

[214] Bian J, Ren J, Li Y, Wang J, Xu X, Feng Y, et al. Discovery of Wogonin-based PROTACs against CDK9 and capable of achieving antitumor activity. Bioorg Chem 2018;81:373–81. https://doi.org/10.1016/j.bioorg.2018.08.028.

[215] Chen H, Chen F, Liu N, Wang X, Gou S. Chemically induced degradation of CK2 by proteolysis targeting chimeras based on a ubiquitin–proteasome pathway. Bioorg Chem 2018;81:536–44. https://doi.org/10.1016/j.bioorg.2018.09.005.

[216] Okuhira K, Shoda T, Omura R, Ohoka N, Hattori T, Shibata N, et al. Targeted degradation of proteins localized in subcellular compartments by hybrid small molecules. Mol Pharmacol 2017;91:159–66. https://doi.org/10.1124/mol.116.105569.

[217] Steinebach C, Lindner S, Udeshi ND, Mani DC, Kehm H, Köpff S, et al. Homo-PROTACs for the chemical knockdown of Cereblon. ACS Chem Biol 2018;13:2771–82. https://doi.org/10.1021/acschembio.8b00693.

[218] Girardini M, Maniaci C, Hughes SJ, Testa A, Ciulli A. Cereblon versus VHL: hijacking E3 ligases against each other using PROTACs. Bioorg Med Chem 2019;27:2466–79. https://doi.org/10.1016/j.bmc.2019.02.048.

[219] Steinebach C, Kehm H, Lindner S, Vu LP, Köpff S, López Mármol Á, et al. PROTAC-mediated crosstalk between E3 ligases. Chem Commun 2019;55:1821–4. https://doi.org/10.1039/c8cc09541h.

[220] Ma D, Zou Y, Chu Y, Liu Z, Liu G, Chu J, et al. A cell-permeable peptide-based PROTAC against the oncoprotein CREPT proficiently inhibits pancreatic cancer. Theranostics 2020;10:3708–21. https://doi.org/10.7150/thno.41677.

[221] Hsu JHR, Rasmusson T, Robinson J, Pachl F, Read J, Kawatkar S, et al. EED-targeted PROTACs degrade EED, EZH2, and SUZ12 in the PRC2 complex. Cell Chem Biol 2020;27:41–46.e17. https://doi.org/10.1016/j.chembiol.2019.11.004.

[222] Potjewyd F, Turner AMW, Beri J, Rectenwald JM, Norris-Drouin JL, Cholensky SH, et al. Degradation of polycomb repressive complex 2 with an EED-targeted bivalent chemical degrader. Cell Chem Biol 2020;27:47–56.e15. https://doi.org/10.1016/j.chembiol.2019.11.006.

[223] Zhang X, Xu F, Tong L, Zhang T, Xie H, Lu X, et al. Design and synthesis of selective degraders of EGFRL858R/T790M mutant. Eur J Med Chem 2020;192:112199. https://doi.org/10.1016/j.ejmech.2020.112199.

[224] Jang J, To C, De Clercq DJH, Park E, Ponthier CM, Shin BH, et al. Mutant-selective allosteric EGFR degraders are effective against a broad range of drug-resistant mutations. Angew Chem Int Ed 2020;59:14481–9. https://doi.org/10.1002/anie.202003500.

[225] Hu J, Hu B, Wang M, Xu F, Miao B, Yang CY, et al. Discovery of ERD-308 as a highly potent proteolysis targeting chimera (PROTAC) degrader of estrogen receptor (ER). J Med Chem 2019;62:1420–42. https://doi.org/10.1021/acs.jmedchem.8b01572.

[226] Roberts BL, Ma ZX, Gao A, Leisten ED, Yin D, Xu W, et al. Two-stage strategy for development of proteolysis targeting chimeras and its application for estrogen receptor degraders. ACS Chem Biol 2020;15:1487–96. https://doi.org/10.1021/acschembio.0c00140.

[227] Lebraud H, Wright DJ, Johnson CN, Heightman TD. Protein degradation by in-cell self-assembly of proteolysis targeting chimeras. ACS Cent Sci 2016;2:927–34. https://doi.org/10.1021/acscentsci.6b00280.

[228] Peng L, Zhang Z, Lei C, Li S, Zhang Z, Ren X, et al. Identification of new small-molecule inducers of estrogen-related receptor α (ERRα) degradation. ACS Med Chem Lett 2019;10:767–72. https://doi.org/10.1021/acsmedchemlett.9b00025.

[229] Popow J, Arnhof H, Bader G, Berger H, Ciulli A, Covini D, et al. Highly selective PTK2 proteolysis targeting chimeras to probe focal adhesion kinase scaffolding functions. J Med Chem 2019;62:2508–20. https://doi.org/10.1021/acs.jmedchem.8b01826.

[230] Burslem GM, Song J, Chen X, Hines J, Crews CM. Enhancing antiproliferative activity and selectivity of a FLT-3 inhibitor by proteolysis targeting chimera conversion. J Am Chem Soc 2018;140:16428–32. https://doi.org/10.1021/jacs.8b10320.

[231] Hines J, Gough JD, Corson TW, Crews CM. Posttranslational protein knockdown coupled to receptor tyrosine kinase activation with phosphoPROTACs. Proc Natl Acad Sci U S A 2013;110:8942–7. https://doi.org/10.1073/pnas.1217206110.

[232] Smalley JP, Adams GE, Millard CJ, Song Y, Norris JKS, Schwabe JWR, et al. PROTAC-mediated degradation of class i histone deacetylase enzymes in corepressor complexes. Chem Commun 2020;56:4476–9. https://doi.org/10.1039/d0cc01485k.

[233] Yang K, Wu H, Zhang Z, Leisten ED, Nie X, Liu B, et al. Development of selective histone deacetylase 6 (HDAC6) degraders recruiting von Hippel-Lindau (VHL) E3 ubiquitin ligase. ACS Med Chem Lett 2020;11:575–81. https://doi.org/10.1021/acsmedchemlett.0c00046.

[234] Li MX, Yang Y, Zhao Q, Wu Y, Song L, Yang H, et al. Degradation versus inhibition: development of proteolysis-targeting chimeras for overcoming statin-induced compensatory upregulation of 3-hydroxy-3-methylglutaryl coenzyme a reductase. J Med Chem 2020;63:4908–28. https://doi.org/10.1021/acs.jmedchem.0c00339.

[235] Tomoshige S, Nomura S, Ohgane K, Hashimoto Y, Ishikawa M. Discovery of small molecules that induce the degradation of huntingtin. Angew Chem 2017;129: 11688–91. https://doi.org/10.1002/ange.201706529.

[236] Nunes J, McGonagle GA, Eden J, Kiritharan G, Touzet M, Lewell X, et al. Targeting IRAK4 for degradation with PROTACs. ACS Med Chem Lett 2019;10:1081–5. https://doi.org/10.1021/acsmedchemlett.9b00219.

[237] Shah RR, Redmond JM, Mihut A, Menon M, Evans JP, Murphy JA, et al. Hi-JAK-ing the ubiquitin system: the design and physicochemical optimisation of JAK PROTACs. Bioorg Med Chem 2020;28:115326. https://doi.org/10.1016/j.bmc.2020.115326.

[238] Bond MJ, Chu L, Nalawansha DA, Li K, Crews CM. Targeted degradation of oncogenic KRASG12Cby VHL-recruiting PROTACs. ACS Cent Sci 2020;6:1367–75. https://doi.org/10.1021/acscentsci.0c00411.

[239] Kargbo RB. Degradation of LRRK2 in the treatment of Parkinson's disease. ACS Med Chem Lett 2020;11:2070–1. https://doi.org/10.1021/acsmedchemlett.0c00453.

[240] Donoghue C, Cubillos-Rojas M, Gutierrez-Prat N, Sanchez-Zarzalejo C, Verdaguer X, Riera A, et al. Optimal linker length for small molecule PROTACs that selectively target p38α and p38β for degradation. Eur J Med Chem 2020;201:112451. https://doi.org/10.1016/j.ejmech.2020.112451.

[241] Li Y, Yang J, Aguilar A, McEachern D, Przybranowski S, Liu L, et al. Discovery of MD-224 as a first-in-class, highly potent, and efficacious proteolysis targeting chimera murine double minute 2 degrader capable of achieving complete and durable tumor regression. J Med Chem 2019;62:448–66. https://doi.org/10.1021/acs.jmedchem.8b00909.

[242] Wang B, Wu S, Liu J, Yang K, Xie H, Tang W. Development of selective small molecule MDM2 degraders based on nutlin. Eur J Med Chem 2019;176:476–91. https://doi.org/10.1016/j.ejmech.2019.05.046.

[243] de Wispelaere M, Du G, Donovan KA, Zhang T, Eleuteri NA, Yuan JC, et al. Small molecule degraders of the hepatitis C virus protease reduce susceptibility to resistance mutations. Nat Commun 2019;10:3468. https://doi.org/10.1038/s41467-019-11429-w.

[244] Zhao Q, Lan T, Su S, Rao Y. Induction of apoptosis in MDA-MB-231 breast cancer cells by a PARP1-targeting PROTAC small molecule. Chem Commun 2019;55: 369–72. https://doi.org/10.1039/c8cc07813k.

[245] Wang S, Han L, Han J, Li P, Ding Q, Zhang QJ, et al. Uncoupling of PARP1 trapping and inhibition using selective PARP1 degradation. Nat Chem Biol 2019;15:1223–31. https://doi.org/10.1038/s41589-019-0379-2.

[246] Bassi ZI, Fillmore MC, Miah AH, Chapman TD, Maller C, Roberts EJ, et al. Modulating PCAF/GCN5 immune cell function through a PROTAC approach. ACS Chem Biol 2018;13:2862–7. https://doi.org/10.1021/acschembio.8b00705.
[247] Li W, Gao C, Zhao L, Yuan Z, Chen Y, Jiang Y. Phthalimide conjugations for the degradation of oncogenic PI3K. Eur J Med Chem 2018;151:237–47. https://doi.org/10.1016/j.ejmech.2018.03.066.
[248] Mu X, Bai L, Xu Y, Wang J, Lu H. Protein targeting chimeric molecules specific for dual bromodomain 4 (BRD4) and polo-like kinase 1 (PLK1) proteins in acute myeloid leukemia cells. Biochem Biophys Res Commun 2020;521:833–9. https://doi.org/10.1016/j.bbrc.2019.11.007.
[249] Song Y, Park PMC, Wu L, Ray A, Picaud S, Li D, et al. Development and preclinical validation of a novel covalent ubiquitin receptor Rpn13 degrader in multiple myeloma. Leukemia 2019;33:2685–94. https://doi.org/10.1038/s41375-019-0467-z.
[250] Tovell H, Testa A, Zhou H, Shpiro N, Crafter C, Ciulli A, et al. Design and characterization of SGK3-PROTAC1, an isoform specific SGK3 kinase PROTAC degrader. ACS Chem Biol 2019;14:2024–34. https://doi.org/10.1021/acschembio.9b00505.
[251] Wang M, Lu J, Wang M, Yang CY, Wang S. Discovery of SHP2-D26 as a first, potent, and effective PROTAC degrader of SHP2 protein. J Med Chem 2020;63:7510–28. https://doi.org/10.1021/acs.jmedchem.0c00471.
[252] Schiedel M, Herp D, Hammelmann S, Swyter S, Lehotzky A, Robaa D, et al. Chemically induced degradation of Sirtuin 2 (Sirt2) by a proteolysis targeting chimera (PROTAC) based on Sirtuin rearranging ligands (SirReals). J Med Chem 2018;61:482–91. https://doi.org/10.1021/acs.jmedchem.6b01872.
[253] Bensimon A, Pizzagalli MD, Kartnig F, Dvorak V, Essletzbichler P, Winter GE, et al. Targeted degradation of SLC transporters reveals amenability of multi-pass transmembrane proteins to ligand-induced proteolysis. Cell Chem Biol 2020;27:728–739.e9. https://doi.org/10.1016/j.chembiol.2020.04.003.
[254] Wang X, Feng S, Fan J, Li X, Wen Q, Luo N. New strategy for renal fibrosis: targeting Smad3 proteins for ubiquitination and degradation. Biochem Pharmacol 2016;116:200–9. https://doi.org/10.1016/j.bcp.2016.07.017.
[255] Kargbo RB. PROTAC compounds targeting α-synuclein protein for treating neurogenerative disorders: Alzheimer's and Parkinson's diseases. ACS Med Chem Lett 2020;11:1086–7. https://doi.org/10.1021/acsmedchemlett.0c00192.
[256] Zhou H, Bai L, Xu R, Zhao Y, Chen J, McEachern D, et al. Structure-based discovery of SD-36 as a potent, selective, and efficacious PROTAC degrader of STAT3 protein. J Med Chem 2019;62:11280–300. https://doi.org/10.1021/acs.jmedchem.9b01530.
[257] Ohoka N, Nagai K, Shibata N, Hattori T, Nara H, Cho N, et al. SNIPER(TACC3) induces cytoplasmic vacuolization and sensitizes cancer cells to Bortezomib. Cancer Sci 2017;108:1032–41. https://doi.org/10.1111/cas.13198.
[258] Silva MC, Ferguson FM, Cai Q, Donovan KA, Nandi G, Patnaik D, et al. Targeted degradation of aberrant tau in frontotemporal dementia patient-derived neuronal cell models. elife 2019;8, e45457. https://doi.org/10.7554/eLife.45457.
[259] Crew AP, Raina K, Dong H, Qian Y, Wang J, Vigil D, et al. Identification and characterization of von Hippel-Lindau-recruiting proteolysis targeting chimeras (PROTACs) of TANK-binding kinase 1. J Med Chem 2018;61:583–98. https://doi.org/10.1021/acs.jmedchem.7b00635.
[260] Gechijian LN, Buckley DL, Lawlor MA, Reyes JM, Paulk J, Ott CJ, et al. Functional TRIM24 degrader via conjugation of ineffectual bromodomain and VHL ligands article. Nat Chem Biol 2018;14:405–12. https://doi.org/10.1038/s41589-018-0010-y.
[261] Zhao B, Burgess K. TrkC-targeted kinase inhibitors and PROTACs. Mol Pharm 2019;16:4313–8. https://doi.org/10.1021/acs.molpharmaceut.9b00673.

[262] Li Z, Pinch BJ, Olson CM, Donovan KA, Nowak RP, Mills CE, et al. Development and characterization of a wee1 kinase degrader. Cell Chem Biol 2020;27:57–65.e9. https://doi.org/10.1016/j.chembiol.2019.10.013.
[263] Montrose K, Krissansen GW. Design of a PROTAC that antagonizes and destroys the cancer-forming X-protein of the hepatitis B virus. Biochem Biophys Res Commun 2014;453:735–40. https://doi.org/10.1016/j.bbrc.2014.10.006.
[264] Demizu Y, Okuhira K, Motoi H, Ohno A, Shoda T, Fukuhara K, et al. Design and synthesis of estrogen receptor degradation inducer based on a protein knockdown strategy. Bioorg Med Chem Lett 2012;22:1793–6. https://doi.org/10.1016/j.bmcl.2011.11.086.
[265] Weng G, Shen C, Cao D, Gao J, Dong X, He Q, et al. PROTAC-DB: an online database of PROTACs. Nucleic Acids Res 2020;49:D1381–7. https://doi.org/10.1093/nar/gkaa807.
[266] Li X, Song Y. Proteolysis-targeting chimera (PROTAC) for targeted protein degradation and cancer therapy. J Hematol Oncol 2020;13:50. https://doi.org/10.1186/s13045-020-00885-3.
[267] Backus KM, Correia BE, Lum KM, Forli S, Horning BD, González-Páez GE, et al. Proteome-wide covalent ligand discovery in native biological systems. Nature 2016;534:570–4. https://doi.org/10.1038/nature18002.
[268] Tong B, Spradlin JN, Novaes LFT, Zhang E, Hu X, Moeller M, et al. A Nimbolide-based kinase degrader preferentially degrades oncogenic BCR-ABL. ACS Chem Biol 2020;15:1788–94. https://doi.org/10.1021/acschembio.0c00348.
[269] Słabicki M, Kozicka Z, Petzold G, Der Li Y, Manojkumar M, Bunker RD, et al. The CDK inhibitor CR8 acts as a molecular glue degrader that depletes cyclin K. Nature 2020;585:293–7. https://doi.org/10.1038/s41586-020-2374-x.
[270] Arora L, Kumar AP, Arfuso F, Chng WJ, Sethi G. The role of signal transducer and activator of transcription 3 (STAT3) and its targeted inhibition in hematological malignancies. Cancers (Basel) 2018;10:327. https://doi.org/10.3390/cancers10090327.
[271] Goedert M. Tau protein and neurodegeneration. In: Seminars in cell & developmental biology, vol. 15. Elsevier Ltd; 2004. p. 45–9. https://doi.org/10.1016/j.semcdb.2003.12.015.
[272] Paul SM, Mytelka DS, Dunwiddie CT, Persinger CC, Munos BH, Lindborg SR, et al. How to improve RD productivity: the pharmaceutical industry's grand challenge. Nat Rev Drug Discov 2010;9:203–14. https://doi.org/10.1038/nrd3078.
[273] Gao H, Sun X, Rao Y. PROTAC technology: opportunities and challenges. ACS Med Chem Lett 2020;11:237–40. https://doi.org/10.1021/acsmedchemlett.9b00597.
[274] Cantrill C, Chaturvedi P, Rynn C, Petrig Schaffland J, Walter I, Wittwer MB. Fundamental aspects of DMPK optimization of targeted protein degraders. Drug Discov Today 2020;25:969–82. https://doi.org/10.1016/j.drudis.2020.03.012.
[275] Pike A, Williamson B, Harlfinger S, Martin S, McGinnity DF. Optimising proteolysis-targeting chimeras (PROTACs) for oral drug delivery: a drug metabolism and pharmacokinetics perspective. Drug Discov Today 2020;25:1793–800. https://doi.org/10.1016/j.drudis.2020.07.013.
[276] Kola I, Landis J. Can the pharmaceutical industry reduce attrition rates? Nat Rev Drug Discov 2004;3:711–5. https://doi.org/10.1038/nrd1470.
[277] Abad-Zapatero C. A Sorcerer's apprentice and the rule of five: from rule-of-thumb to commandment and beyond. Drug Discov Today 2007;12:995–7. https://doi.org/10.1016/j.drudis.2007.10.022.
[278] Edmondson SD, Yang B, Fallan C. Proteolysis targeting chimeras (PROTACs)in 'beyond rule-of-five' chemical space: recent progress and future challenges. Bioorg Med Chem Lett 2019;29:1555–64. https://doi.org/10.1016/j.bmcl.2019.04.030.

[279] Wei M, Zhao R, Cao Y, Wei Y, Li M, Dong Z, et al. First orally bioavailable prodrug of proteolysis targeting chimera (PROTAC) degrades cyclin-dependent kinases 2/4/6 in vivo. Eur J Med Chem 2021;209:112903. https://doi.org/10.1016/j.ejmech.2020.112903.
[280] Waters NJ, Jones R, Williams G, Sohal B. Validation of a rapid equilibrium dialysis approach for the measurement of plasma protein binding. J Pharm Sci 2008;97:4586–95. https://doi.org/10.1002/jps.21317.
[281] Watt GF, Scott-Stevens P, Gaohua L. Targeted protein degradation in vivo with proteolysis targeting chimeras: current status and future considerations. Drug Discov Today Technol 2019;31:69–80. https://doi.org/10.1016/j.ddtec.2019.02.005.
[282] Dobrovolsky D, Wang ES, Morrow S, Leahy C, Faust T, Nowak RP, et al. Bruton tyrosine kinase degradation as a therapeutic strategy for cancer. Blood 2019;133:952–61. https://doi.org/10.1182/blood-2018-07-862953.
[283] ClinicalTrials.gov. A phase 1 clinical trial of ARV-110 in patients with metastatic castration-resistant prostate cancer. ClinicalTrials.gov; 2019. https://clinicaltrials.gov/ct2/show/NCT03888612. [Accessed 10 September 2020].
[284] Benchekroun M. The advent of directed protein degraders in drug discovery. Futur Drug Discov 2019;1:FDD16. https://doi.org/10.4155/fdd-2019-0019.
[285] Mullard A. Arvinas's PROTACs pass first safety and PK analysis. Nat Rev Drug Discov 2019;18:895.
[286] Cacace A. Discovery of brain penetrant PROTAC® degrader molecules that target pathologic tau and alpha-Synuclein protein species. In: Target. Protein Degrad. Summit; 2019.
[287] Wang P, Zhou J. Proteolysis targeting chimera (PROTAC): a paradigm-shifting approach in small molecule drug discovery. Curr Top Med Chem 2018;18:1354–6. https://doi.org/10.2174/1568026618666181010101922.
[288] Khongorzul P, Ling CJ, Khan FU, Ihsan AU, Zhang J. Antibody-drug conjugates: a comprehensive review. Mol Cancer Res 2020;18:3–19. https://doi.org/10.1158/1541-7786.MCR-19-0582.
[289] Birrer MJ, Moore KN, Betella I, Bates RC. Antibody-drug conjugate-based therapeutics: state of the science. J Natl Cancer Inst 2019;111:538–49.
[290] Teicher BA, Chari RVJ. Antibody conjugate therapeutics: challenges and potential. Clin Cancer Res 2011;17:6389–97. https://doi.org/10.1158/1078-0432.CCR-11-1417.
[291] Lu J, Jiang F, Lu A, Zhang G. Linkers having a crucial role in antibody–drug conjugates. Int J Mol Sci 2016;17:561. https://doi.org/10.3390/ijms17040561.
[292] Dragovich PS, Adhikari P, Blake RA, Blaquiere N, Chen J, Cheng YX, et al. Antibody-mediated delivery of chimeric protein degraders which target estrogen receptor alpha (ERα). Bioorg Med Chem Lett 2020;30:126907. https://doi.org/10.1016/j.bmcl.2019.126907.
[293] Dragovich PS, Blake RA, Chen C, Chen J, Chuh J, den Besten W, et al. Conjugation of indoles to antibodies through a novel self-immolating linker. Chem Eur J 2018;24:4830–4. https://doi.org/10.1002/chem.201800859.
[294] Manna D, Verma S. Controlling PROTACs with light. ChemMedChem 2020;15:1–5. https://doi.org/10.1002/cmdc.202000249.
[295] Velema WA, Szymanski W, Feringa BL. Photopharmacology: beyond proof of principle. J Am Chem Soc 2014;136:2178–91. https://doi.org/10.1021/ja413063e.
[296] Broichhagen J, Frank JA, Trauner D. A roadmap to success in Photopharmacology. Acc Chem Res 2015;48:1947–60. https://doi.org/10.1021/acs.accounts.5b00129.
[297] Hüll K, Morstein J, Trauner D. In vivo photopharmacology. Chem Rev 2018;118:10710–47. https://doi.org/10.1021/acs.chemrev.8b00037.

[298] Mayer G, Hechel A. Biologically active molecules with a "light switch". Angew Chem Int Ed 2006;45:4900–21. https://doi.org/10.1002/anie.200600387.
[299] Agostinis P, Berg K, Cengel K, et al. Photodynamic therapy of cancer: an update. CA Cancer J Clin 2017;61:250–81. https://doi.org/10.3322/caac.20114.Available.
[300] Shafirstein G, Battoo A, Harris K, Baumann H, Gollnick SO, Lindenmann J, et al. Photodynamic therapy of non-small cell lung cancer narrative review and future directions. Ann Am Thorac Soc 2016;13:265–75. https://doi.org/10.1513/AnnalsATS.201509-650FR.
[301] Moore CM, Pendse D, Emberton M. Photodynamic therapy for prostate cancer - a review of current status and future promise. Nat Clin Pract Urol 2009;6:18–30. https://doi.org/10.1038/ncpuro1274.
[302] Kunzi-Rapp K. Photodynamic therapy in dermatology. Photon Laser Med 2015;4:378–9. https://doi.org/10.1201/b15582-45.
[303] Kounde CS, Tate EW. Photoactive bifunctional degraders: precision tools to regulate protein stability. J Med Chem 2020;63:15483–93. https://doi.org/10.1021/acs.jmedchem.0c01542.
[304] Fuchter MJ. On the promise of photopharmacology using photoswitches: a medicinal chemist's perspective. J Med Chem 2020;63:11436–47. https://doi.org/10.1021/acs.jmedchem.0c00629.
[305] Liu J, Chen H, Ma L, He Z, Wang D, Liu Y, et al. Light-induced control of protein destruction by opto-PROTAC. Sci Adv 2020;6, eaay5154. https://doi.org/10.1126/sciadv.aay5154.
[306] Beharry AA, Sadovski O, Woolley GA. Azobenzene photoswitching without ultraviolet light. J Am Chem Soc 2011;133:19684–7. https://doi.org/10.1021/ja209239m.
[307] Samanta S, Beharry AA, Sadovski O, McCormick TM, Babalhavaeji A, Tropepe V, et al. Photoswitching Azo compounds in vivo with red light. J Am Chem Soc 2013;135:9777–84. https://doi.org/10.1021/ja402220t.
[308] Matsumura Y, Ananthaswamy HN. Toxic effects of ultraviolet radiation on the skin. Toxicol Appl Pharmacol 2004;195:298–308. https://doi.org/10.1016/j.taap.2003.08.019.
[309] Weissleder R. A clearer vision for in vivo imaging Progress continues in the development of smaller, more penetrable probes for biological imaging. Nat Biotechnol 2001;19:316–7.
[310] Alabugin A. Near-IR photochemistry for biology: exploiting the optical window of tissue. Photochem Photobiol 2019;95:722–32. https://doi.org/10.1111/php.13068.
[311] Yu L, Chen Y, Tooze SA. Autophagy pathway: cellular and molecular mechanisms. Autophagy 2018;14:207–15. https://doi.org/10.1080/15548627.2017.1378838.
[312] Gruenberg J. The endocytic pathway: a mosaic of domains. Nat Rev Mol Cell Biol 2001;2:721–30.
[313] Yin Z, Popelka H, Lei Y, Yang Y. The roles of ubiquitin in mediating autophagy. Cell 2020;9:2025–56.
[314] Ding Y, Fei Y, Lu B. Emerging new concepts of degrader technologies. Trends Pharmacol Sci 2020;41:464–74. https://doi.org/10.1016/j.tips.2020.04.005.
[315] Lim S, Khoo R, Peh KM, Teo J, Chang SC, Ng S, et al. BioPROTACs as versatile modulators of intracellular therapeutic targets including proliferating cell nuclear antigen (PCNA). Proc Natl Acad Sci U S A 2020;117:5791–800. https://doi.org/10.1073/pnas.1920251117.
[316] Ma Y, Gu Y, Zhang Q, Han Y, Yu S, Lu Z, et al. Targeted degradation of KRAS by an engineered ubiquitin ligase suppresses pancreatic cancer cell growth in vitro and in vivo. Mol Cancer Ther 2013;12:286–94. https://doi.org/10.1158/1535-7163.MCT-12-0650.

[317] Hatakeyama S, Watanabe M, Fujii Y, Nakayama KI. Targeted destruction of c-Myc by an engineered ubiquitin ligase suppresses cell transformation and tumor formation. Cancer Res 2005;65:7874–9. https://doi.org/10.1158/0008-5472.CAN-05-1581.

[318] Ghosh P, Dahms NM, Kornfeld S. Mannose 6-phosphate receptors: new twists in the tale. Nat Rev Mol Cell Biol 2003;4:202–12. https://doi.org/10.1038/nrm1050.

[319] Lycia. Therapeutics uncloaks with $50M to tackle protein degradation in a new way. FierceBiotech; 2020. https://www.fiercebiotech.com/biotech/lycia-therapeutics-uncloaks-50m-to-tackle-protein-degradation-a-new-way. [Accessed 18 June 2020].

[320] Feng Y, He D, Yao Z, Klionsky DJ. The machinery of macroautophagy. Cell Res 2014;24:24–41. https://doi.org/10.1038/cr.2013.168.

[321] Filimonenko M, Isakson P, Finley KD, Anderson M, Jeong H, Melia TJ, et al. The selective macroautophagic degradation of aggregated proteins requires the PI3P-binding protein Alfy. Mol Cell 2010;38:265–79. https://doi.org/10.1016/j.molcel.2010.04.007.

[322] Johansen T, Lamark T. Selective autophagy mediated by autophagic adapter proteins. Autophagy 2011;7:279–96. https://doi.org/10.4161/auto.7.3.14487.

[323] Kirkin V, McEwan DG, Novak I, Dikic I. A role for ubiquitin in selective autophagy. Mol Cell 2009;34:259–69. https://doi.org/10.1016/j.molcel.2009.04.026.

[324] Takahashi D, Arimoto H. Targeting selective autophagy by AUTAC degraders. Autophagy 2020;16:765–6. https://doi.org/10.1080/15548627.2020.1718362.

[325] Takahashi D, Moriyama J, Nakamura T, Akaike T, Itto-Nakama K, Arimoto H. AUTACs: cargo-specific degraders using selective autophagy. Mol Cell 2019;76:797–810.e10. https://doi.org/10.1016/j.molcel.2019.09.009.

[326] Sawa T, Zaki MH, Okamoto T, Akuta T, Tokutomi Y, Kim-Mitsuyama S, et al. Protein S-guanylation by the biological signal 8-nitroguanosine 3′,5′-cyclic monophosphate. Nat Chem Biol 2007;3:727–35. https://doi.org/10.1038/nchembio.2007.33.

[327] Li Z, Wang C, Wang Z, Zhu C, Li J, Sha T, et al. Allele-selective lowering of mutant HTT protein by HTT–LC3 linker compounds. Nature 2019;575:203–9. https://doi.org/10.1038/s41586-019-1722-1.

[328] Li Z, Zhu C, Ding Y, Fei Y, Lu B. ATTEC: a potential new approach to target proteinopathies. Autophagy 2020;16:185–7. https://doi.org/10.1080/15548627.2019.1688556.

[329] Kabeya Y, Mizushima N, Ueno T, Yamamoto A, Kirisako T, Noda T, et al. LC3, a mammalian homolog of yeast Apg8p, is localized in autophagosome membranes after processing. EMBO J 2003;22:4577. https://doi.org/10.1093/emboj/cdg454.

[330] Scherzinger E, Lurz R, Turmaine M, Mangiarini L, Hollenbach B, Hasenbank R, et al. Huntingtin-encoded polyglutamine expansions form amyloid-like protein aggregates in vitro and in vivo. Cell 1997;90:549–58. https://doi.org/10.1016/S0092-8674(00)80514-0.

[331] Soto C, Pritzkow S. Protein misfolding, aggregation, and conformational strains in neurodegenerative diseases. Nat Neurosci 2018;21:1332–40. https://doi.org/10.1038/s41593-018-0235-9.

[332] Pardridge WM. Drug transport across the blood-brain barrier. J Cereb Blood Flow Metab 2012;32:1959–72. https://doi.org/10.1038/jcbfm.2012.126.

[333] Good MC, Zalatan JG, Lim WA. Scaffold proteins: hubs for controlling the flow of cellular information. Science 2011;332:680–6. https://doi.org/10.1126/science.1198701.

[334] Gerry CJ, Schreiber SL. Unifying principles of bifunctional, proximity-inducing small molecules. Nat Chem Biol 2020;16:369–78. https://doi.org/10.1038/s41589-020-0469-1.

[335] Stanton BZ, Chory EJ, Crabtree GR. Chemically induced proximity in biology and medicine. Science 2018;359, eaao5902. https://doi.org/10.1126/science.aao5902.
[336] Nakai K. Controlling signal transduction with synthetic ligands. Tanpakushitsu Kakusan Koso 2007;52:1794–5.
[337] Wei LH, Guo JU. Coding functions of "noncoding" RNAs. Science 2020; 367:1074–5. https://doi.org/10.1126/science.aba5319.
[338] Cooper TA, Wan L, Dreyfuss G. RNA and disease. Cell 2009;136:777–93. https://doi.org/10.1016/j.cell.2009.02.011.
[339] Yin W, Rogge M. Targeting RNA: a transformative therapeutic strategy. Clin Transl Sci 2019;12:98–112. https://doi.org/10.1111/cts.12624.
[340] Dey SK, Jaffrey SR. RIBOTACs: small molecules target RNA for degradation. Cell Chem Biol 2019;26:1047–9. https://doi.org/10.1016/j.chembiol.2019.07.015.
[341] Disney MD, Winkelsas AM, Velagapudi SP, Southern M, Fallahi M, Childs-Disney JL. Inforna 2.0: a platform for the sequence-based design of small molecules targeting structured RNAs. ACS Chem Biol 2016;11:1720–8. https://doi.org/10.1021/acschembio.6b00001.
[342] Silverman RH. Viral encounters with 2′,5′-oligoadenylate synthetase and RNase L during the interferon antiviral response. J Virol 2007;81:12720–9. https://doi.org/10.1128/jvi.01471-07.
[343] Velagapudi SP, Cameron MD, Haga CL, Rosenberg LH, Lafitte M, Duckett DR, et al. Design of a small molecule against an oncogenic noncoding RNA. Proc Natl Acad Sci U S A 2016;113:5898–903. https://doi.org/10.1073/pnas.1523975113.
[344] Costales MG, Matsumoto Y, Velagapudi SP, Disney MD. Small molecule targeted recruitment of a nuclease to RNA. J Am Chem Soc 2018;140:6741–4. https://doi.org/10.1021/jacs.8b01233.
[345] Costales MG, Suresh B, Vishnu K, Disney MD. Targeted degradation of a hypoxia-associated non-coding RNA enhances the selectivity of a small molecule interacting with RNA. Cell Chem Biol 2019;26:1180–1186.e5. https://doi.org/10.1016/j.chembiol.2019.04.008.
[346] Costales MG, Aikawa H, Li Y, Childs-Disney JL, Abegg D, Hoch DG, et al. Small-molecule targeted recruitment of a nuclease to cleave an oncogenic RNA in a mouse model of metastatic cancer. Proc Natl Acad Sci U S A 2020;117:2406–11. https://doi.org/10.1073/pnas.1914286117.
[347] Lanyon-Hogg T, Faronato M, Serwa RA, Tate EW. Dynamic protein acylation: new substrates, mechanisms, and drug targets. Trends Biochem Sci 2017;42:566–81. https://doi.org/10.1016/j.tibs.2017.04.004.
[348] Walsh CT. Posttranslational modification of proteins: expanding nature's inventory. Englewood, Colorado: Roberts and Company; 2006.
[349] Ciechanover A. The unravelling of the ubiquitin system. Nat Rev Mol Cell Biol 2015;16:322–4. https://doi.org/10.1038/nrm3982.
[350] Conway SJ. Bifunctional molecules beyond PROTACs. J Med Chem 2020;63: 2802–6. https://doi.org/10.1021/acs.jmedchem.0c00293.
[351] Yamazoe S, Tom J, Fu Y, Wu W, Zeng L, Sun C, et al. Heterobifunctional molecules induce dephosphorylation of kinases-a proof of concept study. J Med Chem 2020;63:2807–13. https://doi.org/10.1021/acs.jmedchem.9b01167.
[352] Tappan E, Chamberlin AR. Activation of protein phosphatase 1 by a small molecule designed to bind to the enzyme's regulatory site. Chem Biol 2008;15:167–74. https://doi.org/10.1016/j.chembiol.2008.01.005.
[353] Siriwardena SU, Munkanatta Godage DNP, Shoba VM, Lai S, Shi M, Wu P, et al. Phosphorylation-inducing chimeric small molecules. J Am Chem Soc 2020;142: 14052–7. https://doi.org/10.1021/jacs.0c05537.

CHAPTER THREE

Automated and enabling technologies for medicinal chemistry

Paula L. Nichols*
Synple Chem AG, Kemptthal, Switzerland
ETH, Zurich, Switzerland
*Corresponding author: e-mail address: nichols@synplechem.com

Contents

Keywords: Automation, Medicinal chemistry, Enabling tools, Synthesis, Synthesis technology, Automated synthesis platforms, Closed-loop drug discovery, Drug discovery, Artificial intelligence, Machine learning

Progress in Medicinal Chemistry, Volume 60
ISSN 0079-6468
https://doi.org/10.1016/bs.pmch.2021.01.003

1. Introduction

The discovery of new drugs is a constantly evolving field, which has always been driven by the need to get more drugs to patients, as quickly as possible. Ever since it was first established, medicinal chemistry has played an important role in the discovery of new therapeutic treatments. Over the years the industry has developed and changed, in part due to the emergence of new enabling technologies. Automating synthetic chemistry, including synthesis, purification and analysis, is attractive for many reasons. Such technologies have the potential to be both time- and resource-efficient by minimising the amount of manual labour and raw materials required, and can result in significantly enhanced productivity when used appropriately. In addition, such technologies often offer both environmental and safety benefits, since human interaction is reduced and chemical processes can be more reliably controlled.

Much progress has been made since the development of the early synthesis, purification and analysis technologies reviewed in this chapter, but there is still much more potential. While these technologies laid the foundations for the automation and enabling tools that are available today, and have undoubtedly impacted medicinal chemistry, it is evident that in this ever-evolving field, most of the current automated synthesis technologies still have very clear limitations. However, enabled by recent developments in synthesis technologies and an increase in computing power, automation in medicinal chemistry now has the potential to cause a step change in how we make new molecules, and in the role that medicinal chemists play in this process.

Looking back, natural substances have been used for their medicinal effects for thousands of years and the field of medicinal chemistry has its roots in naturally derived substances. Following developments in organic chemistry in the latter half of the 19th century, which enabled the identification of the active ingredients in many of traditional remedies, such as quinine from cinchona bark [1], medicinal chemistry became primarily focused on synthetic materials with therapeutic properties. This increased understanding, and the availability of new synthetic techniques, enabled researchers to start making fully synthetic, pure versions of these natural products. However, it was not until the mid-1900s that researchers started making and fully exploring structurally related analogues of medicinal natural products. The ability to systematically make and explore structurally related analogues led to an increased understanding of how the structure of different molecules impacts their therapeutic effects, enabled medicinal chemists to design even more potent analogues, and ultimately led to the establishment of medicinal chemistry as we know it today.

It was evident, even from a relatively early stage, that by making more molecules, it was possible to improve the chance of finding a better drug candidate. As such, there was a desire to make a larger number of compounds in as short a time frame as possible. Despite this significant shift in thinking, and the subsequent evolution of the drug discovery process, during the last century synthetic chemistry has not really evolved much and many of the synthetic processes utilised by medicinal chemists today have not significantly changed. As such, synthesis can still today constitute a significant bottleneck in drug discovery.

It is widely acknowledged that all areas of drug discovery and development need to seek new ways to increase productivity without incurring significant costs. In 2010 an analysis revealed that the pharmaceutical industry was suffering severely from rising costs, a lack of innovation and a poor reputation [2]. At the time, the primary focus was reducing clinical development costs, which accounted for a greater proportion of the costs for launching a new drug than preclinical drug discovery (Fig. 1).

However, even in 2010, it was estimated that preclinical discovery (not including target validation) accounted for approximately 46% of the almost $1.8 billion (capitalised costs) required to launch one new molecular entity (NME) per year, with the average target to hit, hit to lead, and lead optimisation phases costing $94, $166, and $414 million respectively [2]. In the decade that has followed this analysis, R&D costs have increased substantially, with the top 20 pharmaceutical companies believed to be spending a combined total of around $60 billion on drug development each year, making the estimated average cost of bringing a drug to market (including drug failures) in the region of $2.6 billion—a 140% increase [3]. Now more than ever, there is a real need to increase innovation, manage costs and do more, faster at every single stage of drug discovery and development.

The drug discovery process typically requires an extensive series of iterative *Design-Make-Test-Analyse* cycles, in which new molecules are made that have small incremental structural changes compared to a lead structure (Fig. 2). The effect of these changes is then measured in relation to biological activity. The knowledge gained is subsequently utilised to design the next iteration of molecules. All of these cycle iterations are done with the ultimate goal of identifying a drug candidate with the most optimal combination of

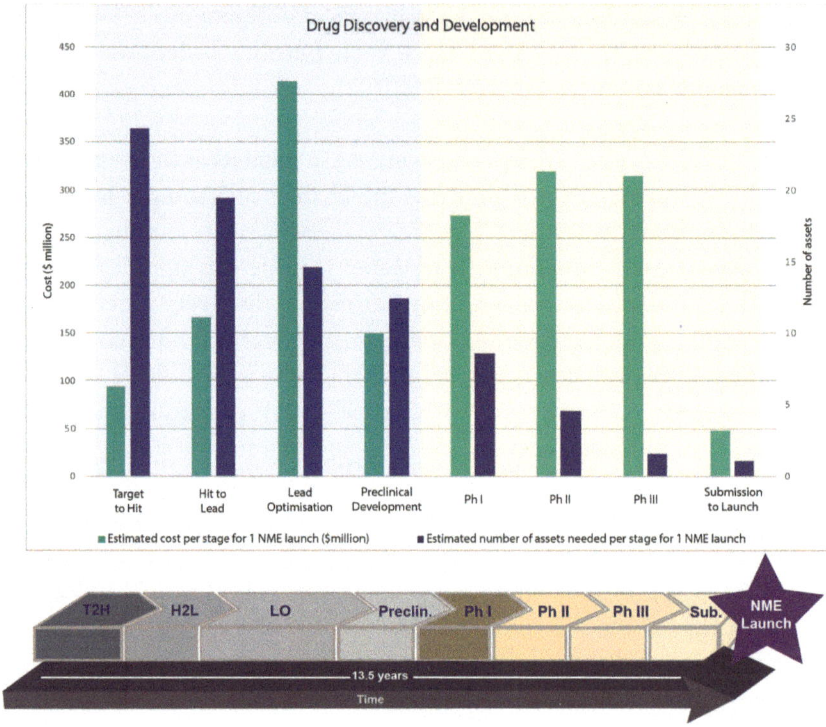

Fig. 1 Modelled R&D costs in 2010 for the successful discovery and development of a single new molecular entity [2]. The model, based on industry benchmarks and data from Eli Lilly and Company, defines the performance of R&D at each stage of drug discovery and development and estimates the number of assets or projects required in each stage, and the capitalised costs* for each stage, which yield a total cost of $1778 million for the launch of one New Molecular Entity (NME) per year (not including the earliest phase of discovery research prior to target selection). *Capitalised costs include out of pocket costs, which includes the cost of failures and 'capitalises' these costs to include the returns required by shareholders to use their money during development.

on-target biological activity and physicochemical properties. For some time now, the goal has been to reduce both the time per cycle and the number of these cycles, through the use of innovative, productivity-enhancing solutions to reduce R&D costs.

For medicinal chemists the greatest bottleneck, and thus the main focus for any improvements, is often the actual preparation of new molecules for biological testing. As such, there is now a trend towards the development and integration of automated equipment, technologies and processes as

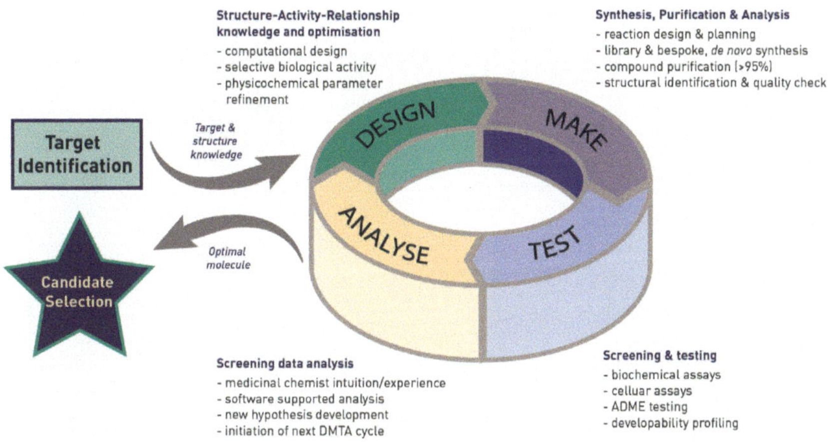

Fig. 2 *Design-Make-Test-Analyse (DMTA)* cycle of medicinal chemistry: the process in which knowledge from the biological target is used to identify hits and optimise leads via a number of iterative cycles, involving the design of new molecules to test hypotheses, probe structure–activity relationships (SAR), and optimise multiple parameters; the synthesis of the newly designed molecules; biological screening and ADME testing; and analysis of the screening data to confirm or disprove hypotheses and initiate the next DMTA cycle.

cost-saving and productivity-enhancing solutions in medicinal chemistry. While a number of automated platforms for evaluating new molecules have been widely adopted in the *Test* stage of the drug discovery process [4–7], this chapter will focus primarily on the advances and innovative solutions applicable to the *Design*, *Make* and *Analyse* stages of the drug discovery cycle, with a particular focus on the development and current availability of automated technologies for expediting and accelerating chemical synthesis.

2. History of automation in medicinal chemistry

From the 1950s, when drug discovery as we know it today first became fully established, the laboratory activities of the medicinal chemist looked quite similar to what they are today. The majority of synthetic work was carried out manually by highly qualified chemists, using a relatively small set of known reactions and standard laboratory equipment, e.g. round-bottom flasks, stirrer hotplates and rotary evaporators (Fig. 3) [8]. In addition, purifications were undertaken manually and were often very time consuming. Standard analytical techniques included TLC, IR, elemental analysis, and eventually NMR. It is really quite astonishing to observe that

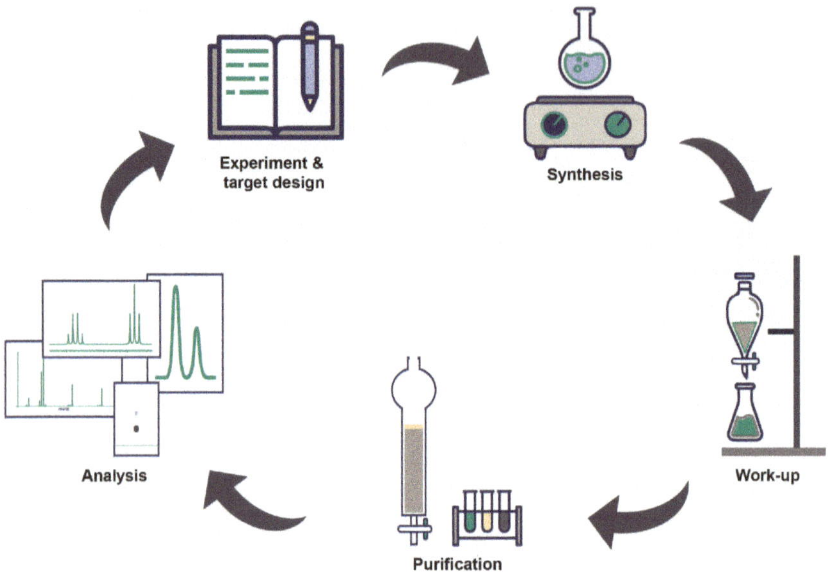

Fig. 3 Traditional process for preparing molecules for biological testing.

over half a century later, the way in which we prepare small molecules has not significantly changed, with almost all of the processes and techniques used then still being used today.

However, these traditional processes and equipment are now used alongside some newer technologies, which have been developed during the last 20–30 years. This section will focus on the history and development of technologies that have had a significant impact on the way in which medicinal chemists work. Both purification and analysis have benefitted from significant developments, which have provided relief from the tedious and time-consuming traditional manual methods. In addition, developments in computer processing power have enabled the use of software and automated processes in every stage of the drug discovery processes. For chemical synthesis, while some high-tech hardware has been developed over the last few decades, the sophistication of such equipment has typically lagged behind purification and analytical technologies. Only very recently have we started to see the emergence of technologies that are not based on the same fundamental practices as traditional synthesis techniques.

2.1 Early synthesis technologies

The automation of organic synthesis has been an on-going goal for several decades, and we have seen the emergence of increasingly complex robotic systems. However, some initial approaches to synthesis automation sought to simply automate traditional techniques and processes. One such example is a system developed in the early 1990s by researchers at Takeda in Japan, who created a fully automated, solution-based synthesis system consisting of modular synthesis units performing various tasks, such as the supply of reagents and solvents, the control of reaction conditions and the purification of products, all of which could be controlled with user-friendly software [9]. Stock reagent solutions and solvents were introduced into reservoirs (Reagent Supply Units) under reduced pressure using a diaphragm pump. From here the materials could flow into a volumetric tube to measure the amount required for the reaction, before being transferred to one of three reaction vessels, which along with a pH adjustment vessel formed the Reaction Unit. Workup and isolation of the products was affected by the reduced pressure transfer of the reaction mixture to the Extraction-Drying Unit, which consisted of a separation funnel equipped with a sensor, two solution reservoirs, and five drying tubes prefilled with desiccants. While such solution-based synthesisers were certainly impressive for the time, solid-phase synthesis techniques were continuing to advance and their distinct advantages, compared to the use of only solution synthesis, emerged. Synthesis and work-up procedures were simplified with solid-phase synthesis, which meant these procedures were more amenable to robotics-based automation.

The development of solid-phase synthesis techniques and robotic automation afforded what was perhaps one of the earliest fundamental changes in organic synthesis and medicinal chemistry [10]. Prior to the early 1990s, new molecules were generally individually synthesised and screened, but rapid developments in molecular biology and target identification, along with the emergence of high-throughput screening technologies, increased the demand for large numbers of small organic molecules for screening. A solution to this challenge presented itself in the form of combinatorial chemistry, in which a small set of building blocks and well-defined, robust chemical reactions are utilised to concurrently prepare large sets of diverse products in high yields by iterative building block assembly, with orthogonal protection group chemistry, and purification. This approach received much attention and was heavily promoted as heralding the ability to prepare large numbers of structurally diverse analogues in parallel (Fig. 4) [11]. While initial work in this area was focused on the preparation of peptide libraries [12–18], the methods developed for these libraries soon proved to be enormously beneficial for the preparation of non-peptidic libraries containing biologically interesting products [19–22].

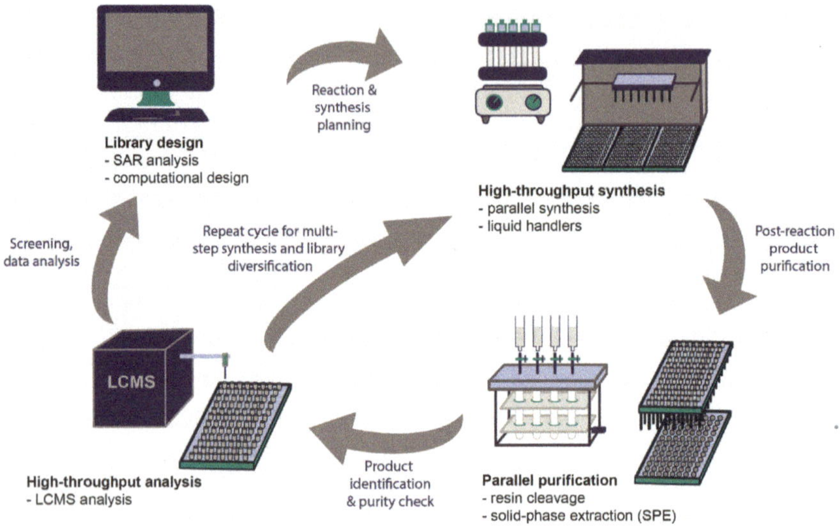

Fig. 4 Combinatorial library synthesis: libraries are designed, based on available technologies and suitable chemical reactions, and prepared, purified and diversified using a sequence of chemical reactions. Biological screening of the compound libraries and the resulting data analysis is utilised in the design of new libraries.

Combinatorial chemistry, and the preparation of large libraries, was enabled by the development of several new methods, including the 'multipin' approach in which polypropylene pins are immersed sequentially in different reagents [23], and the 'tea-bag' strategy in which polypropylene mesh containers hold polystyrene resins [24]. Both facilitated the 'split pool' approach in which the small pool of products from the first synthetic step is then split and reacted further with a selection of new building blocks and the process repeated multiple times to give libraries of structurally diverse products (Fig. 5) [25,26]. In addition, more recently new methods for the generation of biological peptide libraries have been developed, such as phage-display, yeast-display, mRNA-display, and chemically modified phage-display, which offer both advantages and disadvantages compared to traditional combinatorial chemical library generation [27].

Although parallel synthesis of combinatorial libraries can be achieved manually or robotically, both in solution and on solid support, the introduction of techniques such as the techniques described in Fig. 5 offered the potential to simultaneously synthesise libraries containing millions of diverse compounds. While this concept seemed very elegant, it brought with it some problems, namely the small quantity of material produced and the

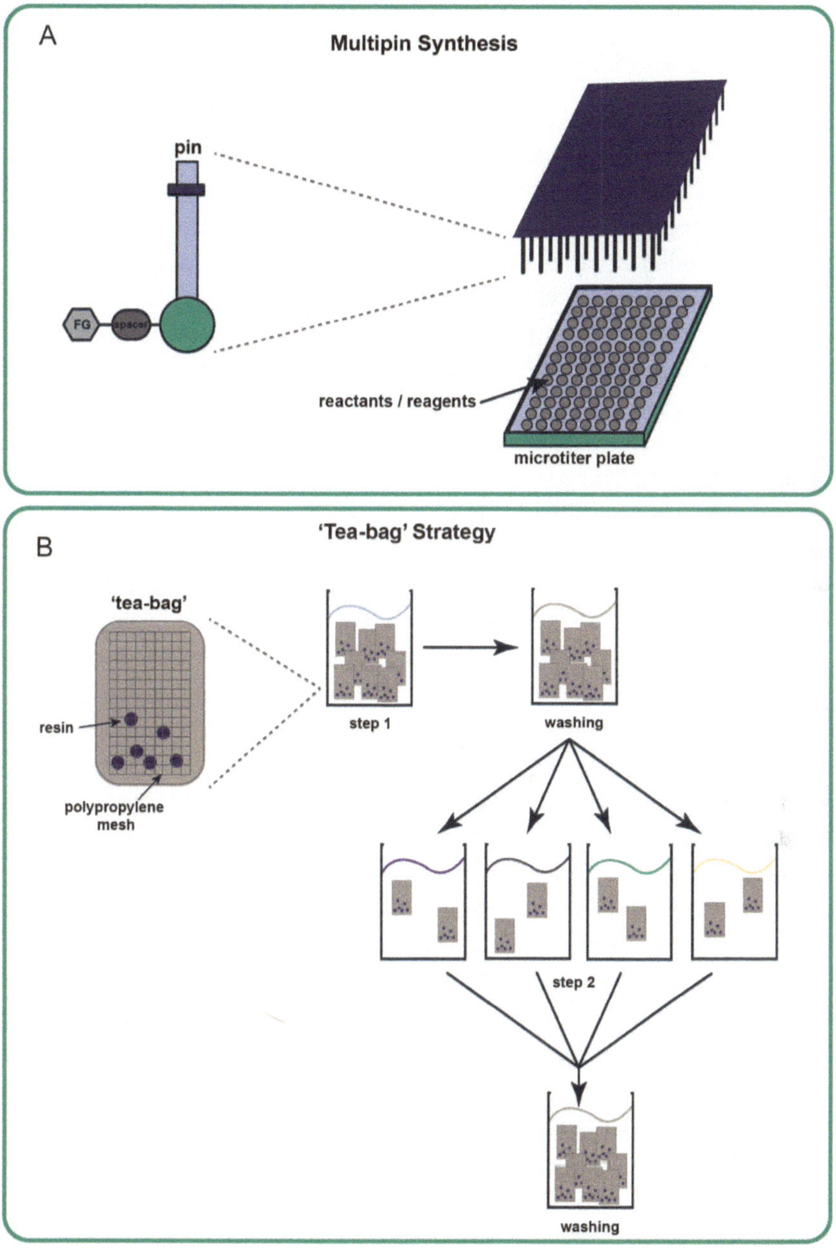

Fig. 5 Early combinatorial chemistry methods for library synthesis. (a) The multipin approach uses an array of polymeric pins that serve as a solid support to which a linker with a reactive functional group is attached. Sequential dipping of the pins solutions of different building blocks effects multistep synthesis and library diversification. (b) The tea-bag approach uses mesh packets encasing resin beads containing a linker and a reactive functional group, which can be added to a solution of reactants to effect the first building block addition. Following a washing step, the bags can be split and immersed in different reactant solutions in order to complete the second sequential step and yield a library of diverse products. Subsequent washing, splitting and building block addition steps yield larger, more diverse libraries.

requirement for structural deconvolution of the compound on each bead. The former issue was addressed through the development of a hybridised approach, using both solution and solid-phase synthesis in combinatorial chemistry. Attempts to address the latter issue led to exploration of different encoding strategies, including the use of DNA strands, and the establishment of DNA-encoded libraries (DELs) as a method for the mass creation and interrogation of libraries via affinity selection, typically on an immobilised protein target. In DEL synthesis, chemical compounds or building blocks are conjugated to short DNA fragments that serve as identification bar codes.

The idea of linking each molecule of a chemically synthesised entity to a particular oligonucleotide sequence (constructed in parallel) and the use of this oligonucleotide sequence as an identifying tag was first hypothesised in 1992 by Brenner and Lerner [28]. A year later, the first practical implementations of this concept began to emerge [29,30]. However, the method did not begin to garner interest until almost a decade later, which was in part due to the restrictions associated with DNA libraries. The greatest limitation of this particular technology has always been the chemical reactions and reaction conditions that are tolerated by the DNA. As such, several other encoding strategies have also been explored [31], but the ease with which DNA can be amplified makes it a great option for screening and encoding. As such, several companies and research groups have sought to push the boundaries and overcome the original limitations of DNA-encoded libraries [32].

The potential to accelerate the identification of new drugs through approaches such as combinatorial chemistry, led to most pharmaceutical companies making significant investments, both in terms of the automation technologies themselves, and the time needed to develop the required chemistry and operating expertise. Many new companies were founded based on novel technologies for library synthesis and screening, as the ability to rapidly synthesise very large numbers of compounds from only a handful of building blocks, for use in high-throughput screening, was very attractive to drug discovery investors due to the potential to drastically improve productivity.

Large investments were required because the simultaneous preparation of large arrays of compounds requires the use of robotics, automated work stations, parallel synthesis equipment, and software for each step in the process: reaction preparation, execution, workup and purification. Up until the 1980s, there were no commercial suppliers of laboratory robotics. Recognising the need for laboratory automation, early pioneers Zymark (acquired in 2003 by Caliper, a PerkinElmer company since 2011), were

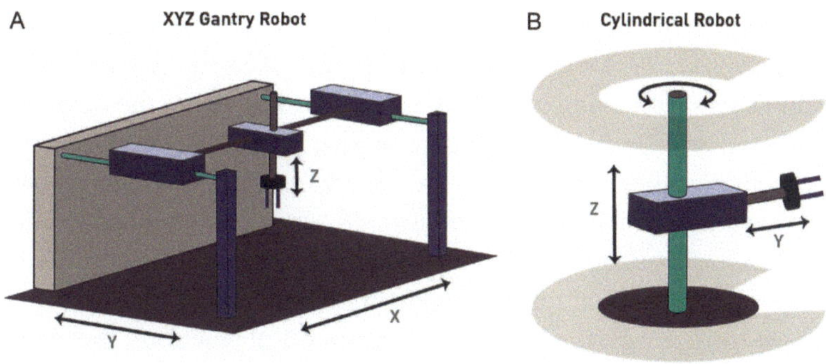

Fig. 6 Basic robot system formats. (a) XYZ Gantry or Cartesian robot systems move on three, orthogonal axes—X, Y, and Z—according to the Cartesian coordinates. (b) Cylindrical robots have three axes of movement—two of which are linear and one that is circular, which enables them can move along Z and Y axes and rotate along Z axis.

one of the first companies to bring small, readily accessible robotics to laboratory scientists, albeit primarily for screening applications [33]. While several other suppliers of robotics existed, including CRS (now part of ThermoFischer), Beckman Coulter, and Mitsubishi, it is Zymark who are often credited for the birth of laboratory robotics. Their work, including the development of robotic systems (Fig. 6), which were compatible with systems from other manufacturers, and their promotion of the laboratory automation industry, encouraged many more companies into the field. Several other companies entered the field and created variants of Zymark's Twister system, a cylindrical robot for handling and sampling microplates, including the Hudson PlateCrane EX® [34], Thermo CataLyst Express™ [35], Peak Robotics (now PAA) KiNEDx/ProNEDx/BiNEDx [36], and Precise Automation PreciseFlex [37].

In some areas of drug discovery, such as high-throughput screening, automated technologies such as sample handling robots, which replaced the manual operation of moving samples, were readily adopted. Liquid handlers are typically XYZ Gantry robots, and are used to automate the task of pipetting by hand. This saves significant time and effort and as such, they have proved to be highly enabling for high-throughput screening. These types of systems now range in complexity, from very simple systems that dispense an allotted volume of liquid from a motorised pipette or syringe, to more complicated machines that can also manipulate the position of the dispensers and containers and/or integrate additional laboratory devices. Today, many manufactures now offer liquid handling workstations and robots, including Hamilton [38], Gilson [39], ThermoFischer [40], INTEGRA [41], Tecan [42], and Andrew Alliance [43]. While these types of technologies have been well established for some time, the development of automated synthesis systems, with more than just liquid handling and sampling capabilities, has been a much greater challenge. This is in part because, unlike screening which can be standardised, synthetic chemistry requires a system that can provide a wide range of reaction environments and can handle a vast array of solvents and reagents with different physical properties.

Due to the complexities of automating chemical reactions, the most popular approach to automating synthesis has been through the concept of self-contained, specific synthesis workstations, in which all the functional components necessary to carry out the synthetic tasks are contained within a single instrument. Using these systems, reagents are automatically added to chemically resistant reactors so that robust reactions can be carried out, worked up and the products isolated. Of the workstations developed, many are based on an overhead gantry XYZ robot (Fig. 6) and a fixed reaction block, containing reaction vessels that can be heated, cooled, agitated, and maintained under an inert atmosphere. Alternatively, some systems use a mobile functional robotic arm, which can much better mimic the actions of a chemist carrying out manual synthesis, and can transport the reaction vessels between the different task areas of the system. A great deal of effort also went into developing disposable consumables such as work-up cartridges and plates that could be used with the system to expedite higher throughput, including those for phase separation/extraction, drying, scavenging and ion-exchange chromatography.

Over the past few decades, a number of automated synthesis systems have been developed. Although not specifically designed for synthesis, early work in the field was led by Zymark who, in collaboration with Zeneca, launched their Zymate Laboratory Robotic System in 1982. This circular system consisted of a compact, microprocessor-controlled central robotic arm with interchangeable hands, operating software, and surrounding laboratory work stations for multistep sample manipulations such as pipetting and dispensing. Over time Zymark further evolved its robotics into other application specific workstations or modules, including the small BenchMate workstation, a fully automated solution-phase reaction work-up facility for multiple parallel synthesis. Zymark also developed an XYZ arm-based automated synthesis workstation, as did many others in this field. In 1997, and in collaboration with several different end-user companies, including Bayer [44], Isra developed and manufactured a one-arm-based system for solution-phase synthesis. The 5 × 2 m system utilised a robotic arm from CRS Robotics (now a ThermoFischer company) and could synthesise, work up, and analyse arrays of up to 50 compounds on a scale of up to 2 g. For synthesis, the system used an XYZ-pipetting robot, heating and cooling reaction blocks with magnetic stirrers, a rack for reagents, and a rack for starting materials. For the workup, it used a filter station, a drying station, a balance, a centrifuge, a solvent evaporator, and a phase separation and phase boundary recognition station. A Shimadzu high-performance liquid chromatography/ultraviolet (HPLC-UV) system was also incorporated for analysis. Several other manufacturers developed similar systems, including Chemspeed, Charybdis Technologies (Iliad), Mettler-Toledo (Discoverer, Myriad, Neptune), PE Biosystems (Solaris 530), Argonaut Technologies (Trident) and Zinsser Analytics (Sophas) [45]. The systems offered varied features for automated parallel synthesis but often needed to be tried and tested to determine if these had the desired synthetic capabilities.

In addition to robotics, alternative reactor systems, such as microwaves, also started to emerge in the 1990s. Despite initial speculation that microwave irradiation may have offered some additional benefit over conventional heating, in reality, these systems simply create a heated, pressurised environment for each reaction. However, they offer ease of use and convenience, which makes microwave reactors appealing to medicinal chemists. While the first application of microwaves in chemical synthesis was reported in 1986, it was not until 2000 that the first commercial reactors, specifically designed for chemical synthesis, became available [46]. The initial reactors, such as the MLS ETHOS system, were adapted versions of microwave digestion units, which were well established within analytical chemistry. However, these systems were not suitable for small-scale synthesis. The first reactor suited to small-scale reactions was launched in 2000 by Personal Chemistry AB and used sealed glass reaction vessels, enabling reactions to be run at elevated pressure and high temperatures. Since then, the technology has been developed to include robotics and as such, microwave reactors also found great utility in high-throughput combinatorial chemistry. In addition to high-throughput applications, microwave reactors are widely used elsewhere in medicinal chemistry because reactions are easy to set up and run. Multiple reaction tubes can be prepared and added to the microwave reactor, which can be programmed to run them either in parallel or sequentially, depending on the system set up, using a single robotic arm to add and remove the reaction tubes from the reactor. A perspective on the development of microwave technologies for synthesis has recently been provided by Professor Oliver Kappe, who was a pioneer in the field of microwave assisted synthesis [47].

Many of these alternative reactor systems and robotic platforms were developed through collaboration between research and development organisations and manufacturers of robotic automated synthesis systems, which resulted in the creation of some bespoke, automated organic synthesis platforms. Zeneca and Chiron both worked with Zymark, the latter using a Zymark robot as part of their computer-controlled, fully automated peptide synthesiser [48]. One outcome of this approach is that, since each drug discovery organisation developed its own bespoke solution, there is now very little standardisation in terms or equipment, interfaces and software, and today instrument manufacturers are being called upon to standardise their systems. However, while efforts to facilitate this are growing [49], this is an enormous task since each organisation has already established its own systems and software. To change elements of this in order to ensure standardisation across the industry could potentially be highly disruptive.

The main advantage of these early fully automated, robot-arm-based systems was clear. Offering a customised solution similar to HTS equipment, large collections of compounds could be simultaneously prepared. However, one big issue was that the individual user needed to define the architecture, the layout, and the processes of the robotic system. Unfortunately, the majority of medicinal chemists do not have experience in programming motor movements, motion controls, and feedback sensors, which makes the technology inaccessible to many. More advanced software is continually being designed in an attempt to make these processes invisible to the average user, but even today such systems typically still require highly skilled operators and thus they are not easily accessible to the vast majority of medicinal chemists.

In contrast to more complex robotic systems, many of the components of robotic synthesisers were developed into non-robotic parallel synthesis equipment. Argonaut's manual 10-position reactor, the Firstmate, incorporated components of their robotic Quest system [50]. Radleys, which has never offered robotic systems, has also played a major role in the provision of this type of 'Personal Synthesis' equipment. The simple 12-position Radleys Carousel Reaction Station, originally developed by GSK [51], sits on a stirrer hotplate and is an effective, space-saving solution that offers an alternative to using multiple round-bottom flasks and multiple magnetic stirrers. Since the first Carousel was launched, Radleys has greatly expanded its range to include compact options for heating, cooling, and maintaining an inert atmosphere for up to 24 different reactions, in addition to complementary parallel work-up stations. JKEM also launched a 24-position parallel reactor block range [52], which utilises an orbital shaker rather than a magnetic stirrer and provides precise temperature control. The Personal Reaction Station from JKEM can be used for up to six reactions, which can be heated, cooled and magnetically stirrer, all under inert conditions. Mettler-Toledo provided medicinal chemists with similar capabilities with their MiniBlock™ range, a flexible, non-robotic, parallel synthesis system that also sits on top of a stirrer hotplate [53]. Since these types of non-robotic, productivity-enhancing parallel synthesis equipment are relatively inexpensive and incredibly easy to use, they have been widely adopted by the majority of medicinal chemists, unlike the complex robotic systems, which have tended to remain only within the realms of the specialist groups.

Despite combinatorial chemistry delivering a plethora of new synthesis methods, reactor and robotic systems, it has unfortunately not really lived up to its hype. This perceived failure is primarily due to the fact that insufficient consideration was given to the 'drug- or lead-likeness' of the compound libraries, and instead those made favoured the use of chemical reactions that were more easily enabled by automation, which ultimately led to an undesirable structural bias in the resulting libraries [54]. At the time, combinatorial chemistry struggled to develop synthetic methods that utilise cheap, readily available building blocks for the synthesis of chemically diverse, high-quality libraries of molecules with desirable features, such as a broad range of functionality and chiral sp^3-centres that provide greater three-dimensional shape diversity, and good pharmacokinetic properties [55].

Although the synthesis of many such molecules is feasible, often the cost of making such molecules, and the timelines required, far outweigh the gain since the vast majority of molecules prepared in medicinal chemistry are unlikely to progress far in the screening cascade and might add little knowledge to the drug discovery process. As such, the ability to produce large numbers of drug-like molecules, in a low-cost fashion, remains a gap in our synthetic capabilities.

Attempts to develop automated systems that circumvent the cost and time issues have been hampered by the fact that there are still too many unknown and unpredictable factors involved in compound design and synthesis. Therefore, the biggest challenge still remains—an easily accessible, low-cost, automated method for the preparation of a broad selection of structurally diverse drug-like molecules. Despite the challenge, interest in automated synthesis platforms remains high within drug discovery organisations because of the potential productivity-enhancing and cost-saving benefits they offer. Although combinatorial chemistry did not fully live up to its promise, automation legacies from this era remain, such as automated purifications, robotic handling of samples, and HTS. In addition, the aims of combinatorial chemistry are still today very compelling. Freeing the chemist from the routine experimental drudgery and enabling them to think more about the science is still a major goal of automation.

2.2 Development of automated purification techniques

In order to be sure that any observed biological activity is actually related to the intended molecule, almost all synthetic products need to be purified in some way and to a level at which one can be confident that a therapeutic effect is derived from that molecule and not an impurity. Purification methods can be broadly classified as either distillation, crystallisation or chromatography. The method of choice for medicinal chemists depends not only on the nature of the substance and the impurities present, but also on the ease and accessibility of the purification method.

Distillation is the combined process of vaporisation and condensation. This method is suitable for the purification of liquids that boil without decomposing and which contain non-volatile impurities. It is most useful for separating liquids that have a significant difference in their boiling points (at least 25 °C). Mixtures with closer boiling points need to be separated by fractional distillation. However, distillation is only suitable for substances that are liquids and will boil at moderate temperatures. It is also relatively low throughput and requires constant observation. As such, it is used infrequently in medicinal chemistry.

In contrast to distillation, crystallisation is a method utilised for purifying solids. It relies on a difference in solubility between the product and the impurities. While it can result in materials of extremely high purity, the greater the amount of impurity, the harder it is to effect the crystallisation process. Also, it can take time to develop the correct conditions for crystallisation, resulting in low throughput. While crystallisation plays a very important role in chemical development and pharmaceutical manufacturing, this technique is used less frequently in medicinal chemistry since it is unsuited to the high-throughput demands of this field.

Chromatography is far more general and can be used for separation of mixtures regardless of whether materials are soluble solids or liquids. Consisting of a stationary and a mobile phase, this technique exploits the difference in the rates at which the components of a mixture move over the large surface area of the stationary phase, e.g. silica, under the influence of the mobile phase, typically a liquid solvent. Chromatography, in a rudimentary form, was first used in the mid-19th century for the separation of plant pigments, and the first chromatography column was developed in 1901 by the Russian botanist Mikhail Tsvet, who discovered that an organic solution of plant pigments could be washed through a vertical glass column packed with an adsorptive material and separated into a series of discrete coloured bands on the column. Subsequently, column chromatography dramatically increased in popularity during the 1930s, to become a generally accepted technique in both Europe and the United States, after chemists Richard Kuhn and Edgar Lederer successfully used the technique to separate a number of biologically important materials [56]. Since this time, there have been a great number of advances in the field.

Most of the early advancements in column chromatography resulted in an increased understanding of both the stationary and mobile phases. The packing materials can vary greatly and chromatography can be classified according to the type of stationary phase used. The options include adsorption materials that differentiate based on how well the components are adsorbed onto the surface (adsorption chromatography Fig. 7a). Ion-exchange resins effect a separation based on the strength of an ionic interaction (ion-exchange chromatography Fig. 7b). Materials can be designed that separate mixtures based on their affinity for the solid phase due to a specific biological interaction (affinity chromatography Fig. 7c).

Of the three different solid phases described above, adsorption chromatography, e.g. using silica, is probably the most popular type used in medicinal chemistry. However, ion-exchange resins, first developed in 1938 by Harold C. Urey and T. I. Taylor [57], are also commonly used. Today, ion-exchange chromatography forms the basis of 'catch and release' purification techniques. Disposable cartridges containing ion-exchange resins are a remnant from the combinatorial chemistry era but these are still regularly used today, e.g. for the isolation of basic or acidic compounds [58]. Affinity chromatography, first developed in 1968 by Pedro Cuatrecasas and co-workers [59], involves the binding of a biomolecule to a substrate that has been immobilised on the solid phase, while the other components remain unbound and are eluted. This is a powerful technique, which is utilised for the purification of biologics such as proteins and antibodies. However, it is not commonly used in medicinal chemistry.

For adsorption chromatography, a glass or plastic tube is packed with the solid phase, typically silica, and saturated with the mobile phase, which is

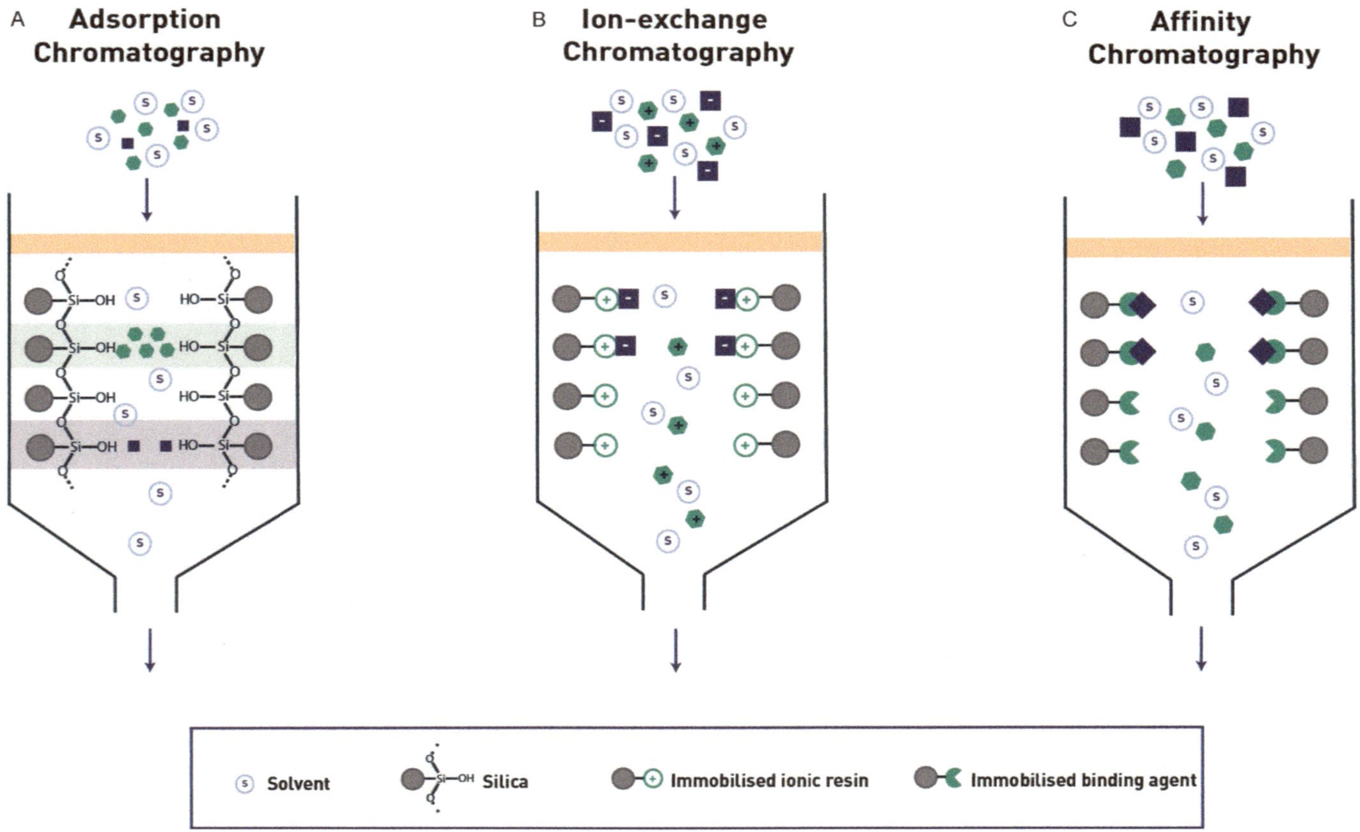

Fig. 7 Variations of chromatography according to the solid phase used. (a) Adsorption chromatography involves passing a concentrated solution of the crude material through a column of the adsorbent, e.g. silica, which has been pre-saturated with the eluent solvent. (b) Ion-exchange chromatography is used to separate either positively or negatively charged materials, e.g. acids or bases, from a crude mixture by passing a solution of the crude material through an oppositely charged resin. Basic materials can be isolated, for example by passing the mixture through an acidic resin. (c) Affinity chromatography involves use of a solid phase with binding sites, to which the desired material from a crude mixture specifically binds.

typically either a mixture of a non-polar organic solvent, such as hexane, and a more polar solvent, such as ethyl acetate (normal phase) or a mixture of water and a polar, water miscible organic solvent, such as acetonitrile (reverse phase). Upon addition of the sample to the top of the column, it is then eluted with more mobile phase. In addition to the stationary phase, the mobile phase also plays a role in determining how quickly the components of a sample mixture progress through a chromatography column, due to differences in solubility.

Given the broad range of solvent mixtures that can be used as the mobile phase, thin layer chromatography (TLC) (discussed in more detail in Section 2.3), where the solid phase is spread thinly on a supporting plate, is often used as means of predicting the correct solvent system to use as the mobile phase for the best separation in column chromatography. While originally the mobile phase was allowed to flow through the packed column under gravity, in 1978, W. Clark Still pioneered a development that led to the establishment of a new column chromatography technique [60]. In his modification, the mobile phase could be passed rapidly through the column using positive pressure, a technique that was known as flash column chromatography. Owing to broad applicability, in part due to the wide range of solid phases and mobile phases that can be employed, flash column chromatography, both manual and automated, is still one of the most popular purification methods in medicinal chemistry.

While flash column chromatography has been a purification method of choice for several decades, the manual preparation and execution of column chromatography can actually be quite a time-consuming process. Traditionally, the preparation of a chromatography column involved packing the solid phase, such as silica, into a cylindrical glass or plastic tube, which was fitted with a filter, e.g. a glass frit, at the bottom to hold the solid phase in place. The solid phase needed to be added either by addition of the dry solid phase, followed by repeated flushing with solvent until it was completely wet and evenly saturated, or by the preparation of a slurry of the solid and mobile phases, and subsequent transfer of this to the glass column. For good separations, it is essential that chromatography columns are well packed, that no air bubbles are present, and that the top of the stationary phase is flat. As such, the manual preparation of good quality columns is a lengthy and laborious process. Additionally, using traditional techniques, the separation of the mixture components itself is also time consuming because the solvent needs to be slowly eluted through the column and fractions collected manually. As a result, traditional column chromatography of a single sample could easily take hours.

Due to the extremely time-consuming nature of column chromatography, the most beneficial advances in purification automation for medicinal chemistry have been in the development of automated flash column chromatography systems to relieve purification bottlenecks. These systems typically include a pump, sample injection ports, a UV detector and a fraction collector to collect the eluent. Many manufacturers, including Biotage [61], Büchi [62], Interchim [63], and Teledyne Isco [64] have developed and commercialised a range of low-pressure automated flash chromatography systems that utilise pre-packed disposable chromatography columns, and are capable of automatically purifying samples on a varying scale, from a few milligrams up to kilogram scale. Given the huge timesaving advantages of these systems, they have almost entirely replaced manual flash column chromatography in industry labs. However, the cost of these systems and the consumables means that they are still out of the reach of some academic groups.

While the resolution on these low-pressure systems is sufficient to get pure enough material in the vast majority of cases in medicinal chemistry, higher resolutions, purities, and increased speed are sometimes required. The solution to this came in the form of high-performance liquid chromatography (HPLC), which utilises smaller solid-phase particles as a means of increasing the surface area and thus improving the resolution. Although HPLC originated with the work of Calvin Giddings in the 1960s [65], it was not until the early 1970s that HPLC columns with very small solid-phase particles, such as microparticulate silica gel, began to emerge. Despite the improved separations, preparative HPLC systems do have limitations. The columns themselves, though re-usable for a time, are expensive. The amount of material that can be injected and separated in each run is low relative to column size. HPLC systems also require far more expensive high-pressure pumps and as such, there are typically some restrictions on the number of such systems available for medicinal chemists to use. As a result, the low-pressure systems remain the preferred technology for routine purification. However, HPLC systems continue to develop, offering faster run times and a more accessible purification method, particularly for small amounts of final compounds that are needed in high purity for biological screening.

2.3 From manual to automated analysis

In medicinal chemistry, analytical tools play an essential role for both monitoring reaction progress and determining the identity and purity of final reaction products. Over the last century, several important analytical techniques have been developed, and ultimately automated, many of which are based around chromatographic processes. One of the earliest forms of such a process was paper chromatography, which was first introduced in 1944 and was then replaced almost entirely by thin layer chromatography (TLC) by the end of the 1950s [66].

The technique of TLC was originally developed in 1937 in the Ukraine by Izmailov and Shraiber, who were seeking to rapidly analyse plant extracts [67]. Even in this pre-automation era, researchers were constantly seeking to develop methods and technologies to speed up research processes. Considering classical column chromatography to be too time consuming, they postulated that the same adsorbent spread thinly on a glass plate would behave in the same way as a chromatography column but that the time taken to run the analysis would be vastly reduced. This hypothesis was confirmed when they were able to spot the extracts onto the plate and observed separation of the mixture into concentric rings of the components, after addition of a spot of solvent, which matched what would have been observed from column chromatography (Fig. 8). The technique was picked up and

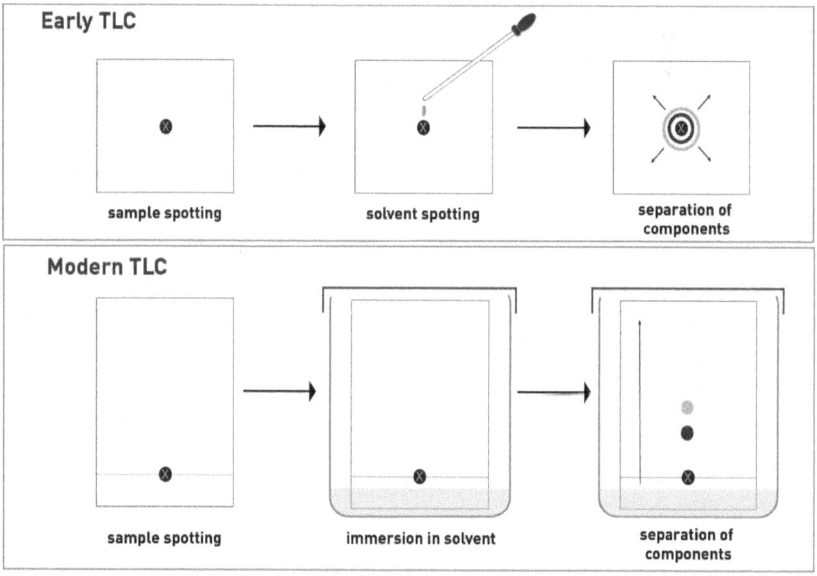

Fig. 8 Early TLC involved addition of drops of solvents to spots of the mixture, which resulted into the separation of the components into concentric rings on the thinly spread absorbent. Modern TLC separates the components of a mixture, according to their affinity for the stationary phase, by immersing the bottom of the TLC plate in the solvent reservoir and allowing the eluting solvent to ascend the TLC plate.

adapted by Crowe in the United States in 1941, who used the technique to scout for the best developing solvent for his column chromatography [68].

TLC as we know it today was developed almost a decade later by Kirchner, who was seeking an analytical method to investigate the composition of fruit juices [69,70]. However, instead of utilising the spot method developed a decade earlier, he instead opted to develop the plates using the ascending mode that was commonly used for paper chromatography, followed by spraying with selective reagents to visualise the components, which is how thin layer chromatography is still carried out today. Kirchner contributed much more to the development of this commonly used analytical technique, including demonstration of two-dimensional TLC, rod variations of TLC plates [71], and the ability to quantitatively analyse mixtures by measuring the light absorbance of the separated component spots [72].

Despite the great advances made by Kirchner, TLC did not become widely accepted as an analytical technique until the end of the 1950s. The eventual acceptance was primarily due to the work of Stahl, who optimised and standardised the method, the equipment and the adsorbent for his work on the analysis of essential oils [73]. In 1958 the new instrumentation, made by Desaga, and the commercially available adsorbent—'silica gel G according to Stahl for TLC'—were introduced at a chemical equipment exhibition in Frankfurt [74]. TLC was further improved by the introduction of new equipment for more accurate spotting of the sample on the plate and by the development of high-performance TLC, which utilised new adsorbents with reduced particle sizes [75]. However, it is clear that the widespread adoption of TLC, like most successful technologies, came about primarily because the technique and equipment were both standardised and made commercially available.

TLC requires little more than a simple glass chamber and a glass capillary tube for spotting. More sophisticated developing chambers are now available, as are automated instruments for spotting samples on plates (dosimeters), which enable a controlled and standardised spotting process. However, these and other advances, such as the use of densitometers to quantitatively measure the separated components, are not routinely used in medicinal chemistry. Recently, the technique has been further developed and TLC can now be combined with techniques such as mass spectrometry (MS) or Fourier transform infrared (FTIR) spectroscopy, to support structural elucidation. However, such sophisticated, automated and complex systems are rarely used in medicinal chemistry laboratories. Instead, TLC is primarily used as a qualitative analytical technique. Applications include: monitoring reaction progress; selecting a suitable solvent system for column chromatography purification and identifying the subsequent product containing fractions; or giving an indication of whether a product contains a significant impurity.

Another analytical technique that was developing around the same time as TLC is gas chromatography. Due to the rapid development of the required instrumentation, this technique, in which an inert carrier gas elutes in continuous flow, was one of the first modern instrumental quantitative chromatography tools [76]. It involves the separation and detection of the chemical components of a sample mixture to confirm the absence or presence of a component and also quantifies the amount of each component in a mixture. Although the technique itself was pioneered in the early 1950s, these developments were built on early exploratory work carried out during the preceding decade [77].

Various instrument manufacturers, such as Spring and Perkin Elmer, played a pivotal role in the rapid advancement of this technique through the development of reliable, standardised instruments and by promoting the technique, increasing awareness, and providing training for novice chromatographers. Over the past few decades, the range of chromatography column adsorbents, such as charcoal, silica gel and alumina, has been greatly expanded and has thus increased the utility of this technique. However, it still has its limitations. It can only be utilised for chemical components that can be vaporised without decomposition. Therefore, its applicability to the less volatile organic molecules typically prepared by medicinal chemists is somewhat limited.

In contrast, liquid chromatography is far more widely applicable as an analytical technique. Standard column chromatography is not suitable for analytical purposes, due to its time-consuming nature and low throughput. However, HPLC is ideally suited, offering both fast analysis and high resolution separation of components. Like gas chromatography, the establishment of this technique was reliant on the development of specialised, high-quality instrumentation. The key components of these instruments are the pump, the column and the detector. Multi-piston pumps are typically used to drive the solvent and sample through the column as they can provide the required pressure and ensure a constant, pulse free, flow rate. HPLC columns are typically narrow metal columns packed with the adsorbent. This is often a solid; such as silica or alumina but the adsorbent can also be a liquid bonded to a solid support. The detector, which needs to be capable of differentiating the components from the solvent, is most commonly a UV absorption detector since many medium to large molecules absorb UV radiation (Fig. 9).

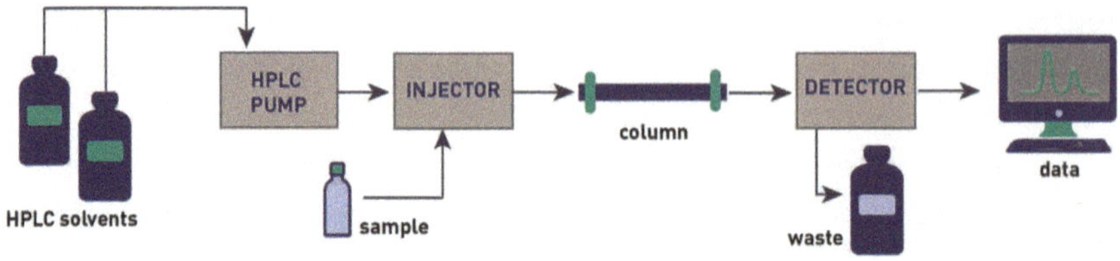

Fig. 9 Schematic of a HPLC system, in which the pump passes a pressurised liquid and a sample mixture through a column filled with adsorbent, leading to the separation of the sample components, which are detected, quantified and reported in terms of the retention time on the column and the amount of each component.

Since its conception, HPLC has undergone almost constant development as a qualitative and quantitative analytical method. Some of the most significant developments relate to the column and the adsorbent. Although silica and alumina are typically used, the incorporation of chiral adsorbents was hugely enabling. Prior to 1979, chiral molecules could only be separated chemically, which was extremely time consuming and not robust. However, in 1979, Okamoto used a stable, enantiomeric form of a helical polymer of triphenylmethyl methacrylate in a HPLC column to make the first chiral chromatography columns [78]. These columns were subsequently commercialised in partnership with Daicel, and their subsidiary Chiral Technologies Inc., who are leading Japanese manufacturers of chiral chromatography media. Over the years, they have successfully developed smaller columns, and also columns containing coated polysaccharide based chiral stationary phases, both of which have resulted in faster, better and more efficient chiral separations. HPLC systems that use a chiral stationary phase in the column are now one of the most effective ways of separating chiral molecules.

The columns, packing materials and processes for non-chiral separations have also undergone significant development over the last few decades. Several companies have played a key role in advancing the HPLC column technology, including Merck KGaA and Waters Corporation, a company that has consistently proved to be a strong player in the HPLC field [79,80]. Since the early 1990s, Waters Corporation has been developing processes for both the manufacture of packing materials and improved column packing procedures, thereby improving reproducibility. In addition, Waters has also made significant improvements to HPLC technology in terms of speed, peak shape and operating pH range.

The development of new column technologies, along with the constant demand for better separations and reduced analysis time, ultimately led to the development of a new form of liquid chromatography. In 2004, Waters Corporation introduced their Acquity Ultra Performance LC (UPLC) system [81], which incorporated a second-generation, pressure-tolerant column containing reversed-phase silica/organosiloxane hybrid particles of extremely narrow size distribution. This column allowed unrivalled separation and resulted in many laboratories replacing their HPLC technology with the Acquity UPLC. Just 2 years later, Agilent Technologies entered the UPLC market with its rival 1200 Series LC system, a high-resolution, high-speed system, which provided 60% higher resolution than an equivalent HPLC column [82]. Two years later again, in 2008, Shimadzu launched its new Ultra-Fast Liquid Chromatography System that doubled separation performance and reduced analysis time by 75% compared to conventional systems [83]. This was achieved by using Shim-pack XR-ODS II series columns, which had optimised particle pore diameters, longer column lengths, increased system pressure resistance and minimised flow path dead volume [84]. In addition, rather than needing to continuously invest in new hardware, some companies focused on new, improved columns that could be used on any existing liquid chromatography instrument. Phenomenex developed their Kinetex® columns, which offered improved speed and separation efficiency compared to traditional columns, and thus were capable of achieving UPLC like results [85].

While these chromatography systems were very effective at identifying the number and amounts of each component, they did not provide much information on the structure of the separated compounds. Coupling of analytical chromatographic separation systems with other spectroscopic analytical techniques, as a means of harnessing the complementary strengths of both, is an idea that first started to gain ground over half a century ago. Initial work focused on coupling gas chromatography with mass spectrometry (MS). In MS, samples are bombarded with electrons to effect ionisation and the resulting ions are then separated according to their mass-to-charge ratio [86].

While the first GC-MS systems were commercialised in the 1970s [87], coupling of MS with LC was operationally more complex and development of the technique only began in the 1970s [88,89]. The biggest hurdle in the development of coupled LC-MS systems was the development of an automated interface that could remove a significant portion of the mobile phase from the LC process, and transfer the maximum amount of analyte to the MS system. While such an interface remained difficult for some time, the advent of electrospray ionisation (ESI), developed by Fenn and collaborators in 1988, changed this [90]. In this method, the eluent from the LC system is pumped through a metal capillary and nebulised at the tip, forming a fine spray of charged droplets. While ESI is one of the most common LC-MS interfaces, other variations have been developed, including atmospheric pressure chemical ionisation (APCI), and atmospheric pressure photoionisation (APPI), for samples that are not readily ionised using ESI.

LC-MS systems are now very popular within medicinal chemistry. They are routinely used for reaction monitoring, product identification and purity determination, due to their speed and simplicity of use. Instrument manufacturers have also developed systems with autosamplers, allowing users to log samples into the system and then simply leave the vials in the sample rack (Fig. 10). The whole process will be carried out automatically and the users receive a report just a few minutes later containing the LC trace and a separate MS analysis for each individual peak in the LC trace.

Aside from LC-MS, the other most commonly used analytical technique in medicinal chemistry for structural elucidation is nuclear magnetic resonance (NMR). The process exploits the specific magnetic properties of atomic nuclei, in particular isotopes with an odd number of protons and/or neutrons that have an intrinsic nuclear magnetic moment and angular momentum, such as ^{1}H and ^{13}C. NMR machines are built around a strong magnet, and when the sample is subjected to this strong magnetic field, the nuclei of some atoms in the molecules will begin to behave like small magnets themselves and start to resonate at their own specific frequencies

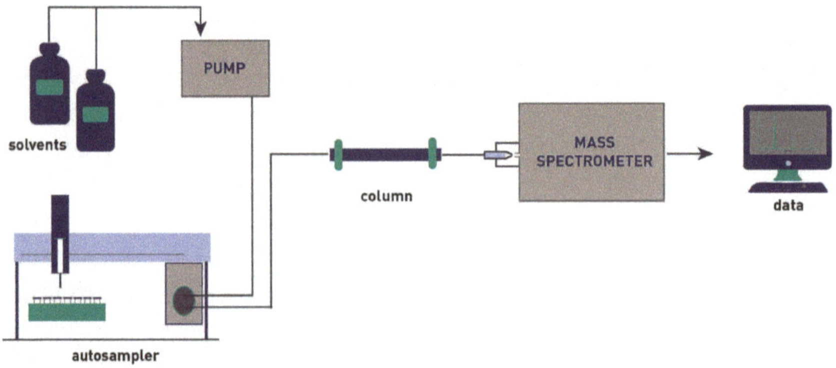

Fig. 10 LC-MS systems select the sample from the autosampler and inject it onto the column, which is eluted with a pressurised solution to effect the separation of the mixture components. Introduction of the separated components into the mass spectrometer results in detection of the mass ions associated with each component to support identification of each component.

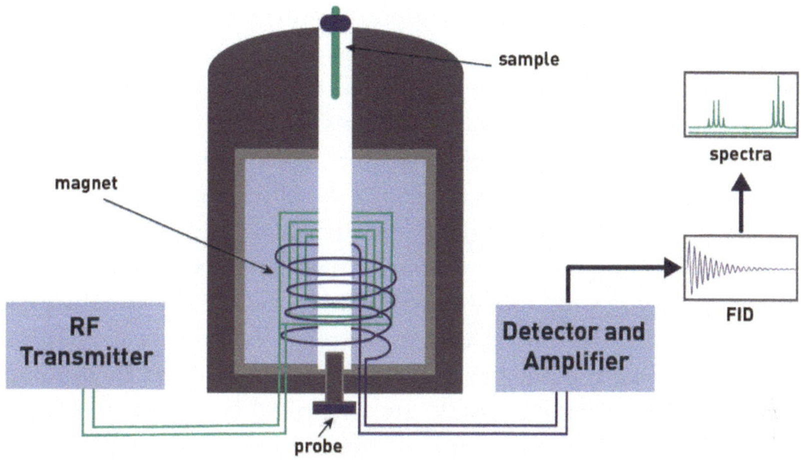

Fig. 11 NMR spectrometers consist of an insulated inner chamber, containing a super cooled, strong magnetic unit wound with super-conducting wire. The sample is spun inside the strong magnetic field and the resulting radio-frequency emitted is detected, amplified, and the resulting free induction decay (FID) is converted into the final NMR spectrum using Fourier Transformation.

(Fig. 11). These frequencies can be measured and converted into individual peaks on an NMR spectrum. Structural information can be derived from the NMR spectrum because the intensity of the peaks correlates with the number of each nuclei, and the frequency at which each nucleus resonates is uniquely dependent on the identity of neighbouring atoms and their molecular connectivity.

NMR was first measured and described in 1938 by Isidor Rabi [91], who later won the Nobel Prize in Physics for his work. The technique was further developed by Bloch and Purcell in 1946, who successfully demonstrated NMR for the first time on liquids and solids. Shortly afterwards in 1951, the first patent for an NMR machine was filed by Russell H. Varian, the owner of Varian Associates, who went on to develop one of the first NMR units in 1952. Recognising the importance of Varian's work, Laukien, another pioneer in the NMR field, developed an alternative NMR method and established his own company, Bruker Physik, in 1960 [92]. These two companies were responsible for the early development of the field and much of the innovation in NMR was driven by the intense competition between the companies [93].

Despite this early work, it took around 30 years for the technique to fully develop. The establishment of NMR as a widely used technique was enabled by three key developments: stronger magnets, which improved both intrinsic sensitivity and peak separation; improved probe technology and design, which enhanced performance; and a rapid increase in computer power to enable software to be developed that simplified data processing and opened up the technique to non-experts. As a result of these key developments, high-resolution NMR machines finally became commercially available in the 1990s. Towards the end of the 20th century, higher dimensional spectroscopy (2D and 3D) was developed and became available. This was made possible by rapid developments in computer technology. Around this time, the cryogenic probes that we find in today's machines were also being developed, which, along with improved electronics, resulted in a significant enhancement in the signal to noise ratio. In addition, incorporation of fluorine in molecules is a rapidly growing trend in medicinal chemistry. As a result, highly sensitive cryoprobes have been developed that are capable of observing this nucleus.

Although the pioneering work in NMR can be credited to both companies, Varian and Bruker, the former was acquired by Agilent in 2010, who exited the NMR field only a few years later, leaving Bruker as the dominant player in the NMR field. Today several other companies have entered the NMR space, particularly with smaller bench-top systems, and much of the process of running an NMR analysis and obtaining a spectrum has now been automated. After logging a sample into the system and adding the sample tube to a carousel, the user can simply leave the whole process to run automatically, before finally receiving and interpreting the NMR spectra at their desk. The technique remains popular because it is highly selective and able to distinguish between the atoms in a molecule, with even subtle differences in their local environment. Through the interpretation of standard ^{1}H and ^{13}C NMR spectra and, where necessary, the use of powerful homo- and heteronuclear 2D NMR experiments, chemists can often unambiguously determine the structure of compounds by studying the peaks of a nuclear magnetic resonance spectrum. The spectrum can also give an indication of the sample purity, further increasing its utility.

In recent years, the utility of NMR in medicinal chemistry has expanded. Not only has it been coupled to liquid chromatography techniques, such as HPLC and UPLC, to enable the separation and structural elucidation of the sample components, but it has also found utility beyond just being a technique for the structural elucidation of products. Now it is also used for the identification of new hits for drug discovery projects [94]. In fragment-based drug discovery, small, low molecular weight chemical fragments, which may bind only weakly to the biological target, are identified and then optimised to produce a lead with a higher affinity. Due to the weak binding affinity of the small fragments, it would be challenging to identify these using traditional screening techniques. However, by using NMR spectroscopy it is possible to observe the three-dimensional experimental binding mode of these fragments when they interact with the target [95].

It is clear that, compared to the relatively basic techniques used in the early 20th century, there have been significant developments in the technologies and tools for synthesis and analysis. Much of the early progress in this area, which occurred during the second half of the 20th century, was made possible as a result of developments in electronics, robotics and computing capabilities. As these key areas have developed, we have observed the concomitant rise of automation. While automated laboratory technologies are well established for some medicinal chemistry applications, e.g. purification, systems for synthesis have typically lagged behind. The very latest developments (see Section 3) and the technologies that are just starting to emerge (Section 4) have now brought us to the point where we are finally starting to realise the full power of automation in chemistry, and where widespread adoption of automation in medicinal chemistry is a realistic possibility in the near future.

3. The current state of automation and established technologies in medicinal chemistry

Medicinal chemistry plays a key role in drug discovery as it links together several different scientific disciplines, including biology, pharmacology, and chemistry. In fact, it is synthetic chemistry that forms the foundation of this discipline since it holds an essential role in three out of the four stages of the iterative *Design-Make-Test-Analyse* cycles that typify drug discovery (Fig. 12). This section will review the current status of automation and the existing enabling technologies for each step in the *Design*, *Make* and *Analyse* stages of drug discovery.

The ultimate goal of most medicinal chemists is to design new molecules to selectively interact with the desired target, and then make these new molecules as quickly as possible, in order to determine whether or not it has the anticipated pharmacological effect on the biological target. Once the design step is complete, the chemist must work out how to make the target, which is normally done via an iterative series of chemical reactions. Upon completion of each of these reactions, there is usually a workup or isolation process, followed by a purification step, and finally an analysis to confirm the identity

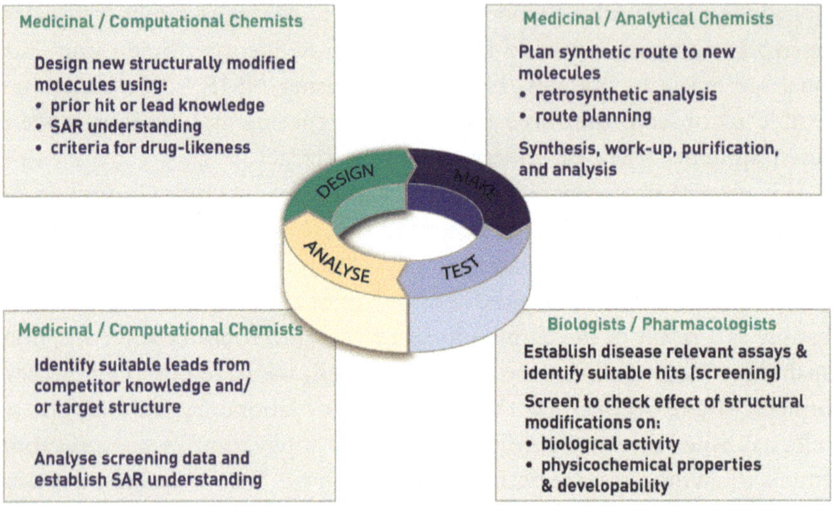

Fig. 12 The impact of medicinal and synthetic chemistry on drug discovery.

and purity of the product. Despite the significant advances already discussed, some of these processes can still constitute a bottleneck in the identification of new drugs.

This section will review the current status of automation tools for every step in the medicinal chemistry process, including molecule design, reaction planning, synthesis, workup, purification, and analysis. It will also highlight the gaps that still need to be filled in the future if we are to fully realise the potential of automation in medicinal chemistry.

3.1 *Analyse and design*—Software tools

Once a biological target has been identified as having a role in a particular disease state, the next challenge is to find a ligand that interacts with the target and that can be suitably redesigned to give a drug candidate. The process of redesign is heavily dependent on the knowledge available on the specific target and its ligands. If there is no knowledge of the three-dimensional structure of the target and no existing ligands that can serve as suitable starting point for drug discovery, then diverse screening must be carried out to identify suitable hits. The knowledge of how structural variants of these hits alter the ligand-target binding interactions then needs to be built using a trial and error process in which data from screening these analogues is amassed and analysed. As this structure–activity relationship (SAR) profile is established and refined, it can be used to aid design of new molecules with improved properties.

If prior knowledge of ligands that interact with the desired target is available, then this can also be used for the rational design of novel molecules (ligand-based drug design, LBDD). Each design iteration again helps to establish and refine the SAR and further refine the designs. If structural information on the target is known, new molecules can be designed to incorporate key features that would appear to favour binding, based on the knowledge of the three-dimensional structure of the target (structure-based drug design, SBDD).

There are several tools available to aid the *Design* of new molecules [96–100], and the tool of choice depends on the approach being taken. However, common to all approaches is the need to be able to assimilate large amounts of screening data and from this identify trends relating to the structure of the compounds tested. Today there are numerous possibilities available for SAR analysis and visualisation. In addition to those developed internally by various drug discovery companies, such as Johnson and Johnson's SAR Maps tool [101], several commercial systems are also available. ICM Molsoft FOCUS was developed in collaboration with Novartis and features customisable spreadsheets, visualisation plots and heat maps, and clustering [102]. It can be readily integrated with external databases and has data sharing options. In addition, it has a 3D ligand editor to aid 3D compound design. Data analytics solutions such as, TIBCO Spotfire®, Dotmatics Vortex or OpenMolecules DataWarrior enable fast data visualisation and exploration. Spotfire® is also supported by a built-in artificial intelligence engine, to enable anyone to explore data and spot trends [103]. StarDrop™ by Optibrium is a complete modular software package for small molecule design, optimisation and data visualisation [104]. Its core features can be extended with a range of add-on plug-in modules for predictive pharmacokinetics, metabolism and toxicity models, to allow scientists to create, manage and analyse chemical structures and related non-chemical data. It serves as a desktop application that gives a convenient and straightforward approach to the analysis of chemical and biological data. ChemAxon's Instant JChem and Dotmatics Browser are database-centric environments that support query and sorting functionality, and handle large volumes of data (hundreds of thousands of structures), both in local and remote databases, without any additional administration [105]. Both also include customisable database views, integration of library enumeration, and dynamic population of columns with singular and combined molecular property and descriptor predictions. Chemically enabled data processing tools include Biovia Pipeline Pilot and open source solutions such as

KNIME. These enable different data gathering and analysis tools to communicate automatically [106].

In ligand-based drug design, computational pharmacophore models, based on overlays and common features of known ligands, can be highly beneficial. Using these models, key binding features can be predicted, as can important electronic and steric considerations. This knowledge can help to focus designs, and focused libraries of novel potential ligands can be screened virtually using a pharmacophore model to enable prioritisation of new molecules for synthesis. Several companies offer commercial software for pharmacophore modelling, including Schrödinger (Phase) [107], Cresset (Blaze, Torch, Spark, Forge) [108], Molecular Discovery Ltd. (Flap) [109], Inte:Ligand (LigandScout) [110], Chemical Computing Group (MOE) [111], and Silicos (Pharao). Some academic groups have also contributed significant efforts to this area [112]. As screening data emerges following the synthesis of molecules, this can be used to optimise the pharmacophore model. However, the process of building these models can be time consuming and it has not been feasible to rebuild the model for each iteration of screening data, which is somewhat limiting. In light of new developments in artificial intelligence, however, machine learning approaches now offer models that adjust themselves automatically based on emerging screening data (see Section 4).

For targets where structural information is known, this can drastically reduce the need for the synthesis of large libraries of compounds, as well as the 'wet' screening of these libraries. Instead, very large virtual libraries can be readily constructed and screened virtually using a computational model based on the known binding site. Promising hits from the virtual screen can then be selected for synthesis. This offers potentially huge savings since the preparation and screening of hundreds or even thousands of compounds is expensive and time consuming. Once hits are confirmed via synthesis and wet screening, the model can be validated and used to aid the rational design and optimisation of the hits.

In the past two decades, there has been increasing consideration given to the 'drug-likeness' of new molecules, as a means of reducing attrition due to poor pharmacokinetic properties or toxicity. The concept of drug-likeness relates to properties that confer a desirable adsorption, distribution, metabolism, excretion, and toxicity (ADMET) profile. Since Lipinski first proposed his Rule of Five (Ro5) as a means of assessing whether a molecule would have good oral absorption, numerous variations of this rule have been described that indicate 'drug-likeness' [113–116].

Today, the concept has become an important consideration in early phase drug discovery and numerous in silico tools are now available to help predict drug-likeness [117]. However, these rules can also be perceived to be too rigid and there is the potential to rule out good compounds because they fail to meet the pre-defined criteria set in these in silico models. Some predictive tools have been developed that try to overcome the lack of flexibility in the rules [118–123], and now artificial intelligence approaches are being employed to try and assess molecules based on specific liabilities, such as hERG inhibition, rather than on the physicochemical properties that make the molecule more prone to these liabilities [124,125]. Ultimately, all these tools are only predictors to aid the design of novel molecules and the output should be taken only as guidance. Since drug discovery involves multiparameter optimisation, some molecules often still need to be made for the purpose of hypothesis testing, even if they do not meet the necessary criteria of the in silico predictor. In the most recent computational approaches to molecular design, AI-based predictive models for biological activity, as well as drug-likeness, are also being utilised. These optimise and select molecules based on multiple different parameters associated with the desired candidate profile. Although these models have yet to be demonstrated as successful in identifying new drugs, they do look promising [126]. However, these technologies are still in their infancy and there are still many challenges to overcome.

3.2 *Make*—Reaction planning tools

Often when a medicinal chemist designs a new molecule, the planning and identification of a suitable synthetic route can take as long as the preparation of the molecule itself. Typically, when faced with a new target compound, one must perform a retrosynthetic analysis to determine the best way to make the compound (Fig. 13). In order to do this, chemists must have extensive knowledge of chemical reactivity and a good understanding of the bond making and breaking processes, which typically takes years of training to achieve.

With a retrosynthetic analysis executed and a theoretical route in hand, the next process is to predict what the ideal reaction conditions might be, for each individual step. For this, chemists typically explore what has already been reported in the literature. Traditionally, chemists needed to manually search an index, provided by the Chemical Abstracts Service (CAS), of all the structures of compounds reported in journals, technical reports,

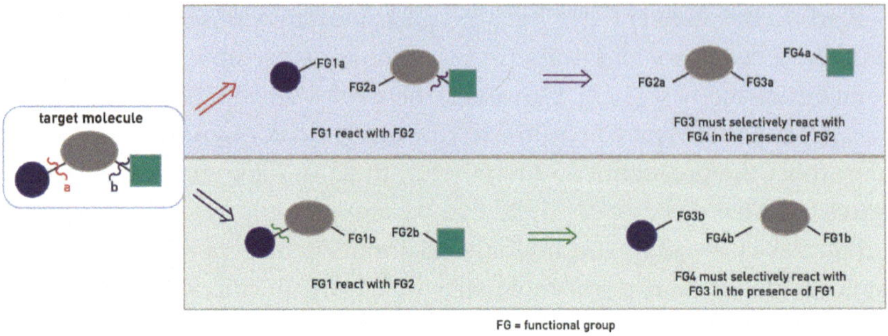

Fig. 13 Simplified retrosynthetic analysis, in which the target molecules are theoretically 'deconstructed' into readily available starting materials by the imaginary sequential breaking of all the potential bonds (disconnections). Alternative retrosynthetic pathways can be explored, e.g. breaking bond **a** vs bond **b** and assessed in terms of both the available starting materials with appropriate functional groups, and what might be the best forward reaction, e.g. reaction of FG1a with FG2a vs reaction of FG1b with FG2b. The same process is repeated until readily available starting materials are reached.

conference proceedings and books. However, this was massively time consuming and laborious. Following the launch of the client application Scifinder in 1995, which utilised the CAS database, the need for manual searching was avoided and printing of Chemical Abstracts actually ceased in 2010. In addition to Scifinder, other automated search tools are also available, including Reaxys (originally the CrossFire Beilstein database) — a web-based tool that draws upon information and data from published literature relating to chemical compounds, chemical reactions, chemical properties, bibliographic data, substance data and synthesis planning information. It provides research chemists with access to current and historical chemistry information, from reliable sources via an easy-to-use interface [127]. Such automated database searching tools are still heavily utilised today by medicinal chemists for reaction planning.

In order for these tools to be successful, they must have an extensive and reliable source of data and information, such as that obtained from peer-reviewed publications. However, there is a great deal of knowledge, particularly within companies, that remains unpublished. Prior to the last decade, it was actually quite difficult to discover what other chemists, even within the same company, were doing since most work is never published externally. One of the biggest challenges in establishing internal systems has been the non-standardised way in which experimental procedures are recorded. However, over the last decade steps have been taken towards standardising procedures. While there is still a great deal to do, the replacement of paper laboratory notebooks with electronic lab notebooks (eLNB), has certainly helped with standardising the way in which experimental procedures are recorded. It has also enabled far greater transparency and has improved the access to, and the retrieval of, experimental data within companies and research groups.

While computer-based tools for the forward planning of reactions have been available for over 20 years, tools that simplify the retrosynthesis step have only fairly recently become a focus of interest. Reaction planning tools, which incorporate both the retrosynthesis and forward reaction planning steps, are one of the most recent developments in enabling technologies for chemical synthesis. Very recently, a plethora of new AI powered tools has been developed and launched for automatic retrosynthesis and identification of the optimal synthetic route to a given target, including Synthia™ from Merck [128], Spaya by Iktos [129], ASKCOS by MIT [130], and ICSynth by Deep Matter [131]. Systems such as SYLVIA [132] and IBM RXN for Chemistry — a state-of-the-art neural machine learning translation/AI web-based tool for retrosynthesis and chemical reaction feasibility predictions [133]. These technologies have a great deal of potential but they still need to be fully demonstrated and proven to be at least as effective as their human counterparts.

3.3 *Make*—Synthesis tools

In the 1980s and 90s, when the first automated synthesis technologies started to emerge, the overarching goal of automation was to increase productivity. Today, that goal is still a valid one, as is the need to increase innovation, develop and embrace automated tools, and free the chemist from routine, mundane work to enable them to focus on greater intellectual contributions. Work towards systems that still embrace the goal of combinatorial chemistry, and which can be used to automatically assemble structurally diverse, interesting arrays of compounds from cheap, readily available building blocks, has continued and many new technologies for synthesis have been developed and launched since the 1980s.

Many of these systems are robotic-based, medium- to high-throughput synthesis platforms based on a number of work stations that carry out specific tasks, and which are capable of performing increasingly complex functions. Academic groups have contributed a great deal by developing systems that can perform new chemistries. In 2015, Burke et al. unveiled their automated building block assembly platform, demonstrated using *N*-methylimidodiacetic boronic acid esters (MIDA boronate) building blocks [134]. While initially restricted to MIDA boronates, the ultimate goal is to expand the scope of the building blocks, resulting in a system capable of making almost any molecule imaginable. This is, however, an enormous task. There are almost infinite ways of combining atoms into molecules and the technology to build any arbitrary molecule has not yet been realised. Inevitably though, broadening the scope of these technologies typically results in an increase in complexity and cost, which is certainly not desirable for medicinal chemistry.

With an eye to trying to retain an element of simplicity, Jensen and Jamison at MIT have developed a modular, automated chemical synthesis system to enable chemists to identify the most optimal reaction conditions for a particular reaction [135]. While reaction optimisation is not the main goal for medicinal chemists, the approach that was taken towards the development of this technology is aligned with the needs of medicinal chemists. The system consists of individual modules that facilitate a specific reaction type. This results in an automated platform that offers the operator greater simplicity and ease of use. As with any automation, it is essential that the technology for chemical synthesis is easy to use and does not create more work for the user than it brings to them in terms of benefits. This is especially relevant in medicinal chemistry where most new molecules will not meet the criteria for further exploration following initial testing. Therefore, unless the technology is very easy and quick to use, it is rarely worth investing time in learning how to use equipment, and setting up and optimising the procedure for each new reaction or molecule.

This consideration is especially important today. This is because, while advances towards meeting the original goal of combinatorial chemistry have certainly been made, the goal posts have somewhat shifted. Perhaps one of the biggest, most recent changes is the nature and needs of the end user. Until recently, most automated synthesis platforms were the domain of specialist groups, who were highly experienced in their operation. As such, many automation tools would appear to have been developed with these end users in mind. However, the complexity of the systems and the time required to programme, operate and trouble shoot the systems, presents too great a hurdle for many medicinal chemists, who have neither the time, skills or inclinations to master these technologies. Now, with the ever-increasing pressure to do more and to do it faster, automated synthesis technologies are something that the majority of medicinal chemists can no longer shy away from. This need for more medicinal chemists to adopt and embrace synthetic automation tools has been noticed by some instrument manufacturers. As such, we have started to see the inclusion of convenient, time-saving optional extras, easy to use interfaces, and supporting software, as a means of increasing the appeal of the technology to the non-specialist.

As automation advances, the number of synthesis tools is ever increasing. Platforms based on individual workstations connected via robotics continue to dominate. However, they can be difficult to set up and operate. Chemspeed® is one such company that has sought to address the challenges associated with these highly complex platform technologies. Their robotic systems are tailored towards specific needs and tasks. In addition, they offer a chemical consumables technology (SMOLE), in which they provide users with the option to purchase sealed, crushable ampules that come prefilled with specific chemicals in various fixed molarities [136]. While such ease of access to pre-weighed reagents is certainly appealing to medicinal chemists, the robotic platforms themselves remain expensive, which precludes the routine use of the technology for the majority of medicinal chemists, despite the addition of user-friendly software.

The subject of operating software is something that is frequently discussed among automation specialist. It is envisaged that tailoring the software and interface to the end user will make the same technology accessible to a wide range of users, from specialist to generalist. There is also a school of thought that believes that in order to provide real value and to simplify their use, automated systems should aim to automate an entire process. In medicinal chemistry, time is best spent actually making compounds, and the time spent programming automated systems can feel like wasted time. Despite the emergence of new reaction planning tools and the increasing availability of automated synthesis platforms, up until very recently the chosen synthetic routes generally still needed to be manually entered in a suitable format in order to instruct the platform exactly how to carry out the reactions, which ultimately reduces the advantages gained in automating reaction planning and synthesis.

This automation gap is one that Cronin has been looking to fill with his Chemify project [137], which aims to develop a method for the conversion of literature synthetic procedures into computer code, to enable the programming of their Chemputer system for the automated synthesis of new molecules [138]. In a related approach, in 2020 IBM also unveiled their 'RoboRXN' system, which combines their 'RXN for Chemistry' AI reaction planning tool with cloud technology and chemistry automation [139]. In this system, cloud-based machine learning algorithms autonomously design and execute the production of molecules, using an automated synthesis platform located in a remote laboratory, with as little human intervention as possible. Such machine learning approaches, and the translation of traditional experimental procedures into computer code, offer a great deal of promise towards the goal of fully automating chemical synthesis. However, it still remains to be seen as to whether removing human involvement completely is either feasible or even desirable.

While the benefits of fully automating synthesis have not yet been convincingly demonstrated, there is clearly a need to develop the option to automate each step, even if all of them are not included in the final process. In traditional synthesis, there are multiple manual operations, many of which involve the transfer of materials from one vessel to another. These transfer steps are repetitive and time consuming, and as such they are the reason that some of the earliest laboratory robotic systems were introduced. However, systems that effect the transfer in a non-robotic fashion have also started to emerge. Many of these utilise the advantages of continuous flow chemistry, which overcomes the need for a human to manually move materials from one step to the next (Fig. 14). As one of the top emerging technologies of the last decade, many companies have established themselves in the field. Vapourtec has a range of versatile, commercially available flow chemistry-based systems [140], as do Uniqsis and Syrris [141,142]. Typically, these systems are flexible enough to conduct a range of chemistries, although solubility of the starting materials and products can be limiting, and they offer the opportunity to safely scale up chemical synthesis and enable chemistries that would be difficult to effect in batch, e.g. the use of unstable intermediates.

While the advantages of flow chemistry systems are clear, namely improved safety and reproducibility, these systems need programming and

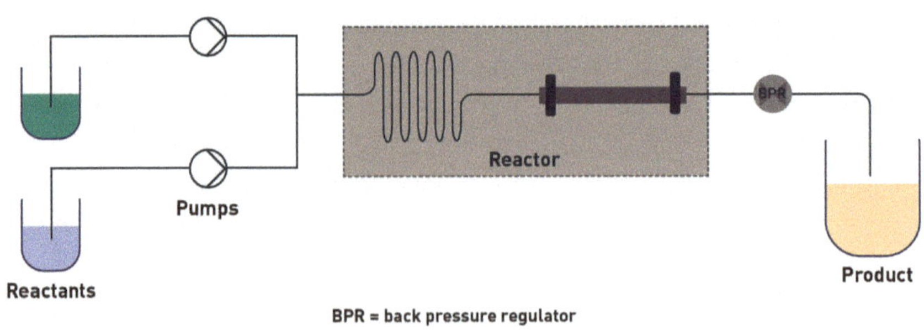

Fig. 14 Flow chemistry systems pump solutions of reagents or reactants, in a continuous flowing stream, into a single tube/coil or column reactor where the reaction takes place. Pressure in the system is controlled by the back pressure regulator (BPR).

time is required to optimise the process for each reaction. Although the option to programme and optimise the systems and processes is highly desirable for some chemists, such as process chemists, it is not so for most medicinal chemists, who have little time for optimising reactions since their primary goal is the preparation of new molecules as quickly as possible for biological screening. As such, medicinal chemists tend to prefer 'kit-like' synthesis technologies, in which the hardware, software and consumables have all been optimised to work together in order to complete a specific task. The manual parallel synthesis technologies developed during the combinatorial chemistry era fall into this classification, and remain very popular today. Therefore, there have been considerable efforts to develop new kit-like automated technologies.

Generally, the more flexible the automated synthesis platform is, the less kit-like it is. As such, continuous flow synthesis systems, which provide the flexibility to make a huge range of molecules under a multitude of reactions conditions, are not usually in a 'kit-like' format and do typically require significant skill and time to programme, operate, and optimise. These systems provide a great deal of benefit for users who need a generic system with which they can optimise multiple parameters, but they can be less appealing to medicinal chemists for these very same reasons. One exception is perhaps the H-Cube and related products from ThalesNano [143]. Although based on flow chemistry processes, they are not generic systems, typically enabling only one function, thus there is little to optimise. Synple Chem, has developed an easy to use, cartridge-based, automated synthesis technology, which has preprogrammed and highly optimised reaction protocols, and thus requires no programming or optimisation for use within discovery chemistry [144].

Despite the number of automated synthesis systems that have been developed to date, not many of these currently offer the easy to use, user-friendly features that medicinal chemists desire. As such, still today a large proportion of synthesis in medicinal chemistry continues to be done using traditional, manual methods, non-robotic parallel synthesis equipment, or simple personal reactors that have a low number of functions. The non-robotic parallel synthesis equipment available today range from the simpler Radleys Carousels, DrySyn Octo reaction stations from Asynt [145], and the StarFish work station from Radleys (a modular, space-saving, general purpose heating and stirring solution) [146], to more sophisticated offerings such as the Mya 4 reaction station, also from Radleys, which has four independent reaction zones and the flexibility to use a range of glassware [147]. More sophisticated still. the EasyMax personal reaction station from Mettler-Toledo, eliminates the need for traditional round-bottom flasks and offers programmable, reproducible methods, with excellent data capture methods [148]. The Amigo Workstation from Amigo Chem, provides automated monitoring of up to 10 reactions [149], and Personal Reactors are also offered by JKEM [150].

Several other types of small personal reactors are worthy of mention here as enabling technologies. Equipment such as the simple non-robotic flow chemistry set ups, including the Asynt fReactor Flow Chemistry Platform, which sits atop a standard stirrer hotplate and utilises an external HPLC pump, can provide a relatively low-cost means of investigating flow chemistry options [151]. The Electrosyn, co-developed and launched by Phil Baran and IKA, is worth a mention as its simplicity appeals to the needs of medicinal chemists. As a low cost, easy to use tool, it enables non-specialists to carry out electrochemistry, which is a synthetic methodology that, until recently, was rarely utilised by medicinal chemists because of its complexity and the lack of general experience with it [152]. Likewise, there is a growing number of simple photochemistry reactors now commercially available. Merck Millipore have an extensive range, including the SynLED Parallel Photoreactor, Penn PhD Photoreactor M2, and the Micro Photochemical Reactor, which can process up to 16 reactions at once [153]. Technologies such as these are not only enabling medicinal chemists to easily explore new or previously minimally utilised chemical methods, but they are also enabling automation to explore a far greater expanse of chemical reaction space than was previously possible.

The on-going demand for this kind of equipment is evidence that medicinal chemists continue to seek the most straightforward and readily applied solution to their synthesis challenges. If one considers the automation technologies that have been successfully integrated into medicinal chemistry practices, they are almost always the technologies that are user friendly and require little in the way of training to use them. One of the greatest successes has been the microwave reactor. While manual input is required to set up, monitor and work up the reactions, several reactions can be prepared in parallel and set up to run sequentially on systems such as the Biotage Initiator, which uses a robotic arm to switch the sample tubes [154]. The systems are so easy and effective to use that they quickly achieved wide spread adoption following their launch and they are still heavily used in medicinal chemistry today.

While these user-friendly technologies are incredibly appealing to medicinal chemists, one of the dangers of easy to use technologies, such as the microwave reactors, is that there inevitably ends up being a bias towards chemistries that are amenable to that particular technology. Suzuki reactions, for example, are conveniently carried out in microwave reactors and arrays of analogues can be readily assembled this way, which can result in a structural bias in the products targeted. This is something that was certainly evident in the era of combinatorial chemistry and to some extent it has persisted even beyond this. While each technology might have its natural bias, if a range of technologies with different capabilities are utilised, significant biases can be avoided.

Given the structural biases that appeared to result from the preparation of traditional combinatorial libraries, it is important to expand diversity by the use of alternative technologies, such as DNA-encoded libraries. While this offers an alternative approach to library generation, it also still has its limitations, particularly in terms of the types of chemistries compatible with DNA. However, in recent years, several groups have reported on their efforts to develop and utilise stabilised forms of DNA [155]. It is still the case though that the available chemical toolbox for DEL synthesis only covers a small fraction of known reactions when compared to classical organic synthesis. In addition, since purification is not normally possible, high yielding reactions with minimal side-products are required, in order to minimise the rate of false positives and negatives [156]. Despite its limitations, DEL synthesis is an example of an enabling technology that is being taken out of the hands of the specialists and presented to the generalists in kit form. Last year, Merck signed an agreement with DyNAbind GmbH to exclusively distribute their fragment-based DNA-Encoded Library kits to accelerate drug discovery applications [157].

New automated synthesis technologies that are capable of performing more challenging, complex chemistries are certainly still needed but so are technologies that automate the mundane, laborious processes in relatively simple chemistry. As summarised by Anat Milo: 'Automation will free us from mundane repetitive tasks, allowing us to focus on the aspects we excel at as humans—providing creative leaps based on curiosity, playfulness, and untraditional thought processes' [158]. However, it is crucial that the accessibility of these future and existing technologies for medicinal chemists is also addressed if we are to truly achieve widespread adoption and relieve the existing bottlenecks in drug discovery.

3.4 *Make*—Work-up tools

Regardless of the manner in which a chemical reaction is conducted, most reactions require some form of workup procedure, in which a series of manipulations are carried out in order to isolate the product or products of the reaction. Typically, this can involve any of the following processes: reaction quenching to destroy unreacted reagents; precipitation and filtration; evaporation; and partitioning of a reaction mixture between organic and aqueous layers in a liquid-liquid extraction process, with the last two being the most time-consuming processes.

In medicinal chemistry today, many workups are still done in the same way that they have always been done, e.g. an aqueous extraction using a separating funnel. Such traditional work-up procedures can be very time consuming and inefficient in terms of solvent usage and waste generation. However, techniques developed during the combinatorial chemistry era are being increasingly applied to non-combinatorial chemistry projects, including use of disposable solid-phase extraction (SPE) cartridges and plates [159].

It is now possible to buy diverse ranges of SPE cartridges and plates for the isolation of a given product from a wide range of reaction mixtures. This range includes those containing ‘catch and release’ ion-exchange resins that capture basic or acidic compounds on a resin while allowing the rest of the reaction mixture to pass through. Isolation of such acidic or basic compounds would traditionally have required a full aqueous work-up procedure. These cartridges offer other advantages, such as avoiding washing of glassware and a reduction in solvent use (Fig. 15).

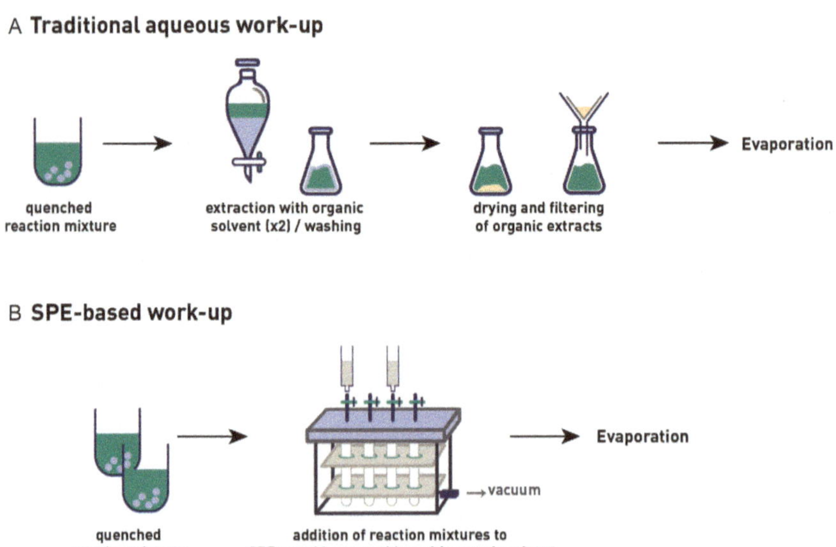

Fig. 15 Acceleration and parallelisation of workups using SPE cartridges.

Given that chemists can spend a large proportion of their lab time on non-productive tasks, there is an increasing trend towards the use of disposable consumables, like SPE cartridges, especially in industry-based laboratories, where it is more cost-effective, for example, to eliminate the time spent cleaning contaminated glassware. Small disposable consumables such as pipettes, syringes and filter cups, are available for many processes in synthetic chemistry. While solid-phase extraction consumables described above can certainly be used with non-robotic parallel synthesis equipment, such as the GreenHouse Work-Up Station™ from Radleys [160], Visiprep™ SPE Vacuum Manifold from Sigma-Aldrich [161], the Biotage VacMaster [162], Radleys' Carousel Work-Up Station™ [163], and the Thermo Scientific™ HyperSep™ [164], other systems have also been developed that are designed for use with automated liquid handling work stations. These include the VacMaster-96 from Biotage [165], Merck Millipore's MultiScreen®$_{HTS}$ Vacuum Manifold [166], and the On-Deck Vacuum Station, which is compatible with Hamilton's automated liquid handling platform [167]. Porvair Sciences also have a vacuum manifold, the MicroLute™, which utilises filter and SPE plates. More recently they have also launched the Universal Robotic Manifold, which can be easily assembled and disassembled by robotic manipulators, thus enabling productive automation of SPE clean-up procedures [168].

Solvent evaporation is an essential component of most work-up procedures but it can be a slow process. The most commonly used traditional evaporation process is rotary evaporation, a type of vortex evaporation in which the sample is swirled to create a vortex and subsequently increase the surface area, which accelerates the rate of evaporation under a partial vacuum. While rotary evaporators require only a few minutes of hands-on time to set up, they are only capable of evaporating one sample at a time. Therefore, sequential evaporation of post-work-up samples from even a small array of compounds could take an inordinate amount of time using a single standard rotary evaporator. As such, systems have been developed to enable solvent evaporation from several samples in parallel. Büchi, one of the leading manufacturers of rotary evaporators have introduced their own range of vacuum-based, parallel vortex evaporator systems, such as the Multivapor™, which can process 12 samples simultaneously [169].

In addition to vacuum-based evaporation, simple manual parallel blow-down systems, in which a flow of nitrogen is passed over the samples, have been used routinely in medicinal chemistry. Although initially these systems released the waste solvent into the environment, which was not desirable, this issue has now been largely addressed with systems such as the GreenHouse Blowdown Evaporator™ from Radleys, which has a condenser attached to collect the evaporated solvent [170]. There is now a plethora of blow-down solvent evaporation products on the market. Like the Radleys equipment, several of these are designed to be compatible with a lab equipment manufacturer's own range of manual parallel synthesis equipment, including the DrySyn Spiral Evaporator developed by Asynt and BioChromato, in which a spiralised flow of gas is applied to effect solvent evaporation in tubes, with evaporation assisted by vacuum suction. This novel spiral technology is reported to prevent 'bumping' during solvent evaporation, which is a common issue with less well controlled evaporation processes, and can result in sample loss and contamination. While many of the evaporation systems discussed above can evaporate relatively small libraries of samples (4–20), systems also exist for medium throughput applications. The MiniVap from Povair Sciences is a simple nitrogen blow-down evaporator that can be utilised to evaporate up to 24 vials or 96-well microplates. While this also needs to be operated manually, it is the simplicity of systems like this that typically suits medicinal chemistry applications [171].

Automated evaporation systems that remove the need for manual operations are also utilised within medicinal chemistry and offer additional benefits. These systems are often based on freeze drying, centrifugal or vortex evaporation. Freeze drying systems, like the Benchtop Pro Freeze Dryers [172], can dry multiple samples simultaneously in a controlled, automated manner. The systems work by freezing the samples and then, at low pressure, subliming the solvent directly from the solid phase to the gas phase to give a 'fluffy' lyophilisate. One advantage is that a high level of dryness can be achieved, resulting in solids from what would otherwise be 'sticky' samples, but on the downside, this process is actually quite slow. The technology is also limited to samples in solvents that can be easily frozen using vacuum alone, such as water/*t*-butanol or water/dioxane mixtures.

One of the biggest challenges for these systems, and indeed all systems that simultaneously evaporate multiple samples, is the avoidance of cross-contamination. Automated evaporation technologies can ultimately handle this much better than the manual parallel evaporators. Centrifugal evaporators, such as the Genevac, carefully control the vacuum and the rotor speed to prevent bumping and cross-contamination [173]. These systems also have good temperature control and are able to automatically detect and cease the evaporation process when the samples are dry. As such, these types of system are able to remove a wide range of solvents. Robot compatible systems, such as the Hettich CombiDance system, which uses a vortex motion and infrared heating, along with an intelligent vacuum pump to remove solvent, have also been developed and can support a variety of sample formats [174].

Due to the wide ranging availability of systems for solvent evaporation, there is essentially a solution to suit almost every need, from basic rotary evaporation to fully automated, robot controlled, high-tech evaporation platforms. Often the biggest challenge is finding a single evaporation technology to suit all users in the laboratory and it is not uncommon to find multiple options within a single lab. Like most laboratory equipment, users tend to have their individual favourites but ultimately, the available space and cost also plays a major role in determining which evaporation tools are selected.

3.5 *Make*—Automated purification systems

After most chemical reactions and workups, the products obtained need to be purified to a certain level in order to be used successfully in subsequent reactions. In medicinal chemistry, it is absolutely essential that the new molecules prepared by medicinal chemists are purified to a high level (typically >95%), prior to being submitted for biological testing. However, standard purification processes, can be very laborious.

'Drug-like' molecules are typically not volatile enough for distillation. Although chemists may well use reagents and prepare intermediates that are suited to purification this way, it is not a process that is used frequently enough to justify a significant investment in any kind of automated system. Distillation for the purification and drying of solvents was previously the most frequent type of distillation used in medicinal chemistry since it is a highly effective means of obtaining dry solvents for moisture sensitive reactions. However, distilling highly flammable solvents is particularly hazardous and this is now longer carried out in most industrial laboratories. Today a large range of anhydrous solvents is commercially available, which meets the needs of the vast majority of reactions carried out by medicinal chemists. Even in academic settings, where the cost of commercial dry solvents may be limiting, solvent distillation is being replaced by much safer solvent purification systems, such as the Pure Solv Micro solvent purification system and other related systems [175–178].

Despite the fact that crystallisation seems to have fallen out of favour in medicinal chemistry, several companies have endeavoured to make this technique easier and more accessible to chemists in this field. Companies, such as Technobis Crystallization Systems, a pioneer in the field of solid-state research, have developed automated platforms for crystallisation, e.g. the Crystal Breeder, a bench-top, multi-reactor crystallisation platform. However, the technique still remains rarely used for purifications in medicinal chemistry [179].

In contrast, flash column chromatography, has benefitted from significant automation in the last 20 years. Since the first low pressure, automated flash chromatography systems were introduced around 20 years ago, and were met with some scepticism, the technology has developed and today's systems have now almost entirely replaced manual chromatography in industrial medicinal chemistry laboratories. Biotage, a pioneer of automated flash chromatography, has the Isolera™ range of systems (Isolera One, Isolera Dalton, Isolera Prime). In addition to time, efficiency and cost benefits, they also offer optional features such as simultaneous detection of product by MS (including both positive and negative ions), and predictions of the best cartridges, solvent systems and gradients for a given mixture [180]. Recently Biotage has also launched its new Selekt system, a compact, flexible, lean system with two column channels [59]. In combination with their Sfär flash columns, which contain spherical silica particles for greater surface area, higher loadings are possible and thus it represents one of the greenest flash chromatography solutions available [181].

Several other automated flash chromatography systems are also commercially available. Teledyne Isco's CombiFlash NextGen range of low-medium pressure systems offer flow rates up to 300 mL/min and can tolerate injection of both liquid and solid samples [182]. In addition, they have a combined flash/prep-HPLC system (CombiFlash EZ Prep) and the CombiFlash Torrent system for larger scale purifications (up to 300 g). Büchi's Pure chromatography systems are very compact and easy to use. They have both an integrated UV and ELSD detector system, and like the CombiFlash EZ Prep, they can be used for both normal and reverse-phase applications [183]. While each system has its own pros and cons, all of them essentially consist of a pump, sample injection ports, a detector, a fraction collector to collect the eluent, and a user-friendly interface, all of which are controlled via a microprocessor (Fig. 16). Generally, the systems are in the same price range and often the selection of which system to buy for general use is based on an individual user's preference, their personal experience with the equipment, and the after sales support.

In addition to the low-medium pressure automated flash chromatography systems and combined low-medium pressure/high-pressure liquid chromatography system, many instrument manufacturers also offer systems

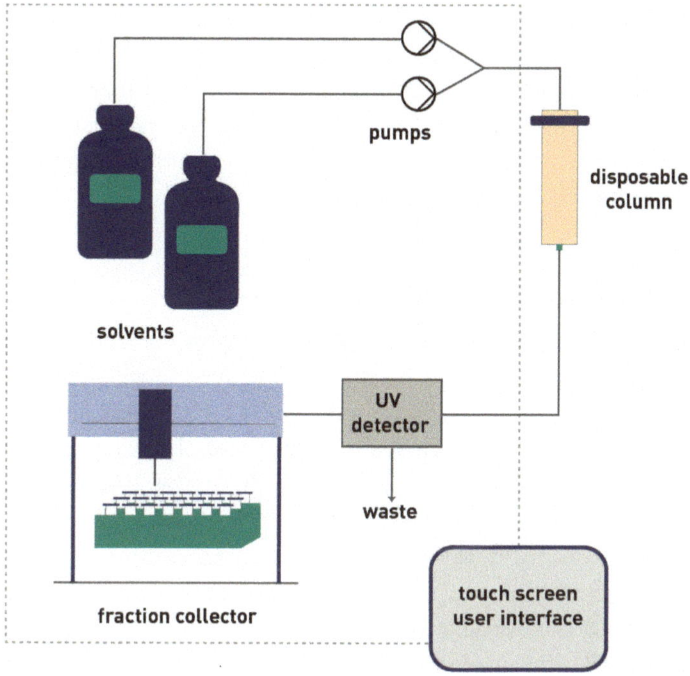

Fig. 16 Automated flash column chromatography systems elute the pre-loaded crude material by pumping the desired solvents through the disposable chromatography column. An integrated detector selects when to collect the eluent as purified fractions and when to send the eluent to waste.

that perform preparative HPLC, which enable compounds to be cleaned to a very high level of purity. Teledyne Isco's ACCQPrep HP150 is an example. It was developed to be as easy to use as their flash systems and utilises the same software [184]. In addition, a range of modules are available for automating column selection and sample injection, which are all compatible with both the ACCQPrep HP150 and the CombiFlash EZ Prep systems [185]. Likewise, Shimadzu has a wide range of products, including the Nexera Prep and the Nexera UC Prep, a preparative supercritical fluid chromatography system that is reported to achieve superior recovery rates for purification [186,187]. For higher throughput needs, systems exist, such as the semi-preparative AZURA Lab Prep LC 50 HPG, which can automatically select and inject from up to 768 positions, depending on the plates or vials that it is equipped with [188]. The Infinity II Preparative LC System from Agilent is also ideal for high-throughput, laboratory-scale purification on a milligram to gram scale [189].

While the manufacturers and instruments listed above are far from an exhaustive list of the systems available, they have been selected to demonstrate the technologies available for a broad range of purification applications that might be encountered by medicinal chemists. Regardless of which system is ultimately selected, it is clear that these technologies offer medicinal chemists significant timesaving benefits.

3.6 *Make*—Automated analysis techniques

Although automation of synthetic processes remains somewhat challenging, the application of automation to analysis has proved to be highly successful. Analytical experiments, such as LC-MS and NMR, generally involve the repeated application of the same standard protocol, thus they are more readily automated than synthetic processes. In addition, because these analytical processes either use only a tiny amount of material or, for those that require larger amounts, do not destroy or alter the original sample, the analysis can always be repeated if the automation fails. As a result, users are more willing to adopt this kind of automation.

Combining liquid chromatography with mass spectrometry was made possible around 20 years ago, and LC-MS systems are a good example of an enabling technology that has seen wide spread adoption. While earlier systems required manual injection of a sample, today's systems are equipped with sophisticated and robust autosamplers, from which the system can select a precise aliquot for sequential injection into the HPLC system under high-pressure flow conditions. As well as the obvious efficiency gains from the use of LC-MS autosamplers, there has been a growing trend of replacing HPLC systems with UHPLC systems, which offer at least as good peak resolution but with much shorter run times. Replacing 5–7-min LC-MS runs with 2–3-min run times greatly increases the capacity of the system and reduces delays in obtaining data, which can be crucial when LC-MS is being used as a reaction monitoring tool.

All of the main instrument manufacturers in the field offer ranges of different HPLC and UHPLC systems, which can be paired with suitable mass spectrometers, and seamlessly controlled with integrated software. Waters have a large selection of systems, including the ACQUITY UPLC System, which utilises high flow rates to reduce the cost and time for each sample run. It can be paired with Waters optical and MS detection technologies to provide a complete end-to-end solution [190]. Similarly, Shimadzu's significant range of products also includes systems that exploit UHPLC for increased speed and productivity [191]. As major contributors to the LC-MS area, Agilent have an extensive portfolio of instruments for LC-MS, including a system for high-throughput LC-MS and one that can detect accurate masses as well as unit masses [192]. Although perhaps better known for their NMR systems, Bruker offer a range of mass-spectrometry systems [193], as well as the Elute and nanoElute HPLC systems [194], which can be seamlessly integrated and automatically controlled using their Compass LC-MS acquisition and processing software [195]. For some applications, further enhancements can be made through the use of automated LC-MS sample preparation systems, which are now readily available. However, these only really add value when there is the requirement to simultaneously prepare large numbers of samples. With the exception of library synthesis, manual preparation of LC-MS samples is unlikely to be a bottleneck in medicinal chemistry so there is not a huge need for such sample preparation systems for routine use.

In addition to the efficiency and productivity gains that automation of LC-MS has brought medicinal chemists, similar benefits have also been achieved with NMR. Prior to the introduction of autosamplers and software for acquisition, each user would need to manually add an individual sample into the NMR probe head, and then manually perform the locking, shimming, tuning, and calibration steps, which was all incredibly time consuming. Upon completion of the sample acquisition, the sample would again need to be manually removed, leaving the system ready for the next user, which meant the user always needed to be there when the acquisition was finished. Finally, the output from the acquisition, the FID spectrum, would then need to be transformed to the actual interpretable spectrum and the necessary phasing- and baseline-corrections carried out along with axis calibration. All of this was required before the user could start to consider analysing the NMR spectrum. Today, almost every step of NMR spectra aquisition can be carried out in an automated manner.

Bruker, leaders in the NMR field, have developed an automation software, TopSpin™, and a graphical user interface, IconNMR™, which enables fully automated acquisition, processing and data analysis, and is free to download for academic users [196,197]. While sample carousels have been in use for some time on NMR machines, originally the user often had to climb steps to reach the carousel, which was located close to the probe head on an NMR machine (top of the system). Now samples can be added automatically to the probe head through the use of user-height level systems, such as the SampleCase [198]. This unit is an automated sample changer system controlled by IconNMR™, for which heating and cooling options are available. Compatible sample preparation robots also exist. However, such systems do not have a general utility in medicinal chemistry.

While acquisition of NMR spectra has been accelerated through automation, assignment and structural elucidation based on the spectra can still be time consuming since for structural elucidation, the chemists need to consider each individual peak in terms of the chemical shifts and splitting patterns. As such, this can still represent a bottleneck for medicinal chemists. However, Goodman and co-workers at the University of Cambridge have recently developed a program, DP4-AI, which automatically analyses ^{1}H and ^{13}C NMR data, assigns the peaks, and suggests which structure is most likely from a set of possible candidates [199,200].

A high field NMR machine is an expensive analytical instrument, requiring a specialised and dedicated location due to the strong magnets employed. However, because of advances in the technology, smaller, bench-top NMR systems are now becoming more accessible and at a fraction of the price of a full sized system. The Fourier from Bruker, for example, is a stand-alone, self-contained 80 MHz high-performance, cryogen-free magnet, nuclear magnetic resonance (NMR) bench-top spectrometer [201]. Like the standard NMR systems, the easy to use software and interface guides users through the pre-defined, dedicated workflows. Magritek have similar 60 and 80 MHz systems, the Spinsolve 60 and 80 [202], as do Buck Scientific [203], Oxford Instruments [204], and Nanalysis [205]. While these bench-top systems may sound ideal, they do have some significant limitations. As they operate at a low field they are lacking both in terms of resolution and sensitivity and thus they are not particularly useful for structural elucidation. They also need very concentrated samples, which is less than ideal for medicinal chemistry where only small amounts of material are made.

While automation of techniques such as LC-MS and NMR have been greatly enabling over the last few decades, as synthesis technologies have moved on thoughts have also shifted to how we can analyse reaction mixtures and products using more modern automated synthesis systems and platforms. Due to the increasing popularity of flow chemistry, there has been more effort focused on the development of analytical techniques that can be employed in a fully integrated and automated manner. As such, techniques such as NMR have been developed to accommodate samples being delivered in flow rather than in pre-prepared glass tubes. Using flow injection probes, which contain a glass chamber with an inlet for the solution to enter the chamber and an outlet for the sample to be evacuated and rinsed out, it is now possible to conduct NMR in flow (Fig. 17). These systems

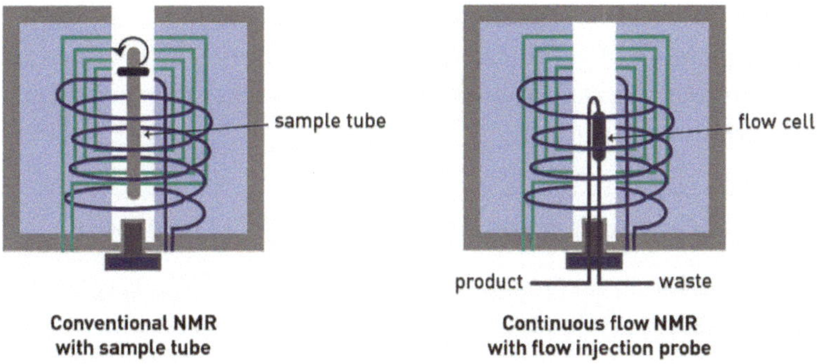

Fig. 17 Conventional NMR has the sample contained within a sealed glass tube, whereas continuous flow NMR uses a flow cell, which has an entry inlet and an exit outlet for the sample.

have several applications, including use in fragment-based NMR screening for drug discovery [206]. The ability to inject samples into the probe has also enabled the coupling of HPLC and NMR. As such, mixtures can be isolated as their individual components and automatically injected into the NMR to characterise the components. It is even possible to go one step further and couple HPLC-NMR with MS. In this case, while the bulk of the sample from the HPLC is directed to the NMR, a small percentage is automatically directed to the mass spectrometer.

While inline/online flow-based HPLC analysis is certainly useful for some applications, in the case of reaction optimisation, it often cannot provide fast enough information for real-time feedback. In these cases, alternative systems can be beneficial, such as the ReactIR Fourier Transform Infrared Spectroscopy (FTIR) instrument from Mettler-Toledo [207]. Allowing for in situ, real-time monitoring of continuous flow chemistry, it enables researchers to save valuable time and materials when screening and optimising their reactions. There are a growing number of online analytical methods but the instruments are still expensive at the moment, and require lengthy calibration, which can only really be justified when a process is routinely used and repeated. However, there is now an observable shift beginning, in which the technologies that were once considered to be the domain of specialists are becoming more mainstream.

Coupling of synthesis with purification and analysis, and automation of the whole process is sure to be an on-going theme in the future. It is likely though that the first major applications will be for reaction parameter optimisation and high-throughput library synthesis. However, it is probable that in the near future, these developments will be adopted by the majority of medicinal chemists. With the advent of AI and more sophisticated robots, it is possible that medicinal chemists will be playing a very different role in the process in the future.

4. Automation gaps and emerging technologies

Automation in medicinal chemistry has come a long way since it was first envisaged as a productivity-enhancing solution. However, the science of drug discovery does not stand still, which continually changes the needs of medicinal chemistry and its requirements for automation. As the targets for therapeutic interventions move away from the traditional receptors and enzymes that can be drugged with small molecules, and into new druggable space such as RNA therapeutics, protein degraders and antibody conjugates, the requirement for new chemical modalities that can interact with these targets is increasing [208]. As such, new methods and automated technologies will be required that enable access to these types of modalities. Therefore, emerging technologies really need to be meeting not just our current needs, but also trying to foresee the needs that could arise over the next few decades.

4.1 Artificial intelligence and machine learning

Due to advances in both computer hardware and in silico methods, artificial intelligence (AI) approaches using machine learning (ML), which are able to improve and accelerate the *Design-Make-Test-Analyse* drug discovery cycle, are becoming a reality today. Utilising algorithms and models, based on sample data, ML processes can make predictions or decisions and also learn from the outcome and refine the original processes. This enables computers or robots to perform more advanced tasks without being specifically programmed to do so. Such systems are increasingly present in everyday life and have been demonstrated to have a positive impact on productivity in a range of different industries.

Progress towards the potential use of intelligent computer systems in medicinal chemistry actually began over 20 years ago. As a result of improved computational processing power, an increasing number of computational tools have been utilised by medicinal chemists for a multitude of different tasks, including molecular modelling and physicochemical property predictors to support the design of new targets, literature searching tools for planning chemical routes to new molecules, and data analysis software for rapid visualisation of screening data. However, there are still many aspects of medicinal chemistry which can further benefit from new computational methods and tools, as well as refinement of the existing ones. AI systems in particular could provide significant benefit to medicinal chemists since these systems are likely to 'think' differently to medicinal chemists, which would be beneficial in terms of broadening the scope of ideas and solutions available.

AI is rapidly developing, in part due to advancements in deep learning (DL), a subclass of ML, which uses more complex artificial neural networks and can handle bigger datasets and utilise a larger number of descriptors. As a result, we have begun to realise its revolutionising potential and are now starting to see the application of machine learning and artificial intelligence in all stages of drug discovery (Fig. 18) [209].

In the initial hit and lead finding stages of drug discovery, computational tools have already had a big impact. Enabling tools for ADMET predictions, as well as virtual library construction, plus quantitative structure activity relationship (QSAR) modelling, are now widely used to help guide molecule design in medicinal chemistry [210]. The ability to make and assess molecules in silico is a powerful technique as it enables potential new molecules to be filtered and prioritised for synthesis and screening, thus avoiding the unnecessary waste of valuable time and resources. Although many of these models and screening procedures have their limitations, due in part to sub-optimal models and poor experimental data sets, AI and ML approaches offer a great deal of promise for the efficient refinement and improvement of these models [126].

Numerous methods already exist for the construction and subsequent screening of virtual libraries using pharmacophore or virtual docking

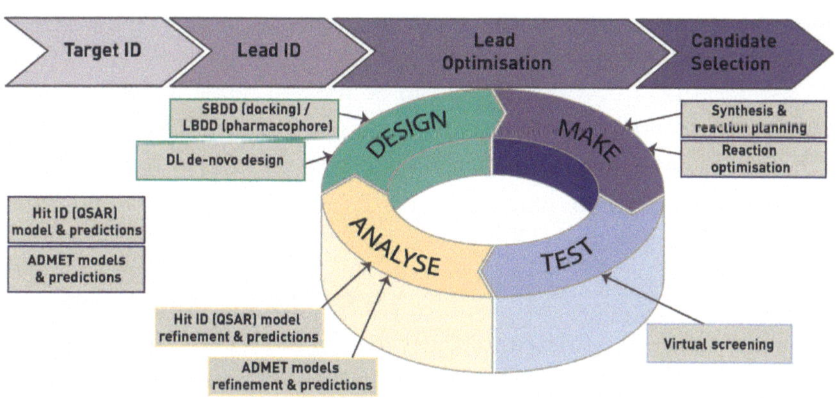

Fig. 18 Potential applications of AI and ML in early drug discovery.

models, which are the bedrock of Ligand- and Structure-Based Drug Design (LBDD and SBDD) [211,212]. However, most of these methods for virtual library assembly use manually defined transformation templates based on a collection of monomers or a starting fragment, which creates some element of rigidity. With new DL approaches, chemical structures can be generated without pre-defined transformation templates. In a recent example of this approach, a SMILES-based model was used to generate a library of novel compounds, which had 97.7% SMILES validity and a 75% pass rate for meeting the criteria for entry into AstraZeneca's screening collection. Following multiple 'fine-tunings' or retrainings of the model for a specific set of desirable properties, it was able to generate molecule sets in which more than 50% of the compounds had predicted activity against the specified target [213]. There are, however, some clear limitations to this approach. The most obvious limitation is that the algorithm does not follow any defined chemistry or transformation rules, and therefore, the synthetic accessibility of the final compounds could be far more limited than products of libraries built using known transformations. While this approach clearly still needs a lot more validation, with this added flexibility it offers a real opportunity to create greater diversity and novelty in virtual libraries and adds a complementary tool to a drug discoverers tool kit.

Virtual library screens often utilise QSAR models, which relate a set of 'predictor' variables, such as physicochemical properties or R-groups, to the biological activity of a given molecule. As such they can be used to identify the compounds that are most likely to interact with the target and have optimal drug-like properties. Although earlier QSAR modelling has been dogged over the years by reports of poor performance, unpredictability, and unreliability, this was often due to poor and incomplete data sets and assumptions of linearity in the modelling. Today, ML techniques offer the opportunity to explore nonlinear SAR relationships with high accuracy and precision [214]. As such, they can be used to model QSAR, or even quantitative structure–property relationships (QSPR). ML can also be used to develop AI systems that predict in silico how changes in chemical structure might alter the biological properties. In addition, other important considerations, such as ADMET, and drug–drug interactions can also be effectively predicted [215].

With the plethora of ML tools to aid hit identification, as well as the design and selection of new molecules for synthesis, attention has recently shifted to the role ML can play in the synthesis of molecules. Given that many ML approaches to compound design generate new molecules through non-reaction based means, it is imperative that a synthetic accessibility assessment is made and used as a selection filter since it is likely that some of these novel chemical entities will not be synthetically feasible. In order to assess synthetic feasibility, one needs to carry out a retrosynthetic analysis to consider how a given molecule might be made [216]. Depending on the complexity of the target, this can be a slow, painstaking process even for experienced chemists looking at a single target. Therefore, it would be an almost unmanageable task for the hundreds or thousands of possible targets generated in virtual libraries. Efforts to try and simplify and speed up this process to address this challenge have been, and still remain, a significant undertaking.

Up until the turn of the century, efforts to develop in silico systems for retrosynthetic and synthesis planning were limited by computational power and the models were hand-coded with a range of pre-set rules for bond making and breaking. It is only in the last two decades that automated reaction planning has become a reality. The development of ML systems, which utilise published literature procedures to train the model, instead of having manually coded, pre-defined reaction rules and templates, has the potential to enable significant time savings and also expand the pool of ideas compared to those generated by medicinal chemists alone. Both approaches to reaction planning tools are valid, have shown demonstrable successes [217,218], and have their own advantages and disadvantages. While the ML approach struggles with predictions when there is a limited amount of data available to train the system, e.g. for the prediction of new reaction classes, hand-coded rules systems fare much better in these situations. In contrast, incorporation of new reactions into a rules-based system is time consuming, whereas this task is simplified and expedited in ML models.

In addition to planning a synthetic route in terms of bond-forming and bond-breaking processes, chemists also need to identify a specific and feasible set of reaction conditions, to ensure that those making and breaking processes occur as expected. Identifying the best conditions can be time consuming and can even require considerable experimental efforts in the lab to optimise the reaction conditions through a trial and error process. To avoid too much wasted time spent optimising reaction conditions, medicinal chemists will often stick with a favourite set of reaction conditions for a particular substrate or series, or select reaction conditions based on the reagents they have available. This can inevitably lead to a bias, which AI systems, such as IBM's RXN AI reaction planning tool, would not necessarily also create and follow. This is not to say that the AI systems will not have other biases, just that these will not necessarily be the same biases that the chemist has. While the AI approach could potentially lead to a greater diversity of reaction conditions and potentially better reactions, in reality some restrictions will have to be placed on the output, either in terms of the available starting materials or the equipment and capabilities required to carry out a specified reaction. Therefore, it is inevitable that regardless of whether a human is making a judgement on the output of the reaction planning software or whether the output is fed directly into an automated synthesis platform, there will need to be an additional practicality and feasibility check.

The usefulness of such reaction and synthesis planning AI systems will always be limited to some extent because of the huge variability in the way that experimental data in publications and patents is recorded. Even within the same laboratory, and sometimes even from the same chemist, there is significant variability in the way that individuals carry out and record their experimental procedures. It is not uncommon to see reactions reported as needing 12-h reaction times, when in reality the reaction may well have been complete in just a few hours but it was left over night for convenience. As a result, the data set upon which these models are built can be poor, incomplete, inconsistent and not always reflective of the actual reaction requirements.

For medicinal chemistry, where the primary goal is the preparation of new molecules for biological screening and the yield is a secondary consideration, this inconsistency does not amount to as big an issue as it would for other chemists, particularly those in process development who want highly optimised reaction conditions. As such, these tools are still of great utility to medicinal chemists since they offer at least a starting point from which the chemist can use their own skill and judgement to determine the best approach. Although it is clear that AI-based reaction and synthesis planning is still very much in its infancy, and there is still much to do to overcome the obvious limitations, there is a great deal of interest in the field. In demonstration of the interest, in 2018, the Machine Learning for Pharmaceutical Discovery and Synthesis (MLPDS) Consortium was formed, consisting of researchers from MIT and representatives from 13 pharmaceutical and chemical companies, with the goal of developing machine learning-based algorithms and tools to accelerate the Make phase of the *Design-Make-Test-Analys*e cycle [219]. Medicinal chemists play a significant role in most stages of drug discovery and through the adoption of AI and ML systems, they have the potential to greatly accelerate the whole process.

There is an increasing shift in the demand for automation, from the specialist technology experts towards the generalist medicinal chemists, who want automation tools that enable them to do their job more efficiently without requiring them to develop an entirely new skill. Consequently, the goal of many companies and research groups now is to find ways to make the same technologies more user friendly and accessible. In light of the advances in AI and ML, one obvious solution is to incorporate intelligent systems that are capable of self-programming and decision making. Such solutions may present as an easier to use option for medicinal chemists, and might also pave the way to the eventual complete automation of discovery research.

4.2 Fully automated integrated synthesis platforms

In order to develop fully automated, integrated synthesis platforms, the automation technology for every stage of the process need to be developed and established. This includes the design of new molecules based on prior knowledge, reaction planning, selection and acquisition of reactants and reagents, as well as reaction set up, execution, workup and purification (Fig. 19). This chapter has so far explored many technologies that can automate one part of the process, e.g. reaction planning systems or synthesis systems, but there are very few examples of fully automated synthesis systems in which all of the stages are automated and fully integrated. This section will review the progress towards this goal to date.

As discussed, early-automated chemical synthesis platforms, led to what many believe to be an unfavourable structural bias, with the products formed typically favouring those with less than ideal physicochemical properties. In order to access structural diversity, a large proportion of the non-library compounds synthesised by medicinal chemists are done so in a highly

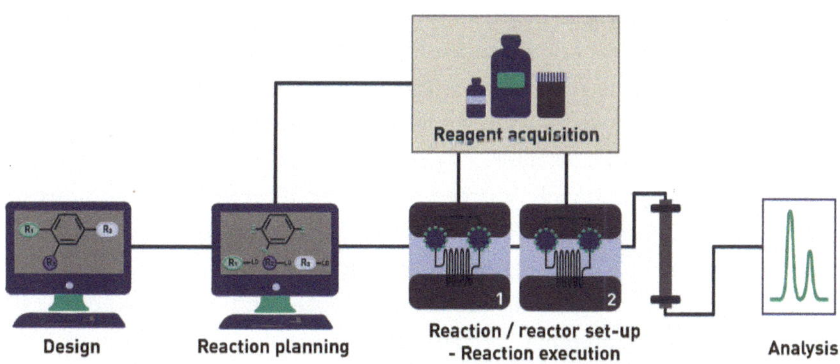

Fig. 19 Components of a fully automated integrated synthesis platform.

customised and predominantly manual way. The challenge now is to develop fully automated, integrated platforms that are easy to set up, programme and operate, and which are flexible enough to carry out the range of chemical reactions typically carried out manually.

The design of new molecules has always been a manual, time-consuming task that requires the expert knowledge of medicinal chemists. However, over the last few years, computational approaches to molecular design have started to emerge. One approach, called multiparameter optimisation, is based on a database of building blocks and known synthetic chemistry, which it uses to generate a library of potential molecules that are then evaluated and scored [220,221]. AI approaches have also been developed that provide the opportunity to fully automate this design process [222,223]. However, given that these approaches are not based on known synthetic transformations, it is highly likely that some of the molecules suggested by the model are synthetically intractable. A suitable means of assessing the synthetic tractability of the molecules designed by these models remains an on-going challenge. Having only started to emerge in the last 5 years, such technologies are still very much in their infancy. Although these approaches have not yet been fully demonstrated to be beneficial in drug discovery, if successful, they would fill the current established automation technology gap in the 'design' stage of drug discovery, and thus represent a step towards the creation of fully integrated platforms for automated synthesis.

Assuming the design stage could be successfully automated and established to successfully predict biologically active, synthetically tractable molecules, the next action for the fully automated, integrated synthesis platform would be AI led reaction planning as discussed in Section 3.2. The next tasks for an integrated automated synthesis platform would be to source the required reagents and reactants and then complete the synthesis. It seems inevitable that the types of reactions that the platform can conduct will always be limited by the availability of the necessary reagents. Therefore, it is foreseeable that this step will act as a significant filter. At present, stocks of reagents, reactants and solvents are usually manually sourced and added to automated synthesis systems. As such, some companies offer solutions that minimise the amount of time chemists need to spend finding chemicals, such as Chemspeed [136], who offer chemical reagents in pre-weighed vials, and Synple Chem [144], who offer a range of prefilled reaction cartridges that contain all the materials to effect the transformation, workup and purification. Merck have also sought to address the issue with their recently launched LANEXO™ Lab Inventory, Safety and Compliance Management System [224].

With sourcing of materials taken care of, the next action would be the synthesis itself. As discussed in Section 3.3, there are now several different types of automated synthesis systems available. A system does not yet exist that could carry out every imaginable chemical reaction in an automated fashion, and it remains an enormous challenge. Given the complexity and functionality that such a machine would need to have, it is questionable whether such a machine would ever be desirable or economically viable.

An alternative approach would be to incorporate a suite of equipment, capable of carrying out chemical reactions and product isolation, into an automated synthesis platform, containing a range of tools with different functionalities that enable the largest possible scope of chemical reactivity. While automated synthesis systems exist that support well established chemistries (Section 3.3), we are now starting to see the emergence of synthesis technologies that support the use of new or previously challenging chemistry methodologies, such as electrochemistry and photochemistry. Even though some of these technologies are not yet automated, they are already hugely enabling in medicinal chemistry. Therefore, it is perceivable that in the near future, fully automated versions of these types of technologies will be available for integration into a synthesis platform.

One potential issue that could arise in an integrated automated synthesis platform is that the proposed 'best reaction conditions' provided by reaction planning software, are only a prediction, and as such there is a reasonable risk that the reactions will not work effectively enough. Therefore, synthesis equipment that could take the suggested reaction conditions and carry out an initial reaction condition optimisation process using high-throughput experimentation, would be beneficial. Such AI controlled, self-optimising reactors already exist, utilising flow chemistry and incorporating inline analysis tools, such as HPLC. The analysis output then feeds back into the algorithm and the next iteration of parameters is automatically executed [225].

Very recently, technology platforms have begun to emerge, which combine large parts of the process into a single integrated platform. While not addressing the molecular design step, the robotic synthesis platform, developed by Timothy Jamison and Klavs Jensen at MIT, was one of the first to integrate a reaction planning tool [226,227]. The planning tool is trained on the Reaxys database and US patents, and predicts the ideal route and reaction conditions. At the moment, the reaction conditions suggested by the reaction planning system still need to be edited by humans, who define parameters such as residence times, stoichiometries, and concentrations and ensure that the proposed conditions are compatible with the selection of the continuous flow chemistry reactor modules present in the platform. A robotic arm is used to assemble the appropriate reactor and separator modules, as required for the reaction. Robots are also employed to connect reagent lines and pumps to reactor inlets. While this is a great step towards fully integrated automated synthesis, some significant challenges remain. Given that flow chemistry is a relatively recent addition to the chemists' tool kit, and that the reaction planning model is trained primarily to batch chemistry, the conversion of the batch conditions to those suitable for flow is still challenging and requires input from highly experienced chemists. The usual limitations of flow chemistry also hinder this system, e.g. solubility. Self-optimisation of the process via inline analysis and feedback into the reaction planning system would be a great addition to this platform, however this is not a simple task for multistep reactions. Despite the existing challenges, this work is a huge step towards the goal of fully automating synthesis.

Several other groups have also started to report on their efforts to combine reaction planning tools with automated synthesis technologies. The US DARPA Make-it program, which funded the work by Jamison and Jensen at MIT [130], has funded several other synthesis technology projects, including the SynFini project, at the not-for-profit research institute SRI International [228]. Although full details are not yet available, the platform consists of an artificial-intelligence-based synthesis planning software tool (SynRoute), an ink-jet-printer-based nanoscale reaction optimisation platform (SynJet), and a reconfigurable flow chemistry-based synthesis platform with integrated LC-MS and ^{1}H NMR analysis (AutoSyn). In October 2020, SRI International also received a further $4.3 million from DARPA to create ProSyn™, an agile, rapidly deployable extension of SynFini for the production of therapeutic small molecules needed for rapid response to biological threats and pandemics, such as COVID-19 [229].

IBM have recently reported a similar approach with their RoboRXN platform, a cloud-based, AI driven robotic platform that is capable of planning and executing the synthesis of a given molecule [230]. The idea behind this is that a chemist located remotely could request a specific molecule to be made, and RoboRXN would then determine the best route to make this and initiate the synthesis using existing automated synthesis technologies. The platform integrates retrosynthetic architecture developed by chemists at the University of Pisa with IBM's own RXN for Chemistry, which uses neural machine learning architectures to predict the most likely outcome of a chemical reaction, and combines these AI tools with cloud technology and chemistry automation.

One of the greatest challenges for IBM was the conversion of the synthesis information compiled by organic chemists, into automation-friendly information that can be read by the system. IBM's approach was to develop a deep-learning model that converts experimental procedures as a whole into a structured, automation-friendly format. In a similar vein, with his Chemify project [231], Cronin has been working on a solution to the same issue in developing a synthesis translator for his Chemputer system, a robotic laboratory platform for synthesis that consists of standard synthesis laboratory equipment, such as round-bottom flasks, separatory funnels, and a rotary evaporators [232]. Since most of the chemistry literature is based on batch chemistry, this approach avoids the issue of needing to convert the majority of the synthesis literature into conditions suited for alternative technologies, which is a problem for the flow chemistry-based platform developed by Jamison and Jensen at MIT.

While the approaches above focus on automation that moves the reagents and reaction mixtures in solution form from one stage of the process to the next, there are alternative systems being developed that utilise robots to mimic the actions a human chemist would make, with the main focus being the replacement of a human carrying out repetitive tasks by the robot. In 2019 Lim and co-workers reported their flexible, automated synthesis system, in which a robot arm mimicked the actions of the human chemist and was able to follow the same work flow—liquid handling to transfer liquids into reaction vial, mixing of chemicals by transferring the vial to the magnetic stirrer via the robotic arm, filtering the reaction mixture through a silica cartridge, and transferring the filtrate to the GC-MS auto-injector rack (Fig. 20A) [233]. More recently, Burger and Cooper adapted a mobile robot, originally designed to work in warehouses or factories, by adding specialised grippers and sample trays so that it could operate liquid and solid chemical dispensers, a gas chromatography system, and other lab instruments required to identify and optimise a new catalyst. Operating 24 h a day, 7 days a week, the robot was able to effect catalyst optimisation around 1000 times faster than a chemist (Fig. 20B) [234].

To be truly successful, these more advanced systems need to be not only highly productive, but also able to learn and make informed decisions. They need to be compatible with operating in the lab environment, be able to locate chemicals and equipment, and they need to be able to communicate effectively with other lab equipment and human researchers alike. The plan to develop such autonomous systems is a grand challenge and there are still many hurdles to overcome and developments to be made before this is a

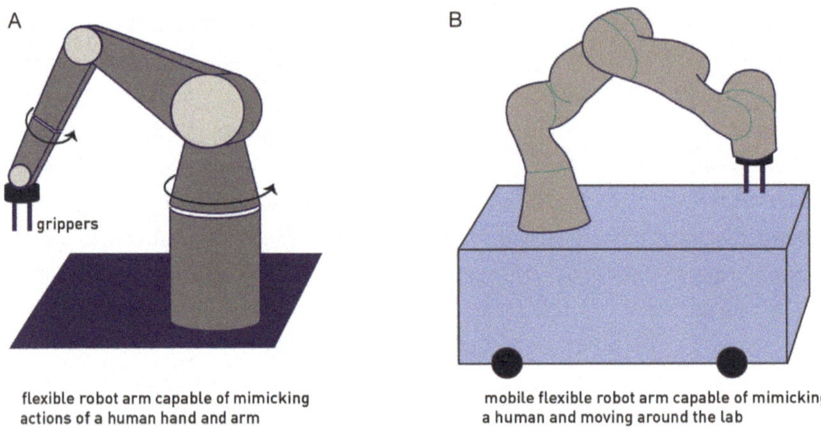

Fig. 20 Automated robots mimicking human chemists.

realistic possibility. Machine learning needs to be far more reliable and effective than it is now because the current systems still rely heavily on human input and direction. However, as evidence by these two examples, big strides are being made towards the realisation of such technologies.

For medicinal chemistry however, where chemists do not often carry out exactly same process repeatedly but instead need to operate with a far greater degree of flexibility, these emerging autonomous robots would need to also be able to perform as efficiently in a flexible manner. At present, this is not a possibility since intelligent robots that can conduct all the operations of humans with the same degree of dexterity have not yet been realised. As such, and despite all the major advances towards fully automated, integrated synthesis platforms, the biggest challenge still remains—an easily accessible, low-cost, automated method that can design, assess and prepare a broad selection of structurally diverse drug-like molecules.

4.3 Closed-loop drug discovery

With R&D costs rising year on year, and early stage drug discovery accounting for around 30% of the costs, productivity-enhancing solutions that do not significantly increase costs are being urgently sought. As a means of improving productivity, the automation of the various processes and stages of drug discovery has been underway for several decades. However, automation and integration of the entire iterative *Design-Make-Test-Analyse* cycle into a single platform—a 'closed-loop drug discovery' platform, is perceived to be the absolute pinnacle of automation goals.

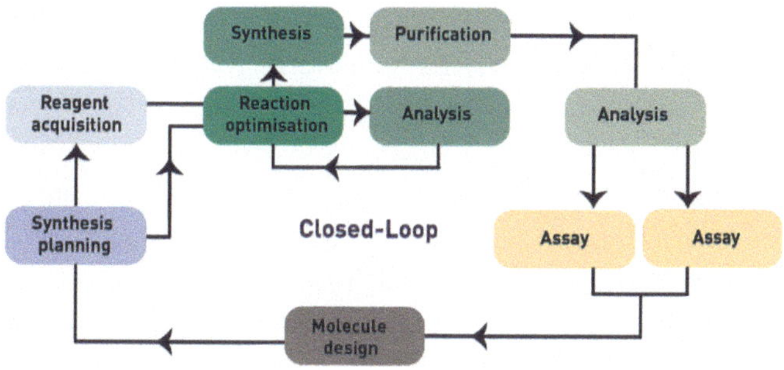

Fig. 21 'Closed-loop' drug discovery is envisaged as a fully automated platform in which new molecules, computationally designed by machine learning (ML) systems, are then assessed for synthetic feasibility, before being synthesised using automated synthesis technologies (with an option for reaction condition optimisation). Following purification and analysis, the new molecules would be screened and the resulting data would then feedback into the ML algorithm, thus initiating the next cycle molecular design and drug discovery.

With great progress being made towards the development of fully automated integrated synthesis platforms that can address both the *Design* and *Make* stages of drug discovery, effective, closed-loop drug discovery platforms are now on the brink of becoming a reality (Fig. 21). Although automation technologies are well established for the *Test* stage of drug discovery and AI technologies are beginning to enable the *Analyse* stage, additional challenges still remain with respect to both individual stages and to the coupling of the stages to each other.

In the *Design* stage, it is essential that the ML algorithms are able to take the feedback from the *Analyse* stage and automatically refine themselves. Examples of this are certainly emerging although the suggestions made by the AI systems do not necessarily correlate well with what a medicinal chemist would design. For closed-loop drug discovery, it is essential that the chemistry used in the *Make* stage is reliable and robust. This is challenging due to the vast complexity and scope of synthetic chemistry. However, if the synthesis fails then the whole loop process falls down. Therefore, a reaction optimisation step would be a beneficial addition to the *Make* stage. Once the optimal conditions have been developed, the platforms could continue with the synthesis of the desired product. Automated systems that could effect this function via the high-throughput optimisation of reaction conditions have recently been reviewed by Nelson [235]. The use of such optimisation systems could also encourage the use of a more diverse range of chemistries.

Linking of the various stages of drug discovery can, in itself, be problematic but despite the challenges, efforts towards linking the *Make* and *Test* stages are now starting to emerge [236–240]. Many of these examples are based on flow or microfluidic synthesis and test platforms, which consist of functional units for specific tasks, and use only small volumes of fluids [241]. Microfluidic-based methods have several advantages, including the ability to manipulate and process small volumes, minimal consumption of reagents, low fabrication costs, reduced analysis times, and the ability to integrate various functional components within a single, monolithic platform [242]. Flow systems offer an additional advantage, in terms of overcoming the challenges associated with linking the *Make* and *Test* stages of drug discovery. Given that synthesis is often carried out in organic solvents, whereas biological screening uses aqueous media, a solvent switch is likely to be required, which would normally be done via evaporation and re-dissolution in the optimal media. However, flow chemistry offers an alternative because some systems allow for slow solvent mixing and thus direct inline testing is possible.

With the progress that has been made towards integration of the *Design-Make* and *Make-Test* stages, AI provides the potential to complete the closed-loop drug discovery cycle by providing a means of analysing the screening data and feeding back the results into the *Design* stage. As discussed in Section 4.3, AI tools are already being developed that can be trained on screening data and then use this to design new molecules for synthesis and screening. Despite the fact that these approaches are very recent and still have significant limitations, it is clear that the basic building blocks of closed-loop drug discovery are already there.

One example, which actually links together all of the *Design-Make-Test-Analyse* stages, is a system reported by researchers at Cyclofluidics, who have developed a flow chemistry platform that can conduct inline synthesis, automated purification, and analysis, which is also coupled with bioassay. Additionally, the system incorporates an activity model that can refine itself after every iteration of synthesis and biological screening [243]. After just 21 compound iterations, the automated process was able to identify a novel Abl kinase binder ($pIC_{50} > 8$). With this example it was clearly demonstrated that integrated microfluidic synthesis and screening coupled with machine learning design, offered real potential to accelerate drug discovery.

It is clear that advances in computational power are fuelling the AI approaches that are critical to the success of closed-loop drug discovery. However, there is still much progress to be made before it can be seen as a robust and reliable alternative. In addition to the AI systems required for molecular design and synthesis planning, machine learning algorithms are also essential in the synthesis, analysis and purification stages, but these are now also starting to emerge. Due to these advances, robotic, cloud-based platforms have now been established, such as the open-access Strateos platform, which enables drug discovery scientists to expand their capabilities beyond their own in-house labs [244]. In collaboration with Strateos, Eli Lilly have incorporated the robotic cloud lab platform into their early discovery automation project: the Lilly Life Sciences Studio Lab. The lab uses conventional commercially available parallel chemistry systems and promises closed-loop discovery, with synthesis and screening automation accessible remotely through a cloud-based interface [245].

From a diversity perspective, the reliance on the existing automated or parallel synthesis platforms is perhaps a concern for closed-loop drug discovery. Many of the systems developed, are based on synthesis happening in flow system. As has already been established in the course of this chapter, while flow chemistry offers many advantages, significant limitations still exist. If closed-loop drug discovery platforms are limited to flow-based systems, this could significantly restrict the number and diversity of the AI generated ideas that can actually be put into practice.

A key question for medicinal chemistry is whether the chemical space accessible using the existing chemistries that are amenable to automation, is sufficient, or whether it is too limiting and would simply result in a bias similar to what was observed during the combinatorial chemistry era. A recent analysis suggests that, due to the availability of many new diverse building blocks, the diversity of the chemical space that is accessible using simple chemistries is not actually restricted compared to that available using more complex chemistries [246]. This observation is good news for synthesis automation and, assuming novel, interesting and diverse building blocks can be made or sourced, brings the goal of closed-loop drug discovery within reach.

4.4 Gaps and outlook for automated technologies in medicinal chemistry

Regardless of the application, almost all forms of automation aim to speed up routine tasks and increase productivity; reduce mistakes and increase robustness; improve safety; and free up humans for higher value tasks that are not so readily automated. Medicinal chemistry is no different. While the work of a medicinal chemist requires a high level of skill that requires years of training to achieve, in reality there is still a significant portion of synthetic chemistry that is routine and mundane, hence automating these elements would free up chemists for more intellectually demanding tasks, and thus enable them to do more and to increase productivity. In addition, with health and safety measures becoming increasingly important, safer working is a clear advantage of automating chemical synthesis since chemists are routinely handing toxic and hazardous materials.

Ever since the rapid developments in high-throughput screening, chemists have been searching for ways to dramatically increase the numbers of compounds available for screening. It was this need to make more with the same amount of resources that drove the development of combinatorial chemistry. However, it soon became clear that there are downsides to being able to make more compounds.

The preparation of large libraries of compounds is expensive, even if it is done in an automated fashion, as is the analysis and subsequent screening. The new compounds also need to be stored somewhere, from which they can be easily retrieved in a suitable format for further screening. In addition, the compounds need to be logged in a searchable database so that they offer value to other drug discovery projects. Therefore, while the automated technologies developed for the synthesis of large libraries certainly increased productivity in terms of the numbers of compounds made, they did so at a huge cost.

As discussed, biases in library chemical diversity has meant that many of the gains achieved from automating the library synthesis were lost because project teams needed to spend years trying to optimise the physicochemical properties of the initial hits and leads. Nevertheless, the goal of doing more, faster has remained. Aided by significant developments in computing power, in silico approaches have become a valid alternative means of automating drug discovery. Predictive models for assessing the physicochemical properties of molecules, such as permeability, solubility and pK_a, enable better triaging of screening hits and progression of only the most promising analogues.

The construction of virtual libraries and analysis of these using pharmacophore models, for example, removes the need to even synthesise almost all but the most promising analogues. However, the virtual libraries often still require the manual input of defined transformations based on available monomers, which is time consuming. Like combinatorial libraries, these virtual libraries can still be a little restrictive since they are based on defined chemical transformations so structural diversity in virtual libraries in an area for improvement. Now AI systems are emerging that take a completely different approach to generating virtual libraries and hold the promise of libraries with increased structural diversity.

While automated computational tools have undoubtedly had a big impact, one weakness of all computational models is that they are based on the data that exist at the time. This presents two potential issues, namely that if the original data is not reliable, robust and standardised, the model will be poor. Also, most models are a snap-shot in time and represent only current knowledge. As such, until very recently, they have needed to be continually updated as new data emerged. Now with the adoption of eLNB and a more standardised approach to recording experimental data, and with the advent of machine learning approaches, there is potential for these models to show their real value.

It is clear that while drug discovery has already benefited immensely from automation, there are still many aspects of medicinal chemistry that need new automated tools and methods. While the new machine learning approaches to virtual library construction and screening hold a great deal of promise, new technologies for automated synthesis are sorely needed. A lot of the mundane, repetitive work can now be done using automated workstations, or platforms based on multiple automated workstations. However, these systems can be highly complex, require skill to operate, and can be time consuming to set up and programme. For the specialist technologists embedded within medicinal chemistry groups, the programming and operation of the systems is not a major issue, and the time invested is worth it when large libraries of compounds are being prepared. However, for many non-specialist users and for those who are making only small sets of analogues, such systems are not advantageous because the effort required to use them far outweighs the potential gain. It is these chemists, who are in fact the majority of medicinal chemists, who have so far been poorly served by automated synthesis technologies.

The challenge in medicinal chemistry is that, unlike production environments, research requires greater flexibility to switch between different reactions and substrates in order to respond to the emerging data and make many different types of target molecules. Only by doing this can each iteration of the *Design-Make-Test-Analyse* cycles be completed and project progression be achieved. While technology that provides this flexibility is certainly already available, there is an additional challenge. In order to identify a drug candidate as soon as possible, medicinal chemists must complete each *Design-Make-Test-Analyse* cycle as quickly as possible. As such, medicinal chemists will only take time to optimise reaction conditions when not doing so means it will be difficult to prepare enough of each compound, or if the compound looks promising following initial screening and will thus be needed in larger quantities.

As a result of the need to make more compounds, and the lack of a desire to 'waste' time optimising reactions, there can be reluctance to use automated synthesis technologies that are perceived to be too difficult to use, require too much time and effort to set up, or offer insignificant benefits. This is perhaps why this sector has been slower to adopt automation compared to other industries. However, with the rise of artificial intelligence and machine learning approaches, suddenly the opportunity for more flexible, easier to use automation is now a real possibility.

5. Medicinal chemists vs machines: What the future holds

Visionary medicinal chemists continually seek ways to relieve drug discovery bottlenecks, by exploring the use of productivity-enhancing tools and equipment. Simply continuing to do work in the same way as was previously done is not sustainable, and trying to do more of what medicinal chemists have always done will have limited effect. As drug discovery evolves to tackle novel biological targets and chemical modalities, new technologies will need to be developed and adopted in order to have a realistic chance of identifying drug candidates for these disease targets. This will undoubtedly drive, or even force, a change in what the role of a medicinal chemist will look like in the future. However, medicinal chemists have the opportunity to shape what this change will look like by working together with laboratory automation equipment manufacturers to develop new automated solutions and to ensure that the optimal systems and processes are developed, to relieve both the existing and future bottlenecks in the identification of new drug candidates.

While innovative automation in drug discovery has already enabled some dramatic shifts in drug discovery practices, there is still room for improvement in the *Make* stage. The automation gaps that remain, with respect to synthesis technologies and reaction methodologies, will be best addressed through the collaboration of medicinal chemists, lab automation equipment manufacturers, and academic research groups working on new methodologies and synthesis techniques.

One of the biggest challenges in developing automation for medicinal chemistry is finding the right balance between complexity and flexibility. In an ideal world, automated systems would exist that can conduct a wide range of synthetic processes, and that are easy to use, low cost, and highly reproducible. However, it is perhaps a little unrealistic to expect that all of these requirements can be achieved simultaneously. Inevitably, systems capable of carrying out a wide scope of chemistry need to have a great deal of functionality. Therefore, such systems are likely to be highly complex and very costly. Conversely, lower cost systems are likely to have reduced functionality and will thus be less flexible. One solution that is trying to bridge the gap between functionality, complexity and usability is software. In theory, with the correct software interface, even the most complex of systems could be rendered easy to use. At present, these ideal software interfaces do not yet exist and there is scope for medicinal chemists to also play a role, via collaborations, in the development of the software interfaces, and thus ensure it actually meets their needs.

Perhaps due to the current difficulty in finding the right balance of complexity and functionality with automated synthesis equipment, the use of traditional, manual techniques persists in medicinal chemistry, and many medicinal chemists are somewhat reluctant to adopt new methods and technologies. Even today, many reactions continue to be run in round-bottomed flasks. The technologies and equipment that have been broadly adopted tend to be simple tools that are easy to use, require little training, and offer an easier, more robust, safer, and crucially, a more convenient version of the traditional processes, or are those which address a significant problem without creating new ones.

A further reason for adoption aversion is that new technologies are complex. Time pressures likely play a role since there is often little time available for evaluating new technologies, making the adoption of new equipment a slow process. There is also, however, a certain element of conservatism and a desire to stay within one's comfort zone, doing what we believe will work because it has done so more often than not in the past. In demonstration of this, a recent analysis revealed that of the current most frequently used synthetic reactions in medicinal chemistry, none were discovered within the past 20 years and only two in the 1980s and 1990s (Suzuki–Miyaura and Buchwald–Hartwig) [247,248].

Another factor contributing to the resistance is cynicism. It seems that regardless of the actual functionality of new technologies, there appears to be an underlying fear in some medicinal chemists that the machines will not being able to function equally as well as the chemist, or a suspicion that the machines will not work at all. Some of this cynicism is understandably grounded in experience with other technologies, but regardless of the basis, this fear is a factor in the slow adoption of new technologies. In an average medicinal chemistry group there will always be those who are keen to try out the latest technologies, but many will be resistant, and these chemists will need to be persuaded, supported, trained and convinced of its value before they are willing to try. Responsibility for accelerating the adoption of automated technologies lies not only with the medicinal chemists but also with the groups and companies developing these technologies. The equipment must add real value and the benefit gained should be significant, or the majority of medicinal chemists will, understandably, not be tempted to try it.

In order to ensure wide spread adoption of new technologies and convince users to step out of their comfort zone, break away from the status quo and find the benefits in these technologies, it is important that automated synthesis technology developers keep in mind that not all technologies are desirable to all chemists. In reality, it is highly unlikely that there will be one single, ideal automation platform for chemical synthesis. The ideal automation tools will always be dependent on the nature of the task. In no other aspect of life do we strive to have a single tool that does every job we can imagine. When cooking, for example, we have a wide range of tools at our disposal and we select the most appropriate tool for the job. The same approach is needed with chemical synthesis technologies. An alternative goal, would be for the medicinal chemist to have a broad tool kit of easy to use, low-cost, enabling synthetic technology equipment, from which they could select the most appropriate tool for the job. This approach might be more palatable since it holds a role for medicinal chemists who are still needed to assess and select the most appropriate tools. If complex highly functionalised equipment is developed, it will be important to create a range of user interfaces that make the same piece of hardware easily accessible.

New technologies should also be capable of being integrated into existing and future technology platforms, to avoid users being restricted in terms of the tools available for them to use. Automated systems should also be readily accessible and affordable to medicinal chemists. This is of course subjective and depends on the user group and the application. The fine balance between functionality, complexity, usability and cost is crucial because the level of complexity of automated technologies and the skills required to operate them also appear to be inversely proportional to a medicinal chemist's willingness to adopt it. However, getting this balance right can be an impossible task when trying to develop a technology that needs to appeal to all chemists, which further lends support to the notion that we should be seeking to develop a selection of automated technologies for chemical synthesis.

Regardless of how well designed and targeted the new automation equipment is, there is another issue to contend with. Among some practitioners of medicinal chemistry, there is a feeling or belief that medicinal chemistry is an art form, and as such, it is not amenable to standardisation and automation. It still remains to be seen whether the complete automation of drug design and discovery will be the death of creativity in medicinal chemistry as feared by some, or whether, in fact, it will free up medicinal chemists to focus on more innovative designs and solutions, and thus result in an overall increase in creativity. Naturally, with the prospect of complete automation, the question of whether chemists need to fear being replaced completely by machines is something that arises [249]. It seems highly unlikely that medicinal chemists will be completely obsolete in the future but it is entirely feasible that the role and requirements of a medicinal chemist will change. Medicinal chemists who are open to working together with these technologies will have the best opportunities to assess the existing tools, and find the most appropriate ones that enable them to spend less time on the mundane aspects of synthesis, and more time contributing intellectually in a way that machines cannot, thus ensuring their role in the future of medicinal chemistry.

In the future, it is realistic to consider that medicinal chemists will work together with automation and AI and utilise it to a much greater extent in reaction planning, route and reaction optimisation, and SAR analysis. It is clear from the efforts towards the development of closed-loop drug discovery platforms that AI can also assist with the task of generating ideas for synthesis. While this does not currently appear to be a bottleneck, automation of this stage of the process perhaps presents a complementary approach to the ideas generated manually by medicinal chemists, and thus expands the diversity of input in the drug discovery project. This complementary approach has the potential to be highly beneficial. While the goal of completely autonomous chemical synthesis is a tempting one, it overlooks the advantages that Man and Machines working together can bring, by each doing what they do best. Over the next decade the focus should perhaps be on making smaller steps towards the greater goal. Even technologies that require some manual input from medicinal chemists can still make a huge difference, as long as the effort of this input does not amount to more than is saved through use of the technology. Over the next 5–10 years, the most significant advancements in drug discovery may well arise as a result of medicinal chemists working together with automation in this way. By harnessing the benefits of both Humans and Machines, it may be possible to truly bring about the acceleration of drug discovery.

References

[1] Achan J, Talisuna AO, Erhart A, et al. Quinine, an old anti-malarial drug in a modern world: role in the treatment of malaria. Malar J 2011;10:144.
[2] Paul SM, Mytelka DS, Dunwiddie CT, Persinger CC, Munos BH, Lindborg SR, et al. How to improve R&D productivity: the pharmaceutical industry's grand challenge. Nat Rev Drug Discov 2010;9(3):203–14.
[3] DiMasi JA, Grabowski HG, Hansen RW. Innovation in the pharmaceutical industry: new estimates of R&D costs. J Health Econ 2016;47:20–33.
[4] Janzen WP. Screening technologies for small molecule discovery: the state of the art. Chem Biol 2014;21:1162–70.
[5] Houston JG, Bank M. The chemical-biological interface: developments in automated and miniaturised screening technology. Curr Opin Biotechnol 1997;8:734–40.
[6] Dittrich PS, Manz A. Lab-on-a-chip: microfluidics in drug discovery. Nat Rev Drug Discov 2006;5:210–8.
[7] Skardal A, Shupe T, Atala A. Organoid-on-a-chip and body-on-a-chip systems for drug screening and disease modelling. Drug Discov Today 2016;21:1399–411.
[8] Campbell IB, Macdonald SJF, Procopiou PA. Medicinal chemistry in drug discovery in big pharma: past, present and future. Drug Discov Today 2018;23:219–34.
[9] Sugawara T, Kato S, Okamoto S. Development of fully-automated synthesis systems. J Anal Methods Chem 1994;16:33–42.
[10] Liu R, Li X, Lam KS. Combinatorial chemistry in drug discovery. Curr Opin Chem Biol 2017;38:117–26.
[11] Fürka Á. History of combinatorial chemistry. Drug Dev Res 1995;36:1–12.
[12] Geysen HM, Rodda SJ, Mason TJ. A priori delineation of a peptide which mimics a discontinuous antigenic determinant. Mol Immunol 1986;23:709–15.
[13] Furka Á, Sebestyén F, Asgedom M, Dibó G. Cornucopia of peptides by synthesis. In: Abstr 14th Int Congr Biochem. Prague, Czechoslovakia, 5; 1988. p. 47.
[14] Furka Á, Sebestyén F, Asgedom M, Dibó G. More peptides by less labour. In: Abstr 10th Int Symp Med Chem. Budapest, Hungary; 1988. p. 288.
[15] Scott JK, Smith GP. Searching for peptide ligands with an epitope library. Science 1990;249:386–90.
[16] Devlin JJ, Panganiban LC, Devlin PE. Random peptide libraries: sources of specific protein binding molecules. Science 1990;249:404–6.
[17] Cwirla S, Peters EA, Barrett RW, Dower WJ. Peptides on phage: a vast library of peptides for identifying ligands. Proc Natl Acad Sci USA 1990;87:6378–82.
[18] Fodor SPA, Read JL, Pirrung MC, Stryer L, Lu AT, Solas D. Light-directed, spatially addressable parallel chemical synthesis. Science 1991;251:767–73.
[19] Gordon DW, Steele J. Reductive alkylation on solid phase: synthesis of a piperazinedione combinatorial library. Bioorg Med Chem Lett 1995;5:47–50.
[20] Chen C, Ahlberg Randall LA, Miller RB, Jones AD, Kurth MJ. Analogous organic synthesis of small-compound libraries: validation of combinatorial chemistry in small-molecule synthesis. J Am Chem Soc 1994;116:2661–2.
[21] Bunin BA, Ellman JA. A general and expedient method for the solid-phase synthesis of 1,4-benzodiazepine derivatives. J Am Chem Soc 1992;114:10997–8.

[22] Zuckermann RN, et al. Discovery of nanomolar ligands for 7-transmembrane G-protein-coupled receptors from a diverse N-(substituted)glycine peptoid library. J Med Chem 1994;37:2678–85.
[23] Geysen HM, Meloen KH, Barteling SJ. Use of peptide synthesis to probe viral antigens for epitopes to a resolution of a single amino acid. Proc Natl Acad Sci USA 1984;81:3998–4002.
[24] Houghten RA. General method for the rapid solid-phase synthesis of large number of peptides: specificity of antigen antibody interaction at the level of individual amino acids. Proc Natl Acad Sci USA 1985;82:5131–5.
[25] Lam KS, Salmon SE, Hersh EM, Hruby VJ, Kazmierski WM, Knapp RJ. A new type of synthetic peptide library for identifying ligand-binding activity. Nature 1991; 354:82–4. [And its correction: Lam KS, Salmon SE, Hersh EM, Hruby VJ, Kazmierski WM, Knapp RJ. Nature 1992;360:768.].
[26] Houghten RA, Pinilla C, Blondelle SE, Appel JR, Dooley CT, Cuervo JH. Generation and use of synthetic peptide combinatorial libraries for basic research and drug discovery. Nature 1991;354:84–6.
[27] Powell Gray B, Brown KC. Combinatorial peptide libraries: mining for cell-binding peptides. Chem Rev 2014;114:1020–81.
[28] Brenner S, Lerner RA. Encoded combinatorial chemistry. Proc Natl Acad Sci USA 1992;89:5381–3.
[29] Nielsen J, Brenner S, Janda KD. Synthetic methods for the implementation of encoded combinatorial chemistry. J Am Chem Soc 1993;115:9812–3.
[30] Gallop MA, et al. Generation and screening of an oligonucleotide-encoded synthetic peptide library. Proc Natl Acad Sci USA 1993;90:10700–4.
[31] Krämer T, Antonenko VV, Mortezaei R, Kulikov NV. Encoding technologies. In: Nicolaou KC, Hanko R, Hartwig W, editors. Handbook of combinatorial chemistry: drugs, catalysts, materials. Weinheim: Wiley-VCH Verlag GmbH; 2002. p. 170–89.
[32] Halford B. How DNA-encoded libraries are revolutionizing drug discovery. Chem Eng News 2017;95:25. https://cen.acs.org/articles/95/i25/DNA-encoded-libraries-revolutionizing-drug.html. accessed October 27, 2020.
[33] Little JN. Advances in laboratory robotics for automated sample preparation. Chemometr Intell Lab Syst 1993;21:199–205.
[34] Hudson Robotics. Lab solutions. https://hudsonrobotics.com/microplate-handling-2/platecrane-ex/. accessed October 27, 2020.
[35] ThermoFischer. Tools. http://tools.thermofisher.com/content/sfs/brochures/D10878~.pdf. accessed October 27, 2020.
[36] Peak Analysis & Automation. KX-2 laboratory robot. https://paa-automation.com/products/kx-2-collaborative-laboratory-robot/. accessed October 27, 2020.
[37] Precise Automation. PF400 sample handler. http://preciseautomation.com/SampleHandler.html. accessed October 27, 2020.
[38] Hamilton. Automation. https://www.hamiltoncompany.com/. accessed October 27, 2020.
[39] Gilson. Liquid handling. https://gb.gilson.com/GBSV/liquid-handling.html. accessed October 27, 2020.
[40] ThermoFischer. Automated liquid handlers. https://www.thermofisher.com/search/browse/featured/ch/en/80012392/Automated+Liquid+Handlers. accessed October 27, 2020.
[41] Integra. Applications—liquid handling. https://www.integra-biosciences.com/global/en/stories/liquid-handling-revolutionized. accessed October 27, 2020.
[42] Tecan. Products—liquid handling. https://lifesciences.tecan.com/products/liquid_handling_and_automation. accessed October 27, 2020.
[43] Andrew Alliance. Products—Andrew+. https://www.andrewalliance.com/pipetting-robot/. accessed October 27, 2020.

[44] Brummer H, Markert RLM, Schwemler C. Robotersysteme zur vollautomatisierten, kombinatorischen Synthese von Wirkstoffen. GIT Labor Fachz 1999;43:598–601.
[45] Cork DG, Sugawara T. Laboratory automation in the chemical industries. New York-Basel: Marcel Dekker, Inc.; 2002. https://epdf.pub/laboratory-automation-in-the-chemical-industries.html.
[46] Sahoo BM. Microwave assisted drug synthesis (MADS): a green technology in medicinal chemistry. J Appl Pharmacol 2016;8:1000e106.
[47] Kappe CO. My twenty years in microwave chemistry: from kitchen ovens to microwaves that aren't microwaves. Chem Rec 2019;19:15–39.
[48] Zuckermann RN, Kerr JM, Siani MA, Banville SC, Santi DV. Identification of highest-affinity ligands by affinity selection from equimolar peptide mixtures generated by robotic synthesis. Proc Natl Acad Sci USA 1992;89:4505–9.
[49] SiLA. Homepage, https://sila-standard.com/. accessed October 27, 2020.
[50] Chemical Online. News; March 15, 2000, https://www.chemicalonline.com/doc/argonaut-launches-parallel-organic-synthesis-0001. accessed October 27, 2020.
[51] Radleys Discovery Technologies. https://www.radleys.com/. accessed October 27, 2020.
[52] J-Kem Scientific Inc. https://www.jkem.com/. accessed October 27, 2020.
[53] Mettler-Toledo. Events and expertise—datasheets, https://www.mt.com/mt_ext_files/Editorial/Generic/1/MBXT_product_brochure_Editorial-Generic_1108489096828_files/mbxtbroch111204.pdf. accessed October 27, 2020.
[54] Kubinyi H. Drug research: myths, hype and reality. Nat Rev Drug Discov 2003;2(8):665–8.
[55] Kodadek T. The rise, fall and reinvention of combinatorial chemistry. Chem Commun 2011;47:9757–63.
[56] Ettre LS. The rebirth of chromatography. In: Hinshaw JV, editor. Chapters in the evolution of chromatography. London: Imperial College Press; 2008. p. 155–66.
[57] Taylor TI, Urey HC. Fractionation of the lithium and potassium isotopes by chemical exchange with zeolites. J Chem Phys 1938;6:429–38.
[58] Biotage. Products and service—sample preparation. https://www.biotage.com/product-page/isolute-scx-2. accessed October 27, 2020.
[59] Cuatrecasas P, Wilchek M, Anfinsen CB. Selective enzyme purification by affinity chromatography. Proc Natl Acad Sci USA 1968;61:636–43.
[60] Still WC, Kahn M, Mitra A. Rapid chromatographic technique for preparative separations with moderate resolution. J Org Chem 1978;43(14):2923–5.
[61] Biotage. Products and services—flash purifications. https://biotage.com/product-page/biotage-selekt-systems. accessed October 27, 2020.
[62] Buchi. Products and solutions. https://www.buchi.com/en/products/preparative-chromatography/pure-chromatography-systems. accessed October 27, 2020.
[63] Interchim. Instruments. http://interchiminc.com/flash-prep-hplc-instruments-flash-instruments.html. October 27, 2020.
[64] Teledyne Isco. Chromatography. https://www.teledyneisco.com/en-us/chromatography. accessed October 27, 2020.
[65] Giddings JC. Dynamics of chromatography, part I. Principles and theory. New York: Marcel Dekker, Inc; 1965. p. 281.
[66] Kalász H. The story of thin-layer chromatography. In: Ettre LS, editor. Milestones in chromatography. LCGC 2001;19(7):712–21. www.chromatographyonline.com.
[67] Shraiber MS. In: Ettre LS, Zlatkis A, editors. 75 Years of chromatography—a historical dialogue. Amsterdam: Elsevier; 1979. p. 413–7.
[68] Crowe MOL. Micro-method of chromatographic analysis. Ind Eng Chem Anal Ed 1941;13:845–6.

[69] Kirchner JG. In: Ettre LS, Zlatkis A, editors. 75 Years of chromatography—a historical dialogue. Amsterdam: Elsevier; 1979. p. 201–8.
[70] Kirchner JG, Miller JM, Keller GJ. Separation and identification of some terpenes by new chromatographic technique. Anal Chem 1951;23:420–5.
[71] Miller JM, Kirchner JG. New type of chromatographic column. Anal Chem 1951;23:428–30.
[72] Kirchner JG, Miller JM, Rice RG. Fungicide determination, quantitative determination of biphenyl in citrus fruit products by means of chromatostrips. J Agric Food Chem 1954;2(20):1031–3.
[73] Stahl E. Thin-layer chromatography; methods, influencing factors and an example of its use. Pharmazie 1956;11:633–7.
[74] Stahl E. Thin-layer chromatography. II: standardization, visualization, documentation, and application. Chem Ztg 1958;82:323–9.
[75] Zlatkis A, Kaiser RE. HPTLC: high-performance thin-layer chromatography. Amsterdam: Elsevier; 1977.
[76] Ettre LS. The development of gas chromatography. J Chromatogr A 1975;112:1–26.
[77] Hesse G, Tschachotin B. Adsorptionsanalyse von Gasen und Dämpfen. Sci Nat 1942;30:387–92.
[78] Okamoto Y, Okamoto I, Yuki H. Optical resolution of atropisomeric poly (triphenylmethyl methacrylate). J Polym Sci Polym Lett Ed 1981;19:451–5.
[79] Merck Millipore. Product 151450. https://www.merckmillipore.com/CH/en/product/Chromolith-SpeedROD-RP-18-endcapped-50-4.6-HPLC-column,MDA_CHEM-151450. accessed October 27, 2020.
[80] Waters Corporation. Products—columns. https://www.waters.com/waters/en_US/Columns-HPLC-UHPLC-UPLC-SFC-GPC-SEC-and-more/nav.htm?cid=134771861&locale=en_US. accessed October 27, 2020.
[81] Anon. https://www.waters.com/webassets/cms/library/docs/720002152en.pdf. accessed October 27, 2020.
[82] Anon. https://www.agilent.com/cs/library/brochures/5989-5200DEE.pdf. accessed October 27, 2020.
[83] Anon. https://www.ssi.shimadzu.com/sites/ssi.shimadzu.com/files/Products/literature/hplc/C196-E070C.pdf. accessed October 27, 2020.
[84] Anon. https://www.shimadzu.eu.com/xr-ods-ii-high-separation-column-0. accessed October 27, 2020.
[85] Anon. https://www.phenomenex.com/kinetex/index. accessed October 27, 2020.
[86] Sparkman OD. Mass spectrometry desk reference. Pittsburgh: Global View Pub; 2000, ISBN:978-0-9660813-2-9.
[87] Hites RA. Development of gas chromatographic mass spectrometry. Anal Chem 2016;88(14):6955–61.
[88] Tal'roze VL, Gorodetskii IG, Zolotoy NB, Karpov GV, Skurat VE, Maslennikova VYA. Capillary system for continuous introducing of volatile liquids into analytical MS and its application. Adv Mass Spectrom 1978;7:858.
[89] Niessen WMA. Liquid chromatography-mass spectrometry. 3rd ed. Boca Raton: CRC Taylor & Francis; 2006. p. 50–90.
[90] Fenn JB, Mann M, Meng CK, Wong SF, Whitehouse CM. Electrospray ionization for mass spectrometry of large biomolecules. Science 1989;246(4926):64–71.
[91] Rabi II, Zacharias JR, Millman S, Kusch P. A new method of measuring nuclear magnetic moment. Phys Rev 1938;53:318.
[92] Kernmagnetische LG. Hochfrequenzspektroskopie (high-frequency nuclear magnetic spectroscopy). In: Flügge S, editor. Handbuch der Physik—Encyclopedia of physics, vol. 38/1. Berlin: Springer; 1958. p. 120–376.

[93] Bruker.com, Anklin C. NMR—past, present and future. https://www.bruker.com/fileadmin/user_upload/8-PDF-Docs/MagneticResonance/NMR/brochures/NMR_Past_Present_and_Future.pdf. accessed October 27, 2020.
[94] Singh M, Tam B, Akabayov B. NMR-fragment based virtual screening: a brief overview. Molecules 2018;23(2):233.
[95] Murray C, Rees D. The rise of fragment-based drug discovery. Nat Chem 2009; 1:187–92.
[96] Jamkhande PG, Ghante MH, Ajgunde BR. Software based approaches for drug designing and development: a systematic review on commonly used software and its applications. Bull Fac Pharm Cairo Univ 2017;55:203–10.
[97] Green DVS, Pickett S, Luscombe C, et al. BRADSHAW: a system for automated molecular design. J Comput Aided Mol Des 2020;34:747–65.
[98] Ajay A, Walters WP, Murcko MA. Can we learn to distinguish between "drug-like" and "nondrug-like" molecules? J Med Chem 1998;41:3314–24.
[99] Leahy DE, Sykora V. Automation of decision making in drug design. Drug Discov Today Technol 2013;10:e437–41.
[100] Harrison S, et al. Extending 'predict first' to the design-make-test cycle in small-molecule drug discovery. Future Med Chem 2017;9:533–6.
[101] Agrafiotis DK, Shemanarev M, Connolly PJ, Farnum M, Lobanov VS. SAR maps: a new SAR visualization technique for medicinal chemists. J Med Chem 2007;50 (24):5926–37.
[102] Molsoft. Technology. https://www.molsoft.com/novartis-focus.html. accessed October 27, 2020.
[103] Tibco. Products. https://www.tibco.com/products/tibco-spotfire. accessed October 27, 2020.
[104] Anon. https://www.optibrium.com/stardrop/. accessed October 27, 2020.
[105] Anon. https://chemaxon.com/products/instant-jchem. accessed October 27, 2020.
[106] González-Medina M, Naveja JJ, Sánchez-Cruz N, Medina-Franco JL. Open chemoinformatic resources to explore the structure, properties and chemical space of molecules. RSC Adv 2017;7:54153–63.
[107] Schrodinger. Platform. https://www.schrodinger.com/phase. accessed October 27, 2020.
[108] Cresset. News, posted April 23, 2018, https://www.cresset-group.com/about/news/pharma-con/. [Accessed 27 October 2020].
[109] Molecular Discovery. Home. https://www.moldiscovery.com/. accessed October 27, 2020.
[110] Inte:ligand. Home. http://www.inteligand.com/. accessed October 27, 2020.
[111] Chemical Computing Group. Products. https://www.chemcomp.com/Products.htm. accessed October 27, 2020.
[112] Schaller D, Šribar D, Noonan T, Deng L, Nguyen TN, Pach S, et al. Next generation 3D pharmacophore modeling. WIREs Comput Mol Sci 2020;10, e1468.
[113] Veber DF, Johnson SR, Cheng H-Y, Smith BR, Ward KW, Kopple KD. Molecular properties that influence the oral bioavailability of drug candidates. J Med Chem 2002;45:2615–23.
[114] Ghose AK, Viswanadhan VN, Wendoloski JJ. A knowledge-based approach in designing combinatorial or medicinal chemistry libraries for drug discovery. 1. A qualitative and quantitative characterization of known drug databases. J Comb Chem 1999;1:55–68.
[115] Egan WJ, Merz KM, Baldwin JJ. Prediction of drug absorption using multivariate statistics. J Med Chem 2000;43:3867–77.
[116] Muegge I, Heald SL, Brittelli D. Simple selection criteria for drug-like chemical matter. J Med Chem 2001;44:1841–6.

[117] Jia C-Y, Li J-Y, Hao G-F, Yang G-F. A drug-likeness toolbox facilitates ADMET study in drug discovery. Drug Discov Today 2020;25(1):248–58.
[118] Sadowski J, Kubinyi H. A scoring scheme for discriminating between drugs and nondrugs. J Med Chem 1998;41:3325–9.
[119] Martin YC. A bioavailability score. J Med Chem 2005;48:3164–70.
[120] Lobell M, et al. In silico ADMET traffic lights as a tool for the prioritization of HTS hits. ChemMedChem 2006;1:1229–36.
[121] Bickerton GR, et al. Quantifying the chemical beauty of drugs. Nat Chem 2012; 4:90–8.
[122] Yusof I, Segall MD. Considering the impact drug-like properties have on the chance of success. Drug Discov Today 2013;18:659–66.
[123] Singh N, et al. A physicochemical descriptor-based scoring scheme for effective and rapid filtering of kinase-like chemical space. J Cheminform 2012;4:4.
[124] Li D, et al. ADMET evaluation in drug discovery. 13. Development of in silico prediction models for P-glycoprotein substrates. Mol Pharm 2014;11:716–26.
[125] Wang S, et al. ADMET evaluation in drug discovery. 16. Predicting hERG blockers by combining multiple pharmacophores and machine learning approaches. Mol Pharm 2016;13:2855–66.
[126] Chan HCS, Shan H, Dahoun T, Vogel H, Yuan S. Advancing drug discovery via artificial intelligence. Trends Pharmacol Sci 2019;40(8):592–604.
[127] White MJ. Chapter 3: chemical patents. In: Currano JN, Roth DL, editors. Chemical information for chemists, a primer. The Royal Society of Chemistry; 2014. p. 84.
[128] Sigma-Aldrich. Products—chemical synthesis. https://www.sigmaaldrich.com/chemistry/chemical-synthesis/synthesis-software.html. accessed October 27, 2020.
[129] Spaya AI. https://beta.spaya.ai/. accessed October 27, 2020.
[130] Anon. https://askcos.mit.edu/. accessed October 27, 2020.
[131] Anon. https://www.deepmatter.io/products/icsynth/. accessed October 27, 2020.
[132] MN-AM.com. Products. https://www.mn-am.com/products/sylvia. accessed October 27, 2020.
[133] IBM RXN for Chemistry. https://rxn.res.ibm.com/. accessed October 27, 2020.
[134] Li J, Ballmer SG, Gillis EP, Fujii S, Schmidt MJ, Burke MD, et al. Synthesis of many different types of organic small molecules using one automated process. Science 2015;47(6227):1221–6.
[135] Bédard A-C, Adamo A, Aroh KC, Russell MG, Jensen KF, Jamison TF, et al. Reconfigurable system for automated optimization of diverse chemical reactions. Science 2018;361(6408):1220–5.
[136] ChemSpeed. Solutions—chemicals consumables. https://www.chemspeed.com/smole/. accessed October 27, 2020.
[137] Chemify. Home. http://www.chem.gla.ac.uk/cronin/chemify/. accessed October 27, 2020.
[138] Steiner S, Wolf J, Glatzel S, Andreou A, Granda JM, Cronin L. Synthesis of many different types of organic small molecules using one automated process. Science 2015;347(6227):1221–6.
[139] Vaucher AC, Zipoli F, Geluykens J, et al. Automated extraction of chemical synthesis actions from experimental procedures. Nat Commun 2020;11:3601. https://doi.org/10.1038/s41467-020-17266-6.
[140] Vapourtec.com. www.vapourtech.com. accessed October 27, 2020.
[141] Uniqsis.com. www.uniqsis.com. accessed October 27, 2020.
[142] Syriss.com. www.syriss.com. accessed October 27, 2020.
[143] ThalesNano.com. https://thalesnano.com/. accessed October 27, 2020.
[144] Synplechem.com. www.synplechem.com. accessed October 27, 2020.

[145] Asynt. Products—DrySyn range. https://www.asynt.com/product/drysyn-octo-reaction-station/. accessed October 27, 2020.
[146] Radleys.com. Our range—StarFish. https://www.radleys.com/range/starfish-work-station/. accessed October 27, 2020.
[147] Radleys.com. Our range—automated reaction stations. https://www.radleys.com/range/automated-reaction-stations/. accessed October 27, 2020.
[148] Mettler-Toledo.com. Automated reactors and in situ analysis. https://www.mt.com/ch/en/home/products/L1_AutochemProducts/Chemical-Synthesis-and-Process-Development-Lab-Reactors/Synthesis-Reactor-Systems/EasyMax-402-Basic.html. accessed October 27, 2020.
[149] Amigochem.com. https://www.amigochem.com/products. accessed October 27, 2020.
[150] JKEM.com. Products—reactors. https://www.jkem.com/product-category/reactors/personal-reaction-station/. accessed October 27, 2020.
[151] Asynt.com. Products—fReactor. https://www.asynt.com/products/freactor/. accessed October 27, 2020.
[152] IKA. Products—electrochemistry kits. https://www.ika.com/en/Products-Lab-Eq/Electrochemistry-Kit-csp-516/. accessed October 27, 2020.
[153] SigmaAldrich.com. Products—chemical synthesis, https://www.sigmaaldrich.com/chemistry/chemistry-products.html?TablePage=124273711. accessed October 27, 2020.
[154] Biotage.com. Products—organic synthesis. https://biotage.com/product-page/biotage-initiator. accessed October 27, 2020.
[155] Flood DT, Asai S, Zhang X, Wang J, Baran PS, Dawson PE. Expanding reactivity in DNA-encoded library synthesis via reversible binding of DNA to an inert quaternary ammonium support. J Am Chem Soc 2019;141(25):9998–10006.
[156] Madsen D, Azevedo C, Micco I, Petersen LK, Hansen NJV. An overview of DNA-encoded libraries: a versatile tool for drug discovery. Prog Med Chem 2020;59:181–249.
[157] Manufacturingchemist.com. Merck partners with DyNAbind to introduce DNA-encoded library kits; April 18, 2019. https://www.manufacturingchemist.com/news/article_page/Merck_partners_with_DyNAbind_to_introduce_DNA-encoded_library_kits/153905. accessed October 27, 2020.
[158] Milo A. Democratizing synthesis by automation. Science 2019;363(6423):122.
[159] Biotage.com. Products and services—workup products. https://www.biotage.com/product-page/work-up-cartridges. accessed October 27, 2020.
[160] Radleys.com. Our range—parallel reaction stations. https://www.radleys.com/range/greenhouse-work-up-station/. accessed October 27, 2020.
[161] Sigma-Aldrich.com. Product 57030-U. https://www.sigmaaldrich.com/catalog/product/supelco/57030u?lang=en&region=US. accessed October 27, 2020.
[162] Biotage.com. Products and services—sample preparation. https://www.biotage.com/product-page/biotage-vacmaster-10-and-20. accessed October 27, 2020.
[163] Radleys.com. Our range. https://www.radleys.com/range/carousel-work-up-station/. accessed October 27, 2020.
[164] Fischersci.com. Products—chromatography. https://www.fishersci.com/shop/products/hypersep-glass-block-manifolds/p-4524468. accessed October 27, 2020.
[165] Biotage.com. Products and services—sample preparation. https://www.biotage.com/product-page/biotage-vacmaster-96. accessed October 27, 2020.
[166] Merckmillipore.com. Products. https://www.merckmillipore.com/CH/en/product/MultiScreenHTS-Vacuum-Manifold,MM_NF-C9206. accessed October 27, 2020.
[167] MerckMillipore.com. Product C9206. https://www.hamiltoncompany.com/automated-liquid-handling/small-devices/on-deck-vacuum-station. accessed October 27, 2020.

[168] Porvair Sciences. Evaporators and manifolds—vacuum manifolds. https://www.microplates.com/vacuum-manifolds/. accessed October 27, 2020.
[169] Buchi.com. Products & solutions—parallel evaporation. https://www.buchi.com/gb-en/products/parallel-evaporation/multivapor-p-6-p-12. accessed October 27, 2020.
[170] Radley.com. Our range—parallel reaction stations. https://www.radleys.com/range/greenhouse-blowdown-evaporator/. accessed October 27, 2020.
[171] Porvair Sciences. Evaporators and manifolds—evaporators. https://www.microplates.com/evaporators/. accessed October 27, 2020.
[172] Biopharmagroup.co.uk. Equipment ranges—freeze dryers. https://biopharma.co.uk/bps/freeze-drying/freeze-dryers/laboratory-freeze-dryers/benchtop-pro/. accessed October 27, 2020.
[173] SPSscientific.com. Products—centrifugal evaporators & sample concentrators. https://www.spscientific.com/ProductCategory/108/Genevac/. accessed October 27, 2020.
[174] Hettich. Products—concentrators, https://www.hettweb.com/products/combidancer/. accessed October 27, 2020.
[175] Sigmaaldrich.com. Products. https://www.sigmaaldrich.com/catalog/product/aldrich/z568031?lang=en&region=GB. accessed October 27, 2020.
[176] Sigmaaldrich.com. Product Z568031. https://www.pureprocesstechnology.com/solvent-purification-systems/. accessed October 27, 2020.
[177] Lctechinc.com. Products—solvent purification systems. https://lctechinc.com/solvent-purification-systems/. accessed October 27, 2020.
[178] Mbraun.com. Products—solvent purification systems. https://www.mbraun.com/us/products/solvent-purifiers.html. accessed October 27, 2020.
[179] Crystallizationsystems.com. Crystal breeder. https://www.crystallizationsystems.com/crystalbreeder. accessed October 27, 2020.
[180] Biotage.com. Products and services—flash purifications. https://biotage.com/product-group/isolera-systems. accessed October 27, 2020.
[181] Biotage.com. Products and services—flash purifications—flash columns. https://www.biotage.com/product-page/biotage-sfr- - -overview. accessed October 27, 2020.
[182] Teledyneisco.com. Chromatography—products. https://www.teledyneisco.com/en-us/chromatography/combiflash. accessed October 27, 2020.
[183] Buchi.com. Products & solutions—preparative chromatography. https://www.buchi.com/en/products/preparative-chromatography/pure-chromatography-systems. accessed October 27, 2020.
[184] Teledyneisco.com. Chromatography—products. https://www.teledyneisco.com/en-us/chromatography/accqprep. accessed October 27, 2020.
[185] Teledyneisco.com. Chromatography—products. https://www.teledyneisco.com/en-us/chromatography/automation-modules. accessed October 27, 2020.
[186] Shimadzu.com. Products—liquid chromatography. https://www.shimadzu.com/an/hplc/nexera/prep/index.html. accessed October 27, 2020.
[187] Shimadzu.com. Products—liquid chromatography. https://www.shimadzu.com/an/hplc/nexera/prep_uc/index.html. accessed October 27, 2020.
[188] Knauer.net. Solutions—preparative HPLC. https://www.knauer.net/en/azura-prep-lab-lc-system-high-pressure-gradient/p15527. accessed October 27, 2020.
[189] Agilent.com. Products—liquid chromatography—HPLC systems. https://www.agilent.com/en/product/liquid-chromatography/hplc-systems/preparative-hplc-systems/1260-infinity-ii-preparative-lc-system. accessed October 27, 2020.
[190] Waters Corporation. Products—instruments & systems. https://www.waters.com/waters/en_GB/Original-UPLC-UHPLC-system-with-sub-2-micron-particle-technology-for-separations/nav.htm?locale=en_GB&cid=514207. accessed October 27, 2020.

[191] Shimadzu.com. Products—liquid chromatography-mass spectrometry. https://www.shimadzu.com/an/lcms/lcms2020/2020.html. accessed October 27, 2020.
[192] Agilent.com. Products—liquid chromatography/mass spectrometry. https://www.agilent.com/en/product/liquid-chromatography-mass-spectrometry-lc-ms/lc-ms-instruments. accessed October 27, 2020.
[193] Bruker.com. Products—mass spectrometry and separations. https://www.bruker.com/products/mass-spectrometry-and-separations.html. accessed October 27, 2020.
[194] Bruker.com. Products—mass spectrometry and separations. https://www.bruker.com/products/mass-spectrometry-and-separations/lc-ms/liquid-chromatography.html. accessed October 27, 2020.
[195] Bruker.com. Products—mass spectrometry and separations. https://www.bruker.com/products/mass-spectrometry-and-separations/ms-software/compass/overview.html. accessed October 27, 2020.
[196] Bruker.com. Products—magnetic resonance—NMR. https://www.bruker.com/products/mr/nmr/software/topspin.html. accessed October 27, 2020.
[197] Bruker.com. Products—magnetic resonance—NMR. https://www.bruker.com/products/mr/nmr/software/iconnmr.html. accessed October 27, 2020.
[198] Bruker.com. Products—magnetic resonance—NMR. https://www.bruker.com/products/mr/nmr/automation/samplecase.html. accessed October 27, 2020.
[199] Kerr H. Program that automatically interprets NMR spectra is boon for structure elucidation. In: Chemistry World News [Internet]; 2020. April 6 https://www.chemistryworld.com/news/program-that-automatically-interprets-nmr-spectra-is-boon-for-structure-elucidation/4011467.article. accessed October 27, 2020.
[200] Howarth A, Ermanis K, Goodman JM. DP4-AI automated NMR data analysis: straight from spectrometer to structure. Chem Sci 2020;11:4351–9.
[201] Bruker.com. Products—magnetic resonance—NMR. https://www.bruker.com/products/mr/nmr/benchtop-nmr.html. accessed October 27, 2020.
[202] Magritek.com. Products. https://go.magritek.com/spinsolve-product-family-brochure-bundle-download#80. accessed October 27, 2020.
[203] Bucksci.com. NMR spectrometers. https://www.bucksci.com/products/nmr-benchtop-spectrometer. accessed October 27, 2020.
[204] Oxinst.com. Products—NMR. https://nmr.oxinst.com/x-pulse. accessed October 27, 2020.
[205] Nanalysis.com. Products. https://www.nanalysis.com/products-overview/. accessed October 27, 2020.
[206] Pellecchia M, Sem D, Wüthrich K. NMR in drug discovery. Nat Rev Drug Discov 2002;1:211–9.
[207] Mettler-Toledo.com. Automated reactors and in situ analysis—ReactIR. https://www.mt.com/us/en/home/products/L1_AutochemProducts/ReactIR/flow-ir-chemis.html. accessed October 27, 2020.
[208] Blanco M-J, Gardinier KM. New chemical modalities and strategic thinking in early drug discovery. ACS Med Chem Lett 2020;11(3):228–31.
[209] Panteleev J, Gao H, Jia L. Recent applications of machine learning in medicinal chemistry. Bioorg Med Chem Lett 2018;28(17):2807–15.
[210] Leelananda SP, Lindert S. Computational methods in drug discovery. Beilstein J Org Chem 2016;12:2694–718.
[211] Kutchukian PS, Shakhnovich EI. De novo design: balancing novelty and confined chemical space. Expert Opin Drug Discov 2010;5:789–812.
[212] Ramírez D. Computational methods applied to rational drug design. Open Med Chem J 2016;10:7–20.
[213] Segler MHS, Kogej T, Tyrchan C, Waller MP. Generating focussed molecule libraries for drug discovery with recurrent neural networks. ACS Cent Sci 2018;4:120–31.

[214] Lo Y-C, Rensi SE, Torng W, Altman RB. Machine learning in chemoinformatics and drug discovery. Drug Discov Today 2018;23(8):1538–46.
[215] Cherkasov A, Muratov EN, Fourches D, Varnek A, Baskin II, Cronin M, et al. QSAR modelling: where have you been? Where are you going to? J Med Chem 2014; 57:4977–5010.
[216] Corey EJ, Wipke WT. Computer-assisted design of complex organic syntheses. Science 1969;166(3902):178–92.
[217] Klucznik T, Mikulak-Klucznik B, McCormack MP, Lima H, Szymkuc S, Bhowmick M, et al. Efficient syntheses of diverse, medicinally relevant targets planned by computer and executed in the laboratory. Chem 2018;4(3):522–32.
[218] Segler MHS, Preuss M, Waller MP. Planning chemical syntheses with deep neural networks and symbolic AI. Nature 2018;555(7698):604–10.
[219] Struble TJ, Alvarez JC, Brown SP, et al. Current and future roles of artificial intelligence in medicinal chemistry synthesis. J Med Chem 2020;63(16):8667–82.
[220] Firth NC, Atrash B, Brown N, Blagg J. MOARF, an integrated workflow for multiobjective optimization: implementation, synthesis, and biological evaluation. J Chem Inf Model 2015;55(6):1169–80.
[221] Brown N, McKay B, Gilardoni F, Gasteiger J. A graph-based genetic algorithm and its application to the multiobjective evolution of median molecules. J Chem Inf Comput Sci 2004;44(3):1079–87.
[222] Matheson R. Automating molecule design to speed up drug development. In: MIT News [Internet]; 2018. July 6 https://news.mit.edu/2018/automating-molecule-design-speed-drug-development-0706. accessed October 27, 2020.
[223] Brown N. Silico medicinal chemistry: computational methods to support drug design. Cambridge, UK: Royal Society of Chemistry; 2016.
[224] Merckgroup.com. https://www.merckgroup.com/en/news/lanexo-system-to-improve-lab-productivity-03-03-2020.html. accessed October 27, 2020.
[225] Fabry DC, Sugiono E, Rueping M. Online monitoring and analysis for autonomous continuous flow self-optimizing reactor systems. React Chem Eng 2016;1:129.
[226] Coley CW, Thomas 3rd DA, Lummiss JAM, et al. A robotic platform for flow synthesis of organic compounds informed by AI planning. Science 2019;365(6453), eaax1566.
[227] Lemonick S. Automating synthesis from planning to execution. Chem Eng News 2019;97:32. https://cen.acs.org/synthesis/Automating-synthesis-planning-execution/97/i32. accessed October 27, 2020.
[228] SRI.com. Synfini. An automated synthetic chemistry platform. https://www.sri.com/case-studies/synfini/. accessed Oct 27, 2020.
[229] Anon. https://www.globenewswire.com/news-release/2020/10/07/2105064/0/en/SRI-International-awarded-4-3M-from-DARPA-to-create-rapid-response-therapeutic-manufacturing-system.html. accessed October 27, 2020.
[230] IBM.com. IBM research—blogs. https://www.ibm.com/blogs/research/2020/08/roborxn-automating-chemical-synthesis/. accessed October 27, 2020.
[231] Chemify. http://www.chem.gla.ac.uk/cronin/chemify/. accessed October 27, 2020.
[232] Steiner S, Wolf J, Glatzel S, Andreou A, Granda JM, Keenan G, et al. Organic synthesis in a modular robotic system driven by a chemical programming language. Science 2019;363(6423), eaav2211.
[233] Lim JX-Y, Leow D, Pham Q-C, Tan C-H. Development of a robotic system for automatic organic chemistry synthesis. arXiv:190607939 2019.
[234] Burger B, Maffettone PM, Gusev VV, et al. A mobile robotic chemist. Nature 2020;583:237–41.
[235] Chow S, Liver S, Nelson A. Streamlining bioactive molecular discovery through integration and automation. Nat Rev Chem 2018;2:174–83.

[236] Baranczak A, et al. Integrated platform for expedited synthesis–purification–testing of small molecule libraries. ACS Med Chem Lett 2017;8:461–5.
[237] Guetzoyan L, et al. Machine-assisted synthesis of modulators of the histone reader BRD9 using flow methods of chemistry and frontal affinity chromatography. Med Chem Commun 2014;5:540.
[238] Guetzoyan L, Nikbin N, Bexandale IR, Ley SV. Flow chemistry synthesis of zolpidem, alpidem and other GAB_{AA} agonists and their biological evaluation through the use of in-line frontal affinity chromatography. Chem Sci 2013;4:764–9.
[239] Werner M, et al. Seamless integration of dose-response screening and flow chemistry: efficient generation of structure–activity relationship data of b-secretase (BACE1) inhibitors. Angew Chem Int Ed 2014;53:1704–8.
[240] Gesmundo NJ, et al. Nanoscale synthesis and affinity ranking. Nature 2018; 557:228–32.
[241] Khalid N, Kobayashi I, Nakajima M. Recent lab-on-chip developments for novel drug discovery. Wiley Interdiscip Rev Syst Biol Med 2017;6, e1381.
[242] Tomazelli Coltro WK, Cheng CM, Carrilho E, de Jesus DP. Recent advances in low-cost microfluidic platforms for diagnostic applications. Electrophoresis 2014; 35:2309–24.
[243] Desai B, et al. Rapid discovery of a novel series of Abl kinase inhibitors by application of an integrated microfluidic synthesis and screening platform. J Med Chem 2013;56:3033–47.
[244] Strateos.com. https://strateos.com/elastic-chemistry/. accessed October 27, 2020.
[245] Nicolaou CA, Humblet C, Hu H, Martin EM, Dorsey FC, Castle TM, et al. Idea2Data: toward a new paradigm for drug discovery. ACS Med Chem Lett 2019; 10(3):278–86.
[246] Tomberg A, Boström J. Can easy chemistry produce complex, diverse, and novel molecules? Drug Discov Today 2020;25:2174–81. https://doi.org/10.1016/j.drudis.2020.09.027.
[247] Brown DG, Boström J. Analysis of past and present synthetic methodologies on medicinal chemistry: where have all the new reactions gone? J Med Chem 2016; 59(10):4443–58.
[248] Boström J, Brown DG, Young RJ, Keserü GM. Expanding the medicinal chemistry synthetic toolbox. Nat Rev Drug Discov 2018;17:709–27.
[249] Chemjobber. Will robots kill chemistry? Chem Eng News 2019;97:15. https://cen.acs.org/careers/employment/robots-kill-chemistry/97/i15?utm_source=Twitter&utm_medium=Social&utm_campaign=CEN&hootPostID=0cd15d64276083de2dc0bc81f1b4cee0. accessed October 27, 2020.

CHAPTER FOUR

Use of molecular docking computational tools in drug discovery

Francesca Stanzione*, Ilenia Giangreco, and Jason C. Cole
Cambridge Crystallographic Data Centre, Cambridge, United Kingdom
*Corresponding author: e-mail address: fstanzione@ccdc.cam.ac.uk

Contents

Keywords: Drug discovery, Molecular target, Homology modelling, Molecular interaction, Virtual screening (VS), Structure-based drug design (SBDD), Scoring functions, Protein databases, Small molecule databases

1. Introduction

Drug discovery is a challenging process and identifying the right lead compound is a determining factor of the overall success of the project. In 2016, the Tufts Center for the Study of Drug Development estimated that the cost associated with developing and bringing a new drug to market has

Progress in Medicinal Chemistry, Volume 60
ISSN 0079-6468
https://doi.org/10.1016/bs.pmch.2021.01.004

increased by almost 145% in the last decade [1]. Furthermore, while the average time to bring a drug to clinical trials has decreased, the success rate of drugs obtaining the US Food and Drug Administration (FDA) approval has dropped to 12% [1]. Computer-aided drug design (CADD) has helped to reduce the costs and the time associated with drug discovery by directing experimental research towards optimal compounds more quickly. Within CADD, techniques such as molecular docking and virtual screening (VS) have provided a valuable complement to the time-consuming and expensive experimental process of high-throughput screening (HTS).

The ability to computationally screen large libraries of compounds that either possess similarity towards known inhibitors (ligand-based) or complementarity towards target structures (structure-based) has proven to be successful at identifying highly focused subsets from which actives can then be experimentally confirmed [2]. To date, generating such 'educated guesses' has provided many examples of lead compounds. Several marketed drugs, such as imatinib [3], zanamivir [4], nelfinavir [5], erdafitinib [6], and several clinical candidates are known to have been discovered or optimised with the aid of computational methodologies [7–9]. We can only speculate as to the number of unpublished examples that reside in corporate collections, but we believe this to be considerable.

In the case of molecular docking, the process of predicting the best position, orientation and conformation of a small molecule (drug candidate) when bound to a protein, provides the additional benefit of simplifying future lead optimisation [10]. Knowing exactly how and where a ligand binds helps to rationally design changes to optimise the protein-ligand interaction, to improve activity, and to avoid changes that could lead to protein-ligand clashed.

2. Molecular docking

Molecular docking is the most common computational structure-based drug design (SBDD) method and has been widely used ever since the early 1980s [11]. It is the tool of choice when the three-dimensional (3D) structure of the protein target is available. Molecular docking popularity has been facilitated by the dramatic growth in availability and power of computers, and the increasing number of and ease of access to small molecule and protein structures.

The main goal of molecular docking is to understand and predict molecular recognition, both structurally (i.e. finding possible binding modes) and energetically (i.e. predicting binding affinity). Molecular docking was originally designed to be performed between a small molecule (ligand) and a target macromolecule (protein) however, in the last decade there has been a growing interest in protein-protein docking, nucleic acid (DNA and RNA)-ligand docking and nucleic acid-protein-ligand docking. In this chapter, we will focus on protein-ligand docking and we will provide an overview on the challenges related to protein-protein docking, protein-peptide docking and nucleic acid-ligand docking.

Molecular docking applications in drug discovery are varied, including structure-activity studies, lead optimisation, finding potential leads by virtual screening, providing binding hypotheses to facilitate predictions for mutagenesis studies and also in assisting X-ray and cryogenic electron microscopy (cryo-EM) crystallography in the fitting of substrates and inhibitors to electron density.

Docking has proved to be extremely successful in SBDD, therefore it has been developed and improved for many years. Over the last 2 decades, more than 60 different docking tools have been developed in academic and commercial settings. Some of these programs will be mentioned in this chapter, but there are several available reviews that detail evaluation and comparison of the different docking programs [12–15].

2.1 Theory of docking

The molecular docking process involves two basic steps: prediction of the ligand (usually a small molecule) conformation as well as its position and orientation within the protein binding site (usually referred to as pose) and assessment of the quality of the pose using a scoring function. Ideally, the sampling algorithm should be able to reproduce the experimental binding mode and the scoring function should also rank it highest among all generated poses.

A further task of the docking procedure would be to score active compounds higher than known inactives (*predictive docking*). However, this level of accuracy is difficult to achieve and it is generally influenced by many factors that are external to the protein. Therefore, primarily a docking algorithm only aims to get the prediction of the ligand pose and assessment of quality of the pose correct (though many scoring functions are developed with active/inactive ranking as a consideration too).

2.2 Searching algorithm

The sampling process is non-trivial. The conformational space involves many degrees of freedom including the rotation and translation of a molecule relative to another, the additional conformational degrees of freedom of both the ligand and the protein and sometimes degrees of freedom related to the solvent [16]. In practice, with current computational resources, it is impossible to explore the search space exhaustively by enumerating all possible conformations and all possible rotational and translational orientations of a single molecule relative to a protein within a second of elapsed time (a time scale that is realistically needed for virtual screening). Hence, efficient sampling of conformational space is still a challenge in molecular docking.

Early approaches to account for these sampling problems treated both the ligand and the protein as rigid bodies thereby reducing the number of degrees of freedom to just six. Such approaches rely on the shape similarities between the ligand and the protein binding site. A very well cited example of a program using this algorithm is DOCK [17].

In a rigid docking approach, the ligand and the protein binding sites are represented as pharmacophore spheres of varying radii and the search algorithm tries to pair the ligand spheres with the protein spheres based on the match of the internal distances of all the ligand's spheres and the internal distances of all the protein binding site's spheres. The ligand is then oriented within the binding site using a least square fit of the atoms to the sphere centres [18,19]. In case of unacceptable orientation (e.g. clashes between the ligand and the protein binding site), the ligand is reoriented until an acceptable orientation is obtained. The orientation is then scored based on the degree of overlap between the ligand and protein pharmacophore spheres (Fig. 1).

Despite its computational efficiency, failing to model molecular flexibility limits effectiveness as the conformation is interlinked with protein-ligand interactions; the optimal binding conformation of a small molecule is a compromise between the best internal geometry of said small molecule and the interactions it forms with the binding site. Rigid docking fails to account for this. This is even more relevant in predictive docking where there is the additional complexity derived from the conformational change of the ligand from its unbound (isolated) conformation and its bioactive (bound) conformation.

To overcome to such limitations, most docking programs reached a trade-off and started to account for the whole conformational space of the ligand while limiting flexibility in the protein to regions of the receptor (e.g. binding

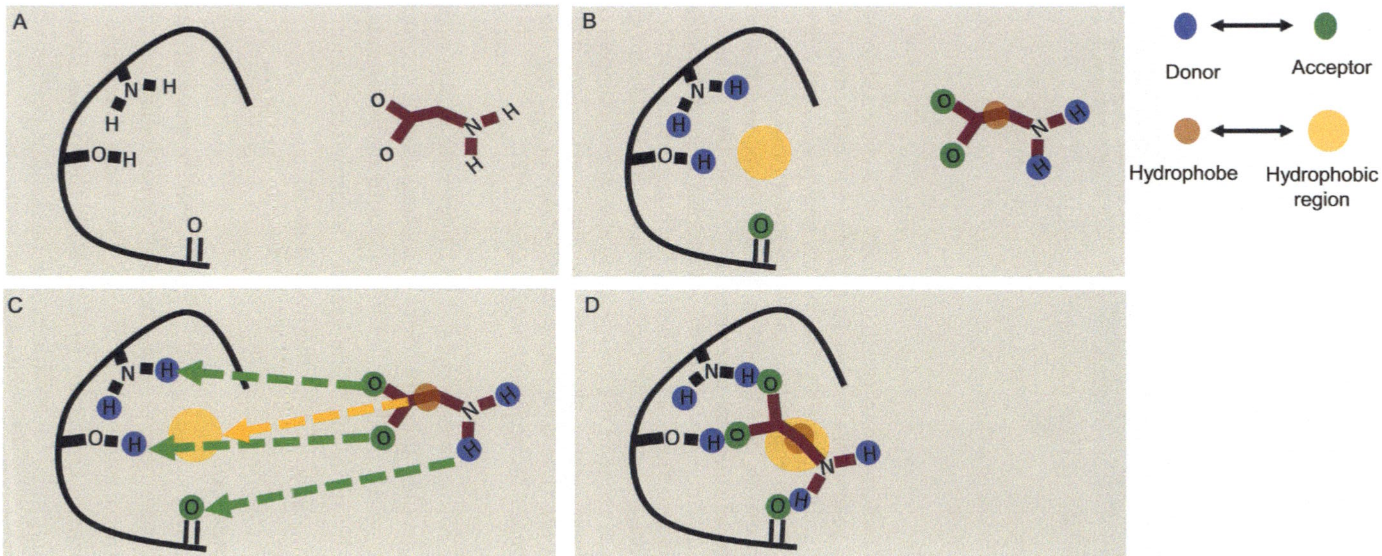

Fig. 1 Rigid docking approach: (A) protein and ligand. (B) The initial pose generation through pharmacophore point matching. Donor–acceptor pharmacophore point and hydrophobic pharmacophore points are added to the protein and the ligand. (C) The searching algorithm tries to match the protein and the ligand fitting points by matching donor with acceptors and hydrophobic atoms with hydrophobic cavities. (D) Different solutions are found.

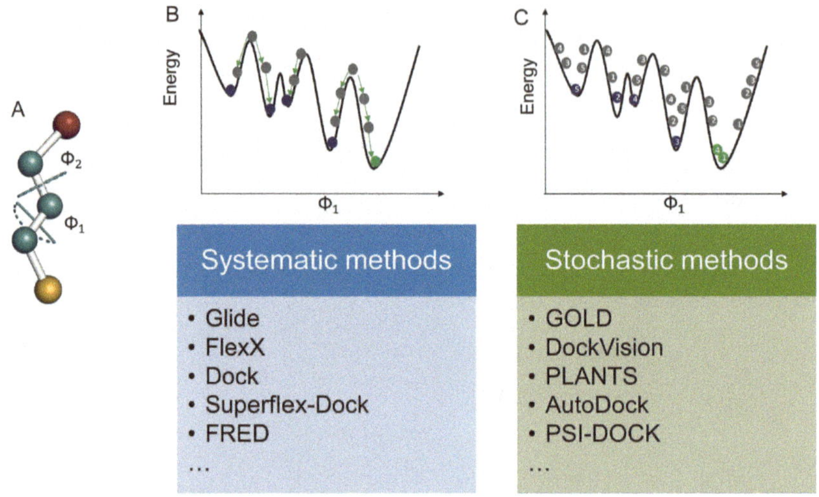

Fig. 2 Small molecule conformational search methods. (A) A molecule containing two bulky groups (red and yellow spheres) has its conformation defined by two internal dihedrals Φ_1 and Φ_2. If we freeze the Φ_2 dihedral, the energy variation due to rotation of Φ_1 can be plotted in a 1D energy landscape. The initial structure (grey spheres) is modified by changing the Φ_1 dihedral. (B) In the systematic search approach, the changes are applied to all structural parameters until local (blue spheres) or global (green sphere) energy minimum is reached; a few examples of docking programs using systematic methods are also listed. (C) Stochastic search approaches explore the conformational space by generating distinct conformations, with an element of randomness, populating a broad range of the energy landscape. A few examples of docking programs using stochastic search algorithms are also listed. *Figure adapted from Ferreira LG, dos Santos RN, Oliva G, Andricopulo AD. Molecular docking and structure-based drug design strategies. Molecules 2015;20(7):13384–13421.*

site), or keeping the entire protein rigid. The underlying reason is that, since ligands are small molecules, they are likely to undergo larger conformational change; furthermore, given their smaller size, accounting for ligand flexibility is also computationally affordable. The ligand conformations can be sampled with both systematic and stochastic methods (Fig. 2).

2.2.1 Systematic methods

Systematic search methods sample the ligand search space at predefined intervals and are deterministic; they can be classified as exhaustive, fragmentation or conformational ensemble methods. The main difference between them is in the approach they take to deal with the ligand flexibility. In exhaustive search methods, for example, the docking is performed by systematically rotating all possible rotatable bonds in the ligand at a given interval. Despite its sampling completeness, the number of possible combinations is huge and increases with the number of rotatable bonds in the ligand (Fig. 2). Therefore, this approach is limited to small, relatively flexible ligands and, in most cases, to make the docking 'practical', it is necessary to apply geometrical or chemical constraints to the initial screening of the ligand poses. A well cited example of a program using the exhaustive sampling method is Glide [20,21].

The fragmentation method is an incremental approach where ligands are divided into modular pieces. One of the fragments is anchored to the protein binding site and then the ligand binding conformation is incrementally grown by placing additional fragments one at the time. The anchor is generally chosen to be the fragment which shows maximum interaction complementarity with the receptor surface (i.e. H-bonds), has the minimum number of alternate conformations, and is fairly rigid (for example, a ring system). One docking program that uses a fragmentation sampling approach is FlexX [22]. A variation of this approach is to dock all the fragments into the binding site and then link them covalently.

As for the exhaustive method, the fragmentation method is restricted to medium and smaller sized ligands and it is not feasible for big ligands where the number of fragments would be too large.

In conformational ensemble methods, the ligand flexibility is represented by rigidly docking an ensemble of pre-generated ligand conformations. Using such an approach removes the computational cost due to the exploration of ligand conformational space, however, it involves additional tools to generate the required ensemble of conformations of the ligand. One limitation in this approach is that the ensemble of generated conformers may not include the bioactive conformation of the ligand.

2.2.2 Stochastic methods

In stochastic algorithms, the ligand binding orientations and conformations are sampled by making changes to the ligand that have some dependence on one or more values generated at random at each step (Fig. 2). The change is then accepted or rejected according to an algorithm-dependent criterion.

The advantage of stochastic algorithms is that they can generate large ensembles of molecular conformations and explore a broad range of the energy landscape increasing the probability of finding a global energy minimum. However, this also means that computational costs associated with this procedure represent an important limitation. Genetic algorithm, Monte Carlo, ant colony optimization (ACO) and tabu search methods are a few examples of stochastic algorithms that differ in the way they generate given moves and the probability criteria of acceptance.

In a genetic algorithm method, a population of potential solutions is set up at random. Each member of the population is encoded as a 'chromosome', which contains information about the mapping of ligand fitting points (e.g. H-bond atoms) onto the complementary protein's fitting points. Each chromosome is assigned a fitness score based on its predicted binding affinity and the chromosomes within the population are ranked according to fitness score. At each step, a point mutation may occur in a chromosome, while the crossover operator exchanges information between two chromosomes of the population. Other operations are also used in some implementations (for example, island migration). GOLD [23] is one of the well cited docking programs that uses a genetic algorithm to explore the ligand conformational space. DockVision [24] is an example of docking program that uses Monte Carlo stochastic method where the probability to accept a random change is calculated by using the Boltzmann probability function. PLANTS [25], instead, is an example of a docking program based on ACO. ACO is inspired by the behaviour of real ants finding the shortest path between their nest and a food source. In the case of protein-ligand docking, an artificial ant colony is employed to find a minimum energy conformation of the ligand in the binding site. These ants mimic the behaviour of real ants and mark low energy ligand conformations with pheromone trails. The artificial pheromone trail information is then modified in subsequent iterations to generate low energy conformations with a higher probability [25].

The tabu search method is a variation of the Monte Carlo approach which maintains a record of the search space of the binding site which has already been visited and thus ensures that the binding site is explored to the maximum. PSI-DOCK [26] is an example of a docking tool that uses a tabu search.

2.2.3 Scoring functions

Scoring functions are fast approximate mathematical methods used to predict the strength of the interaction (or binding affinity) between two or more molecules.

Four aspects should be considered when assessing the reliability of a scoring function [27]: (1) *scoring power*: the ability to produce scores which linearly correlate with experimental binding affinity data, (2) *ranking power*: the ability to correctly rank a given set of ligands that bind to a common target protein by their binding affinities when their binding poses are known,

(3) *docking power*: the ability to identify the native binding pose of a ligand as the one with the best score, and when screening a large set of generated decoy poses, (4) *screening power*: the ability to identify the true binders to a given target protein among a library of random molecules. Ideally, an accurate scoring function would perform equally well on all these four tasks; however, each existing scoring function only perform well on one or two of them at the same time.

Scoring functions can be grouped into four main classes: physics-based, empirical, knowledge-based and machine learning-based scoring functions [28].

The first three types are commonly referred to as 'classical' scoring functions and are based on the assumptions that the change in free energy upon binding of a ligand to its target can be decomposed into a sum of individual energy contributions, and that all these energy contributions are linearly combined. In reality, such linear correlation may not always exist [29]. Two major limitations of classical scoring functions are their minimal description of protein flexibility and the implicit treatment of solvent.

Machine learning-based scoring functions instead use more sophisticated techniques, such as random forests (RF), support vector machines (SVM), and deep learning (DL), to approximate non-linear problems (Fig. 3).

Physics-based or force-field based scoring functions compute the binding energy by summing up the contribution of the bonded interactions (bond stretching, angle bending and torsion angles) and non-bonded interactions (van der Waals and electrostatic interactions) within the protein-ligand complex which accounts for the contribution of enthalpy to energy. Hydrogen bonds are usually considered by adding an additional term to the binding energy. Alternatively, they can be included implicitly in the electrostatic energy term.

Parameters for this type of scoring function are usually derived from both experimental data and ab initio quantum mechanical calculations. The major challenge for physics-based scoring functions is the treatment of the solvent in ligand binding. To overcome this limitation, implicit solvent approaches like Poisson–Boltzmann (PB) or Generalised-Born (GB) continuum solvation models have been widely used [30]. However, more computationally expensive approaches that treat water molecules explicitly are also available (such as free energy perturbation (FEP) and thermodynamic integration (TI) techniques) [31] (Eq. 1).

$$E_{bind} = E_{bond} + E_{no-bond} + [E_{H-bond}] + E_{solv} \tag{1}$$

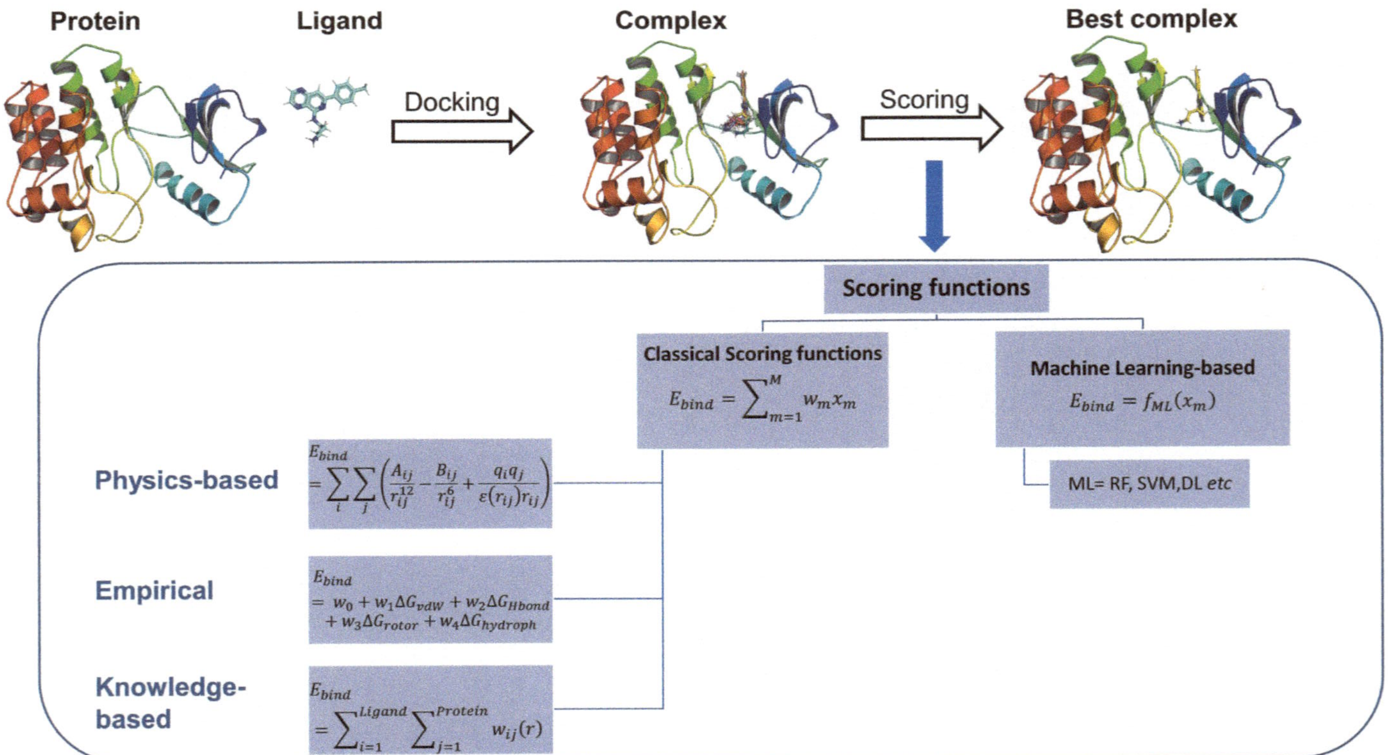

Fig. 3 Scoring functions in docking. Scoring functions can be divided into classical and machine learning scoring functions. The classical scoring functions are physics-based, empirical and knowledge-based. From a mathematical perspective, all classical scoring functions assume an additive functional form; however, they are distinguished by the type of structural descriptors employed. By contrast, non-parametric machine learning scoring functions do not make assumptions about the form of the functional. Instead, the functional form is inferred from training data in an unbiased manner. Machine learning-based scoring functions use more sophisticated machine learning techniques, such as random forests (RF), support vector machines (SVM), and deep learning (DL), to approximate non-linear problems.

In addition to the above, accounting for the entropic effect is an even more severe challenge for physics-based scoring functions. This shortcoming is due to the lack of a reasonable physical model to describe this phenomenon. In principle, the individual terms in Eq. (1) account for the main energetic contributions to the protein-ligand binding however, a separation into individual terms is only possible if the system of interest is divided into mutually independent subsystems [32]. In fact, many of the individual terms are highly correlated with each other and thus can affect the binding affinity in more than one way (i.e. positive or negative contribution) [33]. Moreover, whether the free energy of ligand binding can be decomposed into a linear combination of individual interaction terms without calculating the partition function (ensemble average) also remains in question (the 'non-additive' problem). Despite these approximations, physics-based scoring functions are very appealing as the simplifications result in functions that can be evaluated very rapidly, which is important in a high-throughput docking setting. Another obvious advantage of physics-based methods is that they can ride on the progress of modern force fields, quantum mechanics methods, solvation models, and other developments.

Despite the lucid physical meaning, rigorous physics-based scoring functions are normally computationally expensive. Examples of physics-based scoring functions are GoldScore [34], AutoDock [35], Generalised-Born Volume Integral/Weighted Surface area (GBVI/WSA) [36].

Empirical scoring functions estimate the binding affinity of a protein-ligand complex by summing up different energetic factors involved in the protein-ligand binding, such as hydrogen bonds, hydrophobic effects, protein-ligand clashes, etc. (see Fig. 3). Each factor is multiplied by a coefficient that is obtained from multiple linear regression analyses fitted to a training set of protein-ligand complexes with known binding affinities. Compared to force-field or physics-based scoring functions, empirical scoring functions are much faster in binding score calculations due to their simple treatment of the energy terms. However, the accuracy of empirical scoring functions is directly correlated to the accuracy and the coverage of the protein-ligand training set that is used to develop the model. Examples of empirical scoring functions are ChemScore [37], GlideScore [21] and ChemPLP [25,38].

Knowledge-based scoring functions use statistical analyses to derive the observed interatomic contact frequencies and/or distances in a large database of crystal structures of protein-ligand complexes and employ the Boltzmann law to transform the atom pair preferences into distance-dependent pairwise potentials (Eq. 2).

$$w_{ij}(r) = -k_B T ln\left(\frac{\rho_{ij}(r)}{\rho^*(r)}\right) \quad (2)$$

In this equation k_B is the Boltzmann constant, T is the absolute temperature of the system, $\rho_{ij}(r)$ is the number density of the protein-ligand atom pair at distance r in the training set, and $\rho^*(r)$ is the pair density in a reference state where the interatomic interactions are zero. The protein-ligand density functions $\rho_{ij}(r)$ are constructed by summing the static densities observed in different proteins rather than by averaging different states of the same protein.

The knowledge-based scoring functions assume that favourable interactions occur with higher frequencies than less favourable interactions therefore, the final score is calculated by favouring preferred contacts and penalising repulsive interactions between each atom in the ligand and protein within a given cut-off. Compared to the physics-based and empirical scoring functions, knowledge-based scoring functions offer a good balance between accuracy and speed, because they do not rely on ab initio calculations (physics-based methods) or reproducing binding affinities (empirical methods). Also, because the training protein-ligand dataset database can be large and diverse, they are insensitive to the training set [39].

The main challenge in deriving knowledge-based scoring functions is represented by the calculation of the reference state ($\rho^*(r)$) [40]. Currently, there are two classical strategies used to determine this: traditional atom-randomised reference state and corrected reference state. Traditional methods approximate the reference state by the random distribution of atomic pairs in the training set. Examples include DrugScore [41,42] and GOLD/ASP [43]. The drawback of the atom-randomisation approximation is the neglect of the effects of excluded volume and interatomic connectivity [40]. To overcome these limitations, later approaches introduced correction terms such as the volume factor correction term for the reference state [44,45]. Nevertheless, the accuracy of the reference state remains a challenge for knowledge-based scoring functions. The problem is more relevant for binding mode predictions and virtual screening, as the pairwise potentials, derived from bound structures, are not sufficiently sensitive to different ligand positions and may give good scores even to bad binding modes. A third approach to solving this problem, is to circumvent the accurate calculation of the reference state using iterative methods [39,46]. The basic idea of this method is to adjust the pair potentials by iteration until the interaction potentials reproduce the experimentally determined pair distribution function in the training set, yielding a set of potentials that can discriminate the native structures from decoys. During the iteration procedure, the improvement for the potentials is guided through the difference between the predicted and experimentally observed pair distribution functions rather than through accurate calculation of the reference state [47].

Other challenges for knowledge-based scoring functions include extension of the pairwise interactions to many-body interactions to account for hydrogen bonding and other directional interactions [48,49] and the development of an accurate method that includes the contributions from solvation and entropy [30,50,51].

Over the years several studies have assessed and compared different scoring functions; each one has its virtues and drawbacks. None of the scoring functions available outperforms the others in all tasks but each scoring function may perform better than others in a specific task [27,52]. To compensate for drawbacks in individual scoring functions, the simultaneous use of different scoring functions, or even individual terms of multiple scoring functions, has been widely used to obtain a consensus score [53]. One way to perform consensus scoring is to re-evaluate the best docked pose of each compound with other scoring functions. Only the top scored compounds common to each scoring function will be identified as candidates for bioassay. Consensus ranking is thought to increase hit rates, either by reducing the number of false positives or by statistically reducing the errors in the scores/ranks [34,54,55]. On the other hand, in some cases, single-scoring-function ranking has been shown to outperform consensus-ranking methods [56].

With the abundance of experimental biological, biochemical and biophysical data becoming available, a direct link between energetic and structural information of protein-ligand complexes has become accessible, leading to the design of data-oriented scoring functions using machine learning (ML) techniques [57]. These methods introduced quantitative structure-activity relationship (QSAR) analysis into the protein-ligand interaction evaluation. If the properties of the ligand and the protein, such as atom pairs or structural interaction fingerprints, as well as their interaction patterns (electrostatic interactions, hydrogen bonds, or aromatic stacking), geometrical descriptors (surface or shape properties), and conventional ligand-based descriptors (molecular weight, number of rotatable single bonds, etc.) can be encoded, then ML techniques can be applied to derive statistical models that compute protein-ligand binding scores [47].

Rather than the predetermined functional form, ML-based scoring functions can automatically learn both generalised non-linear functional forms and feature information from the training data. Thus, like empirical scoring functions, ML-based scoring functions also need a training set of protein-ligand complexes with known structures and binding data to derive their final models. Given a set of active and inactive ligands for training, ML-based scoring functions can be trained to distinguish between known ligands by potency with high accuracy. Thus, they have gradually emerged as potential alternatives to classical scoring functions [58,59]. Recently, significant progress has been made on ML-based scoring. In particular a wide range of ML algorithms, such as random forest (RF), support vector machine (SVM), artificial neural network, gradient boosting decision tree (GBDT) and convolutional neural network (CNN), have been applied to the development of new scoring functions [28].

Machine learning scoring functions have been proved to outperform classical scoring functions in such tasks as ranking and screening however, they are rarely directly incorporated into docking software and are mostly used for rescoring [57,58,60]. Examples of machine learning scoring functions are RF-Score [61] and SVM-Score [62].

2.2.3.1 Tailored scoring functions

The intrinsic simplistic nature of the classic scoring functions cannot capture all the features involved in the protein-ligand binding. Some of these scoring functions performs better with one protein target or another based on their structural and chemical features. For example, it was observed that among metalloenzymes, the performance of different scoring functions varied based on the specific metal in the protein, its location in the binding site, type of dominant interactions, exposure to solvent, etc. [63]. Furthermore, the chemical space and properties (such as protonation state, partial charge, number of rotatable bonds) of the small molecules tested in docking has a significant impact on the performance of one scoring function versus another [64].

One can evaluate which scoring function to use for a target protein by experimenting and optimising the choice based on an actives/decoys test set for each specific protein target (Fig. 4).

This strategy is subject to the availability of experimental information on active ligands. Several databases such as BindingDB [65–67], PDBbind [68–70] and ChEMBL [71,72], store experimental binding affinity data of ligands against several protein targets.

This information can be used to select active and inactive ligands for the target of interest that can be used to test each scoring function. Alternatively, services like the DUD-E (database of useful decoys-enhanced) decoys server

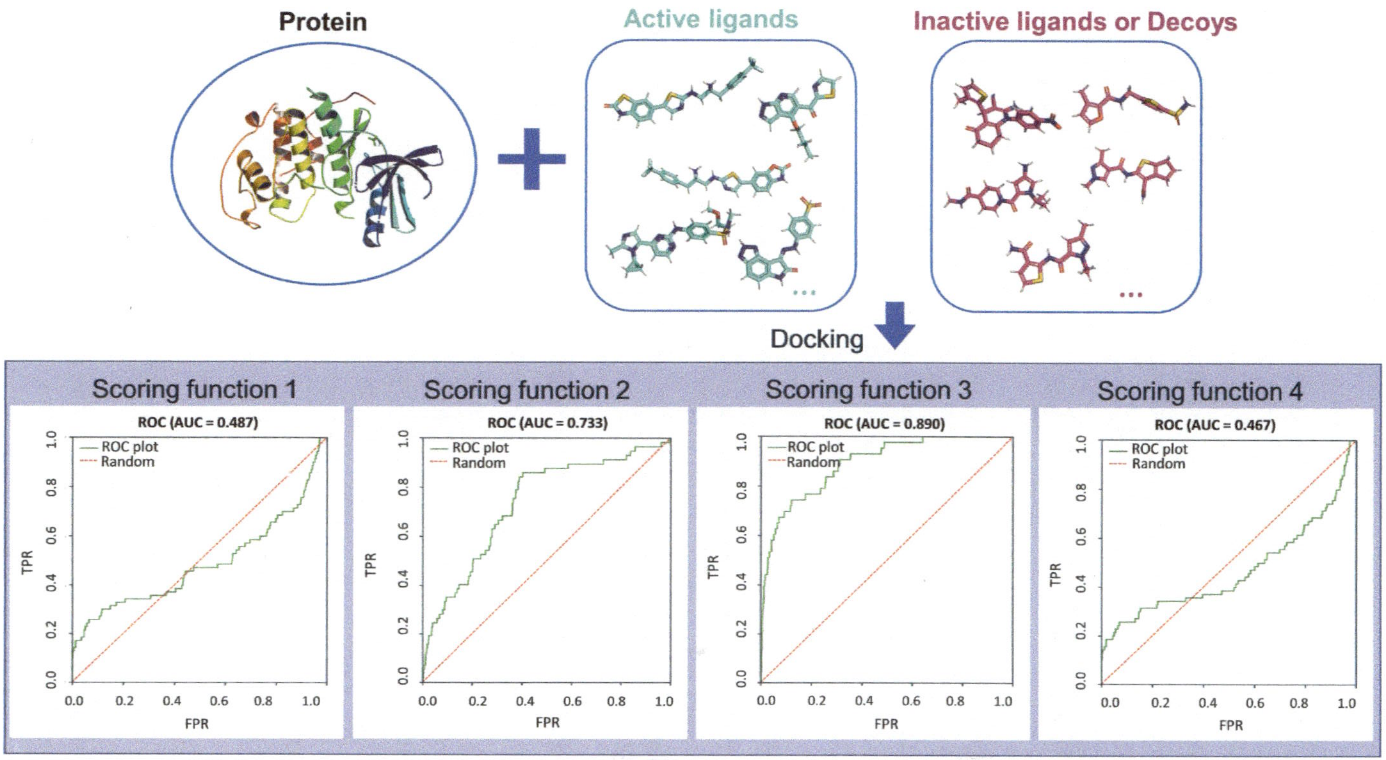

Fig. 4 Approach to evaluate scoring function for a specific protein target. Known active ligands and inactive ligands are docked together to evaluate the performance of different scoring functions. The performance of the docking run can be evaluated by the enrichment of annotated ligands of known binders (True Positive Rate—TPR) from among of non-binding, or decoy molecules (False Positive Rate—FPR). The area under the receiver operating characteristic (ROC) curve is widely used to evaluate its performance, where generally higher is better.

[73] can be adopted to generate decoy molecules with similar physicochemical properties of the actives. The customised actives/decoys test set can therefore be used to assess the performance of the different scoring functions to find the best one available for the specific target for the virtual screening campaign. In the absence of enough data (active ligands), scoring functions are usually selected based on the ability to reproduce the native ligand binding conformation (self-docking). Success of such prediction is defined by the Root Mean Square Deviation (RMSD) value between the top-ranked ligand conformations and the experimentally observed (native) structure. Generally, if the RMSD is 1.5 Å or below, the prediction is considered successful.

Because of its simplicity, the RMSD criterion has been widely used to evaluate the prediction power of scoring functions. However, it may not be reliable for small or symmetrical ligands that are likely to have good RMSD values even when they are randomly placed in a protein binding site. On the other hand, for large flexible ligands, a large RMSD value due to a solvent exposed and/or substituent groups, may hide the correctness in prediction of the overall binding mode. To overcome these limitations, several alternative methods have been proposed for pose evaluations, such as interaction-based accuracy classification (IBAC) [74], real space R-factor (RSR) [75] and Generally Applicable Replacement for RMSD (GARD) [76].

2.3 Practical aspects in molecular docking

Fig. 5 shows the key steps in molecular docking that are common to all protocols. A molecular docking calculation needs 3D structures of the input molecules, both protein and ligand(s). The structure of the protein target is usually determined by experimental techniques such as X-ray crystallography, nuclear magnetic resonance (NMR) or cryogenic electron microscopy (cryo-EM), and can frequently be downloaded from the Protein Data Bank (PDB) (https://www.rcsb.org/) [77]. Other sources of crystallographic structure of proteins and protein-ligands complexes are listed in Table 1.

There are several aspects that should be considered when assessing the quality of the protein target [78].

First is the crystal structure resolution, which is a measure of the degree of measurable diffraction observed in the crystallographic experiment using the protein or nucleic acid. This measure is conveniently represented in a form that expresses the level of details that could be observable when the electron density map is calculated. The crystal resolution depends on the degree of

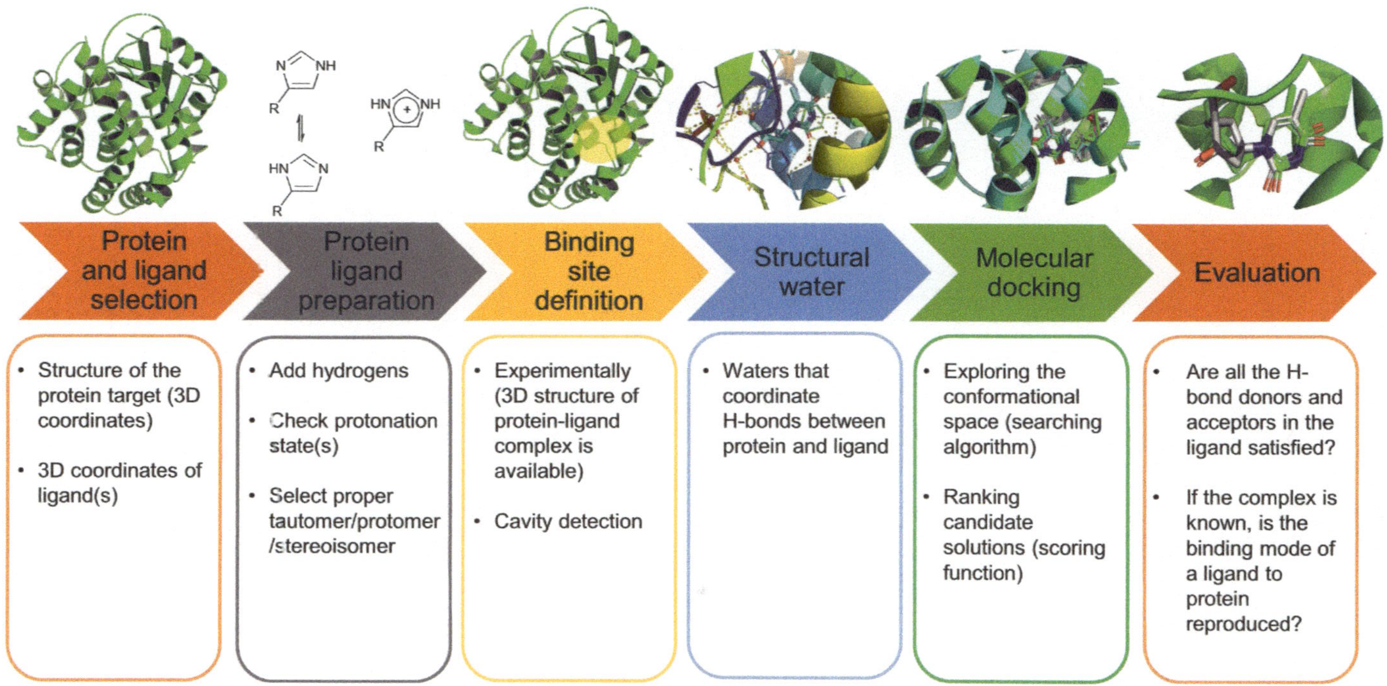

Fig. 5 A typical docking workflow. This flowchart shows the key steps common to all docking protocols. The 3D structures of the target macromolecule and the small molecule must first be chosen, and then each structure must be prepared in accordance with the requirements of the docking method being used. The binding site should be defined using computational tools or using experimental information. Active or structural water should be included as well. Following the docking, the results must be analysed, selecting the binding modes with the best scores, and evaluated.

Table 1 Sources of crystallographic structure of proteins and protein-ligands complexes.

wwPDB: worldwide Protein Data Bank https://www.wwpdb.org/ www.rcsb.org www.pdbe.org www.pdbj.org www.bmrb.wisc.edu	PDB is the central archive of all experimentally determined protein structure data. Today the PDB is maintained by an international consortium known as wwPDB. wwPDB has four members: • Research Collaboratory for Structural Bioinformatics Protein Database (RCSB PDB) • Protein Data Bank in Europe (PDBe) • Protein Data Bank Japan (PDBj) • Biological Magnetic Resonance Data Bank (BMRB) Among these, rcsb acts as archive keeper to ensure that there is only one version of the data which is identical for all users
BindingDB https://www.bindingdb.org/bind/index.jsp	Rich repository of structural and thermocalorimetric data about ligand-protein interactions, including standard free energy, enthalpy and entropy changes upon binding, $\Delta G°$, $\Delta H°$, and $-T\Delta S°$ for a small but growing number of complexes
BindingMOAD—Mother Of All Databases https://bindingmoad.org/	Subset of the PDB containing every high-quality example of protein crystal with clearly identified biologically relevant ligands annotated with experimentally determined binding data extracted from literature. Ligands may be a peptide of 10 amino acids or less; oligonucleotide of 4 nucleotides or less; small organic molecule, and cofactors. Small molecules like crystallographic additives, salts, metals or solvent are not considered as ligands
PDBbind http://www.pdbbind-cn.org/	Comprehensive collection of measured binding affinity data (Kd, Ki, and IC50) exclusively for the protein-ligand complexes available in the PDB. It thus provides an essential linkage between energetic and structural information of these complexes, which is helpful for various computational and statistical studies on molecular recognition occurred in biological systems

Table 1 Sources of crystallographic structure of proteins and protein-ligands complexes.—cont'd

ModBase: Database of Comparative Protein Structure Models https://modbase.compbio.ucsf.edu/	Database of theoretically calculated protein structure models. The models are derived from an automated modelling pipeline relying on PSI-BLAST and MODELLER. In addition to the protein structure models ModBase contains information about putative ligand binding sites, SNP annotation and protein-protein interactions
PDB-REDO databank https://pdb-redo.eu/	Database of optimised existing PDB entries with electron density maps, a description of model changes, and a wealth of model validation data. It is a good starting point for any structural biology project. All the entries are treated with a consistent protocol that reduces the effects of differences in age, software, and depositors
EBI: The European Bioinformatics Institute https://www.ebi.ac.uk/services/structures	EBI provide access to several data resources including: • EMDB, a public repository for electron microscopy (EM) density maps of macromolecular complexes and subcellular structures • EMPIAR-Electron Microscopy Public Image Archive, a public resource for raw, 2D electron microscopy images. It allows users to upload, and download and reprocess the thousands of raw, 2D images used to build a 3D structure • PDBe, the European resource for the collection, organisation and dissemination of 3D structural data (from PDB and EMDB) on biological macromolecules and their complexes)

order of the crystal and the intensity and coherence of the diffraction beam used in the experiment. If a crystal is highly ordered and atoms are in well-defined positions throughout the crystal and remain so over time, then they will scatter X-rays the same way, and the diffraction pattern will show the fine details of the crystal.

In contrast, if atoms move over time or the content of one unit cell differs from the other, the diffraction pattern will not contain as much fine information. The highest resolution structures, with resolution values of 1 Å or so, are highly ordered and it is easy to see every atom in the electron density map. Lower resolution structures, with resolution of 3 Å or higher, show only the basic contours of the protein chain, and the atomic structure must be inferred (Fig. 6). As a rule of thumb, we have more confidence in the location of atoms in structures with resolution values that are lower than 2 Å, called 'high-resolution structures'.

Although the resolution of the data provides information about the theoretical limits on the precision of the model, it does not say anything about the quality or the completeness of the data. Any missing data leads to a deterioration of the model parameters in the same way as reduced resolution does. Therefore, if two data sets are collected at the same resolution, the one with the lower completeness has the poorer data set and will more likely result in a less precise model.

Moreover, the experimental electron density data can be used to calculate several values such as the *R*-factor and R_{free} that can be used to address the quality of the data instead [79].

R-factor is a measure of the difference between measured data and data predicted from the model. A totally random set of atoms will give an

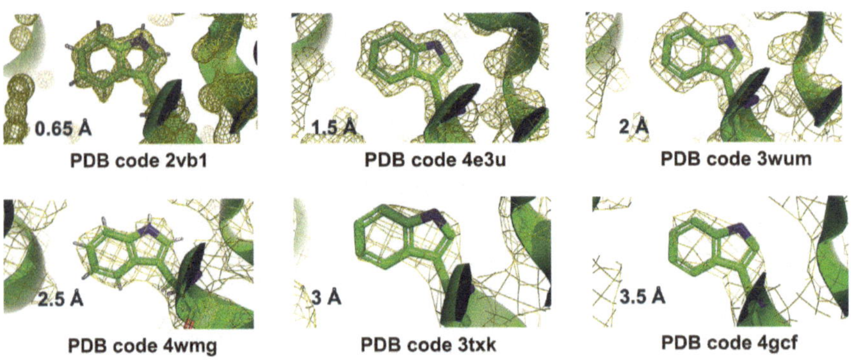

Fig. 6 Electron density maps for structures with a range of resolutions. The three structures on the top panel show tryptophan 28 from Lysozyme C, from PDB entries 2vb1 (0.65 Å resolution), 4e3u (1.5 Å resolution), and 3wum (2 Å resolution), respectively. The three structures on the bottom panel show tryptophan 28 from Lysozyme C, from PDB entries 4wmg (2.5 Å), 3txk (3.0 Å resolution) and 4gcf (3.5 Å resolution), respectively. The protein backbone is displayed as a ribbon cartoon and coloured in green, the tryptophan residue is rendered as green sticks and the electron density map surrounding regions of high electron density is coloured in yellow mesh.

R-factor of about 0.63, whereas a perfect fit would have a value of zero. A small difference indicates a more consistent model and therefore if two structures have similar resolutions the one with lower R-factor is the model that best fits the experimental data.

One potential problem with using R-factor to assess the quality of a structure is that it reflects the fit of the calculated and observed reflection data, the very thing that the refinement process uses to improve the atomic model. In effect, using the R-factor means we are evaluating the refinement model against the data used to train that model and so its value can be artificially low.

A less biased approach is to remove some of the experimental data (usually about 10%) before starting the refinement process, in this way only the remaining data (90%) are used for the refinement while the 10% is used for cross-validation during the refinement process to avoid over-fitting to the data. The R_{free} value represents how well the model predicts the 10% of data that were not used in refinement. However, over-fitting may still occur because there are usually insufficient data to uniquely determine all atom coordinates. For an ideal model that is not over-fitting the data, the R_{free} will be similar to the R-factor with a value of about 0.26 [80]. Another aspect to consider when evaluating the quality of a protein structure is the Debye Waller factors (DWF), or B-factors which are used to describe the relative vibrational motion of different parts of the protein. Atoms with low B-factors belong to well-ordered regions of the protein target while atoms with large B-factors belong to regions of the protein that are very flexible. It is important to ensure that atoms included in the binding site of the protein structure have a low B-factor, as high values imply that their coordinates are less reliable, and the docking experiment could then be affected. Finally, atomic occupancies should be considered. In some structures, regions are so labile that it is impossible to identify atomic positions at all. If prior to refinement a crystallographer knows that a given residue is of a given type from sequence information, they may choose to include a set of model coordinates for it with occupancies set to zero. These atoms will not contribute to the fit and so are, in effect, being placed by the refinement program in an arbitrary single position that may not be accurate.

When the 3D structure of the target protein is not available, protein structure prediction techniques such as homology modelling and protein threading are commonly used to obtain a 3D model of the target protein [81]. Homology modelling or comparative modelling relies on the correlation between the sequence of the target protein and its homologous protein

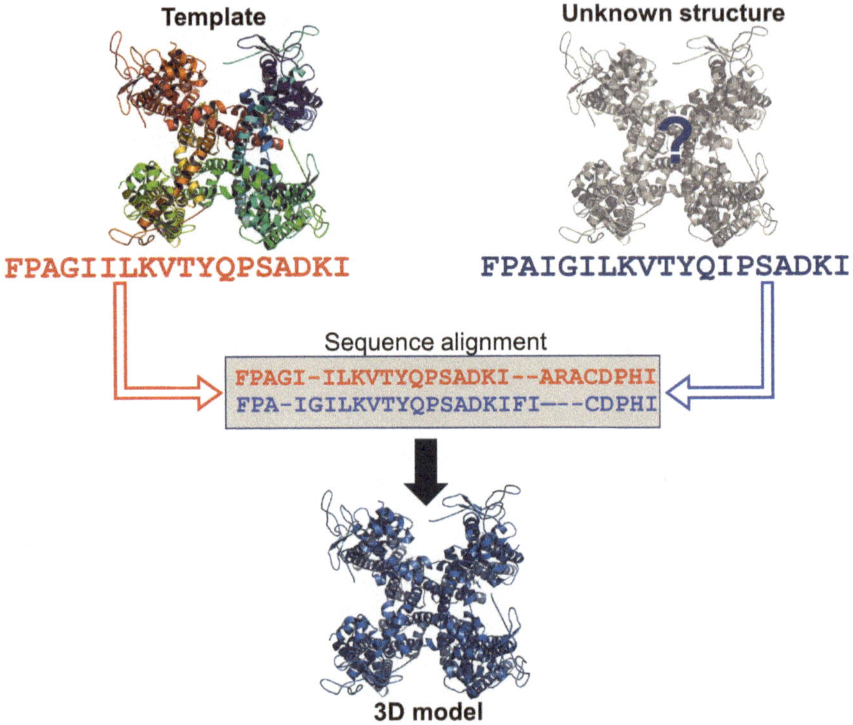

Fig. 7 Outline of the homology modelling process. Given the sequence of a protein with unknown structure, the first step is the identification of a related protein with known 3D structure that serves as a template. An alignment of the target and template sequences is necessary to assign the correspondence between target and template residues. A model is then built for the target based on the alignment and structure of the template, and further refined and validated.

structures (template) available in the PDB (Fig. 7). Such approaches can also be used to generate a variety of receptor conformations using either single-template or multiple-template structures enhancing the understanding of selectivity. Homology modelling can produce high-quality structural models when the protein target and template are closely related. Above 50% sequence identity, structural models tend to be reliable, with only minor errors in side chain packing. In the 30–50% identity range, errors can be more severe and are often located in flexible regions of the proteins (i.e. loop regions). For structures with lower sequence identities (<30%) homology modelling can be extremely difficult and other modelling techniques such as protein threading are then recommended. In protein threading or fold recognition techniques, the protein modelling is not based on the sequence similarities, but on structural fit of the target sequence against different known folds. Protein threading is based on two basic observations. Firstly, the number of different folds in nature is quite small (approximately 1000). Secondly, 90% of the new crystallographic structures added to the PDB have similar structural folds to ones already deposited [82].

2.3.1 *Protein preparation*

Once the 3D structure of the protein target has been obtained (either downloaded from the PDB or generated using protein prediction methods), there are several protein and ligand preparation steps that should be followed before starting a docking run (Fig. 5). Here we will discuss the protein preparation procedure while the procedure for ligand preparation will be covered later in the chapter (see Section 2.3.2).

Due to insufficient resolution, most of the entries in the PDB only contain coordinates of non-hydrogen atoms. To work with these entries, the most common protein preparation task is the placement of the missing hydrogen atoms. This is not trivial, as it should account for the important ambiguities of protein structures, such as rotatable hydrogens, tautomers and protonation states of particular amino acids, alternative water orientations, and terminal side chain flips.

In addition, during protein preparation it is important to ensure that missing side chains are added, missing bonds are detected and fixed, bond orders are assigned, and where alternate locations are present, the atoms with highest frequencies are selected.

Other, more complex, procedures in protein preparation include prediction of protonation states and identification of which water molecules (if any) should be retained in the protein target structure.

The following subsections will look at the methods used to predict the protonation/tautomer states and at how to identify structural water molecules that are known to be vital in mediating hydrogen-bonding interactions, even in some cases key for facilitating tight binding, and hence should be considered part of the protein target structure.

2.3.1.1 Protonation state

Assigning the aqueous protonation state to protein residues is a key preprocessing step when working with crystal structures of protein-ligand complexes, and plays an important role in the prediction of the correct binding mode or binding affinity [64,83,84] of a ligand. This is even more relevant in virtual screening where an incorrect predicted binding mode could lead to the identification of false positives or miss potential true binders.

It is worth mentioning that not all the scoring functions are equally affected by an incorrect protonation; in general, the physics-based scoring functions are likely to be more susceptible in comparison to knowledge-based or empirical scoring functions [85].

Protonation states of ionisable amino acids residues are constantly changing due to the dynamic nature of a protein. This effect is particularly relevant in protein-ligand complexes, where the ionisation states of residues in the binding site could vary, affecting the protein-ligand interactions. To accurately predict the conformation of a ligand in a protein binding site, the protonation state of the protein must be relevant to the bound conformation and in accordance with the pH of the experimental conditions [85].

Inspection of crystal structures and known, experimentally identified active ligands can yield a wealth of knowledge on the protonation state, steric clashes, and hydrogen-bonding networks between ligand and receptor [86]. Assigning protonation states to aspartic acid (Asp), glutamic acid (Glu), arginine (Arg) and lysine (Lys) during the protein preparation is generally straightforward, with deprotonated acids (Asp and Glu) and protonated bases (Arg and Lys) [86]. Histidine (His), however, provides a unique challenge in terms of protonation, as it can be protonated in three different ways. The imidazole ring of the His side chain can be protonated in a neutral form at either the ε-nitrogen or the δ-nitrogen, or in a charged (+1) form where both the ε- and δ-nitrogens are protonated. To further complicate choosing the correct form of the imidazole side chain ring, ambiguities in crystal structures, due usually to poor resolution, often switch the carbon and nitrogen, creating an additional three rotameric conformations, termed 'flipped' [86,87]. In addition, His is a weak base ($pK_a \sim 6.0$) and its protonation state is highly affected by the surrounding environment. To determine the correct protonation state of histidine residues, it is good practice to look at each His in the binding site individually; analysis of hydrogen-bonding networks is likely to yield the most detail about the correct side chain protonation [88]. Glutamine (Glu) and asparagine (Asp) too can be problematic: in poorly resolved crystal structures, the side chain terminal amides can, on occasion, be flipped so that the oxygen and NH_2 groups are misplaced.

Most of the available docking tools provide a set of protocols to prepare the protein target however, because most of these are automatic, it is always good practice to check particularly challenging residues like histidine, asparagine and glutamine.

2.3.1.2 Binding site definition and cavity detection

When targeting a protein, one wants to bind potential small molecule therapeutics to locations that induce a therapeutic effect (Fig. 5). A protein will typically have a key binding site where a given substrate will bind and will possibly have other sites of an allosteric nature. Binding to these sites is desirable, but first such sites need to be identified. They are usually determined experimentally, so that the structure of the protein-ligand complex is available, however, the process of identification of the active site in proteins becomes critical when the bound ligand is absent in the crystal structure.

With protein-ligand docking, a user typically provides guidance to the software to ensure that the predictions generated are placed in the correct regions of the protein. (We note that there is active research into 'blind' docking where no cavity information is assumed up front, which will be discussed later in the chapter.)

To identify likely binding regions, users will typically use cavity detection methods prior to docking. In some cases, the binding sites generated will be further validated experimentally: one can use single point site mutagenesis, for example, to identify whether given residues are essential for binding of known substrates. If a given residue is indeed critical, one can infer that the residue is likely to be in a therapeutically active region of a protein and so researchers will tend to examine the region containing said residue more closely with a view to designing new therapeutic agents.

Programmatic cavity detection is a well-researched field [89]. Many methods using a variety of different algorithmic approaches exist. For example, the classical algorithm LIGSITE [90] uses grid-based ray tracing to identify pockets and GHECOM [91] uses recursive sphere-based detection whereas newer methods, such as CavVis [92], use methods inspired from computer graphics [93]. The best methodology will be open to debate, however, for most docking algorithms the more critical component is identifying the set of atoms to be treated as active in docking.

If there are pre-existing known binding ligands, one can use them as a basis for the docking cavity. A common problem in the evaluation of docking is how to define the cavity in this case; if a cavity definition only encapsulates the close residues to a known binder, it is likely that the performance of the software will be over-represented when redocking the known binder; conversely if the cavity is very large, the search space for the algorithm is made far larger. Experience suggests that a reasonable compromise appears to be to include all the residues within 7 Å of the atoms of a known binder, though a user should visually inspect this cavity: in some cases one can have a larger cavity, particularly if one is starting from a fragment bound to a protein structure. We would recommend surveying the protein using a cavity detection algorithm, even if there are structures of known binders.

2.3.1.3 Protein flexibility

As mentioned earlier in the chapter, due to the large computational resources required for fully sampling both protein and ligand conformations, the standard approach in docking and particularly in structure-based virtual screening is to dock fully flexible ligands in a rigid protein. Moreover, separating the contribution of protein effects on ligand conformation and simultaneous ligand effects on protein conformation, while developing the algorithms to handle such complex interactions, is still a significant challenge in molecular docking.

While conformational variability in different *apo* structures (i.e. without bound ligands) of the same protein suggests intrinsic disorder, variability in different ligand-bound states seems to be related to the presence of the ligand [94]. Ligands with diverse chemical scaffolds and of different sizes can induce different types, or different extents, of changes in protein conformation. Several studies have proven that the incorporation of protein flexibility in automated docking algorithms enables more accurate binding pose prediction and better virtual screening enrichments [95–97], in addition to providing a more realistic description of the physics of the protein-ligand binding interaction. The drawback is the required computational cost; docking with flexible ligands and a flexible protein usually requires the use of supercomputers to be achieved in a reasonable timeframe.

A statistical analysis of the PDB revealed that 85% of the proteins contain only one to three flexible residues in the active site and that single rigid protein dockings predict an incorrect binding pose for 50–70% of all ligands [94,95,98]. These flexible residues are subject to conformational changes ranging from simple side chain movements (rearrangement) to backbone-loop movements, to major domain rearrangements. Combinatorial approaches making use of side chain rotamer libraries and soft potentials are considered very efficient to account for small movement of the protein side chains, while the ensemble of rigid protein structures (ensemble docking) has been proven successful to account for larger changes in the protein [95,96,99,100].

2.3.1.3.1 Soft docking and side chain rotamer libraries Soft docking attempts to treat flexibility by allowing a small degree of overlap between the protein and the ligand. This implicitly models protein accommodation by loosening the criterion for steric fit. This is achieved by reducing the steepness of the repulsion term in the Lennard-Jones potential function [101].

Soft docking has the advantage of being computationally efficient (only the scoring function parameters need to be changed), however, it can take care of only minor side chain movements. Applications of such an approach have been used for both protein-protein docking [102] and protein-ligand docking [103,104].

Side chain rotamer exploration offers an alternative way to model receptor flexibility and is similar in spirit to the approach of exploring multiple ligand conformations for simulating ligand flexibility. Side chain rotamer libraries are usually knowledge-based and generated from experimentally determined structures [105]. Such libraries also tabulate the probability with which different rotamers are observed, therefore in principle, rotamer libraries can also help to decrease conformational search space with the use of rotamers as representative conformations of energetically favourable states [106]. Side chain rotamer exploration can be applied to several amino acid side chains in the binding site. However, based on the conformational space to explore, a computational cost must be added to the docking run. Generally, it is good practice to limit the side chain rotamer library only to those residues that have a large conformational change when comparing unbound and bound forms of the protein structures. One should also note that adding protein flexibility introduces additional scoring terms, extending the necessary approximations made to the scoring functions further than in rigid protein docking.

2.3.1.3.2 Ensemble docking An ensemble docking methodology aims to address protein flexibility by docking flexible ligands against multiple conformations of the target protein rather than just the single rigid protein structure used in standard docking (Fig. 8).

The ensemble of protein conformations mimics the conformational equilibrium which characterises the native state of the target protein and provides a structural degree of freedom by which the conformation of the protein may be matched to fit any particular ligand [107]. Ensemble docking is a two-step process including generation of an ensemble of protein conformations and the actual docking to the selected protein structures.

The ensemble of protein structures can be generated by using available crystal structures of the same protein target that have been isolated and/or co-crystallised with various ligands. Such an approach accounts for induced fit changes occurring upon ligand binding.

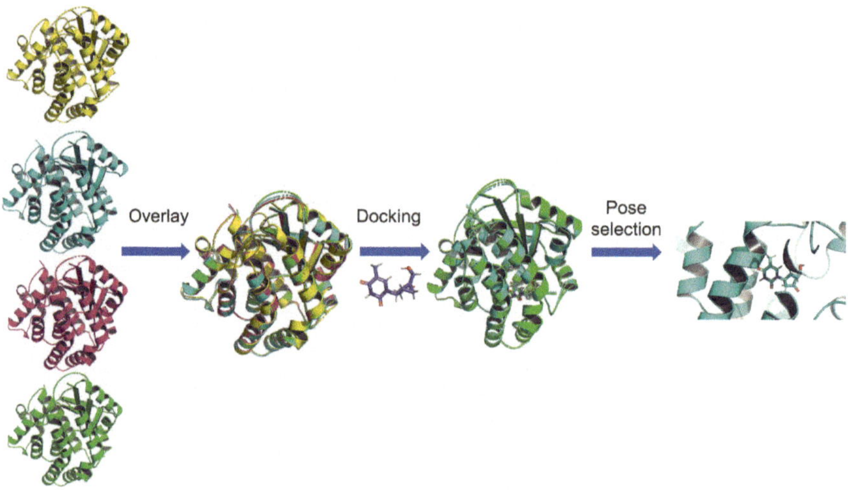

Fig. 8 Ensemble docking workflow for the TK_{HSV1} protein. TK_{HSV1} has been determined in both *apo* form (PDB entry 1e2h displayed in yellow cartoon), and in complex with different nucleoside prodrug ligands (PDB entries 1e2i, 1of1 and 4ivq shown as cyan, magenta, and green cartoon, respectively). Ensemble docking is a two-step process including generation of an ensemble of protein conformations (overlay) and the actual docking to the selected protein structures.

The protein ensemble can also be generated by molecular dynamics (MD) simulations, where different snapshots are isolated from single or multiple trajectories, thus helping to explore possible protein conformation changes across time [100].

To balance the computational efficacy and prediction accuracy of ensemble docking, a relatively small set of representative structures should be selected. Several studies have explored the best way to select experimental protein structures and the best number of structures for ensemble docking, finding in general that the use of many receptor conformations may not necessarily improve the docking performance because a large number of false positives may reduce the enrichment rate in virtual screening [97,107,108]. These studies tend to focus on specific systems however, to our knowledge, there is no general automated solution to the problem of optimal ensemble selection; a key part in ensemble docking [99,109].

2.3.1.4 Structural water molecules

The presence of structural or labile water molecules in the protein binding site plays an important role in protein-ligand binding and therefore can affect the accuracy of protein-ligand docking predictions (Fig. 4). Labile waters can either stabilise a protein-ligand complex by mediating hydrogen bonding between the ligand and the protein, or they can be displaced by the binding ligand. They can also contribute significantly to entropic and enthalpic changes in the protein-ligand complex, where the entropic loss associated with transferring a water molecule from the bulk solvent to protein-ligand binding is compensated for by the enthalpic contribution of the additional hydrogen bonds.

The importance of water molecules is well recognised in structure-based drug design, where displacing, mimicking, and/or targeting of bound water molecules can improve the binding affinity of ligand molecules [110]. For this reason, understanding the role of active water molecules has drawn extensive attention in the past decade, and many tools and strategies have been developed to predict the locations and thermodynamic profile of water molecules in the binding site [111–114].

Most of the commercially-available docking programs are able to deal with active water molecules [115–117]. When water molecules are implicitly or explicitly included in a docking run, the docking algorithm will try to find the best docking pose with a binding site occupied by the water molecule. Among them, GOLD has the additional feature to treat active water molecules as switchable (i.e. the water can be bound or displaced by the ligand) and flexible (i.e. the water can rotate and translate within a given distance to optimise hydrogen bonding). The position of water molecules within an active site can be highly variable. Treating them as static in nature can bias towards ligands that complement the specific orientation and prejudice those that would physiologically replace the given water molecule, leading to an increase in false negatives [86,118]. In GOLD, a simple entropy penalty term has been introduced to account for the unfavourable loss of rotational and translational entropy that accompanies the tight binding of a water molecule to a protein surface; only water molecules with binding affinity that outweighs the loss of rigid-body entropy on binding are considered to be bound [115].

Several studies have proven that docking and virtual screening with active water molecules can improve ligand docking poses and greatly increase the ligand enrichment, however, particularly in a virtual screening context, the addition of water is frequently neglected as the additional computational cost required affects the rapid screening of a large library [119–121]. It is therefore important to determine which water molecules must be kept and exclude those water molecules that are not essential.

2.3.1.4.1 How to recognise active water? One approach to assess which water molecule(s) should be kept for docking is to attempt to replicate the binding mode of experimental ligand structures in the absence of explicit waters. If the accuracy is diminished by the absence of waters in the binding site, then it is important to identify those water molecules which are crucial for binding [86].

In general, waters that are not hydrogen bonded to the protein, and those that are located outside the binding site, have little or no effect on ligand binding and can be removed [86,121]. However, waters that form hydrogen bonds with the receptor, or those with low B-factors, are likely to be highly stable within the protein binding site and should be included in docking studies as they are likely to stabilise the protein and they will be difficult to displace on ligand binding [110]. Waters that form hydrogen bond bridges between the ligand and protein are also likely to be important in ligand binding. However, a hydrogen bond network may be ligand-specific and its importance in virtual screening, where a diverse set of compounds are under study, should be thoughtfully assessed in advance. For this reason, when active water molecules are included in docking, and more importantly in virtual screening, they should, ideally be treated as flexible [121]. As an alternative, several computational methods have been developed to predict the locations of the water molecules in the protein binding site [122–125].

2.3.2 Ligand preparation

Ligand preparation consists of generation, optimisation, and validation of its 3D structure.

3D structures of ligands can be obtained experimentally, for example from protein-ligand co-crystal complexes, or they can be generated using software able to convert 1D and 2D structures (e.g. SMILES, SMARTS, InChi) into 3D molecular structures (Fig. 9).

The 3D structure of the ligand must have realistic bond lengths and angles as these will not usually change during docking. Optimisation of the starting ligand geometry is sometimes required for particularly complex molecules.

Several programs exist to generate and optimise the 3D structure of a ligand (e.g. CSD Conformer Generator [126,127], Omega [128,129], Confab [130], Confect [131], RDKit [132]). They differ in the algorithm used; some systems use force fields to infer intramolecular geometries, whereas others including the CSD Conformer Generator, rely directly on crystal structure data derived from the Cambridge Structural Database (CSD) [133] to produce realistic ensembles of high probability ligand structures.

As in the protein, hydrogens and formal charges must be added to the 3D structure of the ligand. The protonation state should be set according to the physiological pH or the pH of the simulation, and tautomeric states of the

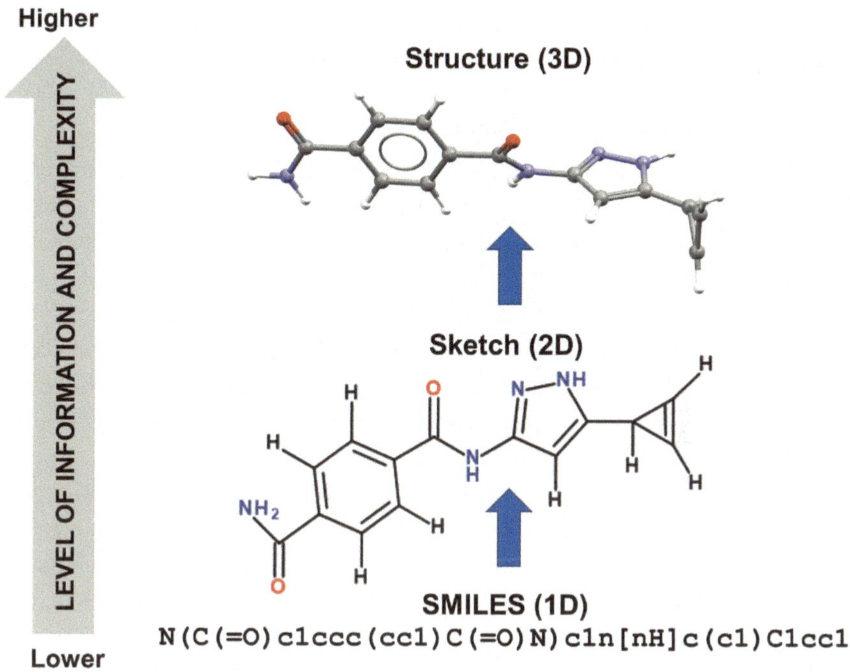

Fig. 9 1D to 3D workflow for a small molecule.

ligand should also be defined. In some cases, it may be worth generating multiple possible protonation states for a given ligand, with a view to docking all forms (in particular, if the pK_a of a given proton dissociation is close in numerical value to the physiological pH).

Tautomers are isomers differing only in the positions of hydrogen atoms and electrons, therefore even a simple molecule can have several different tautomeric forms. Moreover, the protonation and deprotonation of the ionisable sites in the molecule produces additional forms called protomers. Tautomers and protomers differ in shape, functional groups, surface, and hydrogen bonding. Therefore, tautomerism and protonation may result in alternative binding modes that can affect the efficiency of docking and virtual screening [64,84]. For chiral compounds, proper enumeration of the relevant stereoisomers is also necessary to effectively use docking and virtual screening as a drug discovery tool. Since enantiomers of a chiral compound might have different binding affinities for a given receptor, it is important that both enantiomers are used for docking. The docking software should be able to produce different docking scores for enantiomers that bind to the protein receptor. Failure to include appropriate diastereomers and/or enantiomers of compounds could significantly increase the number of false negatives and adversely affect enrichments in virtual screening [134].

2.4 Small molecule databases

Several databases of lead and drug-like small molecules are available for virtual screening purposes. Although a substantial overlap is found among some of the collections, each database has unique features that may make it better than others for a particular virtual screening project. A relevant number of unique compounds is found within each database that makes it worth considering more than one library if possible. Many chemical databases are freely available and may be designed to possess desirable characteristics such as 'drug-likeness', dictated by the 'Lipinski's rule of five' which states that drug-like compounds should have molecular weight lower than 500, lipophilicity (logP) lower than 5, less than 5 hydrogen bond donors, and less than 10 hydrogen bond acceptors [135]. However, over the years, an increasing number of compounds violating some of these rules have been approved as drugs and entered the market (e.g. many natural product drugs as well as 50% of marketed drugs do not comply with the rule of five) [136]. A strict compliance to this rule can strongly limit the variety of chemotypes, indicating that it is advisable to follow it with a certain degree of flexibility [136]. Other available databases contain chemical structures from natural products or approved drugs therefore include compounds that break the rule of five.

Here we are going to provide a brief overview of some of the most common chemical databases that are used for virtual screening. A more exhaustive list is provided in Table 2.

2.4.1 ZINC database

ZINC is a free database of commercially-available compounds developed in the Department of Pharmaceutical Chemistry at the University of California, San Francisco [137]. It contains a constantly growing number of 3D structures ready-to-dock from catalogues of several vendors with annotated relevant information about protonation and tautomeric states, and properties such as size, calculated logP, number of rotatable bonds, etc. Each molecule in the database also contains purchasability and vendor information, making this ZINC's focus on docking and availability the main distinctive characteristic from other databases.

In its latest version, ZINC20 [138] comprises over 736 million lead-like compounds (molecular weight less than 400 g/mol, calculated logP less than 4 and rotatable bonds less than 7), 509 million of these compounds are available for download in 3D ready for docking, together with information

Table 2 List of available small molecule databases.

Database	
ZINC https://zinc.docking.org/	Free database of commercially-available and annotated compounds for virtual screening. ZINC contains over 736 million lead-like compounds of which 509 million are in ready-to-dock, 3D formats. ZINC also contains over 1.3 billion purchasable compounds which you can rapidly search for analogues
ENAMINE https://enamine.net/11-databases	The world's largest collections of novel building blocks (225,000+) and screening compound libraries (2,740,000+)
NCI Open Database https://cactus.nci.nih.gov/download/nci/	NCI database has more than 275,000 small molecules structures, a very useful resource for researchers working in the fields of cancer and AIDS
ChEMBL https://www.ebi.ac.uk/chembl/	ChEMBL provides comprehensive information about 1 million bioactive compounds (small drug-like molecules) with 8200 drug targets
DrugBank https://www.drugbank.com/	The database combines detailed drug data with comprehensive drug target information. It contains 6,712 drug entries including 1,448 FDA-approved small molecule drugs, 131 FDA-approved biotech (protein/peptide) drugs, 85 nutraceuticals and 5,080 experimental drugs
ASINEX Database http://www.asinex.com/	The ASINEX database is a commercial collection of compounds which contains more than 600,000 screening compounds, 27,000 macrocycles, 20,000 fragments and over 22,000 building blocks
Cambridge Structural Database (CSD) https://www.ccdc.cam.ac.uk/solutions/csd-system/components/csd/	The Cambridge Structural Database (CSD) is a repository for small molecule organic and metal-organic crystal structures. With over 1 million structures from X-ray and neutron diffraction analyses, the CSD includes several subsets, such as the CSD Drug Subset with entries that feature in the approved drug list provided by DrugBank and a CSD COVID-19 subset that includes structures of interest in the fight against COVID-19

Continued

Table 2 List of available small molecule databases.—cont'd

Database	
PubChem https://pubchem.ncbi.nlm.nih.gov/	Database of chemical molecules which maintains three types of information namely, substance, compound and BioAssays
SPECS Database https://www.specs.net/	SPECS database (>240,000 compounds) is composed of novel drug-like small molecules obtained from academia and research institutes. SPECS contains available compounds that can be purchased upon request. Every molecule in the collection must fulfil structural characteristics of a biologically active compound and meet ADMET requirements
MAYBRIDGE Database https:/maybridge.com/	Maybridge Screening Hit Discovery collection (over 53,000 compounds) is a commercial library of small hit-like and lead-like organic compounds that covers ~87% of the 400,000 theoretical drug pharmacophores complying with the rule of five and of good ADMET properties. Maybridge also offers a fragment library (30,000 fragments) and a hit-to-lead building block collection
LIFE CHEMICALS Database https://lifechemicals.com/	Contains a commercial compound collection for HTS of 1,213,000 lead-like and drug-like new diverse chemical entities that follow Lipinski's rules. Furthermore, it offers several different diversity libraries on demand: i.e. building blocks, fragment- and scaffold-based libraries, natural product-like compounds, covalent inhibitors, etc.
CHEMBRIDGE Database https://www.chembridge.com/	Contains 1 million drug-like and lead-like compounds in 2 non-overlapping collections of respectively 460,000 and 620,000 compounds, that cover different chemical spaces and that can be customised to create diversity libraries, targeted libraries (KINASet, CNS-Set, and IONSet libraries) and fragment libraries, which can be purchased upon request

regarding target and biological activity, related scaffolds and bioactive and biogenic compounds (with a Tanimoto likeness index of 0.6). ZINC also offers other features such as the possibility to define target-focused libraries and to download subsets of a physical property space (fragment-like, lead-like and drug-like subsets) [139].

2.4.2 ENAMINE database

ENAMINE provides several different commercial collections of compounds for screening. The screening collection currently contains over 2 million low molecular weight organic compounds. The HTS collection (>2,115,000 compounds) represents a highly diverse set of chemotypes developed from in-house research and partner academic organisations, while the Advanced Collection (>493,000 compounds) is intended for lead discovery. This library has been designed according to lead-like properties and/or valuable pharmacophores such as carboxylic, primary amino and amide groups. The Premium collection (>44,500 compounds), instead, contains compounds with the most favourable physicochemical properties (high Fsp^3, low logP and MW).

Enamine also provides a pharmacologically diverse set (10,240 drug-like compounds clustered by activities from biologically relevant chemical space), a 3D diversity set (50,240 compounds from conformational analysis and shape clustering of the HTS collection) and a covalent screening library (10,480 compounds of well validated covalent binders) as well as targeted libraries (e.g. central nervous system (CNS), antibacterial, ion channel, coronavirus, kinase and lipid G protein-coupled receptor (GPCR) libraries) and fragment libraries (e.g. covalent, sp3-rich, Protein-Protein Interactions (PPI), fluorinated and brominated fragments).

The largest Enamine database is called the *REAL* database and is a virtual collection of over 1.36 billion molecules that can be used to find new hit molecules using large-scale virtual screening and to search for analogues of hit compounds. Each molecule in the *REAL* database complies with the rule of five and also the Veber criteria (rotatable bonds ≤ 10, and TPSA $\leq$ 140) [140].

The structure data files of these various collections are regularly updated and can either be directly downloaded from the ENAMINE webpage in MDL SD (.sdf) or MDL ISIS (.db) formats or obtained by request. Along with SMILES and catalogue IDs, the entry for each molecule in the database lists important physicochemical parameters (MW, sLogP, HBA, HBD, etc.), structural alerts (PAINS, Brenk, and Eli Lilly medchem rules), type of chemistry and the difficulty of synthesis ('s', simple chemistry, standard effort, 'm', advanced chemistry, higher effort).

2.4.3 NCI open database

The NCI Open Database is a freely accessible database developed by the Developmental Therapeutics Program of the National Cancer Institute

(NCI). This database contains a set of compounds that have been collected by the NCI, since 1955 for testing in anticancer, and from the 1980s for anti-AIDS screens, that are not covered by confidentiality agreements. The database currently contains more than 260,000 molecules both from organic synthesis and natural source extracts that can be downloaded in .sdf format [141]. The compounds in the database are annotated with information fields including release, structure source and evaluation, calculated/predicted logP, biological activity and commercial availability in addition to 3D atom coordinates, added hydrogens, number of rotatable bonds, stereocentres and bond stereocentres [141].

2.4.4 ChEMBL

ChEMBL is a manually curated database of bioactive molecules with drug-like properties [71,72]. The ChEMBL database contains more than 1.8 million compounds and over 15 million records of their effects on biological systems. It contains information about how small molecules interact with their targets, how the compounds affect cells and the whole organism, and information on absorption, distribution, metabolism, excretion and toxicity (ADMET). Additional data on clinical progress of compounds has been integrated into ChEMBL (ChEMBL Drugs). This highly curated dataset includes marketed compounds and compounds that are or have previously been in clinical development and are annotated with information about their known therapeutic targets and associated indications.

The data in ChEMBL are extracted and curated from the primary medicinal chemistry and pharmacology literature and cover a significant fraction of the SAR and discovery of modern drugs. Additionally, the ChEMBL database contains data deposited by researchers and data extracted from other public databases.

ChEMBL includes 2D structures with calculated molecular properties (e.g. logP, molecular weight, Lipinski parameters) and bioactivity data (such as binding constants, pharmacology and ADMET) with the bioactivity data tagged to show links between molecular targets and published assays.

2.4.5 DrugBank

DrugBank is a web-enabled curated database containing comprehensive molecular information about FDA-approved drugs as well as experimental drugs going through FDA approval [142]. As both a bioinformatics and a cheminformatics resource, DrugBank combines detailed drug data (i.e. chemical, pharmacological and pharmaceutical) with comprehensive drug target (i.e. sequence, structure, and pathway) information. All data in DrugBank is non-proprietary or is derived from a non-proprietary source. It is freely accessible and available to anyone and nearly all data are fully traceable and explicitly referenced to their original source.

To date, DrugBank contains over 13,700 drug entries including approved small molecule drugs, approved biologics (proteins, peptides, vaccines, and allergenics), nutraceuticals and over 6,000 experimental (discovery-phase) drugs. Additionally, over 5,000 non-redundant protein (i.e. drug target, carrier, transporter and enzyme) sequences are linked to the drug entries together with drug-drug and drug-food interactions [143]. All the chemical structures in DrugBank are accessible in canonical SMILES, sdf, .mol, .pdb, InChI and InChIKey formats.

2.4.6 ASINEX database

The ASINEX database is a commercial collection of compounds which contains more than 600,000 screening compounds, 27,000 macrocycles, 20,000 fragments and over 22,000 building blocks [144]. The screening compounds are organised in different libraries that cover different chemical characteristics and try to address different steps in the drug discovery process. The Gold & Platinum Collections of over 260,000 compounds includes diverse and cost-effective coverage of drug-like chemical space. Most compounds have a high degree of drug-likeness, in accordance with Lipinski's rule of five. Other libraries are focused on lead-like compounds are intended for the early stages of drug discovery. For example the ASINEX Synergy and Elite Library of more than 91,000 compounds has been screened against a panel of early ADMET tests to make sure screening hits do not have potential ADMET problems and are amenable for rapid hit-to-lead optimisation.

A subset of over 170,000 compounds (BioDesign) incorporates key structural features of known pharmacologically relevant natural products (e.g. alkaloids and other secondary metabolites) into synthetically feasible medicinal chemistry scaffolds. ASINEX also offers targeted libraries including those focused on CNS disorders, immuno-oncology, PPI, GPCRs, peptide-mimetics, nucleoside-mimetics, etc. All the libraries can be downloaded in SDF format and can be directly purchased.

2.4.7 Cambridge structural database (CSD)

The Cambridge Structural Database (CSD) is the world's leading repository for small molecule organic and metal-organic crystal structures. It contains over 1 million structures from X-ray and neutron diffraction analyses forming a unique database of accurate 3D structures [133,145,146]. Every entry is manually curated and each structure in the CSD is enriched with chemical representations, as well as bibliographic, chemical, and physical property information.

The CSD comes with several subsets including CSD entries that feature in the approved drug list provided by DrugBank. This subset has a wide scope; it contains any solvates, co-crystals or hydrated forms, and currently provides a set of 12,277 entries to help users gather insights into drug-like compound. A single-component CSD drug subset is also available and includes 1,989 CSD entries where a drug molecule is the only modelled component in the crystal structure. Recent additions to the CSD subsets are: the CSD COVID-19 subset with 121 structures of interest in the fight against COVID-19 and the CSD Pesticides subset with 972 entries.

2.4.8 PubChem

PubChem is an open chemistry database managed by the National Institutes of Health (NHI). It contains mostly small molecules, but also larger molecules such as lipids, carbohydrates, nucleotides, peptides and other chemically modified macromolecules. The data in PubChem are organised into three interlinked databases: Substance (as of writing more than 286 million substance descriptions), Compound (over 111 million unique chemical structures) and BioAssay (1.2 million biological assays covering more than 10,000 target protein sequences).

Most of the structures in PubChem are drug-like compounds that satisfy Lipinski's rule of five. Among them more than 10 million are fragment-like compounds which satisfy Congreve's rule of three (molecular weight of a fragment is <300, the cLogP is ≤ 3, the number of hydrogen bond donors is ≤ 3 and the number of hydrogen bond acceptors is ≤ 3) [147]. In addition to bioactivity data, PubChem contains compound information that can be useful for virtual screening. Because of the data integration with DrugBank, PubChem includes comprehensive information on FDA-approved and investigational drugs, including their drug indications, mechanisms of action, target macromolecules, interactions with proteins and genes, ADMET properties and many others. In addition, PubChem provides links to crystal structures available from the CSD. PubChem also offers links between about 6 million patent documents and more than 16 million unique chemical structures, with over 336 million chemical substance-patent links covering the USA.

3. Fragment-based screening

Fragment-based screening aims to identify small chemical fragments which bind weakly to the binding site of a target protein. Although weak, given the limited numbers of interactions those fragments can make with the protein, these are high-quality interactions as they must overcome a substantial entropic barrier to binding [148]. This approach is useful for identifying interaction hot spots for the protein-ligand binding. The key principle in fragment screening is the ability to sample the chemical space more efficiently than when using larger complex molecules. Thus, in theory, this allows sampling of a broader and more diverse chemical space than in currently practical to achieve via standard screening methods. Moreover, smaller fragments are less likely to contain interfering moieties that could block an otherwise favourable ligand–protein interaction, thus optimal binding spots are less likely to be hidden by non-binding elements (Fig. 10). In principle it is possible to use molecular docking for virtual fragment-based screening, however, by definition, fragments are relatively small compared to lead-like or drug-like compounds, which makes molecular docking more challenging. Fragments only form few, key interactions with the binding site, and this usually results in a low docking score, with the risk of missing potential fragment hits if they have weak interactions with the protein. Free energy differences between different binding modes of a fragment are much smaller than those of larger compounds therefore, given the inaccuracies inherent in current scoring functions, they make it more difficult to distinguish the correct/incorrect binding mode of a fragment.

Additionally, due to the relatively small size, docking poses for fragments can be promiscuous, with fragments binding to multiple sub-regions showing similar physicochemical properties inside a binding pocket. In these cases docking results are hard to interpret and post-processing the results can be time-consuming [150].

As for a standard docking screening, to perform a fragment screening one would need access to the 3D structure of the target protein and to the fragment library to be screened. The protein target could be obtained from an experimental or theoretical model and should be prepared as discussed in Section 2.3.1.

Fragment libraries tend to be small because the fragment space is smaller than the chemical space and can be more effectively probed with a relatively

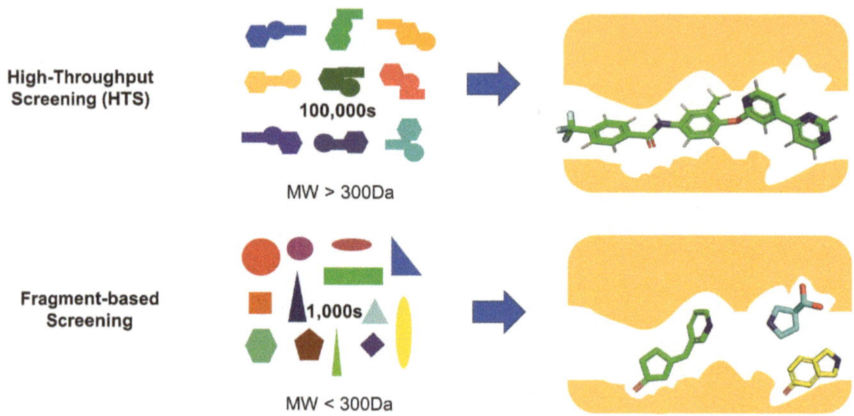

Fig. 10 High-throughput screening (HTS) and fragment-based screening overview. HTS enables testing of a large number of diverse drug-like and lead-like compounds (MW > 300 Da) against protein targets. In order to efficiently cover all the chemical space, the database should contain more than 100,000 compounds. In a fragment screening approach, a database of simple molecules or fragments with MW <300 Da are screened against targets. Due to the small size of fragments, they often have low potency, but high-quality interactions and can be further optimised into potent leads by linkage. Fragment libraries usually are of a small size (~1,000 s) because the fragment space is smaller than chemical space and can be more effectively probed with a relatively small library [149]. *Adapted from https://figshare.com/articles/SSIEM_presentation_Fragment_screening_for_drug_discovery_in_primary_hyperoxaluria_type_1/7057826.*

small library. Screening a few thousand fragments would search a greater fraction of the chemical space than could be achieved by screening over a million full-sized compounds (Fig. 10) [149,151].

Fragment libraries should be chemically diverse, synthetically expandable and should represent a wide range of physicochemical properties, aqueous solubility, molecular diversity, and drug-likeness with medicinal chemistry scaffolds. They contain fragments that comply with the rule of three [147]. Although the validity of this rule is a debatable topic [152], it still remains the preferred reference for fragment selection and fragment library generation. A growing number of commercial companies are now offering well-defined fragment libraries for screening [153]. For instance, the Diamond-SGC Poised Library (DSPL) [154] is a library containing about 760 fragments with at least one functional group that is open to rapid follow-up synthesis whilst maximising chemical diversity. While the majority of fragment libraries are chemically diverse, they still lack shape diversity. To overcome this limitation, the UK 3D Fragments Consortium is working on generating a library focused on fragments that incorporate 3D structures [155].

Once the fragment identification step is completed, these fragments can be grown, linked, or merged to develop the potential lead compounds. Fragment growing is the most used strategy in fragment-based lead design. As the name suggests, the core fragment is modified to increase its size to improve its properties and affinity for the target [150,156]. Fragment growing strategies have been successfully applied to various targets [150,157–160] including Alzheimer's Disease target BACE1 [161] and matrix metalloproteinases [162].

Fragment merging can be used in cases where two distinct fragments partially occupy the same region, or when two binding sites have regions in common and therefore their ligands are partially competitive with respect to the site. In such cases the overlapping parts can be fused into a single molecule [156]. Fragment merging is not only used in novel compound design but also remains a tool for chemical modification and derivative generation [150]. While less common than the growing methodology, a few successful examples of fragment merging have been discussed in the literature with targets including Hsp90 [163], PIγ kinase [164] and mycobacterial transcriptional repressor EthR [165].

Fragment linking describes the process of joining two non-competitive fragments (i.e. fragments that bind in two different sub-pockets of the binding site) with a chemical linker or spacer. Suitable linkers should respect the original conformational constraints of the two initial fragments while making favourable interactions within the protein binding site. Linking strategies have been successfully applied to different targets [160,166,167] including protein kinase ck2 [168] and thrombin [169].

4. Protein-protein docking

Protein-protein docking is an emerging research topic, due to its potential for predicting protein-protein interactions (PPIs) and identifying hot spot residues at the protein-protein interface.

Although protein-protein docking shares the principles of protein-small molecule docking, sampling of the conformational space in protein-protein docking is extremely challenging. Even for relatively rigid proteins, it is difficult to explore the rotational-conformational space of mutual orientations potentially sampled by a pair of proteins as they interact. Due to the huge number of degrees of freedom the computational cost of the search algorithms is considerable for protein-protein docking.

Because of the large size of binding sites in PPIs, the orientational search algorithm often requires strategies for protein-protein docking that aredifferent from those for protein-ligand docking. Additionally, contact surfaces where two proteins interact with each other are significantly different from ligand-protein binding cavities. Protein-protein binding sites are often relatively flat interfaces with no single large and well-defined pocket [170]. Predicting the association of proteins is further complicated by their flexibility. Proteins are dynamic; they constantly interconvert between conformers of varying energies and capturing this flexibility is still a challenge in molecular docking.

Despite the complexity of the problem, a variety of docking methods is currently available for predicting the structures of protein-protein complexes. The choice of the method to use depends on the nature of the docking problem [171]. Rigid docking procedures are generally used when the crystallographic structures of the proteins to dock (or of their very close homologues) and their complex are available. Knowledge of the protein-protein complex makes the prediction of related protein complexes feasible for template-based and homology modelling methods, even when the structures of component proteins are not available. The general idea is to separate the two proteins from the complex structure and use a rigid docking procedure to try to reproduce a near-native approximation of the complex (bound docking) as shown in Fig. 11.

Bound docking is the easiest docking case, because, by definition, it does not involve conformational change and existing rigid-body procedures

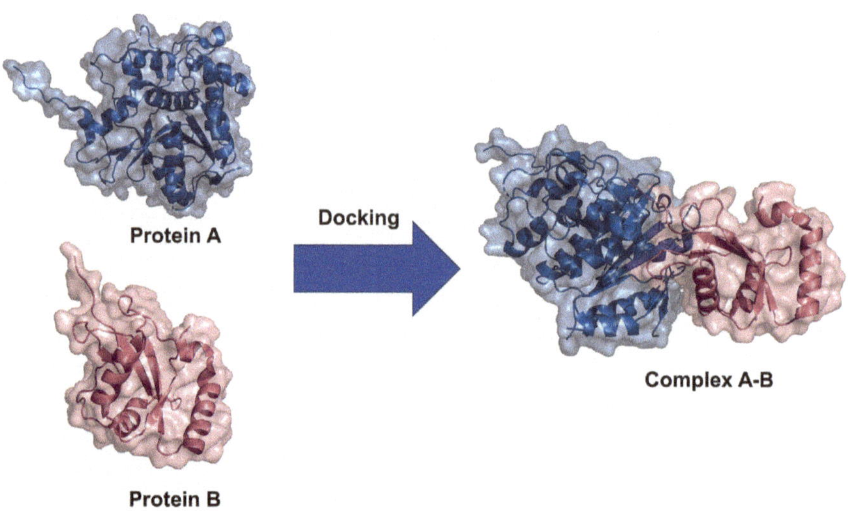

Fig. 11 Example of bound docking procedure. Protein A, in cartoon with the surface coloured in pink, represents the Human Atg4B (HsAtg4B; a mammalian orthologue of yeast Atg4) and protein B, in cartoon with the surface coloured in blue represents the LC3 protein (a mammalian orthologue of yeast Atg8). The complex A–B is HsAtg4B–LC3 complex (PDB code 2z0d).

generally result in good (i.e. near-native) structures of the complex among the top pose predictions.

When the crystallographic structure of the complex is not available, the docking algorithm should be able to predict the complex from the separately determined protein structures (unbound docking). In such cases, the docking algorithm has to deal with the conformational difference between the unbound and the bound structures [172] thus, the landscape of docking solutions could also include many false positives with good surface complementarity that differ considerably from the native complex.

In the case of unbound docking, one approach is to combine a rigid-body search with flexible rescoring and refinement of the docked conformations [173,174]. Other strategies include rigid-body docking of ensembles of structures [175,176], scoring functions that feature soft potentials to allow minor molecular overlaps [177,178], simplified protein representations (coarse-grained), where at such low levels of structural resolution the difference between unbound and bound conformations is less significant [179], and relaxing the interface of docking poses using techniques such as molecular dynamics (MD), Monte Carlo, or simulated annealing [177,179,180]. Additional docking strategies include scoring functions that use a combination of terms to describe physical interactions and penalise models that do not recapitulate the available experimental data. This method allows side chain flexibility at the PPI interface, which is a key aspect to improve the accuracy of the results. Examples of such approaches are HADDOCK [181] and IMP [182].

Protein-protein docking methods have improved substantially over the past few years (as demonstrated by the results of the Critical Assessment of PRedicted Interactions (CAPRI) [183]) but there are still many scenarios that are particularly challenging and pose limitations to current docking approaches. Examples of unsolved challenges are: large movements upon binding; weak or transient binding; and unavailability of 3D structure for one or both subunits. The protein-protein docking field needs to improve not only in optimising and/or developing new docking methods to overcome these challenges but also in identifying those problematic cases and evaluating the reliability of the predictions.

5. Protein-peptide docking

Peptide drugs are gaining attention in drug discovery as a solution for targeting 'undruggable' intracellular protein-protein interfaces characterised by large and relatively featureless interfaces. Peptides can bind large protein interfaces with high potency and great selectivity, which translates into fewer off-target side effects and less potential for toxicity than small molecule drugs [184].

Computational methods, including docking, have proven to be successful in the discovery and design of small molecule drugs as well as in the field of peptide therapeutics.

Protein-peptide docking involves computation steps such as conformational sampling, structural refinement and scoring, similar to traditional protein-small molecule docking techniques. However, peptides are more flexible than small molecules and tend to adopt numerous conformations. Thus, modelling protein-peptide interactions is a challenging and time-consuming task.

Over the past decades a wide range of docking methods has been developed that can be used directly or indirectly for docking peptides on a protein with various degree of success. These methods are described in Fig. 12 and include protein-small molecule docking, protein-protein docking and protein-peptide docking. Due to the limitation of the search algorithm, protein-small molecule docking approaches are currently limited to short peptides (up to 15 amino acids) with a well-structured conformation. However, modelling long peptides can be overcome by docking peptide fragments followed by their merging [185].

Protein-protein docking methods, such as ZDOCK [186], and Hex [187] have also been used to dock peptides onto a protein. However, compared to proteins, peptide molecules are much more flexible and less stable. Protein partners usually have well-defined 3D structures before forming protein-protein complexes while peptides usually do not. Furthermore, peptide-mediated interactions are often transient and weaker than protein-protein interactions due to the smaller interface between peptides and their protein partners.

Therefore, protein docking approaches need to address these challenges before they can routinely be used to predict peptide–protein binding.

The dramatic increase of peptide–protein structures available in the PDB has facilitated the development of more powerful docking and refinement methods for predicting peptide–protein interactions. Peptide-docking approaches can be classified as template-based docking or template-free docking, according to the amount of required input data. Template-based docking methods, or comparative methods, use known structures (templates) as a scaffold to generate a model of the desired complex [188]. These methods are favoured when the template is similar to the investigated

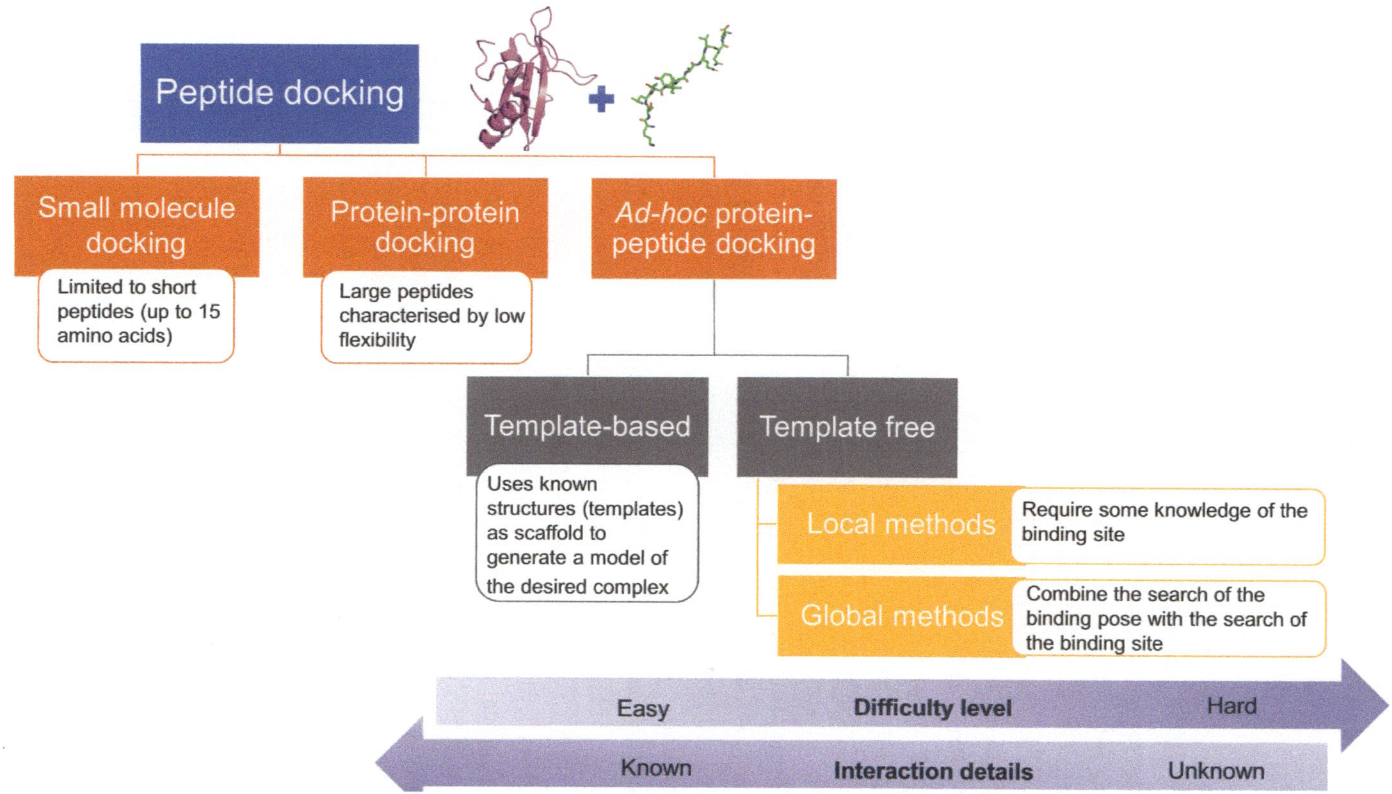

Fig. 12 Different methods in peptide docking. These methods include protein-small molecule docking, protein-protein docking and ad hoc protein-peptide docking. Due to the limitation of the searching algorithm, protein-small molecule docking approaches are limited to short peptides (up to 15 amino acids). Protein-protein approaches instead are limited to large peptides characterised by low flexibility. Ad hoc protein-peptide docking can be divided into two categories: template-based methods that utilise knowledge about the structure of similar complexes (templates) and template-free methods that can be implemented when no templates are available. Template-free methods are further divided in two categories according to the amount of required input data: local docking methods that require some knowledge about the binding site; and global docking methods that assume no knowledge about the peptide beyond its sequence. Different approaches have different level of difficulty and offer different levels of prediction accuracy, often determined by the amount of interaction information provided as input.

complex. Although several successful examples have been published, the main limitation of template-based docking methods is the number of known templates available, thus resulting in an innate limitation for general application [189]. An example of template-based method is GalaxyPepDock, a popular server that performs similarity-based docking by using experimentally determined protein-peptide structures to generate high-resolution complexes [190,191]. A recently developed machine learning-based method, PBRpredict, uses models trained from peptide binding residues of diverse types of domains to build models that robustly predict interacting residues in peptide binding [192].

Alternatively, users can use template-free methods. These methods are further classified based on the amount of information available for the protein and the protein-peptide binding site. Local docking methods require some knowledge about the binding site while global docking methods assume no prior knowledge about the peptide beyond its sequence [193].

Local docking methods are the mostly commonly used strategies and require an initial model of the complex prepared by the user to perform a search for a peptide binding pose in the proximity of a user-defined binding site. DynaDock [194], Rosetta FlexPepDock [195] and PepCrawler [196] are the most popular methods and provide different approaches to defining peptide binding sites.

However, when backbone conformational information of the query peptide is not available, sampling methods that allow acquisition of near-native peptide conformations are essential prior to performing local docking [188]. Rosetta FlexPepDock ab initio, for example, combines ab initio peptide folding with local docking by placing the query peptide into a user-defined binding site from any arbitrary backbone conformation [188,197].

Global docking methods combine the search of the binding pose with the search of the binding site. This makes global docking the method of choice when no prior information is available on the protein binding sites. The simplest approach in such docking is to treat the protein and the peptide input as rigid and to perform exhaustive rigid-body docking. More sophisticated approaches automatically predict the peptide conformation using a sequence provided by the user [188]. Alternatively, global docking can be combined with predictions of the binding site such as in in AnchorDock, which automatically identifies potential binding sites in the target protein and docks a flexible peptide in the proximity of these spots [198].

Although several docking programs have been developed, there is still a lack of a systematic evaluation to reveal the advantages and limitations of these docking programs for protein-peptide systems. A recent effort to systematically assess the performance of docking programs was published by Weng et al. [199] using a large benchmark called PepSet composed of 185 protein-peptide complexes with peptide lengths ranging from 5 to 20 residues. Despite the advances in computational modelling of peptide–protein structures, authors show that major challenges still remain. First among these is flexibility; simultaneously modelling the backbone and side chain conformational change of both peptide and its target protein. Another is the integration of experimental data; combining all available experimental data to solve data ambiguity. For example, NMR experiments to identify native contacts, small-angle X-ray scattering (SAXS) or high-resolution cryo-EM to provide the shape of the bound complex. A third challenge is scoring; many lower-ranked poses were found to be of higher quality in the docking results than the top-ranked poses and vice versa. Interestingly, CAPRI experiments have revealed that hybrid approaches, using energy-based scoring as well as other methods such as mutagenesis, co-evolutionary information, sequence- or structural-clustering function can improve the docking performances and generate accurate peptide–protein docking results that are closer to native models [188].

6. Nucleic acid docking

Nucleic acids (NAs) play important roles in a large number of cellular processes, including cellular reproduction and protein synthesis. Thus DNA binders could interfere with the DNA replication process which affects cell proliferation or regulate the transcription process, and may result in the inhibition of gene expression. Similarly, RNA binders could interfere with the transcription and translation processes. Consequently NAs are potential drug targets for a number of diseases particularly in the area of anticancer, antibacterial and antivirus therapy (Fig. 13) [200].

Small molecules interact with NAs using different mechanisms: intercalation, cross-linkage, strand-cleavage, and reading-molecules. At the time of writing, there are 523 RNA-ligand co-crystalised structures and 730 DNA-ligand co-crystalised structures in the PDB, and the number is increasing year-on-year. These structural data provide not only the opportunity for investigating the molecular interaction between NAs and ligands but they enable structure-based computational methods for the design of nucleic acid-targeting ligands for specific diseases.

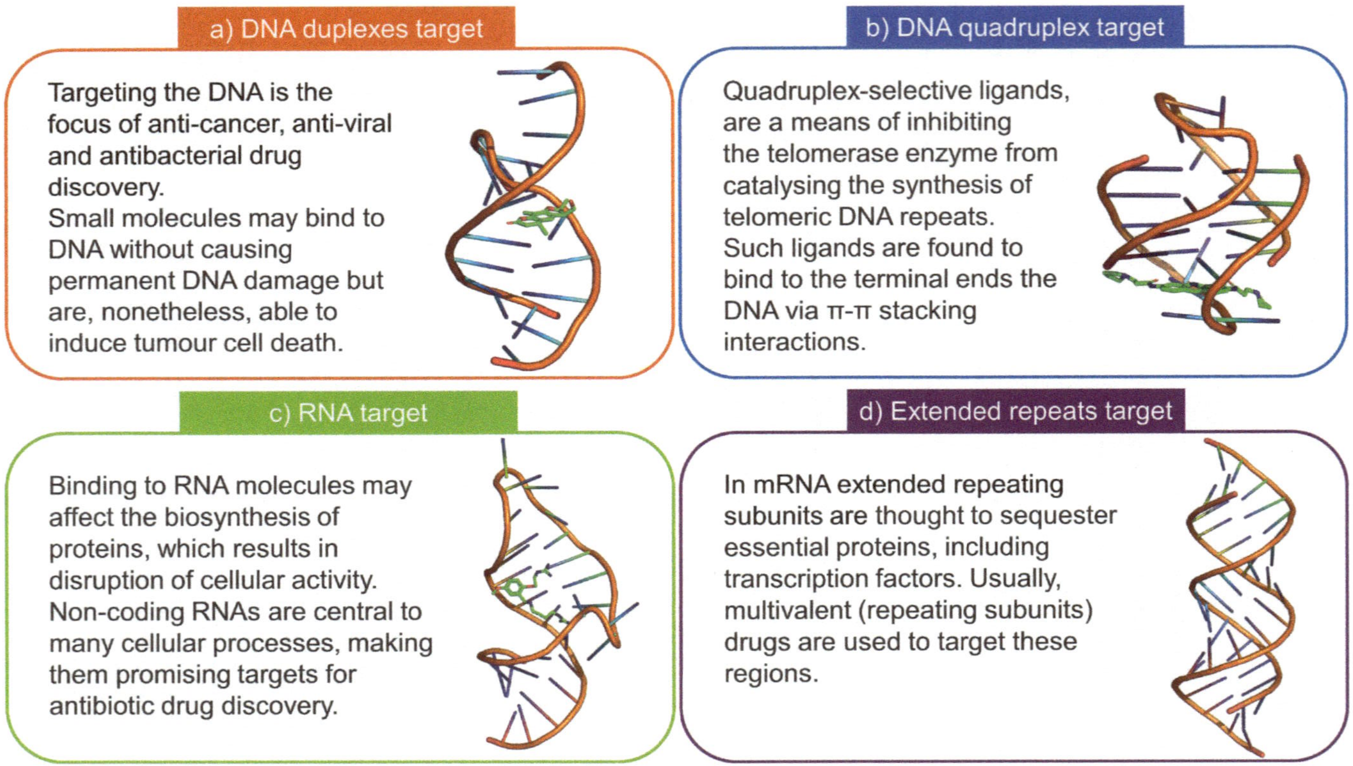

Fig. 13 Example of nucleic acid targets: (A) DNA duplex target; 3D structure of DNA duplex bound by psoralen molecule (PDB code 1fhz). (B) DNA quadruplex targets; 3D structure of DNA quadruplex bound by BSU6039 molecule (PDB code 1l1h). (C) RNA targets; 3D structure of HIV-1 TAR RNA bound to an inhibitor small molecule (PDB code 1uud). (D) Extended repeats target; 3D structure of the RNA duplex containing the CUG repeats associated with several neurodegenerative diseases, including myotonic dystrophy type 1 (PDB code 3gm7).

Generally, most of the protein-ligand docking programs can be used to dock small molecules into NAs, as they follow similar physicochemical binding principles. However, these programs often fail because of incomplete sampling of the conformational space and/or deficiencies of their internal scoring functions, which were not designed for docking of NAs (Fig. 14) [183,201].

While small molecule targeted proteins usually contain a well defined, generally hydrophobic binding site, NAs are characterised by more solvent exposed binding pockets with a high charge density and polarity. These differences require modifications to existing protein-ligand docking programs before they can be used for docking small molecules to NAs. Moreover, most of the current docking methods ignore flexibility and treat NAs as rigid receptors. However, NAs, and RNA in particular, are not rigid; they are usually characterised by induced fit movements and conformational changes in response to ligand binding (e.g. riboswitch) [202,203]. To handle flexibility, specific docking tools such as MORDOR [204] have been developed to allow induced fit binding of small molecule ligands with RNA targets via flexible receptors. MORDOR allows flexibility on both ligand and NAs by applying molecular mechanics minimisation with a restrained conformational search based on the X-ray or NMR experimental structure [205]. Despite the high accuracy, the energy minimisation step is time-consuming and makes this method less satisfactory for screening large libraries [204]. A more computationally efficient approach to deal with NA flexibility is to use ensemble docking, using a set of predetermined NA conformations that can be obtained from different X-ray crystal structures, NMR models, or normal-mode analysis of an MD simulation.

Additionally, scoring functions have been specifically designed for predicting NA-ligand affinities to address the high polarity of NAs compared to proteins and the interactions that require more attention in NAs (e.g. electrostatic interaction, and solvation). RiboDock [206] for example, includes an empirical scoring function with a number of extensions to capture important NA-ligand interaction motifs, such as the interaction between positively charged carbons (e.g. guanidinium) and negatively charged groups (e.g. carbonyl), and an energetic term for parallel π-systems (stacking) to account for electrostatic interactions that characterise NA-ligand complexes.

Two novel knowledge-based scoring functions are DrugScore^{RNA} [207], which was trained from statistical analyses of a set of NA-ligand complexes, and LigandRNA [208] which, in contrast to DrugScore^{RNA} considers the directionality of hydrogen bonds.

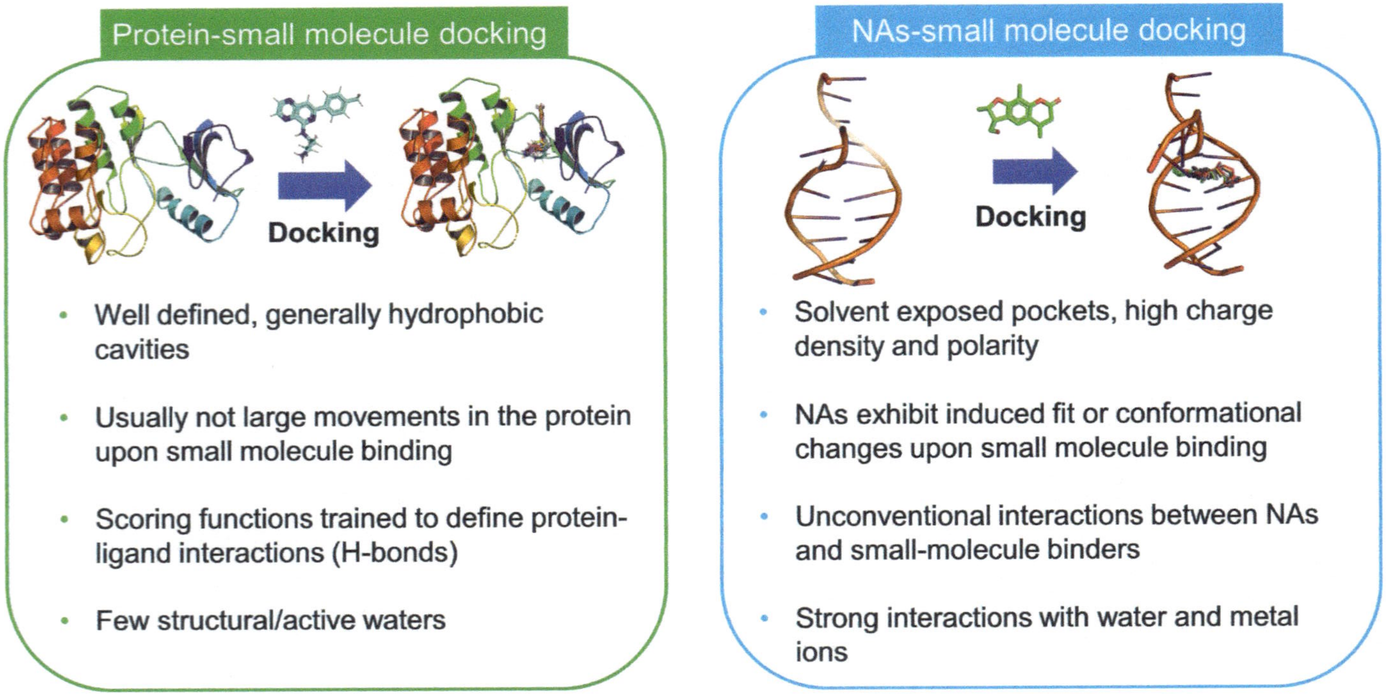

Fig. 14 Main differences between protein-small molecule docking and NAs-small molecule docking.

Another challenge in docking into NAs is the presence of water molecules and metal ions. The highly charged nature of NAs leads to strong interactions with water molecules and metal ions that are not only essential for stabilising and shaping NA structures, but are often necessary for optimal ligand binding. Therefore, docking tools that allow explicit treatment of waters are highly recommended when docking small molecules into NAs [209,210].

Inclusion of water molecules in the docking run has shown better success rate with redocking (or self-docking procedures) [14]. However, caution should be used with cross-docking which may require a different set of water molecules for different ligands. To address this issue, virtually 'displaceable' molecules and a GB/SA model (Generalised-Born model with solvent accessible surface area (SA) term) has been implemented in the solvation module of DOCK [209] to reproduce the electrostatic contribution of waters and ions. The combination with explicit water and counterions in DOCK reaches accuracy values comparable to DrugScore^{RNA} when applied to RNA. Metal ions such as Mg^{2+} and Mn^{2+} are often found at the binding site of NAs, acting as a 'metal bridge' coordinating interactions between ligand and NA residues. When metal ions are included as part of the NA target, it results in improved pose prediction of several targets including TPP riboswitch [211].

Despite the limitations, docking-based screens have already helped medicinal chemists identifying novel NA binders of DNA quadruplex [212,213], triple helix DNA [214] and RNA [215]. These prospective investigations reveal the potential of docking programs in NA binders discovery campaigns [201].

7. Current challenges

Docking is currently in a mature stage of development, but it is still far from perfect.

Most docking programs available are normally able to predict known binding poses with averaged accuracies of about 1.5–2 Å with reported success rates in the range of 70–80% [12]. However, the calculation of accurate binding energies is one of the major limitations in molecular docking, directly correlated with all the approximations made during a docking run (e.g. the treatment of solvent and the flexibility of the macromolecular system).

The lack of a suitable scoring function and searching algorithm, able to efficiently combine both accuracy and speed, are perhaps the most detrimental weakness of docking and have been widely discussed elsewhere in the chapter.

Therefore, despite its invaluable contribution to understanding target-ligand interactions in support of drug discovery projects, the results of a docking experiment should not be taken as the end result, but rather as a good starting point or as part of a workflow for a deeper and more accurate analysis.

In Section 7.1, we will focus on some of the areas where the above-mentioned limitations are likely to impact.

7.1 Blind docking

Blind docking refers to docking a ligand to the whole surface of a protein without any prior knowledge of the target pocket. In blind docking, the entire protein is considered as a region where a ligand might bind leading to a much larger search space with a corresponding increase in the running time. Moreover, the complexity of blind docking grows exponentially according to the number of possible binding sites, which severely limits its use in practice. However, blind docking methods which can predict bound conformations with no a priori knowledge of binding site locations will be needed for fully automated computational approaches for in silico drug design.

The enormous search space is the principal problem blind docking must tackle. Two possible ways to mitigate this are either to reduce the search complexity and split the docking box into multiple boxes, sacrificing the flexibility of some parts of the ligand, or to repeat the search several times using different seeds and then merge the results together, as opposed to one larger blind docking run that covers the complete protein structure.

An alternative approach to decrease the computational complexity of blind docking is to combine binding site prediction tools to identify putative ligand binding sites, with the docking experiment followed by fine-tuning and ranking of the initial solutions using scoring functions and optimisation methods. An example of such approach is BSP-SLIM, an integrated tool in which algorithms for the template-based ligand binding site prediction are incorporated with the SLIM docking method [216].

A more advanced approach for sampling flexibility in blind docking is represented by the Protein Energy Landscape Exploration (PELE), a

Monte Carlo-based technique combined with a protein structure prediction algorithm. Three main steps are performed. Firstly, the ligand and protein perturbation (using a rotamer library), secondly, side chain sampling (via algorithms), and lastly minimisation and acceptance using the Metropolis acceptance criteria. Although such an approach is computationally expensive, it is still lower than for MD simulations [217].

7.2 Covalent docking

Historically, drug discovery mainly focuses on non-covalent drugs due to potential off-target effects and toxicity issues of irreversible covalent drugs. However, in recent years and with the outbreak of Covid-19, we have witnessed the resurgence of covalent drugs [218–221]. Compared to non-covalent drugs, covalent drugs might have extra advantages, including better efficacy, since they are more competitive than non-covalent endogenous substrates. They also offer a lower patient burden and less drug resistance due to lower and less frequent dosing and improved target specificity by careful designs that target specific protein residues [219].

The rational design of covalent ligands is still faced with particular challenges, mostly related to the fact that covalent bond formation, bond breaking and bond rearrangements are quantum mechanical (QM) phenomena which cannot adequately be handled by the force fields or empirical approaches typically used for non-covalent protein-ligand interactions [222].

Historically, to overcome such limitations, many manual interventions and ad hoc solutions have been used to adapt existing docking tools for application with covalent ligands [222]. This has changed more recently, and dedicated workflows and protocols for handling covalent ligands and covalent docking in virtual screening have started to emerge. Even though contributions from QM methods are increasingly incorporated into docking applications, a full QM treatment is still unfeasible in routine applications, given the size of the molecular systems and the number of configurations and compounds to consider. However, the need for QM calculations in covalent docking could be circumvented with faster and simpler modelling approaches and in many cases the QM treatment of docking process may not be required.

Is the binding site known? Is the targeted amino acid and its reactivity known? Is the type of electrophilic warhead of the ligands known? Depending on the answer to these questions different scenarios and different requirements apply. In the simplest case where the target site is well known, particular nucleophilic amino acids featuring in the binding site can be targeted; frequently, the nucleophilic amino acid to target is also known in advance (i.e. from previous experiments). In these cases, the primary task for docking is just to elucidate the binding modes and/or to rank the compounds of interest according to their suitability to fit into the active site after covalent linking to the protein. Therefore, a reactivity assessment is not really required, and classical scoring functions developed on non-covalent interactions are sufficient for the task as are most of the docking programs. For advanced tasks such as the design of covalent ligands for systems without much prior knowledge, more sophisticated approaches are required. A QM-based scoring function has been developed to describe covalent binding of ligands by providing firstly a description of covalent bond breakage using hybrid-QM/semiempirical-QM restrained optimisations starting from the covalent complex (crystal structure or a modelled structure derived thereof) and, second, the addition of a new term ($\Delta G'_{cov}$), to the non-covalent score. This term describes the 'free' energy difference between the covalent and non-covalent complex [223]. Such an approach helps to overcome the common neglection of the energy contributions from covalent bond formation however, it is computationally demanding and more suitable as a post-docking rescoring procedure for selected candidates.

Despite recent advances in QM-based methods [222], an approach in which the covalent docking process itself is driven by QM calculations is not yet on the horizon. Similarly, blind docking remains impossible for covalent ligands because for all approaches available the binding site and the targeted reactive residue must be known in advance.

7.3 Reverse docking

Reverse docking (RD) or inverse docking, as the name suggests, involves docking of a set of one or a few ligands against an array of protein families with the aim of identifying a potential target, their binding affinity or polypharmacology profile (Fig. 15). Additionally, RD can provide a valuable contribution in drug repositioning or drug repurposing and drug rescue, it may reveal targets of drugs with so far unknown mechanism, and contribute to rationally designing less toxic or multitarget drugs [224]. Therefore, clinically approved drugs could be repurposed for other diseases, different from the one they were originally designed for [225,226]; a well-known example is sildenafil [227] a phosphodiesterase-5 (PDE5) inhibitor, used to treat erectile dysfunction but which was first developed for the treatment of angina. Another

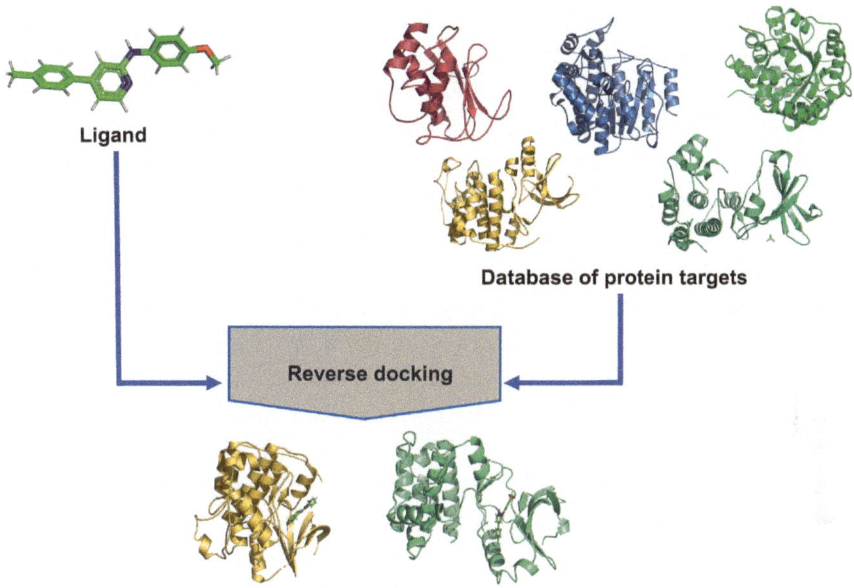

Fig. 15 A flowchart of reverse docking. In a reverse docking a ligand (or multiple ligands) is docked against an array of protein families with the aim of identifying potential targets.

example of a profitable drug repurposing is Minoxidil, which was originally developed for hypertension but was later repositioned to treat male hair loss [228]. While RD was not employed in the discovery of these materials they spurred the application of computational approaches for repurposing [229] that are for example being utilised in the search for treatments for infectious diseases, including Covid-19 [230].

With the development of computing resources, RD has drawn more and more attention in recent years and has had some success in target identification with the predicted results verified by bioassays and crystallographic studies [231,232]. These successful cases show that reverse docking is playing an important role in protein target predictions of small molecules. RD approaches have also proved successful for identifying adverse effects of drugs. For example, RD screenings of a series of anti-HIV drugs highlighted several proteins whose modulation has been associated with adverse reactions in literature [233].

With the increasing amount of crystallographic data available, several databases of protein targets are available to help preform RD screens, providing information about protein structures, diseases, biological functions, and drugs [234]. Moreover, tailored libraries of targets can also be manually built upon publicly available databases of crystal structures and binding pockets, such as PDB [77], sc-PDB [235] and the Therapeutic Target Database (TTD) [236] or can be additionally extended through homology modelling techniques. Because these libraries were not specifically designed for RD, each structure in the databases needs to be properly prepared for the docking calculations. Depending on the library and the number of targets included this step can be time-consuming. These databases are far from exhaustive. Using the data in DrugPort (http://www.ebi.ac.uk/thornton-srv/databases/drugport/) as an example, there are a total of 1664 known druggable protein targets in the database, but only about half of them have 3D structures in the PDB [237]. Furthermore, targets with known structures are not evenly distributed among different superfamilies, due to experimental limitations. For example, the superfamily of membrane proteins, the GPCR, is one of the most important targets in drug design; they account for over a quarter of the known drug targets and about half of the drugs on the market target GPCRs specifically. However, only a fraction of the GPCRs have experimental structures because the structural resolution of membrane proteins like GPCRs is much more complicated and difficult to elucidate than globular proteins such as enzymes [237].

The main limitation of RD approaches is primarily related to the accuracy of current scoring functions to distinguish true target from non-target proteins. Many (if not all) contemporary scoring functions are designed for docking or screening small molecules in protein binding sites without special optimisation for RD, which certainly would be influenced by the properties of protein pockets, resulting in scoring bias to the proteins with particular properties. This bias would produce large number of false positives, interfering with the identification of true targets. In recent years, several attempts have been made to improve the accuracy of docking scores in reverse docking [238–240]. Thus, the docking score or the scoring functions of current docking programs should be rationalised to suit the reverse docking.

Very recently, integration of a docking approach with more sophisticated ML-based methods has also been explored for target prediction [241–243]. Although such methods have achieved notable prediction performances, both in terms of target ranking and on multitarget selectivity predictions [241] they are time-consuming and computationally demanding; moreover, they also need a large amount of bioactivity data to train the models, which will not always be available for some of the targets under study [244].

8. Looking forward

The modalities by which docking is used to assist the different tasks of drug discovery have changed over time. Although it was initially developed and used as a standalone method, docking is now mostly employed in combination with other computational approaches within integrated workflows, to overcome some of the most relevant intrinsic limitations characterising molecular docking.

Applications of combined workflows which include docking, have been explored to assist different tasks of drug discovery. For example, docking has been used in tandem with ligand-based, binding free energy calculations, and AI/ML approaches to improve the prediction performances in de novo virtual screening, as well as to assist target fishing, adverse drug reactions (ADR) prediction and drug repurposing [244]. Likewise, different approaches can also be applied at different phases of the screening workflow to improve docking predictions. For example, MD could be combined with AI-based methods to identify suitable receptor conformations for docking. Then, ligand-based approaches could be applied for rescoring the predicted docking poses.

Considering the number of in silico tools and techniques currently available, there are still countless opportunities for docking to be explored in integrated workflows. Moreover, their integration has also been facilitated by the continuous improvements in hardware and software engineering. For example, parallelisation of molecular docking has enabled the in silico screening of millions of compounds in affordable time by processing the most computationally consuming task (the energy calculation phase) on multiple CPUs using distributed computing infrastructures (DCIs).

In the context of blind docking this computationally expensive task has to be multiplied by the number of binding sites which can also be very large, thus it is of great interest to find different ways to speed-up the whole docking process. A DCI can be a cloud computing resource. Cloud computing is now available on-demand and users are typically charged on a pay-per-use basis. This can make scientific applications, such as VS, more accessible for scientists around the world, lowering the cost of using complex computing infrastructure.

Over the years, there has been rapid progress in developing faster architecture based on graphics processing unit (GPU) clusters, more adequate algorithms for optimised computational analysis, and the tracking of ligand-receptor binding expressed in scoring functions.

Though the initial applications of GPUs were exclusively for creation of computer graphics, GPUs have now matured to a state where they can be successfully applied for non-graphical purposes. For example, GPUs are widely used in drug discovery to deal with demanding computational tasks such as MD simulation or QM calculations. In the context of docking and VS, GPU calculations are particularly appealing to enable one to explore the large conformational landscape potentially accessible to proteins, in shorter times compared to CPUs. GPU-optimised docking and cloud solutions may become the standard soon. Lastly, GPU computing made big-data driven computation tasks more generally accessible, and it is expected to play a prominent role, not only in docking but in future in silico drug design in general.

Another important factor to consider in the further development and optimisation of the docking methodologies is the availability of 3D data. In recent decades, the number of crystallographic data has increased exponentially and new technologies such as cryo-EM and NMR have filled several gaps left by the limitation of X-ray crystallography. Of the three, cryo-EM has emerged as successful for determining the structures of large and dynamic complexes that have proved difficult to obtain by other approaches.

Despite the fact that cryo-EM is a decades-old technique, it has garnered increasing interest since around 2013 due to a series of technological and algorithmic advances that together drove a striking improvement in the resolution obtainable by this technique. Recently, Yip et al. [245] and Nakane et al. [246] reported the sharpest images yet obtained, with a resolution of 1.2 Å, by using a method termed single-particle cryo-EM, enabling the location of individual atoms in a protein to be determined for the first time. Ultimately, these developments will help researchers gain a better understanding, at unprecedented resolution, of how proteins work in health and disease, with the potential to aid the design of better therapeutics. Advancements in cryo-EM have also increased the rate at which structures are solved, such that many general AI-based methods for protein-ligand docking will be applied to modelling protein-ligand recognition in the future. The improved quality of molecular structures has the additional advantage of improving ML and AI-based tools where the main limitation for their application in drug discovery is the lack of large, annotated, unbiased, high-quality data. As a data mining technology, the amount of available data directly affects the performance of the related deep learning models since the successful training of deep neural networks highly relies on a large amount of data.

Developments in computing power coupled to imaging techniques, AI approaches and protein fold prediction tools (i.e. AlphaFold [247]) point to a golden age for the application of computational approaches to drug discovery, where understanding the protein's shape and knowing its role within the cell, provides better protein structures for molecular docking, contributing to the development of specific and selective new therapeutics, while also reducing the costs associated with experimentation.

References

[1] DiMasi JA, Grabowski HG, Hansen RW. Innovation in the pharmaceutical industry: new estimates of R&D costs. J Health Econ 2016;47:20–33.
[2] Ripphausen P, Nisius B, Peltason L, Bajorath J. Quo vadis, virtual screening? A comprehensive survey of prospective applications. J Med Chem 2010;53(24):8461–7.
[3] Druker BJ, Lydon NB. Lessons learned from the development of an Abl tyrosine kinase inhibitor. J Clin Invest 2000;105(1):3–7.
[4] Von Itzstein M, Wu WY, Kok GB, Pegg MS, Dyason JC, Jin B, et al. Rational design of potent sialidase-based inhibitors of influenza virus replication. Nature 1993;363 (6428):418–23.
[5] Kaldor SW, Kalish VJ, Davies JF, Shetty BV, Fritz JE, Appelt K, et al. Viracept (nelfinavir mesylate, AG 1343): a potent, orally bioavailable inhibitor of HIV-1 protease. J Med Chem 1997;40(24):3979–85.
[6] Squires M, Ward G, Saxty G, Berdini V, Cleasby A, King P, et al. Potent, selective inhibitors of fibroblast growth factor receptor define fibroblast growth factor dependence in preclinical cancer models. Mol Cancer Ther 2011;10(9):1542–52.
[7] Talele T, Khedkar S, Rigby A. Successful applications of computer aided drug discovery: moving drugs from concept to the clinic. Curr Top Med Chem 2010;10 (1):127–41.
[8] Kitchen DB, Decornez H, Furr JR, Bajorath J. Docking and scoring in virtual screening for drug discovery: methods and applications. Nat Rev Drug Discov 2004;3 (11):935–49.
[9] Clark DE. What has computer-aided molecular design ever done for drug discovery? Expert Opin Drug Discov 2006;1(2):103–10.
[10] Joseph-McCarthy D, Baber JC, Feyfant E, Thompson DC, Humblet C. Lead optimization via high-throughput molecular docking. Curr Opin Drug Discov Devel 2007; 10(3):264–74.
[11] Kuntz ID, Blaney JM, Oatley SJ, Langridge R, Ferrin TE. A geometric approach to macromolecule-ligand interactions. J Mol Biol 1982;161(2):269–88.
[12] Pagadala NS, Syed K, Tuszynski J. Software for molecular docking: a review. Biophys Rev 2017;9(2):91–102.
[13] Wang Z, Sun H, Yao X, Li D, Xu L, Li Y, et al. Comprehensive evaluation of ten docking programs on a diverse set of protein–ligand complexes: the prediction accuracy of sampling power and scoring power. Phys Chem Chem Phys 2016;18 (18):12964–75.
[14] Li Y, Shen J, Sun X, Li W, Liu G, Tang Y. Accuracy assessment of protein-based docking programs against RNA targets. J Chem Inf Model 2010;50(6):1134–46.
[15] Cross JB, Thompson DC, Rai BK, Baber JC, Fan KY, Hu Y, et al. Comparison of several molecular docking programs: pose prediction and virtual screening accuracy. J Chem Inf Model 2009;49(6):1455–74.
[16] Ferreira LG, dos Santos RN, Oliva G, Andricopulo AD. Molecular docking and structure-based drug design strategies. Molecules 2015;20(7):13384–421.

[17] Kuntz ID, Blaney JM, Oatley SJ, Langridge R, Ferrin TE. A geometric approach to macromolecule-ligand interactions. J Mol Biol 1982;161(2):269–88.
[18] Kabsch W. A solution for the best rotation to relate two sets of vectors. Acta Crystallogr Sect A 1976;32(5):922–3.
[19] Kabsch W. A discussion of the solution for the best rotation to relate two sets of vectors. Acta Crystallogr Sect A 1978;34(5):827–8.
[20] Friesner RA, Banks JL, Murphy RB, Halgren TA, Klicic JJ, Mainz DT, et al. Glide: a new approach for rapid, accurate docking and scoring. 1. Method and assessment of docking accuracy. J Med Chem 2004;47(7):1739–49.
[21] Halgren TA, Murphy RB, Friesner RA, Beard HS, Frye LL, Pollard WT, et al. Glide: a new approach for rapid, accurate docking and scoring. 2. Enrichment factors in database screening. J Med Chem 2004;47(7):1750–9.
[22] Rarey M, Kramer B, Lengauer T, Klebe G. A fast flexible docking method using an incremental construction algorithm. J Mol Biol 1996;261(3):470–89.
[23] Jones G, Willett P, Glen RC, Leach AR, Taylor R. Development and validation of a genetic algorithm for flexible docking 1. J Mol Biol 1997;267(3):727–48.
[24] Hart TN, Read RJ. A multiple-start Monte Carlo docking method. Proteins 1992;13 (3):206–22.
[25] Korb O, Stützle T, Exner TE. PLANTS: application of ant colony optimization to structure-based drug design. In: Dorigo M, Gambardella LM, Birattari M, Martinoli A, Poli R, Stützle T, editors. Ant colony optimization and swarm intelligence. Berlin Heidelberg: Springer; 2006. p. 247–58.
[26] Pei J, Wang Q, Liu Z, Li Q, Yang K, Lai L. PSI-DOCK: towards highly efficient and accurate flexible ligand docking. Proteins 2006;62(4):934–46.
[27] Li Y, Han L, Liu Z, Wang R. Comparative assessment of scoring functions on an updated benchmark: 2. Evaluation methods and general results. J Chem Inf Model 2014;54(6):1717–36.
[28] Li J, Fu A, Zhang L. An overview of scoring functions used for protein–ligand interactions in molecular docking. Interdiscip Sci Comput Life Sci 2019;11(2):320–8.
[29] Ain QU. Machine-learning scoring functions to improve structure-based binding affinity prediction and virtual screening. Wiley Interdiscip Rev Comput Mol Sci 2015;5:405–24.
[30] Zou X, Yaxiong, Kuntz ID. Inclusion of solvation in ligand binding free energy calculations using the generalized-born model. J Am Chem Soc 1999;121(35):8033–43.
[31] Cournia Z, Allen B, Sherman W. Relative binding free energy calculations in drug discovery: recent advances and practical considerations. J Chem Inf Model 2017;57 (12):2911–37.
[32] Mark AE, van Gunsteren WF. Decomposition of the free energy of a system in terms of specific interactions: implications for theoretical and experimental studies. J Mol Biol 1994;240(2):167–76.
[33] Williams DH, Maguire AJ, Tsuzuki W, Westwell MS. An analysis of the origins of a cooperative binding energy of dimerization. Science 1998;280(5364):711–4.
[34] Verdonk ML, Cole JC, Hartshorn MJ, Murray CW, Taylor RD. Improved protein-ligand docking using GOLD. Proteins 2003;52(4):609–23.
[35] Morris GM, Goodsell DS, Halliday RS, Huey R, Hart WE, Belew RK, et al. Automated docking using a Lamarckian genetic algorithm and an empirical binding free energy function. J Comput Chem 1998;19(14):1639–62.
[36] Corbeil CR, Williams CI, Labute P. Variability in docking success rates due to dataset preparation. J Comput Aided Mol Des 2012;26(6):775–86.
[37] Eldridge MD, Murray CW, Auton TR, Paolini GV, Mee RP. Empirical scoring functions: I. The development of a fast empirical scoring function to estimate the binding affinity of ligands in receptor complexes. J Comput Aided Mol Des 1997;11(5):425–45.

[38] Korb O, Stützle T, Exner TE. Empirical scoring functions for advanced protein–ligand docking with PLANTS. J Chem Inf Model 2009;49(1):84–96.
[39] Huang S-Y, Zou X. An iterative knowledge-based scoring function to predict protein-ligand interactions: II. Validation of the scoring function. J Comput Chem 2006;27(15):1876–82.
[40] Thomas PD, Dill KA. Statistical potentials extracted from protein structures: how accurate are they? J Mol Biol 1996;257(2):457–69.
[41] Velec HFG, Gohlke H, Klebe G. Drug score CSD-knowledge-based scoring function derived from small molecule crystal data with superior recognition rate of near-native ligand poses and better affinity prediction. J Med Chem 2005;48(20):6296–303.
[42] Gohlke H, Hendlich M, Klebe G. Knowledge-based scoring function to predict protein-ligand interactions. J Mol Biol 2000;295(2):337–56.
[43] Mooij WTM, Verdonk ML. General and targeted statistical potentials for protein–ligand interactions. Proteins Struct Funct Bioinf 2005;61(2):272–87.
[44] Muegge I, Martin YC. A general and fast scoring function for protein–ligand interactions: a simplified potential approach. J Med Chem 1999;42(5):791–804.
[45] Muegge I. PMF scoring revisited. J Med Chem 2006;49(20):5895–902.
[46] Huang S-Y, Zou X. An iterative knowledge-based scoring function to predict protein-ligand interactions: I. Derivation of interaction potentials. J Comput Chem 2006;27(15):1866–75.
[47] Huang S-Y, Grinter SZ, Zou X. Scoring functions and their evaluation methods for protein–ligand docking: recent advances and future directions. Phys Chem Chem Phys 2010;12(40):12899–908.
[48] Verdonk ML, Ludlow RF, Giangreco I, Rathi PC. Protein–ligand informatics force field (PLIff): toward a fully knowledge driven "force field" for biomolecular interactions. J Med Chem 2016;59(14):6891–902.
[49] Zheng M, Xiong B, Luo C, Li S, Liu X, Shen Q, et al. Knowledge-based scoring functions in drug design: 3. A two-dimensional knowledge-based hydrogen-bonding potential for the prediction of protein–ligand interactions. J Chem Inf Model 2011; 51(11):2994–3004.
[50] Neudert G, Klebe G. DSX: a knowledge-based scoring function for the assessment of protein–ligand complexes. J Chem Inf Model 2011;51(10):2731–45.
[51] Iruela-Arispe ML, Liska DJ, Sage EH, Bornstein P. Differential expression of thrombospondin 1, 2, and 3 during murine development. Dev Dyn 1993;197(1):40–56.
[52] Cheng T, Li X, Li Y, Liu Z, Wang R. Comparative assessment of scoring functions on a diverse test set. J Chem Inf Model 2009;49(4):1079–93.
[53] Charifson PS, Corkery JJ, Murcko MA, Walters WP. Consensus scoring: a method for obtaining improved hit rates from docking databases of three-dimensional structures into proteins. J Med Chem 1999;42(25):5100–9.
[54] O'Boyle NM, Liebeschuetz JW, Cole JC. Testing assumptions and hypotheses for rescoring success in protein−ligand docking. J Chem Inf Model 2009;49(8): 1871–8.
[55] Palacio-Rodríguez K, Lans I, Cavasotto CN, Cossio P. Exponential consensus ranking improves the outcome in docking and receptor ensemble docking. Sci Rep 2019;9 (1):5142.
[56] Verdonk ML, Berdini V, Hartshorn MJ, Mooij WTM, Murray CW, Taylor RD, et al. Virtual screening using protein − ligand docking: avoiding artificial enrichment. J Chem Inf Comput Sci 2004;44(3):793–806.
[57] Khamis MA, Gomaa W, Ahmed WF. Machine learning in computational docking. Artif Intell Med 2015;63(3):135–52.
[58] Wójcikowski M, Ballester PJ, Siedlecki P. Performance of machine-learning scoring functions in structure-based virtual screening. Sci Rep 2017;7(1):46710.

[59] Shen C, Hu Y, Wang Z, Zhang X, Zhong H, Wang G, et al. Can machine learning consistently improve the scoring power of classical scoring functions? Insights into the role of machine learning in scoring functions. Brief Bioinform 2021;22(1):497–514.

[60] Li H, Sze KH, Lu G, Ballester PJ. Machine-learning scoring functions for structure-based drug lead optimization. Wiley Interdiscip Rev Comput Mol Sci 2020;10:1–20. e1465.

[61] Ballester PJ, Mitchell JBO. A machine learning approach to predicting protein–ligand binding affinity with applications to molecular docking. Bioinformatics 2010;26 (9):1169–75.

[62] Li L, Wang B, Meroueh SO. Support vector regression scoring of receptor-ligand complexes for rank-ordering and virtual screening of chemical libraries. J Chem Inf Model 2011;51(9):2132–8.

[63] Çınaroğlu SS, Timuçin E. Comparative assessment of seven docking programs on a nonredundant metalloprotein subset of the PDBbind refined. J Chem Inf Model 2019;59(9):3846–59.

[64] ten Brink T, Exner TE. Influence of protonation, tautomeric, and stereoisomeric states on protein – ligand docking results. J Chem Inf Model 2009;49(6):1535–46.

[65] Chen X, Lin Y, Liu M, Gilson MK. The binding database: data management and interface design. Bioinformatics 2002;18(1):130–9.

[66] Gilson MK, Liu T, Baitaluk M, Nicola G, Hwang L, Chong J. Binding DB in 2015: a public database for medicinal chemistry, computational chemistry and systems pharmacology. Nucleic Acids Res 2015;44(D1):D1045–53.

[67] Chen X, Liu M, Gilson MK. Binding DB: a web-accessible molecular recognition database. Comb Chem High Throughput Screen 2001;4(8):719–25.

[68] Wang R, Fang X, Lu Y, Yang C-Y, Wang S. The PDBbind database: methodologies and updates. J Med Chem 2005;48(12):4111–9.

[69] Wang R, Fang X, Lu Y, Wang S. The PDBbind database: collection of binding affinities for protein–ligand complexes with known three-dimensional structures. J Med Chem 2004;47(12):2977–80.

[70] Grudinin S, Popov P, Neveu E, Cheremovskiy G. Predicting binding poses and affinities in the CSAR 2013–2014 docking exercises using the knowledge-based convex-PL potential. J Chem Inf Model 2015;56(6):1053–62.

[71] Bento AP, Gaulton A, Hersey A, Bellis LJ, Chambers J, Davies M, et al. The ChEMBL bioactivity database: an update. Nucleic Acids Res 2014;42(Database issue):D1083–90.

[72] Gaulton A, Bellis LJ, Bento AP, Chambers J, Davies M, Hersey A, et al. ChEMBL: a large-scale bioactivity database for drug discovery. Nucleic Acids Res 2012;40(D1): D1100–7.

[73] Mysinger MM, Carchia M, Irwin JJ, Shoichet BK. Directory of useful decoys, enhanced (DUD-E): better ligands and decoys for better benchmarking. J Med Chem 2012;55 (14):6582–94.

[74] Kroemer RT, Vulpetti A, McDonald JJ, Rohrer DC, Trosset J-Y, Giordanetto F, et al. Assessment of docking poses: interactions-based accuracy classification (IBAC) versus crystal structure deviations. J Chem Inf Comput Sci 2004;44(3):871–81.

[75] Yusuf D, Davis AM, Kleywegt GJ, Schmitt S. An alternative method for the evaluation of docking performance: RSR vs RMSD. J Chem Inf Model 2008;48(7):1411–22.

[76] Baber JC, Thompson DC, Cross JB, Humblet C. GARD: a generally applicable replacement for RMSD. J Chem Inf Model 2009;49(8):1889–900. Aug 24.

[77] Berman HM, Westbrook J, Feng Z, Gilliland G, Bhat TN, Weissig H, et al. The protein data bank. Nucleic Acids Res 2000;28:235–42.

[78] Warren GL, Do TD, Kelley BP, Nicholls A, Warren SD. Essential considerations for using protein–ligand structures in drug discovery. Drug Discov Today 2012;17(23–24): 1270–81.

[79] Wlodawer A, Minor W, Dauter Z, Jaskolski M. Protein crystallography for non-crystallographers, or how to get the best (but not more) from published macromolecular structures. FEBS J 2008;275(1):1–21.
[80] Read RJ, Adams PD, Arendall WB, Brunger AT, Emsley P, Joosten RP, et al. A new generation of crystallographic validation tools for the protein data bank. Structure 2011;19(10):1395–412.
[81] Kuhlman B, Bradley P. Advances in protein structure prediction and design. Nat Rev Mol Cell Biol 2019;20(11):681–97.
[82] Mishra S, Demo G, Koča J, Wimmerová M. In silico engineering of proteins that recognize small molecules. InTech; 2012.
[83] Sastry GM, Adzhigirey M, Day T, Annabhimoju R, Sherman W. Protein and ligand preparation: parameters, protocols, and influence on virtual screening enrichments. J Comput Aided Mol Des 2013;27(3):221–34.
[84] Kalliokoski T, Salo HS, Lahtela-Kakkonen M, Poso A. The effect of ligand-based tautomer and protomer prediction on structure-based virtual screening. J Chem Inf Model 2009;49(12):2742–8.
[85] Onufriev AV, Alexov E. Protonation and pK changes in protein-ligand binding. Q Rev Biophys 2013;46(2):181–209.
[86] Berry M, Fielding B, Gamieldien J. Practical considerations in virtual screening and molecular docking. Emerg Trends Comput Biol Bioinf Syst Biol; 2015. p. 487–502.
[87] Li S, Hong M. Protonation, tautomerization, and rotameric structure of histidine: a comprehensive study by magic-angle-spinning solid-state NMR. J Am Chem Soc 2011;133(5):1534–44.
[88] Kim MO, Nichols SE, Wang Y, McCammon JA. Effects of histidine protonation and rotameric states on virtual screening of M. tuberculosis RmlC. J Comput Aided Mol Des 2013;27(3):235–46.
[89] Zhao J, Cao Y, Zhang L. Exploring the computational methods for protein-ligand binding site prediction. Comput Struct Biotechnol J 2020;18:417–26.
[90] Hendlich M, Rippmann F, Barnickel G. LIGSITE: automatic and efficient detection of potential small molecule-binding sites in proteins. J Mol Graph Model 1997;15 (6):359–63.
[91] Kawabata T. Detection of multiscale pockets on protein surfaces using mathematical morphology. Proteins 2010;78(5):1195–211.
[92] Simões TMC, Gomes AJP. CavVis—a field-of-view geometric algorithm for protein cavity detection. J Chem Inf Model 2019;59(2):786–96.
[93] Coleman RG, Sharp KA. Protein pockets: inventory, shape, and comparison. J Chem Inf Model 2010;50(4):589–603.
[94] B-Rao C, Subramanian J, Sharma S. Managing protein flexibility in docking and its applications. Drug Discov Today 2009;14:394–400.
[95] Ferrari AM, Wei BQ, Costantino L, Shoichet BK. Soft docking and multiple receptor conformations in virtual screening. J Med Chem 2004;47(21):5076–84.
[96] Cosconati S, Marinelli L, Di Leva FS, La Pietra V, De Simone A, Mancini F, et al. Protein flexibility in virtual screening: the BACE-1 case study. J Chem Inf Model 2012;52(10):2697–704.
[97] Rueda M, Bottegoni G, Abagyan R. Recipes for the selection of experimental protein conformations for virtual screening. J Chem Inf Model 2010;50(1):186–93.
[98] Najmanovich R, Kuttner J, Sobolev V, Edelman M. Side-chain flexibility in proteins upon ligand binding. Proteins Struct Funct Genet 2000;39(3):261–8.
[99] Korb O, Olsson TSG, Bowden SJ, Hall RJ, Verdonk ML, Liebeschuetz JW, et al. Potential and limitations of ensemble docking. J Chem Inf Model 2012;52(5):1262–74.
[100] Campbell AJ, Lamb ML, Joseph-McCarthy D. Ensemble-based docking using biased molecular dynamics. J Chem Inf Model 2014;54(7):2127–38.

[101] Carlson HA, Masukawa KM, McCammon JA. Method for including the dynamic fluctuations of a protein in computer-aided drug design. J Phys Chem A 1999;103 (49):10213–9.
[102] Fernández-Recio J, Totrov M, Abagyan R. Soft protein-protein docking in internal coordinates. Protein Sci 2002;11(2):280–91.
[103] Vieth M, Hirst JD, Kolinski A, Brooks III CL. Assessing energy functions for flexible docking. J Comput Chem 1998;19(14):1612–22.
[104] Hou X, Li K, Yu X, Sun J, Fang H. Protein flexibility in docking-based virtual screening: discovery of novel lymphoid-specific tyrosine phosphatase inhibitors using multiple crystal structures. J Chem Inf Model 2015;55(9):1973–83.
[105] Lovell SC, Word JM, Richardson JS, Richardson DC. The penultimate rotamer library. Proteins Struct Funct Bioinf 2000;40(3):389–408.
[106] Gaudreault F, Chartier M, Najmanovich R. Side-chain rotamer changes upon ligand binding: common, crucial, correlate with entropy and rearrange hydrogen bonding. Bioinformatics 2012;28(18):i423–30.
[107] Bottegoni G, Rocchia W, Rueda M, Abagyan R, Cavalli A. Systematic exploitation of multiple receptor conformations for virtual ligand screening. PLoS One 2011;6(5): e18845.
[108] Tian S, Sun H, Pan P, Li D, Zhen X, Li Y, et al. Assessing an ensemble docking-based virtual screening strategy for kinase targets by considering protein flexibility. J Chem Inf Model 2014;54(10):2664–79.
[109] Sander B, Korb O, Cole J, Essex JW. How to pick a winning team: approaches towards the selection of computationally derived protein structures for ensemble-based virtual screening. J Cheminform 2013;5(Suppl. 1):O7.
[110] Roberts BC, Mancera RL. Ligand–protein docking with water molecules. J Chem Inf Model 2008;48(2):397–408.
[111] Abel R, Young T, Farid R, Berne BJ, Friesner RA. Role of the active-site solvent in the thermodynamics of factor Xa ligand binding. J Am Chem Soc 2008;130 (9):2817–31.
[112] Young T, Abel R, Kim B, Berne BJ, Friesner RA. Motifs for molecular recognition exploiting hydrophobic enclosure in protein–ligand binding. Proc Natl Acad Sci 2007;104(3):808–13.
[113] Jorgensen WL, Chandrasekhar J, Madura JD, Impey RW, Klein ML. Comparison of simple potential functions for simulating liquid water. J Chem Phys 1983;79 (2):926–35.
[114] Imai T, Hiraoka R, Kovalenko A, Hirata F. Locating missing water molecules in protein cavities by the three-dimensional reference interaction site model theory of molecular solvation. Proteins Struct Funct Bioinf 2007;66(4):804–13.
[115] Verdonk ML, Chessari G, Cole JC, Hartshorn MJ, Murray CW, Nissink JWM, et al. Modeling water molecules in protein–ligand docking using GOLD. J Med Chem 2005;48(20):6504–15.
[116] Osterberg F, Morris GM, Sanner MF, Olson AJ, Goodsell DS. Automated docking to multiple target structures: incorporation of protein mobility and structural water heterogeneity in AutoDock. Proteins 2002;46(1):34–40.
[117] Friesner RA, Murphy RB, Repasky MP, Frye LL, Greenwood JR, Halgren TA, et al. Extra precision glide: docking and scoring incorporating a model of hydrophobic enclosure for protein–ligand complexes. J Med Chem 2006;49(21):6177–96.
[118] Kroemer RT. Structure-based drug design: docking and scoring. Curr Protein Pept Sci 2007;8(4):312–28.
[119] Zhong H, Wang Z, Wang X, Liu H, Li D, Liu H, et al. Importance of a crystalline water network in docking-based virtual screening: a case study of BRD4. Phys Chem Chem Phys 2019;21(45):25276–89.

[120] Cheng T, Li Q, Zhou Z, Wang Y, Bryant SH. Structure-based virtual screening for drug discovery: a problem-centric review. AAPS J 2012;14(1):133–41.
[121] Huang N, Shoichet BK. Exploiting ordered waters in molecular docking. J Med Chem 2008;51(16):4862–5.
[122] Hu B, Lill MA. WATsite: hydration site prediction program with PyMOL interface. J Comput Chem 2014;35(16):1255–60.
[123] Verdonk ML, Cole JC, Taylor R. SuperStar: a knowledge-based approach for identifying interaction sites in proteins. J Mol Biol 1999;289(4):1093–108.
[124] Goodford PJ. A computational procedure for determining energetically favorable binding sites on biologically important macromolecules. J Med Chem 1985;28 (7):849–57.
[125] Sridhar A, Ross GA, Biggin PC. Waterdock 2.0: water placement prediction for Holo-structures with a pymol plugin. PLoS One 2017;12(2):1–17.
[126] Cole JC, Korb O, McCabe P, Read MG, Taylor R. Knowledge-based conformer generation using the Cambridge structural database. J Chem Inf Model 2018;58 (3):615–29.
[127] Taylor R, Cole J, Korb O, McCabe P. Knowledge-based libraries for predicting the geometric preferences of drug like molecules. J Chem Inf Model 2014;54(9): 2500–14.
[128] Hawkins PCD, Skillman AG, Warren GL, Ellingson BA, Stahl MT. Conformer generation with OMEGA: algorithm and validation using high quality structures from the protein databank and Cambridge structural database. J Chem Inf Model 2010;50 (4):572–84.
[129] Ebejer J-P, Morris GM, Deane CM. Freely available conformer generation methods: how good are they? J Chem Inf Model 2012;52(5):1146–58.
[130] O'Boyle NM, Vandermeersch T, Flynn CJ, Maguire AR, Hutchison GR. Confab-systematic generation of diverse low-energy conformers. J Cheminform 2011;3(1):8.
[131] Schärfer C, Schulz-Gasch T, Hert J, Heinzerling L, Schulz B, Inhester T, et al. CONFECT: conformations from an expert collection of torsion patterns. ChemMed Chem 2013;8(10):1690–700.
[132] Landrum G, et al. RDKit: Open-source cheminformatics, http://www.rdkit.org/. 2018; 2018.
[133] Groom CR, Bruno IJ, Lightfoot MP, Ward SC. The Cambridge structural database. Acta Crystallogr Sect B Struct Sci Cryst Eng Mater 2016;72(2):171–9.
[134] Brooks WH, Daniel KG, Sung S-S, Guida WC. Computational validation of the importance of absolute stereochemistry in virtual screening. J Chem Inf Model 2008;48(3):639–45.
[135] Lipinski CA, Lombardo F, Dominy BW, Feeney PJ. Experimental and computational approaches to estimate solubility and permeability in drug discovery and development settings. Adv Drug Deliv Rev 2001;46(1–3):3–26.
[136] Shultz MD. Two decades under the influence of the rule of five and the changing properties of approved oral drugs. J Med Chem 2019;62(4):1701–14.
[137] Irwin JJ, Sterling T, Mysinger MM, Bolstad ES, Coleman RG. ZINC: a free tool to discover chemistry for biology. J Chem Inf Model 2012;52(7):1757–68.
[138] Irwin JJ, Tang KG, Young J, Dandarchuluun C, Wong BR, Khurelbaatar M, et al. ZINC20—a free Ultralarge-scale chemical database for ligand discovery. J Chem Inf Model 2020;60(12):6065–73.
[139] Sterling T, Irwin JJ. ZINC 15—ligand discovery for everyone. J Chem Inf Model 2015;55(11):2324–37.
[140] Veber DF, Johnson SR, Cheng H-Y, Smith BR, Ward KW, Kopple KD. Molecular properties that influence the Oral bioavailability of drug candidates. J Med Chem 2002;45(12):2615–23.

[141] Anon. Downloadable Structure Files of NCI Open Database Compounds. Available from https://cactus.nci.nih.gov/download/nci/.
[142] Wishart DS, Knox C, Guo AC, Shrivastava S, Hassanali M, Stothard P, et al. DrugBank: a comprehensive resource for in silico drug discovery and exploration. Nucleic Acids Res 2006;34(Database issue):D668–72.
[143] Wishart DS, Feunang YD, Guo AC, Lo EJ, Marcu A, Grant JR, et al. DrugBank 5.0: a major update to the DrugBank database for 2018. Nucleic Acids Res 2018;46(D1): D1074–82.
[144] ASINEX. Screening libraries. Available from: http://www.asinex.com/libraries-html/.
[145] Cole JC, Wiggin S, Stanzione F. New insights and innovation from a million crystal structures in the Cambridge structural database. Struct Dyn 2019;6(5):054301.
[146] Taylor R, Wood PA. A million crystal structures: the whole is greater than the sum of its parts. Chem Rev 2019;119(16):9427–77.
[147] Congreve M, Carr R, Murray C, Jhoti H. A 'rule of three' for fragment-based lead discovery? Drug Discov Today 2003;8(19):876–7.
[148] Verdonk ML, Giangreco I, Hall RJ, Korb O, Mortenson PN, Murray CW. Docking performance of fragments and druglike compounds. J Med Chem 2011;54(15): 5422–31.
[149] Leach AR, Hann MM, Burrows JN, Griffen EJ. Fragment screening: an introduction. Mol Biosyst 2006;2(9):429–46.
[150] Bian Y, Xie X-QS. Computational fragment-based drug design: current trends, strategies, and applications. AAPS J 2018;20(3):59.
[151] Fink T, Reymond J-L. Virtual exploration of the chemical universe up to 11 atoms of C, N, O, F: assembly of 26.4 million structures (110.9 million stereoisomers) and analysis for new ring systems, stereochemistry, physicochemical properties, compound classes, and drug discovery. J Chem Inf Model 2007;47(2):342–53.
[152] Jhoti H, Williams G, Rees DC, Murray CW. The "rule of three" for fragment-based drug discovery: where are we now? Nat Rev Drug Discov 2013;12(8):644.
[153] Anon. Fragment collections. Available from https://www.cambridgemedchemconsulting.com/resources/hit_identification/fragment_collections.html.
[154] Cox OB, Krojer T, Collins P, Monteiro O, Talon R, Bradley A, et al. A poised fragment library enables rapid synthetic expansion yielding the first reported inhibitors of PHIP(2){,} an atypical bromodomain. Chem Sci 2016;7(3):2322–30.
[155] Morley AD, Pugliese A, Birchall K, Bower J, Brennan P, Brown N, et al. Fragment-based hit identification: thinking in 3D. Drug Discov Today 2013;18 (23):1221–7.
[156] de Souza Neto LR, Moreira-Filho JT, Neves BJ, Maidana RLBR, Guimarães ACR, Furnham N, et al. In silico strategies to support fragment-to-lead optimization in drug discovery. Front Chem 2020;8:93.
[157] Kumar A, Voet A, Zhang KYJ. Fragment based drug design: from experimental to computational approaches. Curr Med Chem 2012;19(30):5128–47.
[158] Congreve M, Chessari G, Tisi D, Woodhead AJ. Recent developments in fragment-based drug discovery. J Med Chem 2008;51(13):3661–80.
[159] Chessari G, Woodhead AJ. From fragment to clinical candidate—a historical perspective. Drug Discov Today 2009;14(13):668–75.
[160] Hung AW, Silvestre HL, Wen S, Ciulli A, Blundell TL, Abell C. Application of fragment growing and fragment linking to the discovery of inhibitors of *Mycobacterium tuberculosis* pantothenate synthetase. Angew Chemie Int Ed 2009;48(45):8452–6.
[161] Cheng Y, Judd TC, Bartberger MD, Brown J, Chen K, Fremeau RT, et al. From fragment screening to in vivo efficacy: optimization of a series of 2-aminoquinolines as potent inhibitors of beta-site amyloid precursor protein cleaving enzyme 1 (BACE1). J Med Chem 2011;54(16):5836–57.

[162] Taylor SJ, Abeywardane A, Liang S, Muegge I, Padyana AK, Xiong Z, et al. Fragment-based discovery of indole inhibitors of matrix metalloproteinase-13. J Med Chem 2011;54(23):8174–87.
[163] Brough PA, Barril X, Borgognoni J, Chene P, Davies NGM, Davis B, et al. Combining hit identification strategies: fragment-based and in silico approaches to orally active 2-aminothieno[2,3-d]pyrimidine inhibitors of the Hsp90 molecular chaperone. J Med Chem 2009;52(15):4794–809.
[164] Hughes SJ, Millan DS, Kilty IC, Lewthwaite RA, Mathias JP, O'Reilly MA, et al. Fragment based discovery of a novel and selective PI3 kinase inhibitor. Bioorg Med Chem Lett 2011;21(21):6586–90.
[165] Villemagne B, Flipo M, Blondiaux N, Crauste C, Malaquin S, Leroux F, et al. Ligand efficiency driven design of new Inhibitors of *Mycobacterium tuberculosis* transcriptional repressor EthR using fragment growing, merging, and linking approaches. J Med Chem 2014;57(11):4876–88.
[166] Istyastono EP, Kooistra AJ, Vischer HF, Kuijer M, Roumen L, Nijmeijer S, et al. Structure-based virtual screening for fragment-like ligands of the G protein-coupled histamine H4 receptor. Med Chem Commun 2015;6(6):1003–17.
[167] Friberg A, Vigil D, Zhao B, Daniels RN, Burke JP, Garcia-Barrantes PM, et al. Discovery of potent myeloid cell leukemia 1 (Mcl-1) inhibitors using fragment-based methods and structure-based design. J Med Chem 2013;56(1):15–30.
[168] De Fusco C, Brear P, Iegre J, Georgiou KH, Sore HF, Hyvönen M, et al. A fragment-based approach leading to the discovery of a novel binding site and the selective CK2 inhibitor CAM4066. Bioorg Med Chem 2017;25(13):3471–82.
[169] Howard N, Abell C, Blakemore W, Chessari G, Congreve M, Howard S, et al. Application of fragment screening and fragment linking to the discovery of novel thrombin inhibitors. J Med Chem 2006;49(4):1346–55.
[170] Huang S-Y. Search strategies and evaluation in protein–protein docking: principles, advances and challenges. Drug Discov Today 2014;19(8):1081–96.
[171] Porter KA, Desta I, Kozakov D, Vajda S. What method to use for protein–protein docking? Curr Opin Struct Biol 2019;55:1–7.
[172] Ruvinsky AM, Kirys T, Tuzikov AV, Vakser IA. Side-chain conformational changes upon protein-protein association. J Mol Biol 2011;408(2):356–65.
[173] Jackson RM, Gabb HA, Sternberg MJ. Rapid refinement of protein interfaces incorporating solvation: application to the docking problem. J Mol Biol 1998;276(1):265–85.
[174] Król M, Tournier AL, Bates PA. Flexible relaxation of rigid-body docking solutions. Proteins 2007;68(1):159–69.
[175] Król M, Chaleil RAG, Tournier AL, Bates PA. Implicit flexibility in protein docking: cross-docking and local refinement. Proteins 2007;69(4):750–7.
[176] Grünberg R, Leckner J, Nilges M. Complementarity of structure ensembles in protein-protein binding. Structure 2004;12(12):2125–36.
[177] Gray JJ, Moughon S, Wang C, Schueler-Furman O, Kuhlman B, Rohl CA, et al. Protein-protein docking with simultaneous optimization of rigid-body displacement and side-chain conformations. J Mol Biol 2003;331(1):281–99.
[178] Zacharias M. Protein-protein docking with a reduced protein model accounting for side-chain flexibility. Protein Sci 2003;12(6):1271–82.
[179] Zacharias M. ATTRACT: protein-protein docking in CAPRI using a reduced protein model. Proteins 2005;60(2):252–6.
[180] Fernández-Recio J, Totrov M, Abagyan R. ICM-DISCO docking by global energy optimization with fully flexible side-chains. Proteins 2003;52(1):113–7.
[181] Dominguez C, Boelens R, Bonvin AMJJ. HADDOCK: a protein – protein docking approach based on biochemical or biophysical information. J Am Chem Soc 2003;125 (7):1731–7.

[182] Russel D, Lasker K, Webb B, Velázquez-Muriel J, Tjioe E, Schneidman-Duhovny D, et al. Putting the pieces together: integrative modeling platform software for structure determination of macromolecular assemblies. PLoS Biol 2012;10(1):e1001244.
[183] Janin J, Henrick K, Moult J, Ten Eyck L, Sternberg MJE, Vajda S, et al. CAPRI: a critical assessment of PRedicted interactions. Proteins 2003;52(1):2–9.
[184] Tsomaia N. Peptide therapeutics: targeting the undruggable space. Eur J Med Chem 2015;94:459–70.
[185] Liao J, Wang Y-T, Lin CS. A fragment-based docking simulation for investigating peptide–protein bindings. Phys Chem Chem Phys 2017;19(16):10436–42.
[186] Pierce BG, Wiehe K, Hwang H, Kim B-H, Vreven T, Weng Z. ZDOCK server: interactive docking prediction of protein-protein complexes and symmetric multimers. Bioinformatics 2014;30(12):1771–3.
[187] Macindoe G, Mavridis L, Venkatraman V, Devignes M-D, Ritchie DW. HexServer: an FFT-based protein docking server powered by graphics processors. Nucleic Acids Res 2010;38(suppl_2):W445–9.
[188] Lee AC-L, Harris JL, Khanna KK, Hong J-H. A comprehensive review on current advances in peptide drug development and design. Int J Mol Sci 2019;20(10):2383.
[189] Agrawal P, Singh H, Srivastava HK, Singh S, Kishore G, Raghava GPS. Benchmarking of different molecular docking methods for protein-peptide docking. BMC Bioinf 2019;19(13):426.
[190] Shin W-H, Seok C. GalaxyDock: protein–ligand docking with flexible protein side-chains. J Chem Inf Model 2012;52(12):3225–32.
[191] Lee H, Heo L, Lee MS, Seok C. GalaxyPepDock: a protein-peptide docking tool based on interaction similarity and energy optimization. Nucleic Acids Res 2015;43 (W1):W431–5.
[192] Iqbal S, Hoque MT. PBRpredict-suite: a suite of models to predict peptide-recognition domain residues from protein sequence. Bioinformatics 2018;34(19):3289–99.
[193] Ciemny M, Kurcinski M, Kamel K, Kolinski A, Alam N, Schueler-Furman O, et al. Protein–peptide docking: opportunities and challenges. Drug Discov Today 2018;23 (8):1530–7.
[194] Antes I. DynaDock: a new molecular dynamics-based algorithm for protein-peptide docking including receptor flexibility. Proteins 2010;78(5):1084–104.
[195] Raveh B, London N, Schueler-Furman O. Sub-angstrom modeling of complexes between flexible peptides and globular proteins. Proteins 2010;78(9):2029–40.
[196] Donsky E, Wolfson HJ. PepCrawler: a fast RRT-based algorithm for high-resolution refinement and binding affinity estimation of peptide inhibitors. Bioinformatics 2011;27(20):2836–42.
[197] Raveh B, London N, Zimmerman L, Schueler-Furman O. Rosetta FlexPepDock ab-initio: simultaneous folding, docking and refinement of peptides onto their receptors. PLoS One 2011;6(4):e18934.
[198] Ben-Shimon A, Niv MY. AnchorDock: blind and flexible anchor-driven peptide docking. Structure 2015;23(5):929–40.
[199] Weng G, Gao J, Wang Z, Wang E, Hu X, Yao X, et al. Comprehensive evaluation of fourteen docking programs on protein–peptide complexes. J Chem Theory Comput 2020;16(6):3959–69.
[200] Wang M, Yu Y, Liang C, Lu A, Zhang G. Recent advances in developing small molecules targeting nucleic acid. Int J Mol Sci 2016;17(6):779.
[201] Luo J, Wei W, Waldispühl J, Moitessier N. Challenges and current status of computational methods for docking small molecules to nucleic acids. Eur J Med Chem 2019;168:414–25.
[202] Bao L, Zhang X, Jin L, Tan Z-J. Flexibility of nucleic acids: from {DNA} to {RNA}. Chin Phys B 2016;25(1):18703.

[203] Stagno JR, Liu Y, Bhandari YR, Conrad CE, Panja S, Swain M, et al. Structures of riboswitch RNA reaction states by mix-and-inject XFEL serial crystallography. Nature 2017;541(7636):242–6.
[204] Guilbert C, James TL. Docking to RNA via root-mean-square-deviation-driven energy minimization with flexible ligands and flexible targets. J Chem Inf Model 2008;48(6):1257–68.
[205] Tessaro F, Scapozza L. How "protein-docking" translates into the new emerging field of docking small molecules to nucleic acids? Molecules 2020;25(12):2749.
[206] Morley SD, Afshar M. Validation of an empirical RNA-ligand scoring function for fast flexible docking using RiboDock®. J Comput Aided Mol Des 2004;18(3):189–208.
[207] Pfeffer P, Gohlke H. DrugScoreRNA—knowledge-based scoring function to predict RNA-ligand interactions. J Chem Inf Model 2007;47(5):1868–76.
[208] Philips A, Milanowska K, Lach G, Bujnicki JM. LigandRNA: computational predictor of RNA-ligand interactions. RNA 2013;19(12):1605–16.
[209] Lang PT, Brozell SR, Mukherjee S, Pettersen EF, Meng EC, Thomas V, et al. DOCK 6: combining techniques to model RNA-small molecule complexes. RNA 2009;15 (6):1219–30.
[210] Moitessier N, Westhof E, Hanessian S. Docking of aminoglycosides to hydrated and flexible RNA. J Med Chem 2006;49(3):1023–33.
[211] Chen L, Calin GA, Zhang S. Novel insights of structure-based modeling for RNA-targeted drug discovery. J Chem Inf Model 2012;52(10):2741–53.
[212] Chan DS-H, Yang H, Kwan MH-T, Cheng Z, Lee P, Bai L-P, et al. Structure-based optimization of FDA-approved drug methylene blue as a c-myc G-quadruplex DNA stabilizer. Biochimie 2011;93(6):1055–64.
[213] Kaserer T, Rigo R, Schuster P, Alcaro S, Sissi C, Schuster D. Optimized virtual screening workflow for the identification of novel G-quadruplex ligands. J Chem Inf Model 2016;56(3):484–500.
[214] Holt PA, Ragazzon P, Strekowski L, Chaires JB, Trent JO. Discovery of novel triple helical DNA intercalators by an integrated virtual and actual screening platform. Nucleic Acids Res 2009;37(4):1280–7.
[215] Daldrop P, Reyes FE, Robinson DA, Hammond CM, Lilley DM, Batey RT, et al. Novel ligands for a purine riboswitch discovered by RNA-ligand docking. Chem Biol 2011;18(3):324–35.
[216] Lee HS, Zhang Y. BSP-SLIM: a blind low-resolution ligand-protein docking approach using predicted protein structures. Proteins 2012;80(1):93–110.
[217] Grebner C, Iegre J, Ulander J, Edman K, Hogner A, Tyrchan C. Binding mode and induced fit predictions for prospective computational drug design. J Chem Inf Model 2016;56(4):774–87.
[218] Singh J, Petter RC, Baillie TA, Whitty A. The resurgence of covalent drugs. Nat Rev Drug Discov 2011;10(4):307–17.
[219] Li Q, Wang Z, Zheng Q, Liu S. Potential clinical drugs as covalent inhibitors of the priming proteases of the spike protein of SARS-CoV-2. Comput Struct Biotechnol J 2020;18:2200–8.
[220] Liu S, Zheng Q, Wang Z. Potential covalent drugs targeting the main protease of the SARS-CoV-2 coronavirus. Bioinformatics 2020;36(11):3295–8.
[221] Hoffman RL, Kania RS, Brothers MA, Davies JF, Ferre RA, Gajiwala KS, et al. Discovery of ketone-based covalent inhibitors of coronavirus 3CL proteases for the potential therapeutic treatment of COVID-19. J Med Chem 2020;63:12725–47.
[222] Sotriffer C. Docking of covalent ligands: challenges and approaches. Mol Inf 2018;37 (9–10):1800062.
[223] Fanfrlík J, Brahmkshatriya PS, Řezáč J, Jílková A, Horn M, Mareš M, et al. Quantum mechanics-based scoring rationalizes the irreversible inactivation of parasitic *Schistosoma*

mansoni cysteine peptidase by vinyl sulfone inhibitors. J Phys Chem B 2013;117 (48):14973–82.
[224] Kharkar PS, Warrier S, Gaud RS. Reverse docking: a powerful tool for drug repositioning and drug rescue. Future Med Chem 2014;6(3):333–42.
[225] Ashburn TT, Thor KB. Drug repositioning: identifying and developing new uses for existing drugs. Nat Rev Drug Discov 2004;3(8):673–83.
[226] Keiser MJ, Setola V, Irwin JJ, Laggner C, Abbas AI, Hufeisen SJ, et al. Predicting new molecular targets for known drugs. Nature 2009;462(7270):175–81.
[227] Terrett NK, Bell AS, Brown D, Ellis P. Sildenafil (VIAGRATM), a potent and selective inhibitor of type 5 cGMP phosphodiesterase with utility for the treatment of male erectile dysfunction. Bioorg Med Chem Lett 1996;6(15):1819–24.
[228] Zappacosta AR. Reversal of baldness in patient receiving minoxidil for hypertension. N Engl J Med 1980;303:1480–1.
[229] Jarada TN, Rokne JG, Alhajj R. A review of computational drug repositioning: strategies, approaches, opportunities, challenges, and directions. J Cheminform 2020;12(1):46.
[230] Mohamed K, Yazdanpanah N, Saghazadeh A, Rezaei N. Computational drug discovery and repurposing for the treatment of COVID-19: a systematic review. Bioorg Chem 2020;106:104490.
[231] Chen X, Ung CY, Chen Y. Can an in silico drug-target search method be used to probe potential mechanisms of medicinal plant ingredients? Nat Prod Rep 2003;20 (4):432–44.
[232] Erić S, Ke S, Barata T, Solmajer T, Antić Stanković J, Juranić Z, et al. Target fishing and docking studies of the novel derivatives of aryl-aminopyridines with potential anticancer activity. Bioorg Med Chem 2012;20(17):5220–8.
[233] Ji ZL, Wang Y, Yu L, Han LY, Zheng CJ, Chen YZ. In silico search of putative adverse drug reaction related proteins as a potential tool for facilitating drug adverse effect prediction. Toxicol Lett 2006;164(2):104–12.
[234] Tanoli Z, Seemab U, Scherer A, Wennerberg K, Tang J, Vähä-Koskela M. Exploration of databases and methods supporting drug repurposing: a comprehensive survey. Brief Bioinform 2020;1–23.
[235] Kellenberger E, Muller P, Schalon C, Bret G, Foata N, Rognan D. Sc-PDB: an annotated database of druggable binding sites from the protein data bank. J Chem Inf Model 2006;46(2):717–27.
[236] Chen X, Ji ZL, Chen YZ. TTD: therapeutic target database. Nucleic Acids Res 2002;30(1):412–5.
[237] Xu X, Huang M, Zou X. Docking-based inverse virtual screening: methods, applications, and challenges. Biophys Rep 2018;4(1):1–16.
[238] Schomburg KT, Bietz S, Briem H, Henzler AM, Urbaczek S, Rarey M. Facing the challenges of structure-based target prediction by inverse virtual screening. J Chem Inf Model 2014;54(6):1676–86.
[239] Kellenberger E, Foata N, Rognan D. Ranking targets in structure-based virtual screening of three-dimensional protein libraries: methods and problems. J Chem Inf Model 2008;48(5):1014–25.
[240] Luo Q, Zhao L, Hu J, Jin H, Liu Z, Zhang L. The scoring bias in reverse docking and the score normalization strategy to improve success rate of target fishing. PLoS One 2017;12(2):e0171433.
[241] Nogueira MS, Koch O. The development of target-specific machine learning models as scoring functions for docking-based target prediction. J Chem Inf Model 2019;59 (3):1238–52.
[242] Luo H, Fokoue-Nkoutche A, Singh N, Yang L, Hu J, Zhang P. Molecular docking for prediction and interpretation of adverse drug reactions. Comb Chem High Throughput Screen 2018;21(5):314–22.

[243] Bagherian M, Sabeti E, Wang K, Sartor MA, Nikolovska-Coleska Z, Najarian K. Machine learning approaches and databases for prediction of drug–target interaction: a survey paper. Brief Bioinform 2020;22(1):247–69.
[244] Pinzi L, Rastelli G. Molecular docking: shifting paradigms in drug discovery. Int J Mol Sci 2019;20(18):4331.
[245] Yip KM, Fischer N, Paknia E, Chari A, Stark H. Atomic-resolution protein structure determination by cryo-EM. Nature 2020;587(7832):157–61.
[246] Nakane T, Kotecha A, Sente A, McMullan G, Masiulis S, Brown PMGE, et al. Single-particle cryo-EM at atomic resolution. Nature 2020;587(7832):152–6.
[247] Senior AW, Evans R, Jumper J, Kirkpatrick J, Sifre L, Green T, et al. Improved protein structure prediction using potentials from deep learning. Nature 2020;577(7792):706–10.

CHAPTER FIVE

An industrial perspective on co-crystals: Screening, identification and development of the less utilised solid form in drug discovery and development

Thomas Kendall[b,*], Sam Stratford[a], Adam R. Patterson[a], Ruth A. Lunt[a], Dyanne Cruickshank[a], Thierry Bonnaud[a], and C. Daniel Scott[a]
[a]Johnson Matthey, Pharmorphix, Cambridge, United Kingdom
[b]Technobis Crystallization Systems, Alkmaar, The Netherlands
*Corresponding author: e-mail address: thomaskendall508@gmail.com

Contents

Progress in Medicinal Chemistry, Volume 60
ISSN 0079-6468
https://doi.org/10.1016/bs.pmch.2021.05.001

Keywords: Co-crystals, Solubility, Bioavailability, Crystallisation, SCXRD, Phase diagrams, Formulation, Mechanochemistry

1. Introduction

In the research and development of a novel active pharmaceutical ingredient (API) for human consumption through an appropriately formulated drug product, there are many considerations to undertake in the development pipeline. Medicinal chemists design synthetic pathways to the isolation of new chemical entities, often with the incorporation of specific functional groups that provide a level of specificity towards the required mechanism of action or to allow for transport across specific membranes of cellular environments.

After identification of the candidate API and in subsequent development it is critically important to not only understand the solid form properties of your API but also the correlation of your selected solid form with the final formulation and drug product. Past examples have shown the importance of recognising and understanding the solid form of your API and the impact this will have on the physiochemical properties the solid form of the API and the final drug product. One of the most quoted examples highlighting this scenario is that of Ritonavir, Norvir® by Abbott Laboratories [1]. Norvir® is a capsule based medicine for the oral delivery of Ritonavir. The drug product was successfully manufactured in separate lots; however, manufacture was quickly halted when the physical properties of lots started failing. In the case of Ritonavir, a significant drop in the solubility of the API was observed which led to a concomitant drop in bioavailability and reduced the therapeutic effect of the drug product at the dosing levels tested during clinical trials. The cause of this change in physicochemical properties was due to a change in solid form, specifically there was a polymorph change of the API in the semi-solid capsule formulation.

Within the pharmaceutical industry, interest in multicomponent systems such as salts, solvates and co-crystals, lies primarily in their ability to change the physicochemical properties of a potential drug, without having to inherently change the molecular structure of the API. By incorporating different molecules (e.g., guest molecule or co-former) into the solid form, one can alter the intermolecular interactions within the crystalline state. This in turn can provide increased drug solubility, bioavailability, stability, purity or improved processability compared to the API entity alone [2,3]. Despite the possibilities in co-crystal development, the response from the pharmaceutical industry until recently has been moderately low. The slow uptake has been attributed to regulatory ambiguity and difficulties with co-crystal formation during scale-up, among other challenges, and a long established practice of moving from conventional salt to amorphous formulations [3–5].

In this review we aim to provide an overview of the important aspects of co-crystals, covering the design and methods used for co-crystal screening and the different analytical techniques used for identification and differentiating a co-crystal from other solid forms. The importance of characterisation and the legal implications of forming and developing co-crystals will be discussed. There are important considerations with respect to the scale up and formulation of co-crystals too. The applications of co-crystals will be covered addressing issues with solubility, bioavailability, purification, processability, stability and offering new and alternative formulations.

2. What are co-crystals?

The many solid forms available are visualised in Fig. 1 and are the focus of this initial discussion.

A solid chemical entity can primarily exist as a crystalline solid or a non-crystalline, amorphous solid. This discussion focusses on the engineering

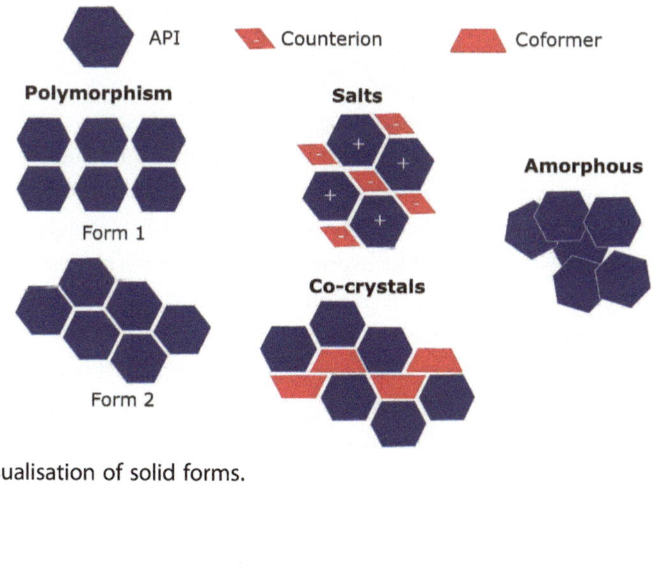

Fig. 1 Visualisation of solid forms.

of crystalline products not amorphous products. If we consider a crystalline solid, the individual molecules of the chemical entity arrange and pack in a repeating fashion within a 3D lattice. This repeating arrangement of molecules can be defined by a unit cell, the minimum size unit that contains all chemical elements to describe the crystalline structure when the unit cell is repeated ad infinitum along the 3-directions of 3D space, typically the x-, y- and z-axes. It is equally possible that the same chemical entity upon crystallisation may arrange in a different arrangement within the crystalline lattice and would therefore be a different crystalline form, a *polymorph* of the chemical entity. Fig. 1 provides a visual representation of two different polymorphic forms of the same chemical entity. When a chemical entity crystallises with no solvent or water molecules in the crystalline lattice, the resulting polymorph may be called a non-solvated form or an anhydrate. However, the interpretation for an anhydrate may differ between researchers, and for some specifically describes a solid form that has been prepared through the dehydration of a hydrate. Examples of anhydrous polymorphs would include Form 1 and Form 2 of paracetamol [6]. Both forms are layered crystal structures, Form 1 crystallises in the monoclinic space group, $P2_1/n$ and is the thermodynamically stable and well known marketed form. The more elusive Form 2 crystallises in the orthorhombic space group *P*bca, which displays better compaction properties and would be the preferred form for tabletting but is difficult to obtain [7].

A *hydrate* is a crystalline solid that incorporates water molecules within the crystalline lattice. Hydrates can exist as stoichiometric hydrates (i.e., monohydrate, dihydrate and other higher order hydrates) or as non-stoichiometric hydrates. In these cases, the water content within the structure can vary and this may be related to the crystal structure. For example in channel hydrates, where water is populating channels propagating within the crystal structure, the water occupancy of the channels can vary with varying ambient humidity conditions [8]. A *solvate* is a crystalline solid where solvent molecules are retained within the structure, much like a hydrate. Similarly, solvates can exist as stoichiometric and non-stoichiometric. In addition it is possible and indeed relatively commonplace to have two or more solvents present within a crystalline lattice, of which one may be water, often referred to as mixed solvate species.

Many APIs developed are acidic or basic in nature. This characteristic can be exploited in the formation of pharmaceutical *salts*. A salt relies on the formation of the charged species of the drug molecule and that undergoing reactive crystallisation with a corresponding acid or base. The driving force in salt formation is the electrostatic attraction of the two entities resulting in a neutral complex. To achieve charge balance upon formation, the salt will be inherently stoichiometric in nature. The salt former may be inorganic (e.g., HCl to form hydrochloride-, sulfuric acid for sulphate- or NaOH to form sodium-salts) or the salt former may be organic (e.g., citric acid to form citrate- or mesylic acid to form mesylate-salts). Salt formation is often considered in early development as one of the primary routes for increasing the aqueous solubility of poorly soluble drug molecules. Other physicochemical reasons for targeting a salt form include decreasing the hygroscopicity of an API (the propensity for absorbing water), increasing the thermal stability, as a route to crystallise poorly crystalline or amorphous drug molecules and to identify a salt form with a low propensity for polymorphism, where the free form of the API may have a complicated polymorphic landscape.

Whereas pharmaceutical salts have been accepted in the pharmacopoeia and many blockbuster drugs are on the market are salt forms; there is another group of solids, pharmaceutical co-crystals, which are an alternative approach to addressing many of the above challenges and often overlooked by the pharmaceutical innovator space. Draft guidance on the use of co-crystals released by the FDA in 2011 caused consternation in the assignment of co-crystals as a drug intermediates. Subsequently, a collaboration of industry and academic experts suggested that co-crystals should be classified as salts and provided the definition of a co-crystal as '*solids that are crystalline single phase materials composed of two or more different molecular and/or ionic compounds generally in a stoichiometric ratio which are neither solvates nor simple salts*' [9]. If one of these entities is an API, we may consider the resulting phase to be a pharmaceutical co-crystal [10]. Following this definition, co-crystals may be sub-divided into two groups; molecular co-crystals (MCC) and ionic co-crystals (ICC). Molecular co-crystals are single crystalline phases resulting from the interaction of two neutral organic entities by non-covalent interactions (hydrogen bonding, π-π stacking and van der Waals interactions) and includes the majority of pharmaceutical co-crystals. The term ionic co-crystals came more into use from circa 2010 [11] for uses where there is an ionic coformer present as a salt. This may be an inorganic salt (e.g., NaCl) or acid salts (e.g., sodium benzoate). In these cases, the single crystalline phase is driven by the formation of, for example, hydrogen bonds between the neutral coformer with the salt, visualised in Fig. 2.

With the benefit of an additional component present, ICCs offer even greater diversity in tuning physicochemical properties and can potentially display a new set of physicochemical properties superior to the pure drug and/or other multicomponent forms.

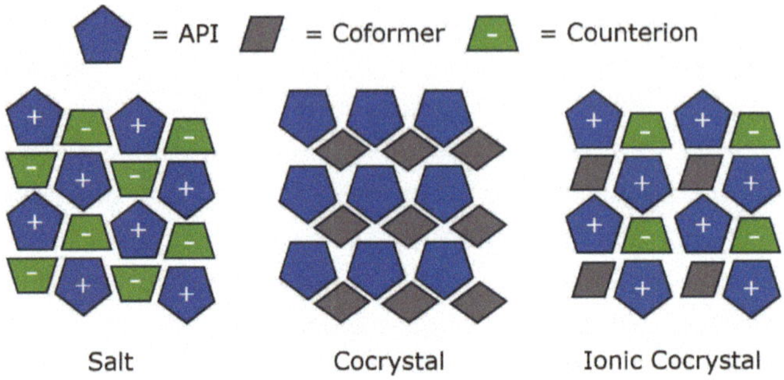

Fig. 2 Visualisation of an API Salt, co-crystal and ionic co-crystal (Left, Center, Right).

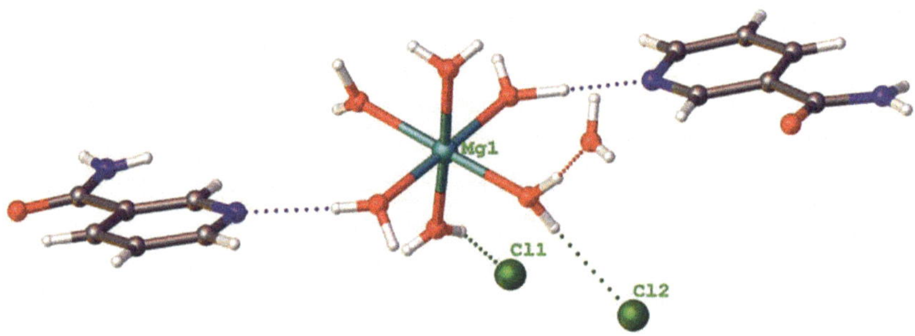

Fig. 3 Single crystal of $MgCl_2$-nicotinamide ionic co-crystal.

Shown in Fig. 3 is an example of an ionic co-crystal nicotinamide-magnesium chloride heptahydrate prepared by storing nicotinamide and $MgCl_2$ at elevated humidity. The Mg^+ ion is coordinated to 7 water molecules which in turn are hydrogen bonded to 2 Cl^- ions, and the neutral nicotinamide.

A further example of ICCs is provided by Braga et al., who successfully produced a series of organic–inorganic ICCs with barbituric acid. While barbituric acid is not pharmaceutically active in its own right, it is the parent structure of a family of API derivatives with hypnotic and anticonvulsant properties [12]. In the ICCs reported barbituric acid acts as the neutral molecule, while a selection of bromide and iodide salts (LiBr, NaBr, KBr, RbBr, CsBr and CsI) form the cation/anion pair. In the set reported, researchers showed that the intrinsic dissolution rates (IDRs) of three ICCs were higher than for pure barbituric acid (Form 2). Braga et al. further provided evidence that the thermal stability of the ICC could be increased or reduced in comparison to pure barbituric acid, depending on the cation/anion pair incorporated [12].

While the vast range of conceivable solid forms that ICCs and co-crystals can offer is very exciting, the added diversity may have the knock-on effect of increasing complexity for the crystallisation process. Selection of the right solid form is paramount to a potential drug successfully reaching the market. Therefore, a robust understanding of co-crystal screening and selection will form an essential aspect in identifying the most suitable multicomponent form for subsequent scale-up.

Co-crystals may be viewed as multi-component molecular crystals containing (net) neutral molecules/entities in a stoichiometric combination. To facilitate the crystallisation of such phases, one can foresee that interactions will inherently involve weaker non-covalent interactions and co-crystal chemistry is often recognised as a branch of supramolecular and crystal engineering. Through co-crystal formation, the hydrogen bond plays a key role in structure. However, further supramolecular interactions including π-π stacking, van der Waals interactions and halogen–hydrogen interactions may also be crucial in stabilisation of the crystal structure.

Pharmaceutical co-crystals are attractive and appealing to innovators and generics producers alike in their application of lifecycle management, where co-crystals offer a solid form that is non-obvious.

2.1 History and discovery

2.1.1 Crystal engineering

The term 'crystal engineering' was first used by Pepinsky in 1955 in his attempts to exploit complex ions for structure solution using direct methods with optically active ions [13]. The first reported co-crystal, however, between quinone and hydroquinone (known as quinhydrone, Fig. 4), was studied by Friedrich Wöhler in 1844 [14]. He correctly identified this material to be made up of a 1:1 mixture of the two components, however, it was only in 1958 that the crystal structure (and therefore its co-crystal nature) was confirmed by single crystal X-ray diffraction [15].

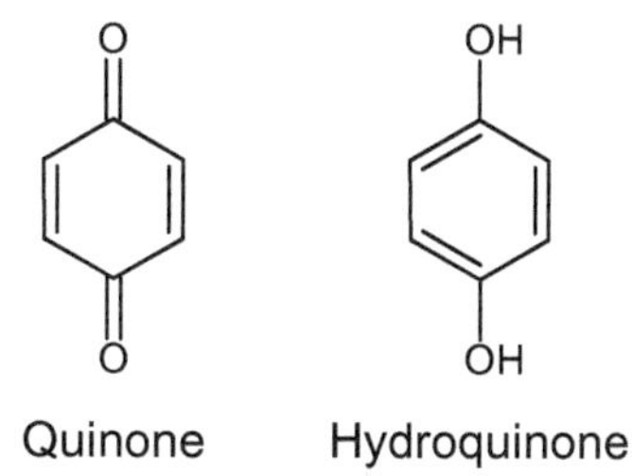

Fig. 4 Structures of first reported co-crystal.

While the first application of crystal engineering, in the modern day sense, was in 1964 by Schmidt in his work on cinnamic acid derivatives (Section 2.1.2), [16] it was not until the late 1980s that the scientific discipline developed further. From this stage efforts to move towards the design of organic solids from molecular building blocks, using robust intermolecular interactions, began to develop.

2.1.2 Cinnamic acid example

Different crystal forms can have different physicochemical properties and molecules packing in a different fashion can also potentially alter reaction pathways. This is exemplified in the classic example of cinnamic acid, whose different polymorphs can form a centro-symmetric dimer, a mirror symmetric dimer or have no reaction when their crystals are exposed to UV light (shown in Fig. 5) [17,18]. Studies such as this, whereby controlling the crystal packing directly affects the reactivity, have given rise to the field of crystal engineering—a scientific discipline that concerns the design and synthesis of solids with predefined architectures and properties [19].

Following Wohler's initial discovery of a co-crystal in 1844 (albeit its nature was not known at the time), numerous co-crystals were discovered in the early 1900s. A collection of these was published by Paul Pfeiffer in 1922 [20]. The first patent for a co-crystal was published in 1937 (described in Ref. [21]) but the term was not used again until 1967 for adenine and thymine derivatives. Its use in general scientific parlance was greatly influenced by Margaret (Peggy) Etter and her work during the 1990s [22].

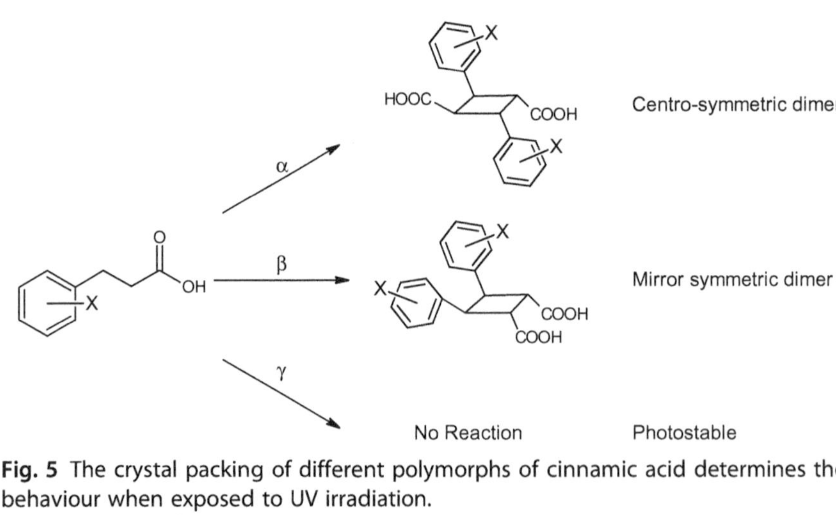

Fig. 5 The crystal packing of different polymorphs of cinnamic acid determines their behaviour when exposed to UV irradiation.

There has been much debate as to the specific definition of co-crystals, initialised when Desiraju expressed his preference for it to be 'a multi-component system held together by non-covalent interactions'. This was disputed by Jack Dunitz who realised that this would also encompass solid solutions and amorphous solids among other systems not typically thought of as co-crystals. A consensus began to appear in 2012 with the publication of the highly cited 'Polymorphs, Salts, and Cocrystals: What's in a Name?' co-authored by many of the leading names in the solid-state chemistry area [9]. However, many still questioned the regulatory stance of bodies such as the FDA, mainly as there is such a great deal of overlap between salts, co-crystals, hydrates and solvates. The academic community believed that 'co-crystals should naturally be grouped with salts' whereas the FDA have since expressed that they will treat co-crystals as polymorphs following the guidance issued in February 2018. Somewhat confusingly, the European Medicines Agency (EMA) have grouped salts, co-crystals and polymorphs together which appears to better fit with the academic communities' general consensus [23].

These aspects will be discussed in greater detail in later sections.

2.2 The salt co-crystal continuum

While the guidance given by the FDA allows for pharmaceutical salts and co-crystals to be distinguishable from a regulatory perspective, in reality, and from a scientific standpoint there exists a continuum between salt and co-crystal formation [24].

Both salts and co-crystals are considered multicomponent crystal systems; they are generally distinguished by the degree of proton transfer between the acid and base components in the crystalline state. At one end of the continuum, proton transfer from an acid to the base molecule is complete. In this case, the system is classified as a salt. At the opposing end, proton transfer between the proton donor and acceptor is absent and the multicomponent system corresponds to a co-crystal.

One common method used to predict whether a multicomponent drug forms a salt or a co-crystal in the crystalline state is to compare the pKa values of the API and the guest molecule. In the so-called 'pKa rule' it is generally accepted that a salt will form where the ΔpKa ($\Delta pKa = pKa_{base} - pKa_{acid}$) is greater than 3 [24]. If the ΔpKa is less than 0, formation of a co-crystal is most likely [25]. This criterion is often used as a rule of thumb when selecting potential counterions or coformers for screening purposes.

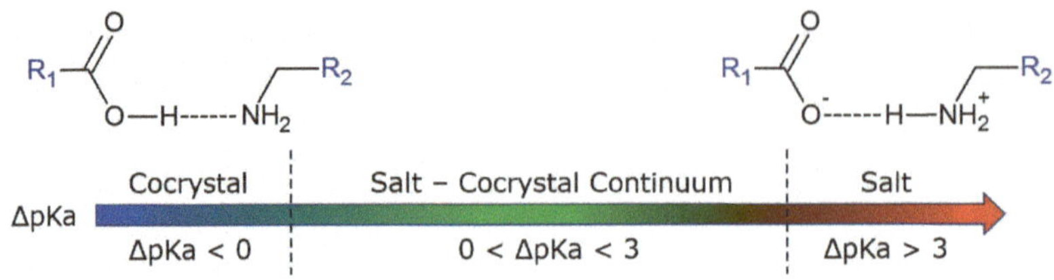

Fig. 6 The salt-co-crystal continuum.

While most crystalline acid–base complexes are expected to fall into one category or the other, a ΔpKa between 0 and 3 results in some uncertainty. Within this region, the pKa rule cannot be used to predict whether a salt or co-crystal will form (Fig. 6).

Under these circumstances, the formation of a salt or co-crystal will be dependent on both the ΔpKa and the crystalline environment. [24,26] Examples of this can be seen in Table 1 which shows the multicomponent systems obtained with pyrimethanil generated in house [27]. It can be seen that when the coformer has a pKa in the region of 3.57–2.91 that both salts and co-crystals can be formed, and with 3-nitrobenzoic acid, both can be formed. It is also possible to form a co-crystal/salt, where both a neutral and charged coformer are present and interacting with the API.

Classification of multicomponent crystals which exist inside the 'continuum' can be particularly problematic. For instance, the extent of charge transfer in a crystal can be variable and in some cases external stimuli (such as temperature) can influence the degree of proton transfer, resulting in ambiguous multicomponent crystals that cannot be easily classified, such as the case of urea-phosphoric acid [28]. Some crystals in the continuum also show switchable behaviour, where a co-crystal may transform to a salt form, or vice versa [29]. While sorting systems which exist within the salt-co-crystal continuum poses a significant challenge, analytical techniques such as single-crystal X-ray diffraction (SCXRD), powder X-ray diffraction (XRPD), neutron diffraction and solid-state NMR (SSNMR) have been used to interpret the structure and intermolecular interactions within the solid form. However, preparation of samples suitable for study by these techniques is not always feasible [30]. As such, characterisation of such crystals remains a difficult task and an active area of study.

Multicomponent solid forms are not limited to salts, co-crystals and solvates (as well as hydrates) which are a commonplace in the solid state but are not mutually exclusive. Other classes of crystalline materials include solvated co-crystals [31,32], solvated salts [33,34] and ionic co-crystals [35] It should also be noted that multicomponent crystals are as capable of displaying

Table 1 Pyrimethanil salts and co-crystals.

Coformer	Ratio	Product	Solvent	Coformer pKa
Benzoic acid, Form 1	1:1	Co-crystal	IPA	4.20
Benzoic acid, Form 2	1:1	Co-crystal	DMSO	4.20
Succinic acid	2:1	Co-crystal	DMSO	4.20
2-Fluorobenzoic acid	1:1	co-crystal	DMSO	4.20
Biphenyl carboxylic acid	1:1	Co-crystal	IPA	4.19
2-Methoxybenzoic acid	1:1	Co-crystal	DMSO	4.09
Pyrimethanil, Form 1	N/A	Freebase	IPA	4.06
Pyrimethanil, Form 2	N/A	Freebase	IPA	4.06
4-Iodobenzoic acid	1:1	Co-crystal	IPA	4.02
4-Chlorobenzoic acid	1:1	Co-crystal	DMSO	3.97
Diphenylacetic acid	1:1	Co-crystal	IPA	3.94
3-Chlorobenzoic acid	1:1	Co-crystal	DMSO	3.83
Formic acid	1:1	Co-crystal	Neat	3.77
1-Naphthoic acid	1:1	Co-crystal	IPA	3.69
Thiazole-4-carboxylic acid	1:1	Salt	DMSO	3.57
3-Nitrobenzoic acid, Form 1	1:1	Co-crystal	IPA	3.45
3-Nitrobenzoic acid, Form 2	1:1	Salt	IPA	3.45
4-Nitrobenzoic acid	1:1	Intermediate	DMSO	3.44
4-(Methylsulfonyl)benzoic acid	1:1	Intermediate	IPA	3.42
R/S-Mandelic acid	1:1	Salt	EtOAc	3.41
R-Mandelic acid	1:1	Salt/Co-crystal	EtOAc	3.41
2,4 Dihydroxybenzoic acid	1:1	Salt	IPA	3.32
2,5 Dihydroxybenzoic acid	1:1	Salt	IPA	3.01
2,5 Dihydroxybenzoic acid: DMSO	1:1:1	Salt/Solvate	DMSO	3.01
DL- Tartaric acid:IPA	2:1:0.5	Salt/Co-crystal	IPA	2.96/4.24
2-Chlorobenzoic acid	1:1	Co-crystal	IPA	2.97
Pyrazinecarboxylic acid, Form 1	1:1	Co-crystal	DMSO	2.91
Pyrazinecarboxylic acid, Form 2	1:1	Co-crystal	DMSO	2.91
2-Bromobenzoic acid	1:1	Salt	DMSO	2.85
2-Nitrobenzoic acid	1:1	Salt	IPA	2.16
Maleic acid	1:1	Salt	IPA	1.92
Oxalic acid	4:3	Salt	IPA	1.27

polymorphism as single component crystals, each with unique physicochemical properties [36–38]. Published work by Grothe et al. provides a good guide for categorising a multicomponent crystal, but classification of solid forms remains hotly contested among the scientific community and a consensus has yet to be reached [26].

3. Intellectual property and regulatory perspective on co-crystals

Screening for the optimum solid form is now a routine part of the drug development process [39] from a clinical, regulatory, and legal aspect.

In an industry associated with risks and high development costs, pharmaceutical companies rely on strategic IP protection through patents. As soon as a molecule is found useful for a therapeutic treatment, its chemical structure is typically protected with the filing of a patent application. During the development phase, new opportunities for IP protection can later arise particularly related to the synthesis and the solid states of the API.

From a regulatory point of view [40], the FDA defines co-crystals as crystalline materials composed of two or more molecules within the same crystal lattice with a defined stoichiometry interacting non-ionically. Co-crystals can be viewed as a special case of solvates and hydrates, wherein the second component, the coformer, is not a solvent (including water), and is typically non-volatile [41]. The recommendations in this guidance apply to materials that the FDA has not previously evaluated and determined to be pharmaceutical co-crystals.

Co-crystals can be considered as novel solid entities which cannot be predictable with regard to their existence, structure and properties and, therefore, they may qualify as an invention if they are shown to be novel, inventive and industrially applicable [8,9]. The term novelty will apply to co-crystals that have never been made public in any way, this includes not being described before in the literature or at conferences. Co-crystals may be inventive if they display an attribute that provides a benefit or if their formation or existence is not obvious [42]. The utility could be linked to the improvement of properties that the co-crystals can provide over the current known polymorphs or salts of the API.

Pharmaceutical co-crystals now have a proven record of patentability and some have been successfully marketed: Depakote® [43], Lexapro® [44], Entresto® [45], Suglat® [46], Steglatro® [47], Odomzo® and agomelatine-citric acid co-crystal [48], which are summarised in Table 2. There are

Table 2 Pharmaceutical marketed co-crystal products.

Brand drug	Company	Marketed co-crystals	Marketed year	Therapeutic indications
Depakote®	Abbott Laboratories	Co-crystal of valproate sodium and valproic acid	1983	Treatment of epilepsy
Lexapro®	Lundbeck	Co-crystal of escitalopram oxalate salt and oxalic acid	2002	Treatment of major depressive and anxiety disorders
Suglat®	Astellas Pharma/Kotobuki Pharmaceuticals	Co-crystal of ipragliflozin and L-proline	2014	Treatment of type 2 diabetes
Entresto®	Novartis	Co-crystal of 2 actives: sacubitril and valsartan	2015	Treatment of symptomatic chronic heart failure
Odomzo®	Novartis/Sun Pharma	Co-crystal of sonidegib and phosphoric acid	2015	Treatment of cancers
Steglatro®	Pfizer	Co-crystal of ertugliflozin and L-pyroglutamic acid	2018	Treatment of type 2 diabetes
Generic	Generic Companies	Co-crystal of agomelatine and citric acid	2018	Treatment of major depressive and anxiety disorders

additional co-crystals that are currently being evaluated in clinical phases further supporting their applicability to patentability [49].

Interestingly Depakote®, which consists of a sodium valproate-valproic acid co-crystal, was originally described [43] as being an oligomer having a 1:1 M ratio of sodium valproate and valproic acid, the preparation of which gained further protection in 2000 (Fig. 7) [50].

The co-crystal nature of Escitalopram was only described in 2007, much later after its introduction to the market [51]. It may have been originally been regarded as a salt but due to the advancements in characterisation, it

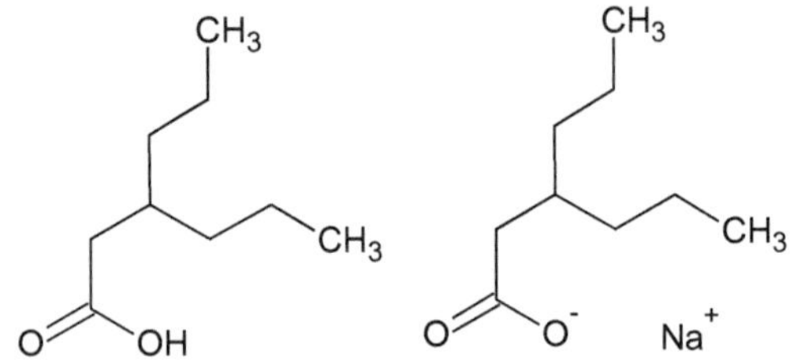

Co-crystal of valproate sodium and valproic acid

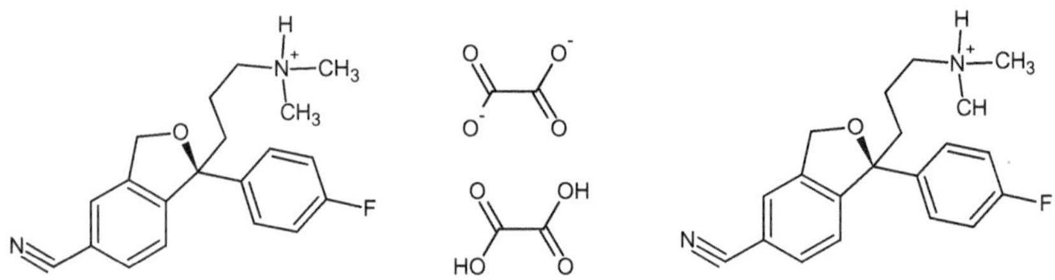

Co-crystal of escitalopram oxalate salt and oxalic acid:

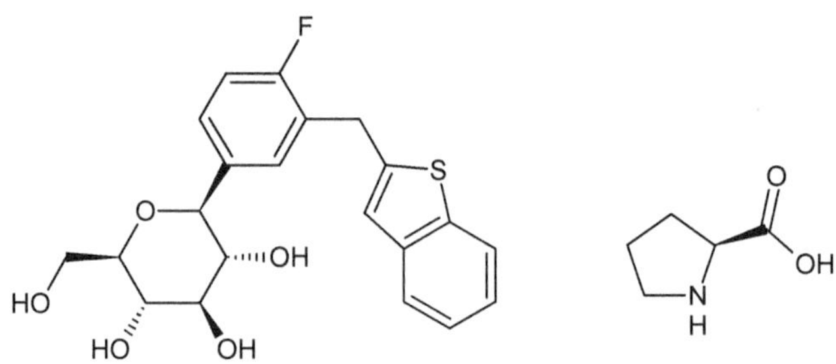

Co-crystal of ipragliflozin and L-proline:

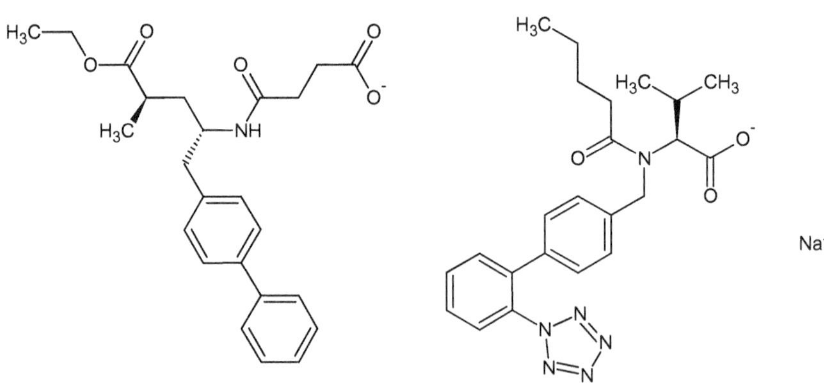

Co-crystal of sacubitril and valsartan

Fig. 7 Structures of co-crystals on the market.

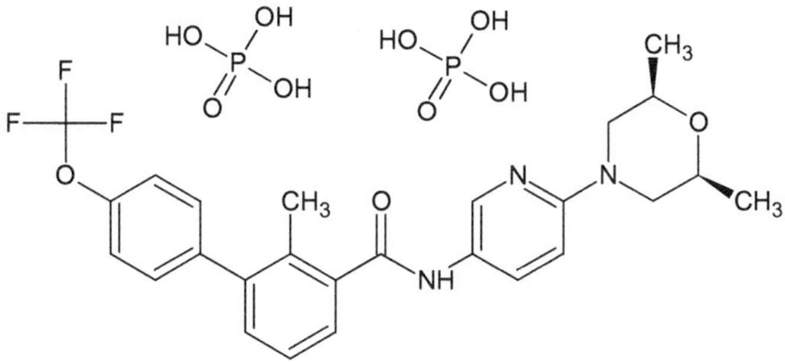

Co-crystal of sonidegib and phosphoric acid

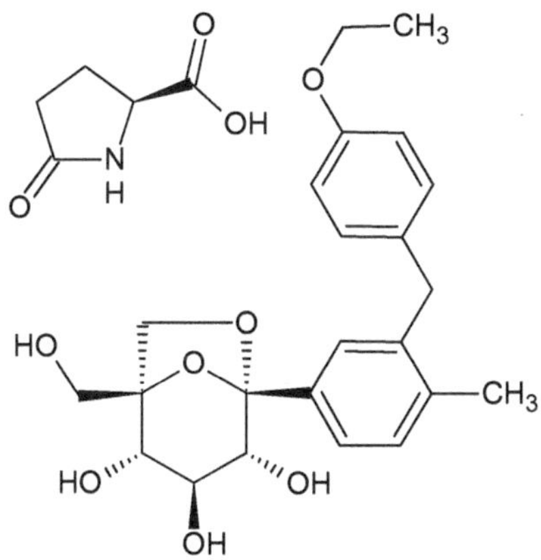

Co-crystal of ertugliflozin and L-pyroglutamic acid

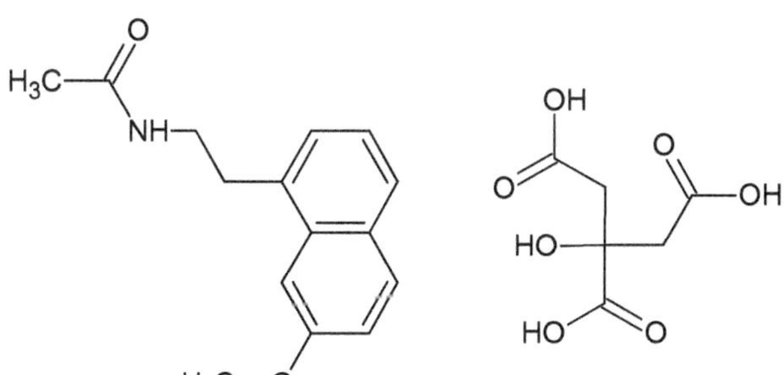

Co-crystal of agomelatin and citric acid

Fig.7—Cont'd

can now be assigned as a co-crystal. The crystalline API sold under the brand name Odomzo® is described as a phosphoric acid co-crystal in the EMA documentation [52] although no evidence on the nature of the crystal has yet been published.

When a co-crystal offers an improvement over the current solid form properties of the API, an additional patent protection may be obtainable even after the expiration of the patent of the chemical entity. For example, in patent US 8,097,592 B2 [46] the co-crystal between ipragliflozin, a C-glycoside derivative, and L-proline is claimed. Ipragliflozin has the tendency to form a clathrate hydrate which can also incorporate the organic solvent from the manufacturing process. This clathrate can contain variable amounts of water and its weight can vary by up to 3% w/w depending on the humidity conditions. A screen with 33 amino acids was performed and only one co-crystal was identified with L-proline. The co-crystal displayed improved hygroscopic behaviour when compared with the ipragliflozine clathrate (summarised in Table 3) and was the form selected for the drug Suglat®, commercialised by Astellas.

A solid form identified as a co-crystal with a pharmaceutically acceptable coformer will have a regulatory classification similar to that of a polymorph of the API, as a salt will have a different regulatory pathway.

For the EMA, co-crystals are considered to share many conceptual similarities with salts and therefore similar principles for documentation should be applied [53].

The selection of the coformers for a co-crystal screen is crucial. It could be based on previous experience on co-crystal formation for an API or related analogue and/or guided by in silico tools [54] with a supramolecular

Table 3 Comparison of properties of ipragliflozin clathrate and ipragliflozin L-proline co-crystal.

Solid form	Hygroscopicity	DSC	Variable humidity XRPD
Ipragliflozin clathrate	3.1% w/w change between 5 and 95% RH with step change between 30% and 35% RH—hysteresis	Sharp endotherm at 158°C (peak temperature)	Change in XRPD with humidity (peak shifts)
Ipragliflozin-L-proline co-crystal	0.7% w/w gradual change between 5% and 95% RH—no hysteresis	Sharp endotherm at 211°C (peak temperature)	No change in XRPD with humidity

synthon approach, for example, data mining/analysis from the Cambridge Structural Database (CSD) [55]. Alternatively, the coformers can be selected by considering the Hansen solubility parameter, which can be a useful tool in matching the solubility of the API with prospective coformers for potential screening [56]. For a screen targeting a final API, the coformers should be pharmaceutically acceptable and be recognised as Generally Recognised As Safe (GRAS) [57] or be included in the Substances Added to Food Inventory [58]. A pool of around 3000 substances is available [4], however a much smaller subset of chemicals will be suitable and appropriate for co-crystal formation. Pharmaceutically acceptable acids or bases can be a good starting point [59] for coformer selection if the API is not expected to form a salt. The Inactive Ingredient Database (IID) from the FDA contains information on the highest level allowed of each excipient in approved drugs. It is a resource that is regularly updated and can be used to refine the number of coformers for a screen, based on the anticipated amount of coformer in the final drug product. From the patent literature, the most common coformers are small bis-carboxylic acids such as maleic-, malonic-, oxalic- and succinic-acids. A non-exhaustive list of successful coformers found in the patent literature is provided in Table 4.

As the FDA has defined co-crystal as similar to polymorphs of an API for a regulatory point of view, this has opened new opportunities for generic companies for generating their own IP. In recent years, the number of patents related to co-crystals has increased dramatically to allow API manufacturers or generic companies to attempt to protect their own 'polymorphs' of an API before the brand polymorph patents have expired [60].

In addition to gaining IP protection, a co-crystal can allow the generic solid form to be approved through the submission of an ANDA. A new drug application(NDA) is a typical pathway for bringing a new drug to the market whereas an abbreviated new drug application (ANDA) is utilised for the development of generic drug products and should provide evidence to demonstrate that both the API and coformers are present in the unit cell as well as evidence that API and coformer interact non-ionically if ionisable functional groups are present (spectroscopic data could be provided) and assurance that substantial dissociation of the API from its co-crystal form occurs before reaching the site of pharmacological activity.

For the solid form to qualify for approval by the ANDA process, the final (generic) drug needs to display a similar dissolution profile [61] to the Reference Listed Drug (RLD) therefore the co-crystal may need to be formulated appropriately in the drug product. The added weight due to the

Table 4 Typical FDA approved coformers.

1-Hydroxy-2-naphthoic acid	Lactose
2,5-Dihydroxybenzoic acid	L-aspartic acid
3-Hydroxy-2-naphthoic acid	L-malic acid
4-Hydroxybenzoic acid	L-phenylalanine
Adipic acid	L-pipecolic acid
Ammonium chloride	L-proline
Benzenesulfonic acid	L-pyroglutamic acid
Benzoic acid	Maleic acid
Caffeine	Malonic acid
Camphoric acid	Meloxicarnimaltol
Cholic acid	Nicotinamide
Citric acid	Nicotinic acid
DL-malic	Orotic acid
DL-pipecolic acid	Oxalic acid
D-pipecolic acid	p-toluene sulphonic acid
D-Proline	Pyrazine-2-carboxylic acid
Ethanesulfonic acid	Pyrazole
Ethyl maltol	R-mandelic acid
Fumaric acid	Saccharin
Glycolic acid	Salicylic acid
Hydrocinnamic acid	S-mandelic acid
Hydroxy-2-naphthoic acid	Succinic acid
Isonicotinamide	Urea

coformer must also be taken in account as the drug product (tablet, capsule, etc.) is required to have similar attributes to the RLD [62]. The stability of the co-crystal will also need to be demonstrated and particular attention should be taken to ensure co-crystal retention during subsequent formulation processes.

A non-exhaustive list of API co-crystals patented by generic companies can be found in Table 5.

Table 5 Examples of generic co-crystals on the market.

API names	Structure	Applicant	Patent reference	Coformer listed	First API approval year
Agomelatine	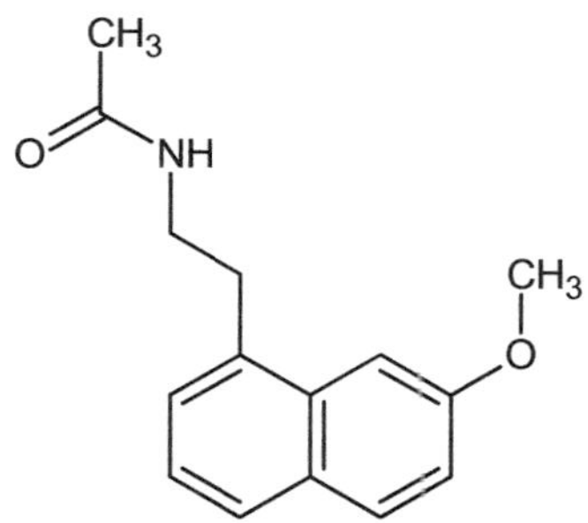	Zentiva	EP2517700A1	Citric acid, benzenesulfonic acid and maleic acid	2009
		Ratiopharm	WO2013082302A1	Urea	
Apremilast	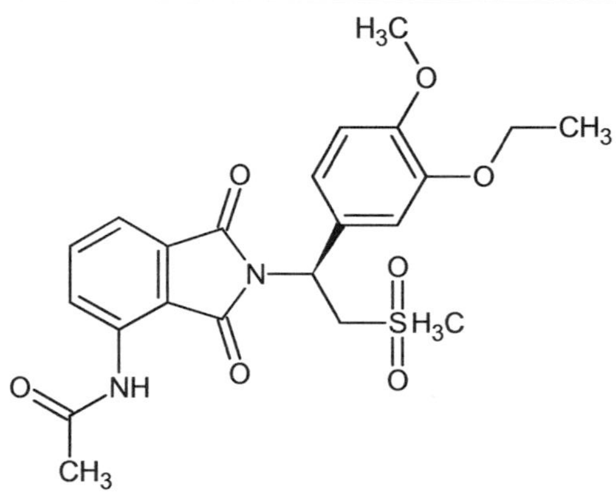	Zentiva	CZ2017390A3	Benzoic acid, nicotinic acid, nicotinamide	2014

Continued

Table 5 Examples of generic co-crystals on the market.—cont'd

API names	Structure	Applicant	Patent reference	Coformer listed	First API approval year
Baricitinib	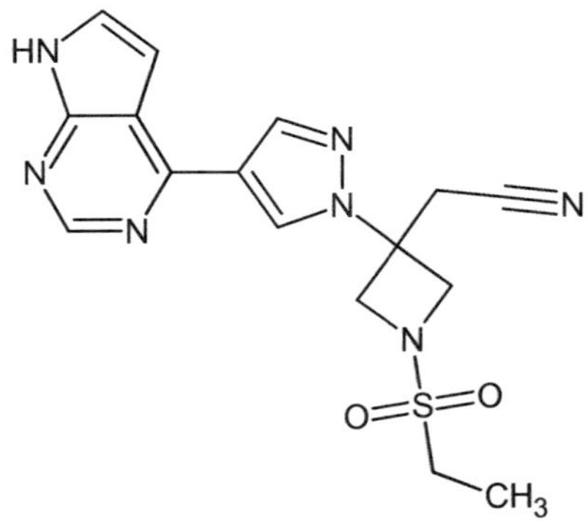	Zentiva	CZ31155U1	Benzenesulfonic acid, citric acid, maleic acid, isonicotinamide, saccharin and oxalic acid	2017
Bexagliflozin	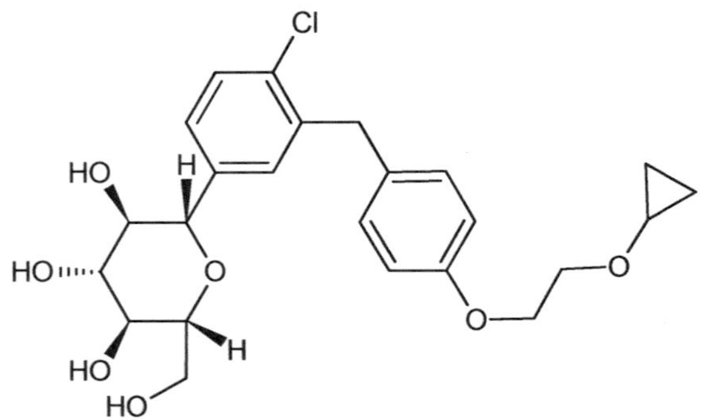	Solipharma	CN108239055A	L-aspartic acid	Under investigation

Canagliflozin	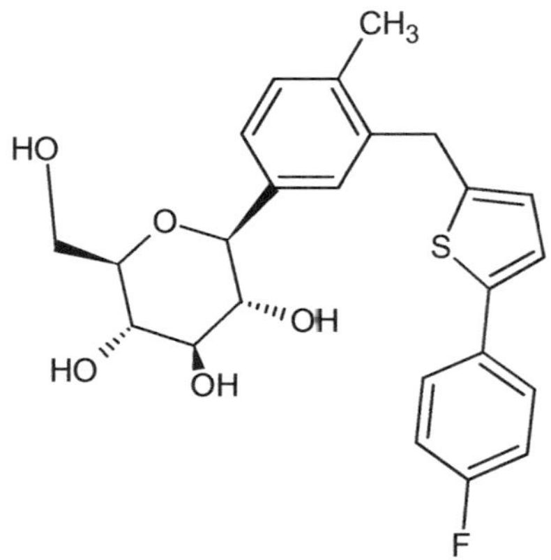	Laurus Labs	WO2017046730A1	DL-pipecolic acid, L-pipecolic acid, D-pipecolic acid, pyrazine-2-carboxylic acid and pyrazole	2013
Dapagliflozin	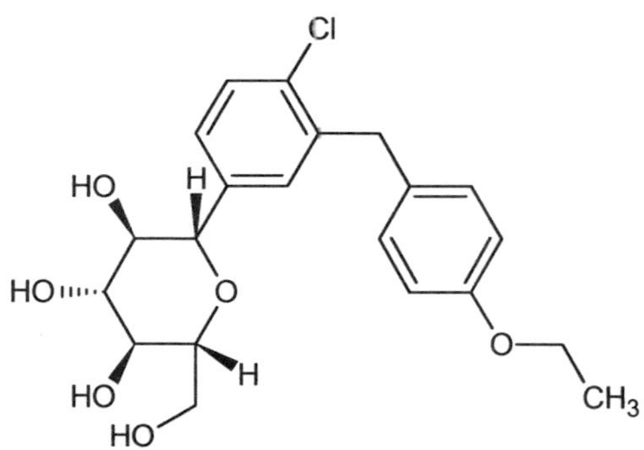	Zentiva	CZ2015834A3	L-proline and L-phenylalanine	2012
		Shanghai Sun-Sail	CN105524033A	Fumaric acid	
		Sun Yat-Sen Univ	WO2015198227A1	Citric acid	
		Laurus Labs	WO2017046730 A1	DL-pipecolic acid, L-pipecolic acid and D-pipecolic acid	

Continued

Table 5 Examples of generic co-crystals on the market.—cont'd

API names	Structure	Applicant	Patent reference	Coformer listed	First API approval year
Empagliflozin	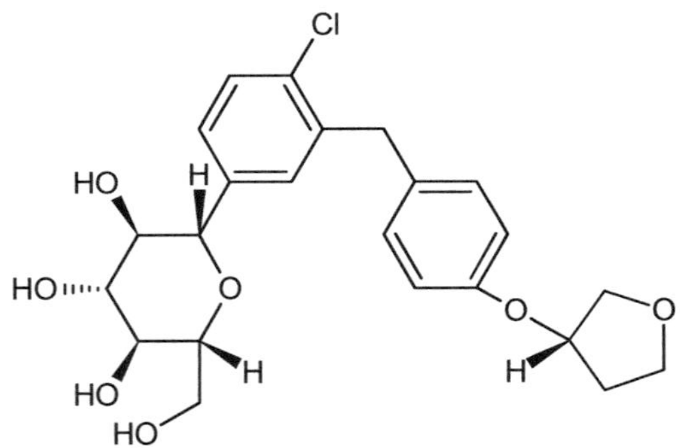	Cadila Healthcare	US2017247356A1	L-proline	2014
		Zentiva	CZ2015110A3	L-proline	
		Laurus Labs	WO2017046730AI	DL-pipecolic acid	
Entinostat	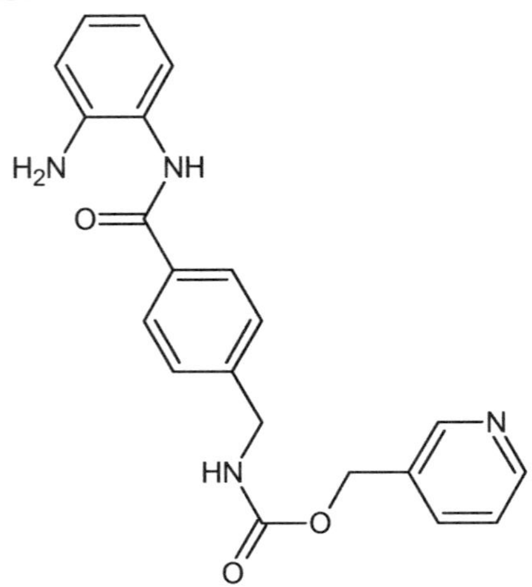	Solipharma LLC	WO2019019130A1	Fumaric acid, maleaic acid, succinic acid and adipic acid	Under investigation

Etoricoxib	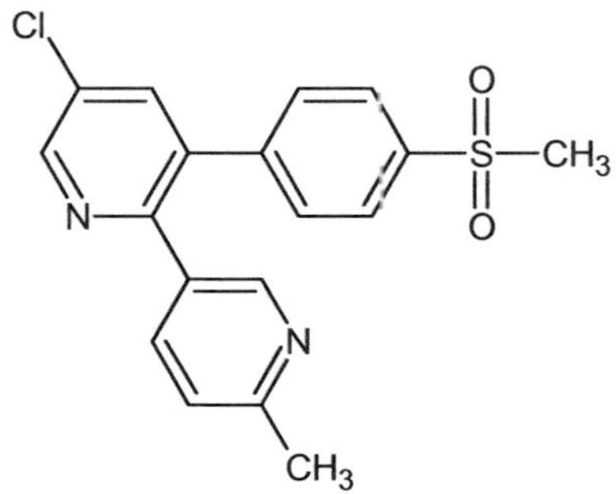	Zentiva	EP2601952A1	Succinic acid, adipic acid, fumaric acid, benzoic acid, salicylic acid, 1-hydroxy-2-naphthoic acid, 3-hydroxy-2-naphthoic and saccharin	2002
Fruquintinib	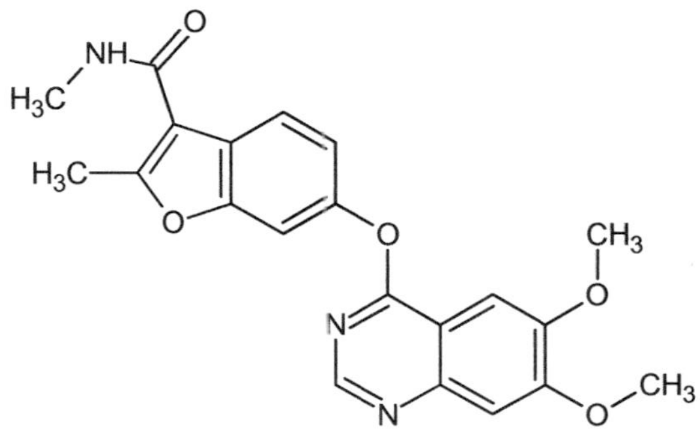	Solipharma	WO2019183916A1	Maleic acid, malonic acid and saccharin	2018
Galunisertib	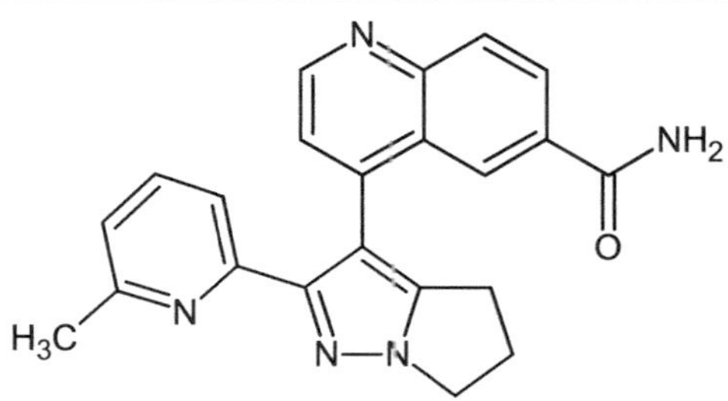	Solipharma	US2020017470A1	Succinic acid and fumaric acid	Under investigation

Continued

Table 5 Examples of generic co-crystals on the market.—cont'd

API names	Structure	Applicant	Patent reference	Coformer listed	First API approval year
Ibrutinib	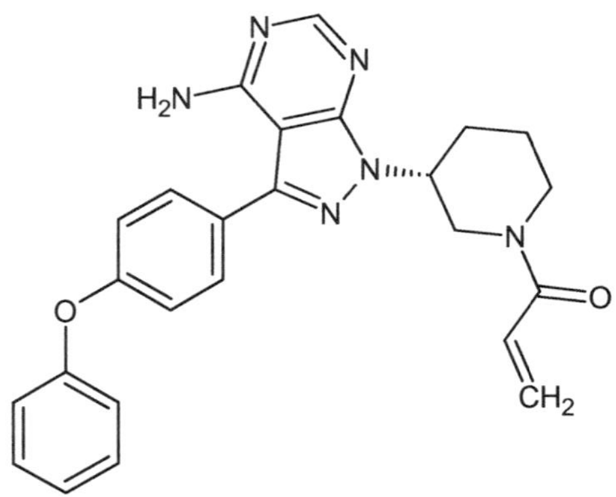	Ratiopharm	WO2016156127A1	Benzoic acid, fumaric and succinic acid	2013

Ivabradine. HCl	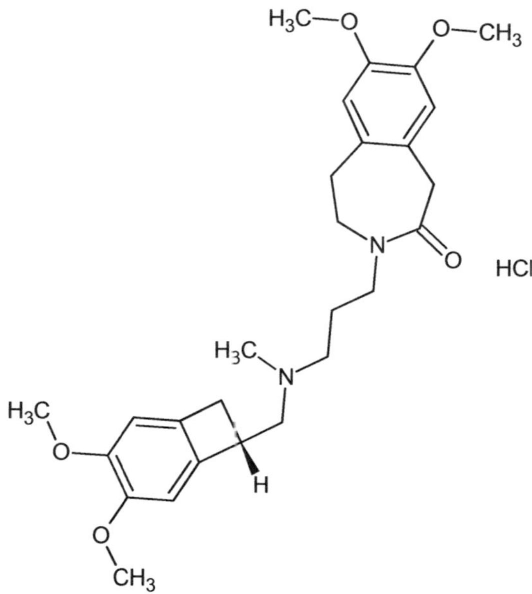	Zentiva	WO2016004911A1	R-mandelic acid	2005
			WO2015048937A1	S-mandelic acid	
Lenalidomide	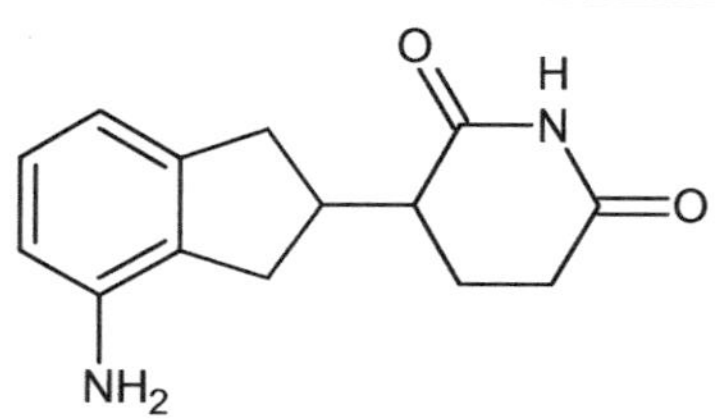	Amplio Pharma	WO2013012485	Ammonium chloride	2005

Continued

Table 5 Examples of generic co-crystals on the market.—cont'd

API names	Structure	Applicant	Patent reference	Coformer listed	First API approval year
Lesinurad	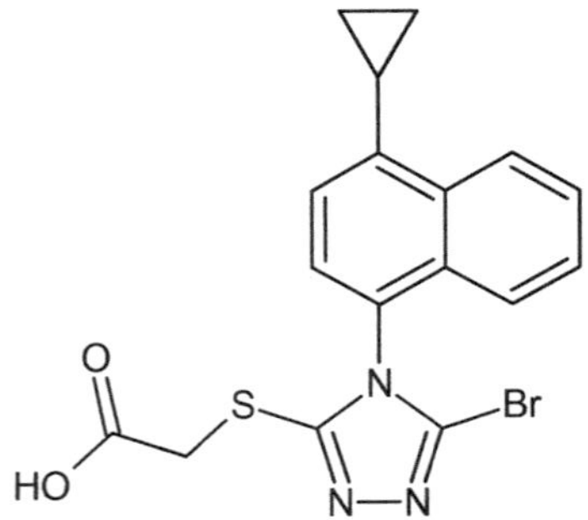	Zentiva	EP3372592A1	Urea	2015
Meloxicam	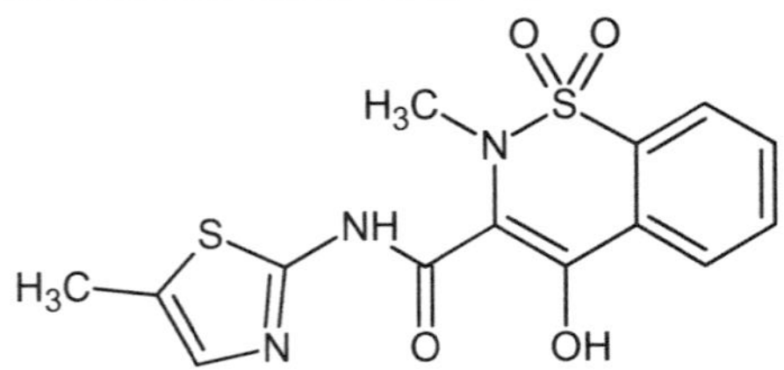	Mylan Laboratories	WO2020095316A1	2,5-dihydroxybenzoic acid, 4-hydroxybenzoic acid, salicylic acid, benzoic acid, succinic acid and hydrocinnamic acid	1998
		Thar Pharmaceuticals	US8,124,603B2	Fumaric acid, succinic acid, adipic acid, benzoic acid, DL-malic acid, L-malic acid, glutaric acid, acetylsalicylic acid, salicylic acid, 1-hydroxy-2-naphthoic acid, maleic acid, 4-hydroxybenzoic acid, malonic acid, glycolic acid, 2,5-dihydroxybenzoic acid, camphoric acid, camphoric acid, ethyl maltol and hydrocinnamic acid	

Obeticholic acid	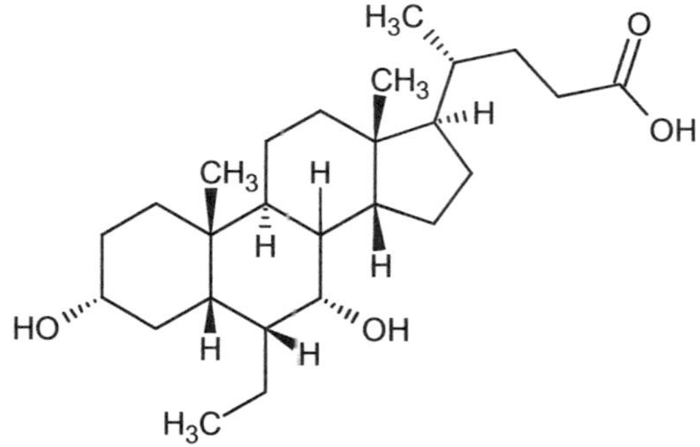	Zentiva	CZ2017646A3	Benzoic acid	2016
Ribociclib	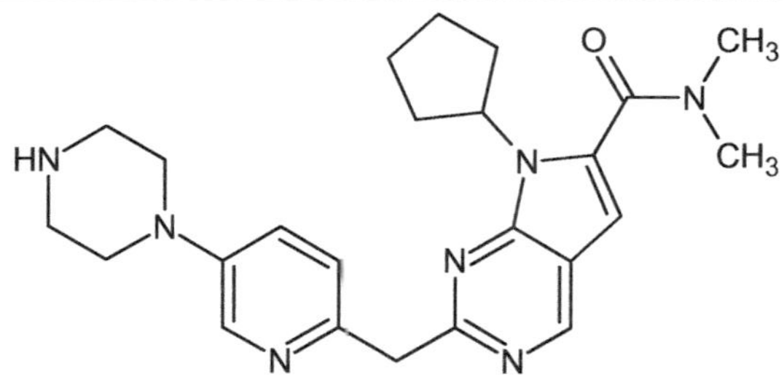	Solipharma	WO2019062854A1	Saccharin, cholic acid, orotic acid, succinic acid and citric acid	2017
Roxadustat	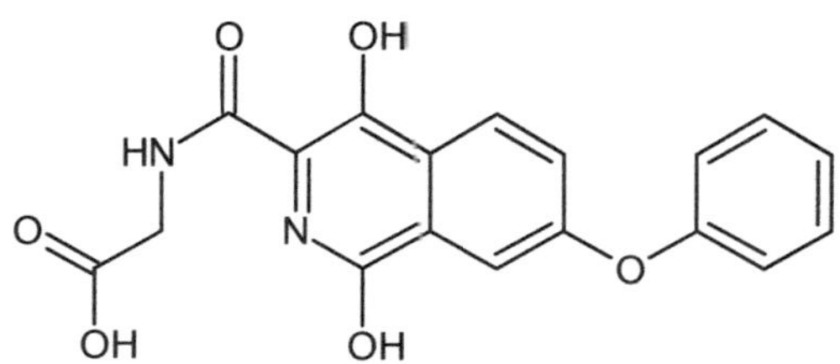	Zentiva	WO2019042485A1	Caffeine	2018
		Sandoz	WO2019042641	L-proline	

Continued

Table 5 Examples of generic co-crystals on the market.—cont'd

API names	Structure	Applicant	Patent reference	Coformer listed	First API approval year
Sofosbuvir	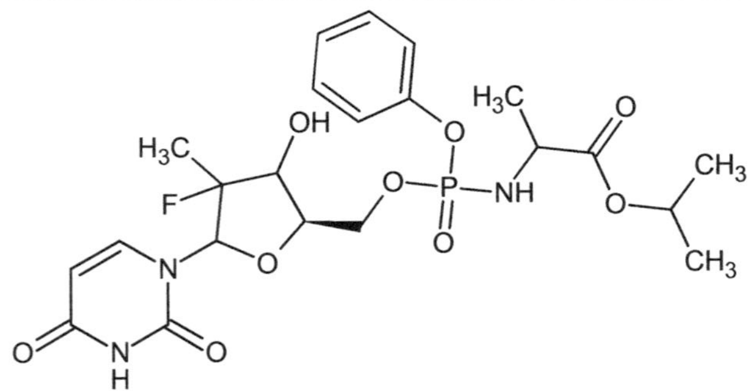	Cadila Healthcare	WO2016042576A1	L-proline	2013
		Ratiopharm	WO2017029408A1		
Sorafenib	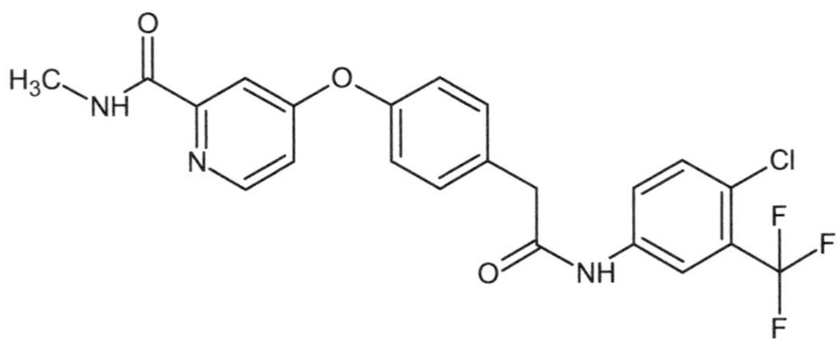	Ratiopharm	WO2010142678A2	p-toluene sulphonic acid and benzenesulphonic acid	2005

Tenofovir disoproxil	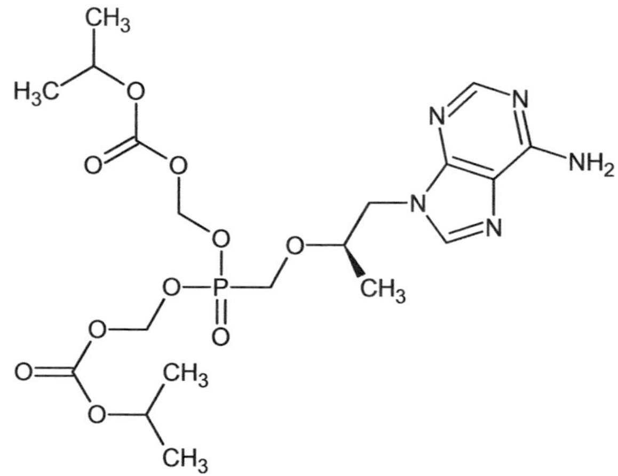	Ultimorphix Technologies	US20090270352A1	Fumaric acid	2001
Ticagrelor	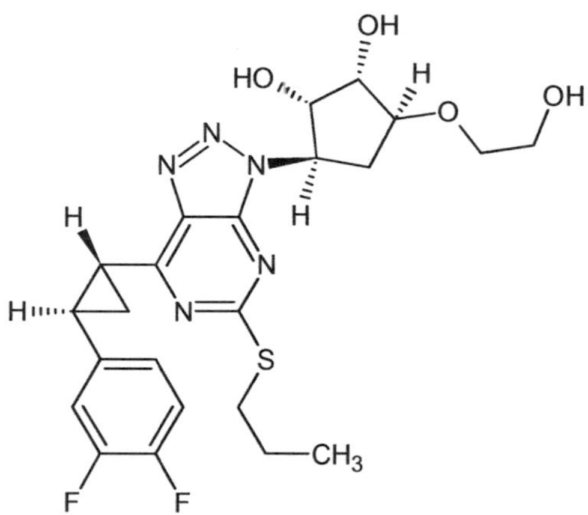	Zentiva	WO2014000719A1	3-hydroxy-2-naphthoic acid	2010

Continued

Table 5 Examples of generic co-crystals on the market.—cont'd

API names	Structure	Applicant	Patent reference	Coformer listed	First API approval year
Tiotropium bromide	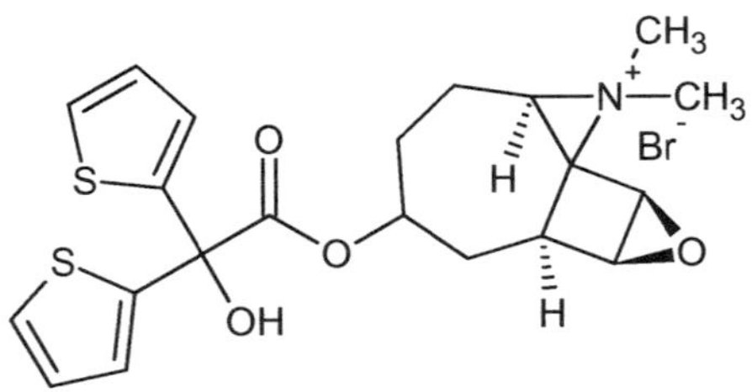	Euticals SpA	EP3110814A1	Lactose	2002
Rac-tramadol hydrobromide and celecoxib	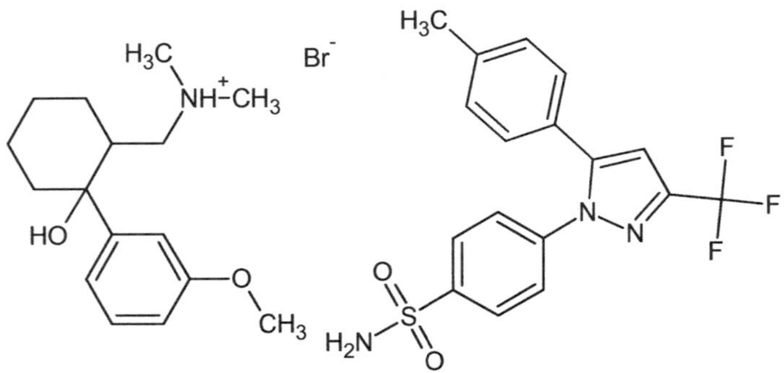	Mylan Laboratories	WO2020039456A1	(rac)-tramadol hydrobromide and celecoxib	Under investigation

Vadadustat	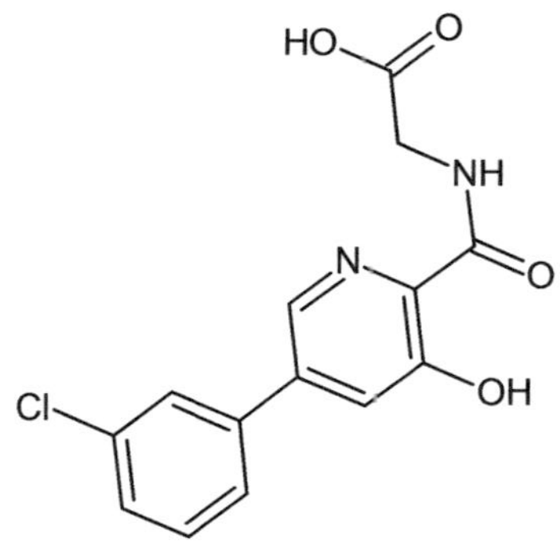	Mylan Laboratories	WO2020075199A1	Caffeine, nicotinamide, isonicotinamide, L-proline and D-proline	2020

In Europe, a generic version of Valdoxan® containing the active agomelatine in the form of a citric acid co-crystal (which is different from the branded product marketed by Servier) has been marketed since 2018 by several generic companies (Zentiva, Accord-UK, Aristo Pharma, Neuraxpharm Arzneimittel).

4. Generation, characterisation, and development of co-crystals

4.1 Designing screens and synthons

Co-crystals form as a result of directed and specific intermolecular interactions between an API and a suitable coformer. Several different types of intermolecular interactions exist such as halogen bonding, π–π interactions and weaker van der Waals forces, however, hydrogen bonding interactions dominate. Predicting hydrogen bonding interactions can be exceptionally challenging and thus several model compounds have been studied to understand the hierarchy of various hydrogen bonding interactions. Crystal engineering is used to design and predict connections between an API and coformer using the concept of supramolecular synthons. By establishing and possibly designing the connections between two or more molecules the supramolecular structure can be controlled to produce co-crystals with desired properties.

In principle, forming a co-crystal may seem trivial; identifying the potential hydrogen bond donors and acceptors of an API; and choosing a suitable coformer with a complementary hydrogen bonding motif. This unfortunately does not always work and other factors such as the location of the hydrogen bonding groups, the conformation of the molecule and competing interactions between like molecules need to be considered. Polymorphic APIs that display several conformational forms of the same molecule are more likely to form co-crystals as they are not locked into one conformation and packing arrangement [63]. They also typically have intermolecular interactions between adjacent molecules that are not particularly strong, making them more likely to interact with a second different chemical molecule. Knowledge of the intermolecular interactions that could occur between an API and coformer forms the basis for designing a pharmaceutical co-crystal.

The intermolecular interactions that occur between two molecules and intramolecular interactions that take place within the same molecule ultimately direct the overall packing arrangement of a crystalline solid.

This is made up of repeating units and thus the interactions between the molecules are also repeated. Hydrogen bonding is a highly directional type of intermolecular interaction and common hydrogen bonding motifs between different functional groups have been extensively studied over the last 25 years [64]. Herein, the focus will be on hydrogen bonded co-crystals but it is important to note that other interactions such as weaker C–H⋯N, C–H⋯O, C–H⋯F, C–H⋯Cl, C–H⋯π and π–π interactions are also important and can have a huge influence in terms of the overall packing arrangement and properties of a solid form.

Etter applied graph set notation to describe hydrogen bonded motifs and proposed rules on various outcomes depending on the hydrogen bond donors and acceptors present within a system [65]. Following these initial proposed rules, many different systems were studied to try understand the hierarchy of hydrogen bond donors and acceptors and build up an understanding of what is termed 'supramolecular synthons' to produce multicomponent complexes with favourable physical properties [66–68].

4.1.1 Supramolecular synthons

Desiraju coined the term 'supramolecular synthon' in the 1990s [69]. These are structural units within a 'supermolecule' (complexes of two or more molecules (often macromolecules) that are not covalently bonded), which can be formed by intermolecular interactions. They can be thought of in a similar fashion to conventional synthons in synthetic chemistry, such as those used for Corey's retrosynthetic analysis [70]. Supramolecular synthons can be used by the crystal engineer to design new complexes comprising more than one molecular entity, such as co-crystals. Both homo- and hetero-synthons are used; where homosynthons are entities sustained by identical chemical functional groups and heterosynthons are comprised of complimentary functional groups.

Common synthons include carboxylic acid dimers, amide dimers, oxime dimers and carboxylic acid-pyridine dimers. The first three listed are known as homodimers, while the latter is a heterodimer (Fig. 8). Homodimer is when there bonding between two identical functional groups, a heterodimer is bonding between two different functional groups. As well as dimers, single hydrogen bond motifs exist where a single hydrogen bond is present seen in Fig. 2E, and are referred to as heteromeric (acid-pyridine, phenol-acid) or homomeric interactions. Ideally, heterodimers and heteromeric interactions need to dominate in a crystalline solid in order to bring two different molecules together to build 3-D structures with directionality

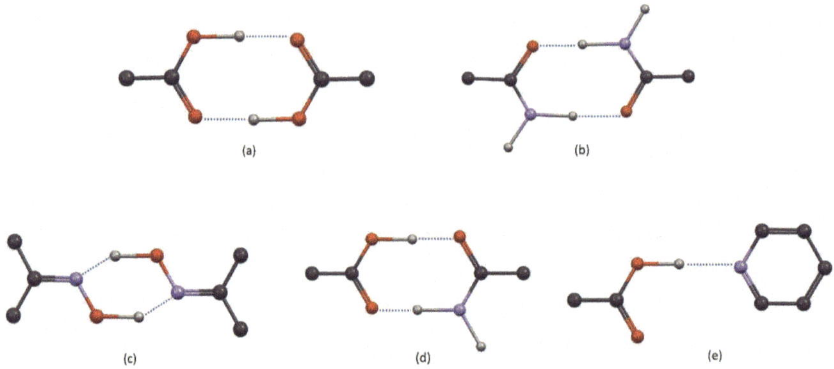

Fig. 8 Representative supramolecular synthons. (A) carboxylic-acid dimer, (B) amide dimer, (C), oxime dimer, (D) carboxylic acid—amide dimer, (E) acid-pyridine heterosynthon.

and dimensionality. Homodimeric interactions are not favoured as these interactions would form between like molecules and potentially provide fewer hydrogen bond donor/acceptor groups for a second molecule to hydrogen bond with.

The Cambridge Structural Database (CSD) contains vast amounts of crystallographic data and is often the initial source scientists studying co-crystals refer to for information on crystal structures and determining the preferred interactions that occur between different functional groups. Several computational techniques based on molecular modelling [71,72], molecular descriptors [73], hydrogen bond propensities [74,75] and machine learning have been developed using the CSD to help guide the process of forming a suitable co-crystal. One promising method designed by Devogelear et al., using the crystal structures in the CSD is a data-driven co-crystal prediction method, based on a network of coformers and link-prediction algorithms. This method led to the discovery of 10 new co-crystal structures, which were subsequently confirmed experimentally. Devogelear et al. also showed how the method can be applied to molecules not in the CSD [76]. This type of prediction method provides great opportunities for the development of pharmaceutical co-crystals and could provide a more directed approach by limiting the number of coformers to be assessed and allowing for more time to be spent on alternative methods used for co-crystal formation or increasing the number of solvents used during a co-crystal screening experiment.

Most of the literature that focusses on designing co-crystals deals with model compounds and coformers that have favourable hydrogen bond donor/acceptor capabilities. These studies have been exceptionally valuable as they have provided information on what intermolecular interactions may be favoured in the presence of other functional groups. It is common for APIs to have either carboxylic acid or phenol groups in their chemical structure which can form hydrogen bonds with suitable coformers [77]. One of the earliest reported pharmaceutical co-crystals was that of paracetamol with cyclic amines [78] and this was investigated by Walsh et al. who reported co-crystals of ibuprofen, aspirin and flurbiprofen with another amine [79]. The amine coformers are not considered as *Generally Recognised as Safe* (GRAS) molecules by the FDA and cannot be found on the '*Substances Added to Food*' list, but the principal of forming a co-crystal and disrupting the common carboxylic acid homodimer synthon with a stronger O–H⋯N hydrogen bond was demonstrated.

Crystal engineering is an evolving area of research and selecting a coformer with the knowledge that it will form a pharmaceutical co-crystal with a particular API is not straightforward and at present cannot be wholly achieved without practical experimentation. Crystal engineering strategies discussed including supramolecular synthons serve only as a guide and the field still requires intensive experimental screening and characterisation to confirm the successful formation of a pharmaceutical co-crystal.

4.2 Screening by grinding

4.2.1 Synthesis of pharmaceutical co-crystals

Pharmaceutical co-crystals can be prepared by a wide range of methods including solution (evaporative crystallisation, quench cooling), spray drying, slurrying and mechanochemistry. Solution preparation can often be found to yield multiple crystal forms, especially if there is a large difference in solubility of the co-crystal former and API. The process can also be laborious and wasteful owing to the relatively large volumes of solvent required for preparation. Mechanochemistry, on the other hand, has emerged as a method of choice to readily screen for pharmaceutical co-crystals. When thinking about phase diagrams, in particular, ternary phase diagrams, grinding also offers the advantage that it is easier to be in the region of stability for co-crystals. In essence, the solvent amounts are so small as to render it almost a binary system. This can then improve the chances of obtaining a co-crystal even if the API and coformer have drastically different solubilities.

4.2.2 Mechanochemical co-crystal screening

Mechanochemistry has long been used for the synthesis of solids and molecules with Faraday initially demonstrating the reduction of AgCl to Ag by grinding silver chloride with metallic zinc. Traditionally mechanochemistry referred to chemical reactions induced by the impact of mechanical energy such as in ball mills. However, following on from the synthon approach developed by Desiraju and others as previously described, it can be seen that mechanochemistry need not refer solely to the making/breaking of covalent bonds but also to the non-covalent bonds used by crystal engineers which form the basis of co-crystallisation.

Mechanochemistry offers several advantages over traditional solution growth, e.g., control over ratio of reactants (leading to different stoichiometry's of co-crystals), co-crystal formation using insoluble compounds, and is quick, easy and reproducible (a typical grinding experiment takes 30 min compared to evaporative co-crystallisation that could take a week or more). Grinding can be performed 'neat' or 'liquid-assisted' (LAG) in which catalytic amounts of solvent (typically 30 μL used in 100 mg co-crystal synthesis are used, with LAG generally yielding co-crystals of a higher degree of crystallinity than neat grinding, which can cause partial to complete amorphisation of the sample. More recent advances include the use of polymers to assist grinding (POLAG) as well as ion-assisted grinding [80]. Not only can grinding yield bimolecular co-crystals but termolecular synthesis is also possible. Polymorphic interconversion and co-crystal fragmentation can be achieved by using slightly different conditions. Some of these areas will be discussed in further detail in the following section.

4.2.3 Liquid-assisted grinding (LAG)

The addition of a small amount of liquid to a grinding experiment has been shown to catalyse potential reactions (chemical or non-covalent) but can also help to dictate the polymorphic outcome. Trask et al. showed in 2005 that the solvent selected for the LAG experiment could reproducibly influence the polymorph obtained for both succinic acid and anthranilic acid. For succinic acid, an hypothesis that the resulting form was possibly due to the polarity of the solvent was put forward, but the exact mechanism was still an area of further research [81].

More recent work by Hasa et al. in 2016 shed further light on the form control aspect of LAG [82]. While previous studies (such as Trask et al.) had focussed on the chemical nature of the liquid, this piece of work explored the amount of liquid added to the grinding jar. A 1:1 co-crystal of caffeine with anthranilic acid was selected as the model system. The system consists of three known polymorphic forms with previous studies identifying conditions to obtain each. Forms I and II had been obtained through grinding reactions (Form 1 both neat and with 16 other solvents; Form 2 using acetonitrile, acetone and DMF), while Form 3 could only be obtained by desolvating a group of aromatic solvates. A conclusion from a study by Fischer et al. was that the polymorphic outcome was strongly dependent on the dipole moment of the liquid (low dipole moment gave Form 1 while high dipole moment gave Form 2) [83]. No grinding experiments had yet yielded Form 3. In the Hasa study, all three polymorphs were obtained as phase pure materials following grinding studies. Only 4 of the 15 investigated liquids gave one polymorphic form regardless of volume added. The remaining 11 yielded different behaviours but all were found to give different results depending on the volume. This adds to the variables available to the crystal engineer for screening purposes which is possibly particularly important when considering the implications for scaling up any mechanochemical, non-covalent synthesis.

Another example by Caira et al. used a sulfadimidine system and a range of aromatic, carboxylic acids [84]. Kinetics were measured using XRPD with certain conditions found to be selective despite the presence of competing acids. They also observed that the resulting products from co-grinding were identical to those obtained from solution.

While the observation of identical products from both grinding and solution based methods is not uncommon, LAG has been shown to sometimes yield different outcomes to crystallisation from solution. Trask et al. showed three different outcomes when looking at co-crystals with caffeine: identical products, different stoichiometric product by milling compared to solution and, different polymorphic forms [81].

Ionic liquid-assisted grinding, using ionic liquids in a similar manner to the use of solvents in LAG, has recently been shown by Myerson et al. to be a further promising approach to tune conditions to obtain otherwise inaccessible co-crystal polymorphs [85]. They investigated two well studied co-crystal systems, caffeine-citric acid and caffeine-glutaric acid using imidazolium-based ionic liquids demonstrating polymorphic control.

An earlier study by Hasa et al. illustrated the effect of using polymers to improve the reaction rate and to increase product diversity during mechanochemical co-crystallisation reactions. The polymer-assisted grinding approach (POLAG) was found to yield advantages when compared to the conventional LAG approach. One such advantage was to eliminate the chance of unwanted solvate formation as well as enabling particle size control. They also demonstrate polymorphic control of caffeine-glutaric acid system suggesting that the number of monomer units (and consequently the polarity of the polymer) was important in obtaining the desired polymorph [86].

4.2.4 *In situ monitoring*

A study in 2013 by Friščić et al. gave some insight as to the mechanism of mechanochemical reactions. In this case the mechanosynthesis of metal–organic frameworks reveals that mechanochemistry is highly dynamic and has reaction rates comparable or greater than those in solution [87]. More recently, in 2020, Julien et al. used a similar protocol to analyse the synthesis of an agrochemical ionic co-crystal [88]. This enabled direct and quantitative evidence of chemical autocatalysis in a mechanochemical reaction. An observation from this study was that reaction acceleration was seen due to water generated from the reaction at amounts up to three orders of magnitude lower than in a typical LAG system. Additionally, the agrochemical ionic co-crystal was highly hygroscopic so using such minute amounts of water (even by the standards of mechanochemistry) to effect catalysis was highly beneficial.

Furthermore, the first real-time, in situ X-ray diffraction study of the mechanochemical synthesis of organic co-crystals was reported by Halasz et al. using the carbamazepine and saccharin system [89]. Crystalline co-crystal formation was not observed when the grinding was performed in the absence of solvent (amorphisation) but when acetonitrile was used in the reaction mixture the crystalline co-crystal was obtained (CSD ref UNEZAO). Time resolution was reported in seconds and showed that the phase pure co-crystal could be obtained in as little as 4 min at a 0.5 g scale.

4.3 Screening by solution methods

Co-crystals can be accessed via several different solution-based methods, such as evaporative, cooling and anti-solvent crystallisation, or solution-mediated phase transformations (slurry/suspension conversion). Each of these methods has its own merits when it comes to its utilisation in co-crystal screening, with no one method being superior. The ideas and methods utilised in co-crystal screening are the same as those employed in both polymorph and salt screening. This means that co-crystal screens are conducted largely in the same way as one would conduct a polymorph or salt screen, but with the addition of potential coformers. If the formation of a co-crystal is possible, the co-crystal must therefore be more stable, and hence less soluble, than the individual counterparts in the selected solvent system; it is this solubility relationship that allows the co-crystal to be accessed from solution. As the relative solubility relationship of the API, coformer and co-crystal will vary depending on the solvent being used, solution-based screening methods must encompass a wide range of solvent systems to increase the probability of discovering a novel co-crystal form. The coformers themselves and the stoichiometric ratio of API to coformer must also be considered as a factor in these screens along with other variables such as temperature and stirring rate/agitation. The combination of these factors leads to a combinatorial screening approach.

Solution based screening methods may be considered more practical than other techniques as crystallisation from solution is the main method utilised in large scale manufacture of pharmaceuticals [90]. These methods can give early indication of the solvent system and stoichiometric ratios of API to coformer required to produce the co-crystal that other methods such as solid state grinding do not. These screening techniques do, however, require some prior knowledge of the solubilities of both the API and coformer.

4.3.1 Evaporative crystallisation

Evaporative crystallisation screening techniques are both simple and effective. Both API and coformer are dissolved in the solvent to create an undersaturated solution. The solvent is then removed from the system either by leaving it undisturbed and allowing the solvent to naturally evaporate (slow evaporation) or by removing the solvent quickly by techniques such as evaporation at elevated temperature, rotary evaporation or blow down evaporation. As solvent is removed the system becomes supersaturated resulting in spontaneous nucleation and subsequent crystal growth. If the formation of a co-crystal is possible this form will likely nucleate, however, kinetics will play a significant role resulting in the possibility that the pure API and/or coformer will also nucleate forming a physical mixture.

Slow evaporation screening can be conducted easily on a small scale and doesn't require any sophisticated platforms or techniques; however, it can be time consuming and is not generally regarded as efficient for fast screening purposes, especially if solvents with low volatility are used. This being said, slow evaporation is one of the most popular screening methods [91]. The temperature at which the evaporation is conducted, however, can lead to an increased rate of evaporation, reducing the experimental timeframe.

The evaporation temperature can also play a significant role in what solid form is accessed due to the changes in both kinetics and thermodynamics with temperature; if a simple evaporative screening process is to be utilised, temperature should become one of the screening factors along with coformer, solvent and the API to coformer ratio.

Rapid evaporation can be achieved by use of a rotary evaporator offering a screening route for the quick identification of novel API co-crystals. This fast evaporation technique enhances crystallisation kinetics and can result in the isolation of metastable forms of an API co-crystal that may otherwise have remained unseen in other screening techniques. Bag et al. demonstrated this by accessing the metastable Form 2 of the saccharin-carbamazepine co-crystal (previously only captured via polymer induced nucleation techniques) using rotary evaporating screening techniques [92].

Screening via evaporative techniques may lead to the discovery of multiple co-crystal forms. However, due to the difficulty and impracticability of scaling up evaporative techniques [93], developing a route for the large scale manufacturing of the co-crystal can be difficult; a co-crystal accessed under evaporative crystallisation techniques may not be directly accessible using more scalable techniques such as cooling or anti-solvent crystallisation.

4.3.2 Cooling crystallisation

Screening techniques based on cooling crystallisation, as with evaporative crystallisation, also start from an undersaturated solution of API and coformer. However, unlike evaporative crystallisation, these solutions are prepared at elevated temperatures and subsequently cooled in the hope that solid will precipitate out of solution due to a reduction in solubility with temperature [94]. This method therefore relies on increasing the supersaturation within the system enough to surpass the metastable limit for the putative co-crystal, before nucleation of the pure components is induced. This kinetic barrier may be high in some cases resulting in slow nucleation kinetics which has the potential to result in long induction times; for some systems precipitation from solution may not occur without an external input. High throughput methods using cooling crystallisation can be developed utilising automated equipment to prepare small scale experiments, however, even with the use of sophisticated equipment, the wait time for crystallisation can be weeks resulting in a relatively slow output [95,96].

As the relative solubilities of the API and coformer can have a significant effect on the stability of the putative co-crystal and its subsequent dissolution behaviour, (whether it dissolves congruently or incongruently discussed further in Section 4.5.2) a wide range of solvent systems need to be screened in combination with different stoichiometric ratios of API to coformer. It is generally assumed that solvent systems for which both the API and coformer have similar solubilities will result in a congruently dissolving co-crystal (if the formation of said co-crystal is thermodynamically feasible). The crystallisation kinetics also play a significant role in what solid form will be accessed during a cooling crystallisation process and as a result, parameters such as cooling rate, final temperature and stirring rate/agitation should all be considered when conducting a cooling screen [97]. This can result in more sophisticated equipment being required than for slow evaporative techniques. However, with cooling crystallisation being one of the most common methods for large scale pharmaceutical manufacture, screening using this technique can provide a directs developable co-crystal crystallisation method for subsequent scale up and manufacture [91].

4.3.3 Anti-solvent crystallisation

As with both evaporative and cooling techniques, this method requires some prior knowledge and understanding of the solubility behaviour of the starting components as an undersaturated solution of API and coformer must first be created. Once solutions are prepared potential anti-solvents will be screened in the hope that upon their addition crystallisation of a co-crystal form will be induced. Potential anti-solvents are chosen based on prior solubility knowledge of the API and coformer, or by estimation of solubility based on the structures of the two components. Screening a wide range of solvent/anti-solvent systems alongside a wide range of coformers will maximise the potential of accessing novel co-crystal forms, including any polymorphic or solvated forms of these co-crystals. The addition of anti-solvents can lead to very quick crystallisation and as a result this screening method has the potential to be developed into an automated high throughput screening technique which can be conducted at a very small scale on a short timescale [94].

As with cooling crystallisation, anti-solvent crystallisation is one of the most common crystallisation techniques utilised in large-scale pharmaceutical manufacturing and as a result screening with this technique can lead directly to scalable routes to large-scale co-crystal formation. This, combined with its ability to be developed into a high throughput technique, makes it a desirable screening method.

4.3.4 Solution-mediated phase transformation (slurry conversion) methods

Screening based on solution-mediated phase transformations (slurry conversion), is the only solution-based technique that does not require full dissolution of the individual components. This method works on the premise that, if the putative co-crystal is more stable and hence less soluble in a specific solvent than the individual API and coformer, the starting components will transform to the co-crystal when suspended in this solvent if given enough time and mobility. Upon the addition of solvent to an excess of solid API and coformer to form a slurry/suspension, the solution phase becomes saturated with respect to both components. As the putative co-crystal form is less soluble, the solution will therefore be supersaturated with respect to this form, resulting in the co-crystal crystallising out of solution. The formation of the co-crystal removes the API and coformer from the solution rendering it undersaturated with respect to these components and as a result more of the solid API and coformer dissolve. This process continues until full conversion of API and coformer to co-crystal has occurred [98].

If given enough time and mobility this method tends to drive the system to the thermodynamic minimum. However, crystallisation kinetics still play a significant role in the rate of this conversion. Using a solvent system where both the API and coformer have low solubility can reduce the rate of conversion; potentially resulting in metastable forms being captured as the system moves towards the thermodynamic form. The temperature at which the samples are slurried/suspended can also have a significant impact on what solid form is accessed. Increased temperature will increase the transformation kinetics and subsequent rate of conversion, resulting in the system moving towards the thermodynamic form of the putative co-crystal more quickly; this may result in any metastable forms that the system may pass through not being captured. However, at varying temperatures the relative stability of forms may change, and what is a metastable form at low temperature may become the thermodynamically stable form at an elevated temperature (enantiotropic relationship). Thus additional forms being captured using this method.

Solution-mediated phase transformation screening methods have been shown to be successful in accessing co-crystals not accessed using other screening techniques. For example, a 'hidden' co-crystal of caffeine-adipic acid which had not been previously accessed during solvent drop grinding and fast evaporation screens was discovered by Bučar et al. using this technique [99].

Each of the techniques discussed above has its own merits, with each one offering a method of accessing novel pharmaceutical co-crystals. Slow evaporation is a very simple and easy screening method; however, it can be very slow, sometimes taking several months for any successful hits to crystallise out of solution. Fast evaporation screening has a much shorter timeframe and can result in the discovery of metastable forms, increasing the potential for intellectual property. Solution-mediated phase transformation screening (slurry conversion) has been shown to have an increased success rate in comparison to evaporative techniques [93,94]. However, this method is less likely to access metastable forms and can be an inherently slow process if the transformation kinetics are slow. Cooling and anti-solvent techniques are much more applicable to the subsequent development and manufacturing scale up of a pharmaceutical co-crystal; if these techniques are used in screening it is likely that these screening conditions can be directly developed, optimised and scaled up. Screens conducted using cooling crystallisation, however, can be time consuming, whereas anti-solvent screening can be conducted as a high throughput screening method [94].

Overall, it is unlikely that a single solution based screening technique will yield all possible co-crystal forms of an API and in some cases these screens can have relatively low success rates [100]. It is therefore advisable to implement a combination of screening techniques when searching for potential pharmaceutical co-crystal forms. For example, the combination of initial mechanochemical grinding to produce 'leads' on potential coformers, followed by solution based screening methods with both these and structurally similar coformers has been shown to allow for fast identification of co-crystal candidates as well as offering insight into their scalability and robustness [101].

4.4 Characterisation of co-crystals

Generally, the methods used to identify and characterise a pharmaceutical co-crystal are the same methods used to characterise any solid crystalline form. Single crystal X-ray diffraction is, however, the preferred method that provides conclusive evidence of co-crystal formation. When single crystals of a pharmaceutical co-crystal are unattainable, a combination of techniques including structure solution from powder X-ray diffraction with advanced spectroscopic techniques can also be used to determine whether a co-crystal has been produced.

4.4.1 Single crystal X-ray diffraction (SCXRD)

Lab-based instruments developed to determine crystal structures of crystalline solids have come a long way in the last century. Increasingly more powerful X-ray sources and more sensitive detectors, coupled with new experimental protocols and modern software, mean that smaller and more complex crystal systems can be evaluated with greater accuracy and within minutes as opposed to days. The current ease of structure determination using single crystal X-ray diffractometers means potential co-crystals can be identified quickly and further structural characterisation tools may not be required. Furthermore, the absolute structure and chirality of the components comprising the pharmaceutical co-crystal can be determined unequivocally using SCXRD.

Single crystals of suitable size and quality for SCXRD are normally obtained by using solution-based methods to prepare pharmaceutical co-crystals. SCXRD is used to determine the spatial arrangement of atoms in pharmaceutical co-crystals and the intra- and inter-molecular interactions between the API and coformer themselves and between the API and coformer molecules. The FDA and EMA require confirmation of the presence of both the API and coformer in the unit cell and, where applicable, whether proton transfer has or has not occurred. SCXRD is the definitive tool to answer these regulatory questions, especially for the latter where, if a salt rather than a co-crystal has been formed, an entirely different development strategy may be required. Other methods such as SSNMR and complementary spectroscopic techniques can also determine the position of a proton, however, the time it takes can be extensive [102] and the results may not be as unambiguous. SCXRD is currently the most reliable technique to establish the formation of a pharmaceutical co-crystal [103].

Finally, on knowing the crystal structure of a co-crystal, the geometry of each entity comprising the co-crystal is known, the preferred interactions between the molecules can be explained and the overall arrangement of the molecules in the crystal structure which allows for a better understanding and rationalisation of a co-crystal's physical properties. Controlling Levofloxacin's hygroscopicity and physical stability can be achieved by forming a co-crystal with metacetamol. Recently the SCXRD structure of a drug–drug co-crystal comprising levofloxacin-metacetamol was obtained. Levofloxacin exists as a hemihydrate under ambient conditions and has four anhydrous polymorphic forms. When comparing the single crystal structure of the levofloxacin hemihydrate with the co-crystal form, it was shown that metacetamol formed an important hydrogen bond with levofloxacin which prevents a water molecule from associating, and thus improves levofloxacin's overall hygroscopicity and stability under various temperatures and humidity conditions [104].

4.4.2 Powder X-ray diffraction (XRPD)

XRPD is a non-destructive technique that is used extensively as a screening tool and quickly provides information on whether a new chemical entity may have formed, whether it be a pharmaceutical co-crystal or salt. Depending on the polymorphic nature of an API and the selected coformer molecule, they can both have one or more fingerprint XRPD patterns, which should all be considered when evaluating the XRPD results from a co-crystal screening experiment. The resulting XRPD patterns from a co-crystal screening experiment could demonstrate three main outcomes. Firstly, the XRPD pattern may contain diffraction peaks that correspond to both the API and coformer and this would suggest a physical mixture of the two starting materials and a new chemical entity has not been formed. Secondly, a new XRPD pattern, with diffraction peaks not corresponding to either of the starting materials may be obtained, suggesting co-crystal or salt formation, and thirdly, it could be a mixed phase containing new peaks as well as peaks that correspond to the API and/or coformer. The latter result could indicate that a different API to coformer ratio is required or that the complexation between the two components is not complete and longer experiment times are required.

Additional characterisation methods such as solution and solid-based ^{1}H NMR spectroscopy, IR spectroscopy and thermal analysis, can assist with further characterisation of potential co-crystal hits. A standard powder diffractogram cannot provide chemical information on intermolecular interactions and thus it cannot confirm whether a pharmaceutical salt or co-crystal has been formed. Structure determination from a high-resolution powder diffractogram data can be undertaken although in an industrial setting this exercise is not routinely performed [105].

One of the most common ways to form co-crystals, through grinding methods, produces microcrystalline powders [106,107]. Recrystallisation of the resulting powder to generate larger crystals is often unsuccessful, and as a result these microcrystalline powders cannot be examined using SCXRD. Structure determination from powder diffraction data remains an option.

Structure solution from powder diffraction data has evolved rapidly in the last two decades, however, there are still problems and challenges, especially for light atom molecular complexes such as pharmaceutical co-crystals [108]. Powder diffraction patterns essentially contain the same information as single crystal diffraction patterns, but the data is condensed into one dimension for XRPD data. This means there is a large amount of peak overlap in the XRPD patterns which obscures information on peak position and intensities. As pharmaceutical co-crystals have large unit cells and low symmetry the problem of peak overlap makes solving structures from powder exceptionally challenging [109] and time consuming. There are continual developments of new and improved techniques for structure solution from powder. Harris et al. have demonstrated the use of powder XRPD to determine the crystal structure of a three component organic system comprising bis-β-naphthol, benzoquinone and anthracene [110]. Several other crystal structures of organic co-crystals, comprising an API and coformer, have been determined from powder data, and the crystal structure has been confirmed with an already determined single crystal X-ray structure [38,111].

If the approximate structure of a pharmaceutical co-crystal, is obtained from powder data, the model may be optimised computationally to a minimum-energy configuration. From such models the predicted solid state NMR (SS-NMR) spectra can be generated and compared to the experimental SS-NMR spectra of the pharmaceutical co-crystal/salt. Using these two techniques in parallel, a greater understanding of the molecular arrangements of the molecules involved in the co-crystal can be obtained and potentially the proton position can be located. Progress in computer-based structure prediction of co-crystals will also be exceptionally useful in this area.

4.4.3 Solid-state NMR

SS-NMR spectroscopy is a useful technique to use when probing molecular structures and the location of distinct protons of co-crystals, salts and pure APIs. ^{1}H NMR chemical shifts in the solid state are a good indication of hydrogen bonding interactions taking place, and the change in bond lengths could give useful information of the protonation taking place between two molecules in a crystalline sample. ^{13}C solid state cross polarisation magic angle spinning (SS CPMAS) NMR experiments are used to evaluate the different environments of carbon atoms in a crystalline solid.

^{15}N nuclei are very sensitive to changes in proton environments and ^{15}N SS CPMAS experiments can be used to determine the protonation states of basic nitrogen atoms of co-crystal and salts [102]. A limitation encountered when conducting ^{15}N solid-state NMR experiments is the exceptionally long experiment times due to the low natural isotropic abundance of ^{15}N (0.37% of nitrogen isotopes). Different types of ^{15}N SS-NMR experiments can be performed, such as solid state ^{1}H–^{15}N scalar attached proton tests (APT)/J-resolved experiments, which have been used to determine the number of protons attached to a nitrogen atom and readily differentiate between a neutral and charged amine derivative [13]. Conducting ^{15}N solid-state CPMAS NMR with short and long contact times can also help identify whether a specific nitrogen atom has been protonated or not. The short contact time spectra are expected to show only protonated nitrogen atoms, while the long contact time spectra are expected to show all nitrogen atoms [30].

More advanced 2D SS-NMR experiments are being used to probe the proximities between two protons or between a nitrogen atom and proton nuclei of a pharmaceutical co-crystal. Magic Angle Spin ^{1}H–^{14}N dipolar heteronuclear multiple-quantum coherence (D-HMQC) experiments have been used to investigate the proximity of hydrogen and nitrogen atoms in a nicotinamide-palmitic acid co-crystal, without relying on the abundance of the ^{15}N isotope [112,113]. 2D ^{1}H–^{1}H homonuclear and 2D ^{1}H–^{13}C heteronuclear correlation experiments have also been useful when determining the location of particular proton atoms in the solid state [114–116].

4.4.4 Infrared and Raman spectroscopy

Raman and infrared spectroscopy, like the use of XRPD, are utilised as screening techniques that allow screening a large number of coformer molecules with a particular API. It has, however, also shown promise in distinguishing co-crystal formation from salt formation.

Co-crystal formation is initially assessed by IR and Raman spectroscopy by analysing the resulting spectra and comparing them to the IR or Raman spectra of the individual components. A physical mixture will appear as an addition of the bands from the individual components, while a shift of the bands to higher or lower frequencies can indicate salt or co-crystal formation. Accurate assignment of bands affiliated with different functional groups is, however, difficult because salt and co-crystal formation is often accompanied by frequency shifts and peak overlapping.

The most notable IR bands that have been monitored in terms of co-crystal or salt formation between an API and coformer molecules include the NH stretching bands of an amine base, carbonyl bands from a carboxylic acid, O–H stretching bands of a carboxylic acid and the H–N stretch of

COOH···N. These IR vibrations are all related to groups that are typically involved in hydrogen bonding [117].

Lamivudine, which is used for the treatment of AIDS and hepatitis B, forms a multicomponent crystal form with both a neutral and ionised form of mandelic acid. Using IR spectroscopy, bands affiliated with the neutral COOH functional group and the charged COO^- group could be assigned. These assignments were further supported by density functional theory calculations as well as other experimental methods such as thermal analysis [118]. Wang et al. studied the interactions between 6-mercaptopurine (6-MP) with isonicotinamide and found that the vibrational modes corresponding to the N–H stretching frequency in the 3300–3500 cm^{-1} range were red shifted when compared to 6-MP·H_2O, which suggested a change of H-bonding interactions on the amine functional groups [119]. The actual position of the proton, which was used to confirm co-crystal formation, was determined using SCXRD.

In situ Raman spectroscopy has been used as a screening tool for high throughput co-crystal experiments. The first example of this being employed was on indomethacin, which was screened with 46 different coformers. Raman microscopy was able to identify 5 new co-crystals and two salts. The extent to which the C=O stretching band shifted from the 1,4-dioxane solvate of indomethacin was used to determine salt and co-crystal formation [120]. In situ Raman spectroscopy is a method that can quickly determine whether or not a new multicomponent system has been formed in solution, but it requires a good understanding of the system including solvent effects. Information both on the equilibrium of co-crystal/salt formation and the time it takes for co-crystal/salt formation can also be obtained using this method.

4.5 Importance of phase diagrams

Phase diagrams provide information on the thermodynamically stable phases in equilibrium under a certain set of conditions and can offer a route to further understanding the thermodynamic relationships between co-crystals and their respective components; they define the region where the co-crystal is the thermodynamically stable solid phase. This knowledge is therefore paramount for the successful design and operation of a manufacturing process to access the desired co-crystal [121,122].

4.5.1 *Binary phase diagrams*

For ease of understanding, an example of a temperature vs composition binary phase diagram will firstly be discussed for a two-component system

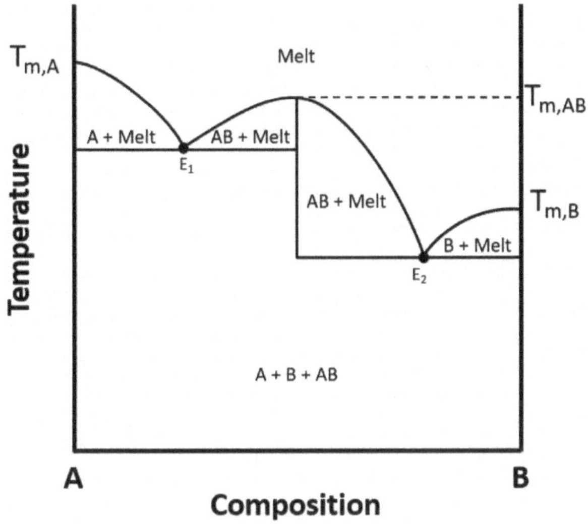

Fig. 9 Binary phase diagram for components A and B.

(A and B), where a co-crystal is incapable of being formed (Fig. 9). Within this phase diagram the intercepts with the *y*-axis demonstrate the melting temperatures of the pure components A($T_{m,A}$) and B ($T_{m,B}$). The upper region of the phase diagram is monophasic and represents the miscible melt phase of both A and B. Directly below this phase there are two biphasic domains: the left represents the conditions at which solid A co-exists with the melt; with the right representing the conditions where solid B exists alongside the melt. The solid lines separating these biphasic phases from the upper phase are called the liquidus and the point at which these lines intersect one another is known as the eutectic point (E), with the corresponding temperature denoted as the eutectic temperature, T_e. At the eutectic point, three phases are in equilibrium with each other, <A>, <B> and the melt. Below the eutectic temperature, T_E, (the rectangular phase) only solid exists (a mixture of A and B), with no melt phase present [123,124].

The binary phase diagram in Fig. 9 can also be viewed with regard to a solute solvent system, with A acting as the solute and B acting as the solvent. For this case, the monophasic upper region represents the undersaturated solution. The left biphasic region represents crystallised solute in equilibrium with a saturated solution, with the right biphasic region representing crystallised solvent in a saturated solution. The left liquidus demonstrates the solubility curve of the solute in the solvent, starting at the melting point of the solute and ending at the eutectic point (E) [125]. Traditional solubility

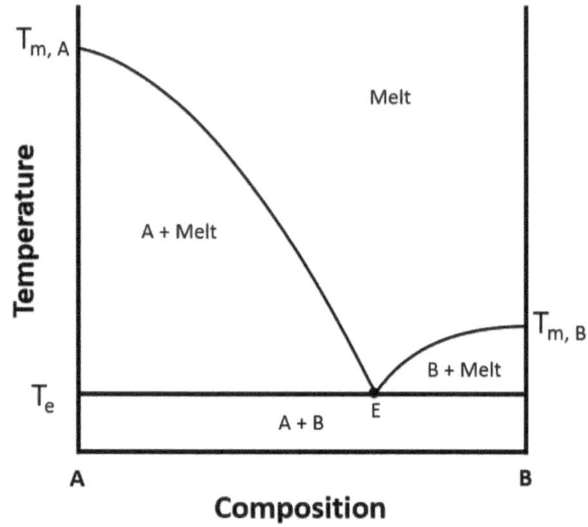

Fig. 10 Binary phase diagram for components A, B and co-crystal (AB).

curves (solute concentration vs temperature) are small sections of these binary phase diagrams with the term solvent being used for the component in excess [126].

If a stable co-crystal can be accessed from a mixture of A and B, however, the binary phase diagram looks quite different, now displaying a stable co-crystal region (Fig. 10).

As before, the upper region is monophasic and represents the melt of both A and B. However, there are now three biphasic regions that lie directly below the melt phase: the left represents a mixture of solid A with the melt, the middle region represents the conditions where the co-crystal exists in equilibrium with the melt, and the right represents the region where solid B exists in equilibrium with the melt phase. This then leads to two eutectic points (E_1 and E_2), where either solid A (E_1) or solid B (E_2) exists in equilibrium with solid co-crystal (AB) and the melt phase. Under these eutectic points a single solid phase exists, consisting of a mixture of A, B and AB [104,123,127].

The construction of binary phase diagrams is relatively simple, involving investigation into the melting behaviour of physical mixtures of A and B in varying compositions [128,129].

4.5.2 Ternary phase diagrams

Phase diagrams can also be constructed for ternary systems (A, B and solvent) and identify the stability regions for a co-crystal and its counter parts with

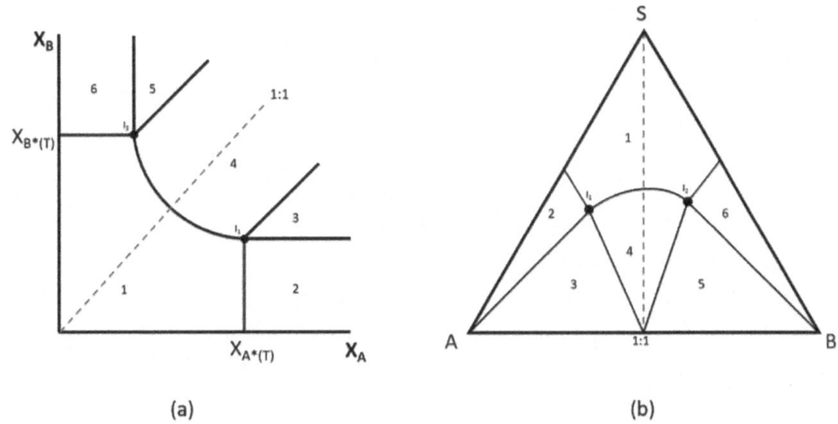

Fig. 11 (A) Orthogonal phase diagram representation for ternary system of A, B and solvent at temperature, T (the solubility values of both B and A at T are denoted by XB*(T) and XA*(T), respectively) (B) triangle phase diagram representation for the same system.

respect to solvent composition (i.e., solute concentration) at a specific temperature, T. These phase diagrams can be depicted as either orthogonal Fig. 11A or triangle representations Fig. 11B. The information contained within these different representations is identical.

In both representations, the red line indicates the solubility of component B with increasing mole fraction of A, with the blue line indicating the solubility of component A with increasing B (for simplicity, it has been assumed that the solubilities of A and B are unaffected by the presence of the other component until the co-crystal is formed). For both diagrams, region 1 represents the undersaturated solution and is bound by the solubility curves of both the individual components (A, red and B, blue) and the co-crystal (black). All other regions represent areas where solid is in contact with saturated solution. Regions 2 and 6 represent solid phase equilibrium solutions of A and B, respectively, with region 4 representing the solid phase equilibrium solution of the co-crystal, AB. Regions 3 and 5 represent the coexistence of the co-crystal with A and B, respectively. The solution concentrations at which the co-crystal is in equilibrium with either A or B are known as the invariant points (I_1 and I_2, respectively) [130,131].

The size and shape of the co-crystal region provides information on the stability and dissolution behaviour of the co-crystal. For example, if the 1:1 stoichiometry line (dashed line) intercepts the solubility of the co-crystal, the system is said to dissolve congruently. If it does not, however, and the co-crystal region is skewed and lies either side of this line, the co-crystal will display incongruent solubility. In this case, solid–liquid equilibrium between the solid co-crystal phase and stoichiometric solution cannot be achieved with the least stable phase recrystallising out of solution upon dissolution of the co-crystal [121,132,133].

The solvent system used can, in most cases, have a significant impact on the phase diagram and the resulting region where the co-crystal is stable. The size, shape and area in which this region falls will be largely dominated by the relative solubility of the individual components. Fig. 12 shows ternary phase diagrams for a co-crystal system in three different solvent systems. In S_2, B has a much higher solubility than A and as a result the co-crystal region is skewed towards B and results in an incongruent system. If the co-crystal was to be left to equilibrate in a saturated solution the composition would move towards the invariant point resulting in the solid phase becoming a

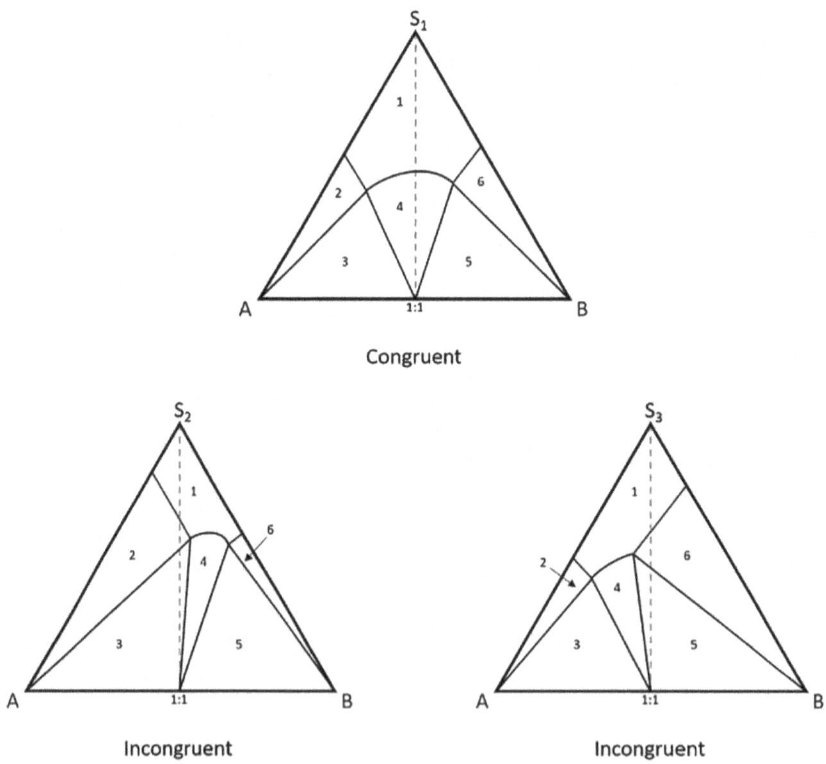

Fig. 12 Examples of phase diagrams for congruent (top) and incongruent (bottom) co-crystal systems (at a specified temperature, T).

mixture of co-crystal and A (the less soluble component). In S_3, the opposite is true, with component B having a much higher solubility than A, resulting in the co-crystal region tilting towards the more soluble component B. In S_1, the solubility ratio of A and B is much closer, resulting in the co-crystal region becoming more symmetrical and also wider [131,134]. It is generally accepted that having a small solubility ratio will result in a congruent system and that the width of the co-crystal region is inversely proportional to the solubility ratio, however, this is not always the case [121,132,135,136].

For a successful manufacture of a co-crystal an understanding of the phase diagram and hence the stability region for the co-crystal is vital. Understandably, congruent systems offer an easier route to manufacture as a simple cooling crystallisation can be utilised, however, if the system is incongruent sophisticated control and moderation of the solution composition must be employed in order to access the co-crystal [121,122,132].

The construction of ternary phase diagrams is relatively simple and can be conducted using two different methods. One method involves slurring mixtures of set compositions at a specified temperature until equilibrium is reached (usually a few days). The samples are then isolated with an aliquot of the solution being analysed for individual component concentration by methods such as HPLC or gravimetric solubility. The solid is then analysed for solid form via XRPD and usually DSC. The concentration values gained give points along the solubility line for each composition and this, in combination with the isolated solid form, is then used to build the ternary phase diagram [131]. Alternatively, different mixtures of A and B can be diluted with solvent until a clear solution is accessed giving the solubility value for this composition [137].

4.6 Scaling up of co-crystals

4.6.1 Mechanochemistry

Grinding methods [107,138,139] are becoming increasingly more popular as part of the screening strategy to identify novel co-crystals. During grinding, both the API and the coformer are simultaneously mixed and ground together by mechanical forces typically generated by a mortar and pestle or a ball mill. The grinding process can either be performed dry or with the aid of a solvent [140]. Multiple mechanisms of co-crystal formation using dry powder methods have been discussed [107] over the last 10 years and due to little or no solvent being used, these methods can be considered as green routes for obtaining co-crystals in high yield [141].

Some of the co-crystals accessed using grinding methods may be difficult to obtain by traditional solution-based methods [142] and scale-up procedures which generate similar conditions to the grinding methods are therefore required.

4.6.2 Twin screw extrusion (TSE)

Twin screw extrusion (TSE) relies on high shearing forces generated by screws to combine individual components together; when combined with heating the process is considered as Hot Melt Extrusion (HME). As a physical mixture of components is added through a feeder, the components are simultaneously sheared and heated together as they progress through the different heating and mixing zones of the barrel resulting in the formation of solid dispersions or multicomponent forms, such as co-crystals (Fig. 13). The process is continuous, and the product is extruded through a die into a filament or simply as an aggregate.

The use of TSE in the pharmaceutical industry to formulate an API with polymers [143] in order to achieve a desired dissolution profile [144] or achieve taste masking [145], as well as accessing multicomponent forms such as co-crystals, is becoming increasingly popular. Hot Melt Extrusion (HME) was used at Amgen to produce AMG 517-sorbic acid co-crystals continuously [146] using a Pharma 16 Twin-Screw Extruder from Thermo Scientific™. This continuous process was heat-mediated without decomposition of the API and was considered efficient, scalable and environmentally friendly.

Alternatively, co-crystal formation may need to be performed with the aid of a small amount of solvent as demonstrated by the formation of a theophylline-citric acid co-crystal [147]. The advantage of these operating conditions is that the formation can be performed at room temperature, with the drying of the co-crystal being performed in the last zone of the extrusion barrel if required. More recently, polymers (up to 15% w/w) have been added to the process to facilitate the co-crystal formation [80] with a similar

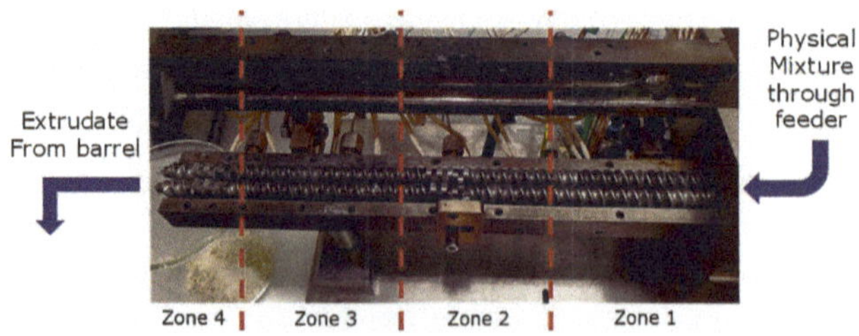

Fig. 13 Twin screw extruder.

effect as achieved previously with a solvent. One such example is a flufenamic acid-nicotinamide co-crystal which was obtained by HME in the presence of five commonly used and pharmaceutically acceptable polymers (Poloxamer P407, PEG-PVA, Soluplus®, PVPVA64 and HPMCAS) [148].

Process Analytical Technology (PAT) tools such as FT-IR or NIR can also be introduced into HME to allow for inline monitoring of the solid state transformations [149]. This was demonstrated in the optimisation of ibuprofen–nicotinamide co-crystal formation by monitoring specific peaks and peak shifts attributed to the co-crystal in the 4800–5200 cm^{-1} region [150].

A recent review [151] by Gajda et al. highlights the tremendous progress in hot melt extrusion over the last 10 years. Within the pharma industry, TSE currently sits within the formulation expertise, however, this is expected to move gradually towards API manufacturing due to it fulfilling considerations of both continuous manufacturing and achieving green chemistry routes. The equipment discussed above is readily available to GMP standard and the HME processes are demonstrated to be scalable with the aid of modelling [152]. The impetus at present is for the industry to seriously consider these processes as viable tools for use in co-crystal manufacture after which the use of extrusion practices may become more ubiquitous in the years ahead (Fig. 14).

4.6.3 Resonant acoustic mixing (RAM)

Resonant Acoustic Mixers (LabRam, Resodyn Acoustic Mixers) subject a reaction vessel to extreme acceleration (100 G); where the samples are placed in a container which is subjected to intense vibration. To date, the technology has been intensively employed for the mixing of powders, paint and viscous material, but the use of RAM for a co-crystal formation was patented in 2012 [153]. The co-crystal formation of the carbamazepine-saccharin co-crystal (Fig. 15) was initially demonstrated in small scale RAM experiments (1.5–22.5 g) with the aid of small quantities of solvent (ca. 15% w/w) [154]. A handful of other articles [155–157] have been published on additional co-crystal systems. Continuous versions of RAM also exist, however, use in scale-up has not been demonstrated yet.

4.6.4 High shear granulation (HSG)

High-shear granulation (HSG) is very commonly used in pharmaceutical formulation practices to blend fine powders of API and excipients into small granules to achieve good drug content uniformity and flowability properties.

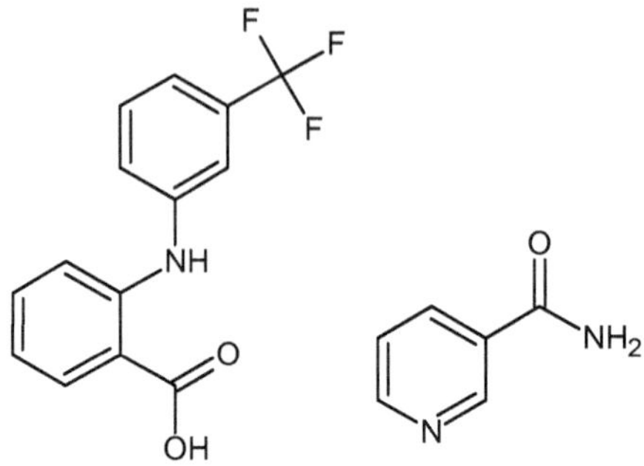

Co-crystal of flufenamic acid and nicotinamide

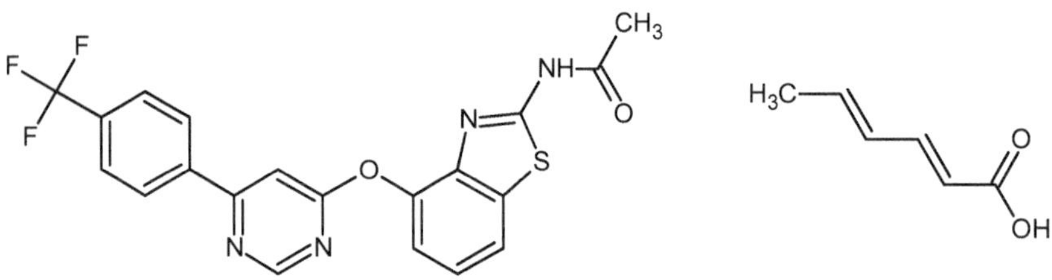

Co-crystal of AMG 517 and sorbic acid

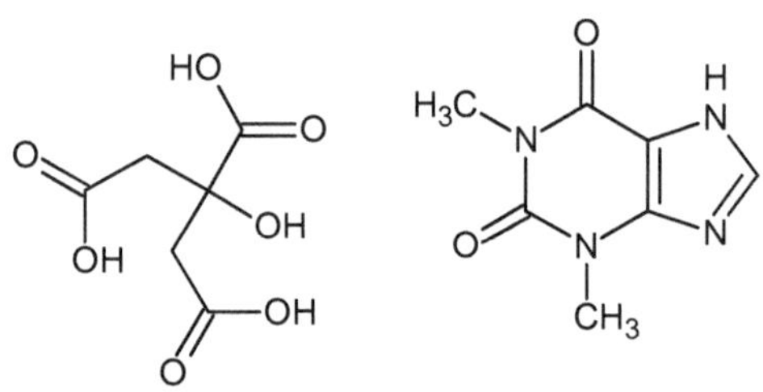

Co-crystal of citric acid and theophylline

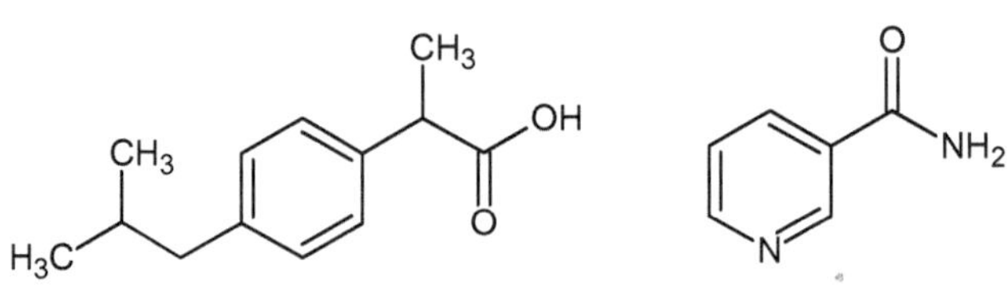

Co-crystal of iboprufen and nicotinamide

Fig. 14 Co-crystals achieved by hot melt extrusion.

High-shear granulators consist of a mixing bowl, a three-bladed impeller (with speeds of up to 3000 rpm being achieved) and a chopper which breaks apart the wet agglomerates to produce granules [158]. The principal manufacturers for HSG are Diosna, Freud Vector and Glatt. The interchangeable bowls (beginning at 250 mL) make the process easy to evaluate in the lab and during scale-up. Large units are also used commercially by formulators.

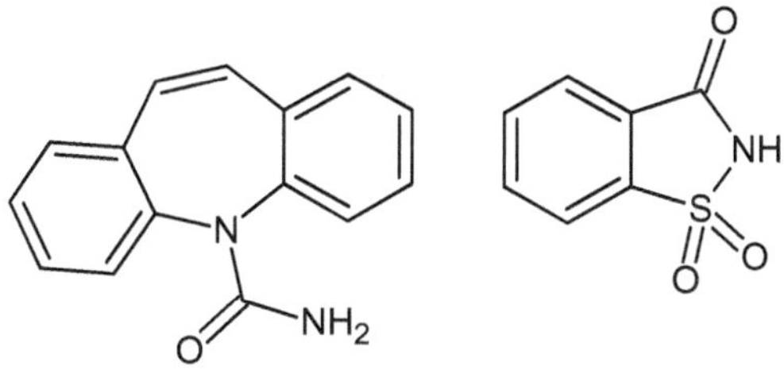

Co-crystal of carbamazepine and saccharin

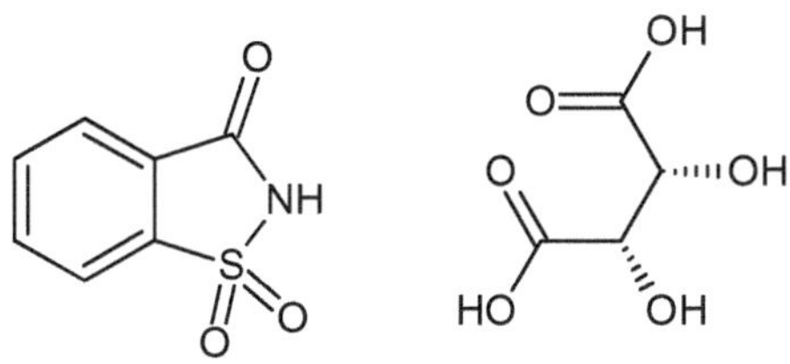

Co-crystal of piracetam and L-tartaric acid

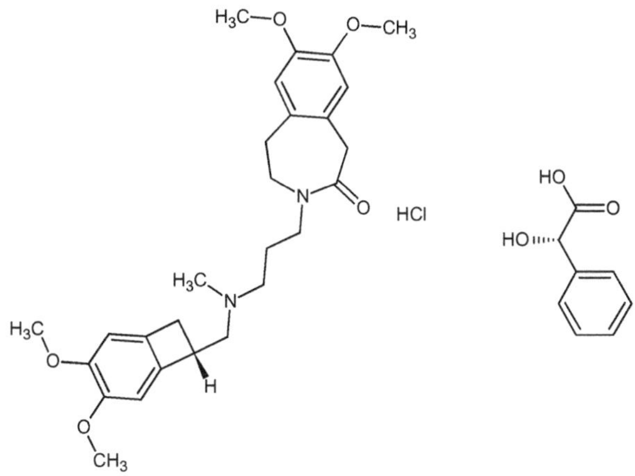

Co-crystal of ivabradine hydrochloride and (S)-mandelic acid

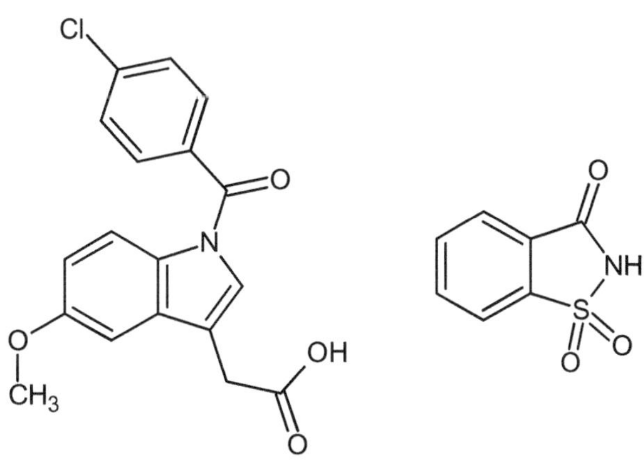

Co-crystal of indomethacin and saccharin

Fig. 15 Examples of co-crystals obtained by RAM and HSG.

HSG has been successfully used to produce the piracetam-L-tartaric acid co-crystal (Fig. 15), alongside commonly used excipients, in the presence of water [159]. The formation of the co-crystal was monitored by XRPD and Raman spectroscopy and parameters such as the amount of water, impeller speed and excipient type were investigated. The best results showed that granules of co-crystal could be obtained within 5 min on a 30 g scale in a 300 mL jar. Some excipients were observed to retard the rate of the co-crystallisation process, however, unfortunately no data on the formation of the co-crystal without excipient was presented in this study.

Zentiva, a pharmaceutical company specialising in generics, have also reported the use of HSG to produce a co-crystal [160]. In order to telescope the processes, the ivabradine hydrochloride-(S)-mandelic acid co-crystal (Fig. 15) was successfully produced in situ during formulation with excipient using HSG with the aid of ethanol. The input ivabradine hydrochloride was not detected and the complete conversion to the co-crystal was confirmed by SS-NMR [161]. This demonstrated an elegant way to produce the co-crystal during drug production.

A similar approach was followed for an indomethacin-saccharin co-crystal [162]. The choice of the polymer was found to be crucial in influencing the kinetics of co-crystallisation; the best results were obtained using polyethylene oxide which was found not to interact with indomethacin or saccharin.

4.6.5 Solution based scale up

Co-crystals are typically identified through methods that are efficient for screening but not conducive to a controlled scale up. Methods such as ball milling and evaporation are efficient methods of screening a large number of samples to identify API and coformer combinations, however, these methods do not address factors important in scale up and manufacturing such as yield, purity or particle size and shape [136,160,161].

Crystallisation using solution-based methods such as cooling and antisolvent are the most common ways of manufacturing APIs on an industrial scale and have been shown to be an effective way of controlling final particle properties. However, the scale up of a selected co-crystal through solution-based methods can be a more challenging process than the crystallisation of single component APIs.

Uncontrolled and not fully understood crystallisation processes can be riddled with problems such as oiling out [163], encrustation and fouling [164], or out-of-specification purity profiles [165], all of which can necessitate the rework or disposal of batches. Furthermore, through the application of crystal engineering it is possible to optimise the properties of the particles produced to improve their secondary processing performance in filtration and drying [166].

Many of the considerations that go into designing a robust and scalable process are the same regardless of the process: solubility curve generation, understanding crystallisation parameters and tracking these parameters using PAT tools. However, the introduction of the coformer as an additional component adds to the level of complexity in the process. This section aims to give an overview on how to build and design a scale up co-crystal synthesis with monitoring that can be employed to ensure the integrity of the scale up.

4.6.6 Solubility

The foundation of any robust crystallisation process is a strong understanding of the solubility landscape. Solubility measurements can be collected in several ways, including aliquot addition and gravimetric determination. Measuring solubility by aliquot addition is conducted by adding solvent to a known mass of API in incremental additions isothermally with agitation until dissolution is observed. Gravimetric solubility is measured by creating a slurry of API in solvent and stirring isothermally for at least 24 h. This suspension is filtered into a pre-tared vial and the known volume of mother liquor is allowed to evaporate. The remaining solid after evaporation is then weighed and the solubility concentration can be calculated. From the resultant solubility curves that can be produced one can determine if a neat cooling crystallisation is possible or, if the solubility is still high at low temperatures, determine if the addition of antisolvent is required to achieve a higher yield. Using the preferred method of solubility assessment, the measurements should be taken at several temperatures and/or solvent-antisolvent compositions to fully understand the solubility landscape, allowing a controlled process to be designed.

There are several factors in a crystallisation process, but the two most significant competing factors are nucleation and crystal growth. These two factors are driven by supersaturation (S) which is defined as the concentration (C^*) divided by the solubility concentration (C_s). When the concentration is below the solubility the system is undersaturated ($S < 1$), if the concentration is greater than the solubility it is considered as supersaturated ($S < 1$). If a sample is supersaturated but remains as a stable solution

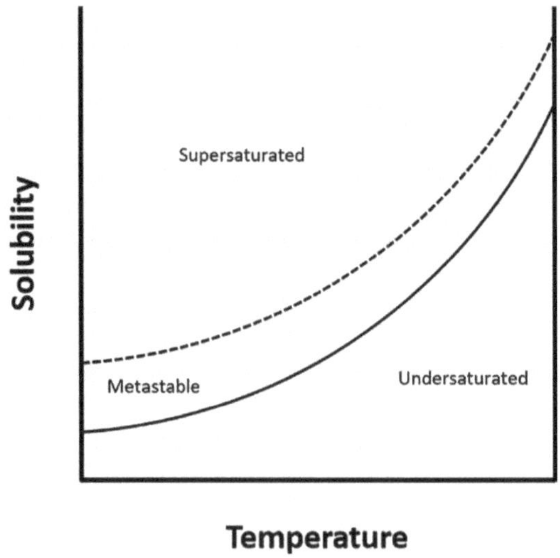

Fig. 16 Solubility and metastable curves and landscape.

then it is called metastable, once it moves out of the metastable zone the sample becomes unstable (labile) and will spontaneously nucleate. This is summarised in Fig. 16.

In solution-based crystallisations supersaturation is commonly achieved in one of 2 ways: either by lowering the solubility concentration by cooling (Fig. 17, blue line), or by introducing a second solvent (antisolvent) reducing the solubility concentration (Fig. 17, red line).

In a single component crystallisation, solubility points are measured across a range of temperatures and solubility curves are determined for a variety of solvent:antisolvent ratios, if required. However, the inclusion of a second component for co-crystallisation adds an additional layer of complexity and the system must be supersaturated with respect to the co-crystal rather than the individual components, as summarised in Fig. 18. This reduces the likelihood of crystallising mixtures of the co-crystal with the starting components.

4.6.7 Nucleation and seeding

Nucleation is the formation of crystalline material from solution and can occur either by primary nucleation (spontaneous), or secondary nucleation (induced by the presence of a heterogeneous source such as seeds or impurities). Primary nucleation can be problematic at scale due to its stochastic nature, making it difficult to design a controlled process. Furthermore, if

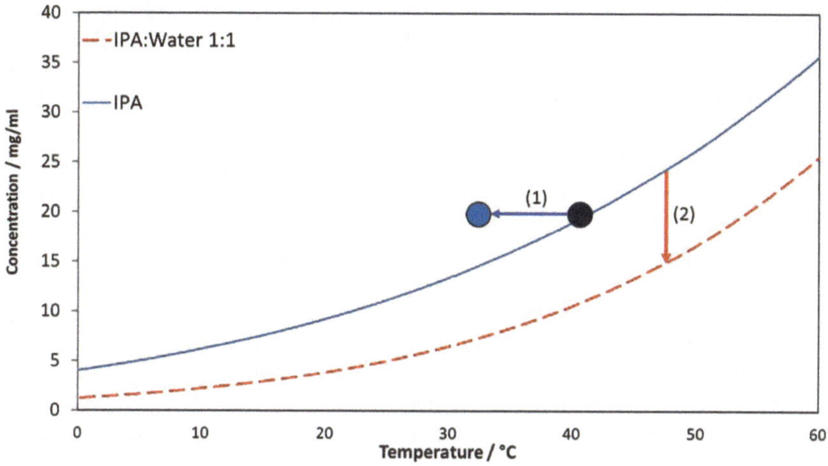

Fig. 17 Supersaturation generation (1) cooling crystallisation, (2) anti-solvent crystallisation.

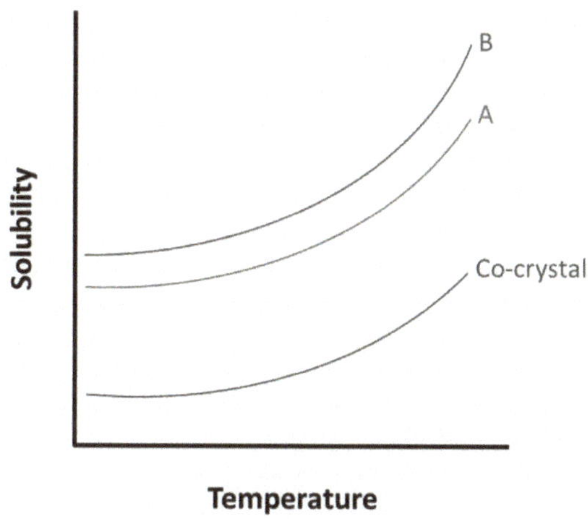

Fig. 18 Solubility curves of single components A and B, and the solubility curve of co-crystal (AB).

there are multiple forms of the product there is always a possibility of nucleating the undesired form due to the relative nucleation rates of competitive polymorphs [167]. A common strategy is to induce nucleation with control of both the solid form and the point of nucleation is through the introduction of seeds [168]. Seed particles of the desired form are introduced when the system is sitting within the metastable zone, inducing secondary nucleation. A large volume of work has been conducted highlighting the benefits of seeding in crystallisation, particularly surrounding the ability to obtain metastable forms which can be difficult to access through spontaneous nucleation [169,170]. By exerting control on solid form, the point of nucleation and number of crystals introduced into the system through a tight seeding strategy, supersaturation generation can be geared towards optimising for crystal growth.

4.6.8 Crystal growth

Crystal growth is the transfer of molecules from solution onto existing crystal surfaces. It is a kinetically driven process and is dependent on several factors such as the solvent system, particle size and surface, and temperature. A way of understanding crystal growth rates with respect to a crystallisation is through de-supersaturation experiments. These experiments consist of monitoring the rate at which supersaturation is consumed upon the addition of a fixed amount of seeds at different temperatures. The information produced from these measurements informs the rate at which supersaturation can be safely generated. Generating supersaturation at a faster rate than the existing solids can deplete it via crystal growth increases the risk of excessive secondary nucleation. Uncontrolled nucleation events during a crystallisation process have the potential to result in nucleation of undesired forms or impurities, entrapment of unwanted solvent molecules, or the formation of crystals that are difficult to isolate due to their particle size, shape and surface properties. Typically, to generate larger crystals a slow generation of supersaturation is employed via slow cooling or antisolvent addition rates. Ripening can be also be employed to generate larger crystals through temperature cycling; dissolving smaller crystals upon the heating steps and growing on the remaining larger crystals during the cooling stages. This has been successfully applied to change the aspect ratio of succinic acid from plates into diamond like crystal habits [171]. This technique could also be applied to needle habits which are known to be problematic during isolation [172].

Understanding the growth rate also allows for modelling to be conducted in order to further optimise the supersaturation generated during a process to prevent loss of control. Understanding and controlling supersaturation can yield a robust and reproducible crystallisation process that gives the predicative and consistent results required of industrial manufacturing.

4.6.9 Process analytical technologies

The development of Process Analytical Techniques (PAT) allows for in situ monitoring of the crystallisation process, creating new avenues of process control by allowing the opportunity for corrective actions to be taken in real time if the material falls out-of-specification. Technologies such as Focussed Beam Reflection Measurements (FBRM™), Particle Vision Measurement (PVM™) or Blazemetric™ have introduced the ability for real time particle size and habit tracking during a crystallisation. Shown in Fig. 19 are examples of data collected using a Blaze 900 probe during a cooling crystallisation of the pyrimethanil-L-tartaric acid co-crystal [27]. The co-crystal form has a block-like crystal habit, however, if the crystallisation process is uncontrolled the form can change to the salt which has an acicular needle morphology. Using the in situ imaging it is possible to observe this transformation in real time. Through a robust and controlled crystallisation process, it is possible to reliably obtain the phase pure co-crystal. While not directly measuring particle size, FBRM measures chord length distributions which can be a strong indicator of particle sizes. Probes utilising Attenuated Total Reflection (ATR) FTIR allow for the tracking of concentration in solution ensuring that you can maintain the desired growth driven process to achieve the optimal particle size.

Tracking the nature of the solid form being produced during a crystallisation is also possible using Raman spectroscopy probes. As long as there are differences in the Raman spectra between solid forms it is possible to monitor any possible form transitions in real time.

By using PAT tools it is possible to ensure the integrity of the crystallisation process through scale up to manufacturing and the application

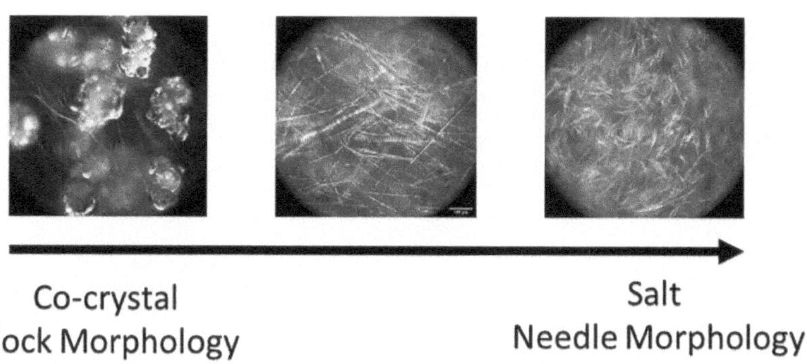

Fig. 19 In-situ imaging showing transformation from co-crystal to salt.

of PAT to monitor and track crystallisation processes has now begun to be incorporated into the crystallisation of co-crystals.

4.6.10 Case studies

Work by Leung et al. [173] from Merck demonstrated the development of a co-crystal form of an API for the treatment of diabetes from screening to scale up. In this study they presented an initial screen of the compound with 12 carboxylic acid coformers using solvent drop assisted sonication in 7 solvents. From screening, co-crystals with benzoic acid and maleic acid were selected for scale up by a solution-based method. Through modelling of the solubility profiles and generation of the ternary phase diagrams, both co-crystals were successfully modelled and scaled using a seeded antisolvent crystallisation from a THF/heptane solvent/antisolvent system. From the ternary phase diagram, they were able to determine the amount of antisolvent that could be added to maintain a safe operating window within the critical co-crystal region.

Yu et al. [174] demonstrated the scale up and polymorphic control of a caffeine-glutamic acid co-crystal through supersaturation control in a cooling crystallisation at a 1 L scale using ATR-FTIR and PVM to observe any polymorphism changes. The caffeine-glutamic acid co-crystal has two known polymorphs: Form 1, the kinetic metastable form, and Form 2, the desired thermodynamic form. In an unseeded crystallisation it was found that the metastable Form 1 crystallised first and then converted to the stable Form 2 during the cooling process. This can be a problematic scenario during manufacturing as trace amounts of an undesirable form requires the batch to be reworked to ensure a phase pure material is isolated. Through investigating both seed loading and cooling rate, it was found that a high supersaturation (measured by ATR-FTIR) resulted in the secondary nucleation of crystals of Form 1 when seeded with Form 2. By implementing a high seed loading and following a crystal growth driven process by carefully reducing the level of supersaturation generated by cooling throughout the crystallisation, they were able to limit the amount of secondary nucleation of Form 1 and also achieved large crystals of the desired Form 2.

An antisolvent crystallisation for a carbamazepine-saccharin co-crystal was designed in work outlined by Wang et al. [175] Carbamazepine is an anticonvulsant used in the treatment of seizures and epilepsy. As part of the investigation five solvents were investigated (methanol, ethanol, acetone, methyl acetate, and ethyl acetate) with water used as an antisolvent.

From these solvents, methanol was selected being the only solvent that yielded a phase pure co-crystal, with a final yield of 84%. The effect of kinetic factors such as antisolvent addition rate and stir speed on the polymorphic outcome of the crystallisation were subsequently investigated [36]. Two polymorphs of the carbamazepine-saccharin co-crystal are known, Form 1 and Form 2. It was found that increasing kinetic factors (faster antisolvent addition and increased agitation rate) favoured the formation of Form 1, whist the inverse (slower antisolvent addition and lower agitation rate) favoured the formation of Form 2.

Svoboda et al. [176] demonstrated the selective continuous reactive crystallisation of a benzoic acid-isonicotinamide co-crystal by rapid mixing of benzoic acid in ethanol directly into a stream of isonicotinamide in water. In the benzoic acid-isonicotinamide system, there are two known co-crystals, one forming in a 1:1 stoichiometry and another forming in a 2:1 stoichiometry, so in order to selectively isolate each form a comprehensive understanding of the crystallisation process is required. Ternary phase diagrams were generated in water, ethanol and water/ethanol mixtures to identify how the stable co-crystal regions changed with mixing. By identifying safe regions to operate, the flow rates of each stream were varied to reach the correct region in the phase diagram to selectively form either the 1:1 or 2:1 co-crystal.

Another example of continuous manufacturing of co-crystals includes work conducted by Powell et al. [177]. This focussed on a cooling crystallisation of the urea-barbituric acid (1:1) co-crystal which has multiple polymorphs. Two of the known polymorphs, Form 1 and Form 3, often form together due to similar stabilities and exhibit very similar solubilities making their isolation difficult. Using a periodic mixed suspension mixed product removal (MSMPR) crystalliser consisting of a cascade of multiple stir tanks held at different temperatures, where transfer between vessels is such that the system remains in a steady state, they were successfully able to selectively make Form 1 (the most stable form) of the co-crystal with high yield. Throughout the continuous process PVM and FBRM were utilised to observe particle size and Raman and ATR-UV/Vis absorption spectroscopy were used to monitor solid form.

4.7 Formulating co-crystals

The formulation stage of pharmaceutical development involves multiple processing steps and the introduction of several chemicals used as excipients to optimise the physiochemical properties of the API. Excipients are used for a multitude of reasons including as binders, lubricants and sweeteners, all to ensure the API is effective and reaches the point of use. Furthermore, formulation involves many processes such as milling, tableting and mixing, and the solid form of the API needs to be stable to the processes it is subjected to throughout, to ensure that the desired form persists in the final formulated product. However, the formulation of co-crystals adds an extra layer of complexity which must be considered, and that is to avoid the potential dissociation of the co-crystal to the individual components. This dissociation can be directly related to the excipients used, as shown by Duggirala et al. [178] where the stability of a caffeine-oxalic acid co-crystal was investigated. The co-crystal, previously reported as being stable, would dissociate during the formulation process when common excipients were used. It was found that the acidic/basic nature of the excipients selected was the cause of the co-crystal dissociation due to the provision of new hydrogen bonding sites (from the excipients) that formed stronger interactions than those found in the co-crystal.

Additional research conducted by Koranne et al. [179] investigated the ionic and hygroscopicity properties of excipients and their effect on co-crystal stability. When a theophylline-glutaric acid co-crystal was tableted without any excipients, within 2 weeks, the co-crystal would significantly dissociate when stored at elevated temperature and humidity (40°C/75% RH), indicating that the dissociation process was water mediated. Several commonly used pharmaceutical excipients were investigated with different hygroscopic profiles and ionizabilities, and it was found that excipients with high hygroscopicity and ionizability, such as sodium starch glycolate, resulted in significant dissociation of the co-crystal in storage. A neutral excipient with high hygroscopicity, such as microcrystalline cellulose, also resulted in the co-crystal dissociating. However, when a neutral molecule with low hygroscopicity, such as lactose, was used as an excipient, dissociation of the co-crystal was not observed. This research demonstrates the importance of careful consideration in excipient selection and a need to investigate their effect on the integrity of the co-crystal form during the formulation stage.

Thermodynamically unstable co-crystals which often exhibit the greatest improvement in dissolution rate over pure API, can also undergo rapid phase transformation in aqueous formulations [180]. It is suspected that the mechanism involves the co-crystal dissolving, dissociating into its constituent components (API and coformer), which then precipitate as separate components (each of lower solubility) which negates any dissolution benefit the co-crystal may offer. Therefore, any formulations utilising these unstable co-crystals must use excipients and additives that prevent such phase transformations from occurring. A possible approach to stabilising co-crystals is through the use of polymers, as demonstrated by Wang et al. [181]. Co-crystals of dihydromyricetin, a Biopharmaceutical class IV drug, were investigated separately with caffeine and urea. It was shown that the IDR of the caffeine co-crystal was 17 times higher than pure dihydromyricetin, however, after 2 min the IDR decreased to match the same rate as the known dihydrate form, indicating a rapid phase transformation. Various polymers were also investigated, and it was found that the presence of PVP K30 in the formulation was successful in inhibiting this transformation, resulting in a higher IDR being maintained for a longer period. The study highlighted how careful formulation design can successfully exploit the use of unstable co-crystals in a drug product where the active molecule displays poor solubility and bioavailability.

5. Application of co-crystals

5.1 Processability

The physiochemical properties of an API are intrinsic to the form being developed. Properties such as bioavailability, solubility and dissolution rate determine the biological performance—how well the drug performs in the body. Others such as processability, stability and hygroscopicity determine the economic performance of the molecule—will the development and manufacturing be profitable? In certain cases, the physiochemical properties can be poor and may result in many additional steps and measures being required to process and protect the molecule to mitigate these poor properties. This in turn incurs additional costs and requires extra time. Co-crystals can offer a route of tailoring the physiochemical properties by using different coformers, removing these additional steps.

A low melting point of an API can be problematic during drug development and manufacturing owing to the pressures and temperatures produced during secondary processing stages; such as milling, granulation and tableting. This can limit the formulations available for use with the molecule. Utilising coformers that have high melting points can make it possible to access co-crystals with a range of melting points. Fig. 20 demonstrates the variations in melting point possible for co-crystals of the agrochemical pyrimethanil [27]. The melting point of pyrimethanil is around

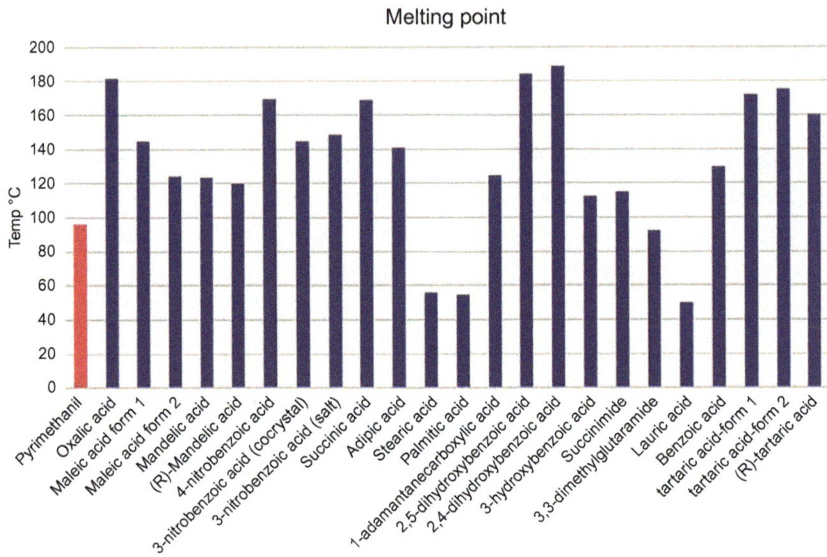

Fig. 20 Melting point of pyrimethanil co-crystals.

96°C, but by forming stronger hydrogen bonds with dicarboxylic acid coformers, such as oxalic acid, it is possible to increase the melting point by some 85°C. Using benzoic acids with multiple hydrogen bonding sites such as 2,4- or 2,5-dihydroxybenzoic acid can result in an increase of about 100°C.

The improvement in melting point is suspected to be the result of increasing the crystal density, packing efficiency, and the number of strong hydrogen bonds within the crystal lattice. This behaviour has also been observed in co-crystals of 4-nitrophenol with aza-compounds [182].

The ability of a molecule to be efficiently processed from a powder to the final tablet is related to the particle morphology and the internal crystal structure. Tabletability is defined as the ability of a powder to form a tablet through compression and is governed by both the compressibility and compactability of the material. Poor performance can require extensive pressures and the addition of additional excipients in order to counter the poor properties of the API powder, resulting in increased cost and development time.

One such example is paracetamol, where the thermodynamically stable form expresses poor tableting performance. However, by changing the crystal structure through the formation of an oxalic acid co-crystal [183] it is possible to significantly improve the tableting performance. Similar trends have also been observed with the metformin-salicylic acid co-crystal, where the co-crystal displays significantly improved tableting compared to pure metformin [184]. The improved performance is attributed to the increased availability of slip planes increasing compression within the crystal lattice.

Co-crystals also offer a unique way of enhancing the chemical stability of a molecule. The improved stability can be explained by the change in the bonding within the crystal structure providing protection to a molecule susceptible to reaction. This has been demonstrated in the study of brexpiprazole and its co-crystal with catechol [185]. Brexpiprazole is used in the treatment of atypical psychotic disorder, however, it is prone to undergo photo-oxidation of the nitrogen of the piperazine ring during granulation with the commonly used excipient PVP. Photo-oxidation of structurally similar phenylpiperazine analogues was seen in the presence of peroxides, which are used in the production of PVP. This suggests the source of the photo-oxidation are the peroxides left over from the PVP manufacturing process. The co-crystal of brexpiprazole and catechol showed significant stability in elevated conditions and in the presence of PVP; analysis of the crystal structure shows bonding at the nitrogen which is susceptible to oxidation, offering protection from the photo-oxidation.

Similar improvements in photostability have been demonstrated in co-crystals of the supplement ubiquinol [186]. The photo-oxidation of ubiquinol, which occurs at room temperature, results in the formation of ubiquinone which has lower bioavailability. Therefore, stabilising the molecule is necessary for the supplement to be effective. Co-crystals were formed using coformers with acceptable toxicity levels and antioxidative properties, with the 3,4-dihydroxybenzoic acid co-crystal showing high stability to oxidation under stressing conditions. It was observed that hydrogen bonds form between the coformer directly to the oxygen centre that undergoes oxidation, thereby protecting the ubiquinol.

Chemical stability can also be improved with the formation of co-crystals that prevent transformation to less chemically stable solid forms, as is demonstrated in the case of levofloxacin [187]. This broad-spectrum antibacterial agent shows no photostability issues as an anhydrous form. However, it is hygroscopic and will readily convert to a hemihydrate under ambient conditions due to channels in the crystal lattice which take up and hold water. Unfortunately, the hemihydrate form has been shown to degrade when exposed to light due to photo-oxidation. Levofloxacin is typically prescribed alongside antipyretics such as paracetamol analogues and a co-crystal was successfully developed with metacetamol.

The co-crystal exhibits significantly lower hygroscopicity than amorphous levofloxacin and is stable under conditions of elevated temperature and humidity, thus preventing the formation of the hemihydrate and subsequent photo-degradation. The improved stability is attributed to the absence of water accessible channels within the co-crystal structure, thus preventing the transformation to the hemihydrate. These examples demonstrate the unique opportunities that co-crystals allow to improve chemical stability of different molecules. Formation of a co-crystal offers the chance to tune the crystal structure and therefore enables the crystal engineer to protect the sites of the molecule that are susceptible to degradation pathways.

5.2 Solubility

Often, pharmaceutical APIs display poor solubility in their free form. A common approach to improving the solubility is by salt formation. However, this relies on the presence of ionisable centres in the molecule. More recently, co-crystal formation has been shown to have a similar effect and can be a viable alternative to salt formation. This has been demonstrated in numerous studies including Thakuria et al. and Bavishi et al. [188,189].

Fig. 21 shows the variation in the aqueous solubility behaviour of pyrimethanil and its co-crystals while Fig. 22 shows the IDR generated in house [27].

Both increases and decreases are observed in solubility and significant changes in the IDR are observed in phosphate buffered saline (PBS, pH 7.4) depending on the coformer used. Free form pyrimethanil was found to have a solubility of 0.08 mg/mL, and an IDR of 18.5 μg/min/cm^2. The co-crystals tested showed between a 30-fold increase to a sevenfold decrease in dissolution rate compared to pyrimethanil itself. Increases in IDR open the possibility for immediate release formulations and low IDR present the opportunity for sustained release.

Similar trends can be seen in research conducted on the broad spectrum antifungal ketoconazole [190]. The API is administered topically and orally, however, the hydrophobic nature of the molecule causes it to exhibit poor aqueous solubility, limiting its application. Work by Martin et al. generated four complexes with ketoconazole: an oxalate salt, and three co-crystals with fumaric, succinic and adipic acid, all prepared by slow evaporation with their nature confirmed by SCXRD. The IDR of these complexes was determined, and it was found that all four had higher IDRs and solubility than the free form of Ketoconazole. The IDR of the co-crystals was also

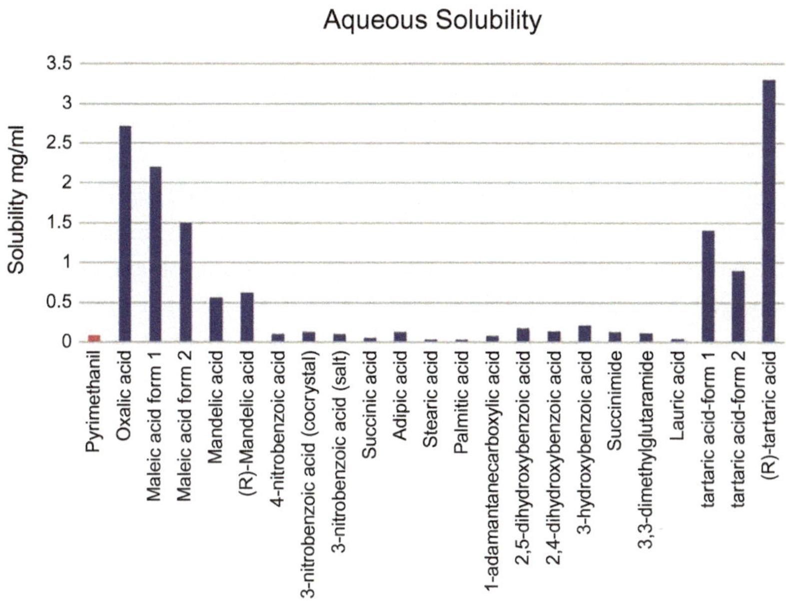

Fig. 21 The aqueous solubility of various co-crystals of pyrimethanil.

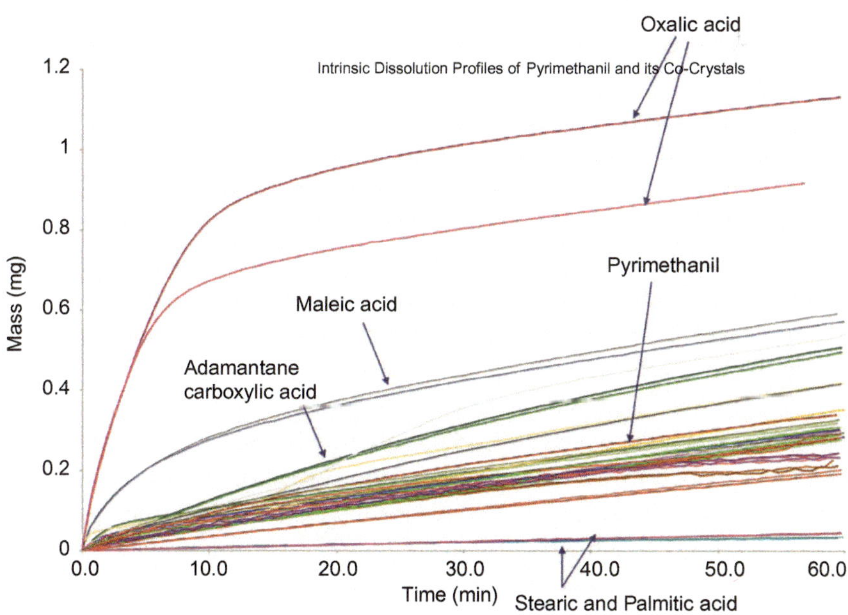

Fig. 22 IDR analyses of pyrimethanil and co-crystals in PBS pH 7.4.

demonstrated to be higher than the respective salt, with the adipic acid co-crystal having an IDR 75 times greater. This work further demonstrates the improvement in solubility of poorly soluble molecules that co-crystal approaches can achieve. However, the opposite effect can also be achieved and was demonstrated by co-crystals of zonisamide [191]. The anticonvulsant API has high solubility and a fast dissolution profile, which result in a short half-life within the body, affecting the dosage requirements. In their study, Aitipamula et al. generated four co-crystals by solvent evaporation and by grinding with L-proline, catechol, 5-methyl-2-pyridone and caffeine. All four co-crystals showed a lower sustained release by IDR with caffeine showing the greatest promise for further development as a sustained release formulation product.

An example of the application of co-crystals to improve solubility on a pharma compound resulting in an improved solubility that is now on the market is aripipozole. An atypical antipsychotic, aripipozole is used in the treatment of schizophrenia, and has 12 anhydrous polymorphs which possess low solubility [192]. A co-crystal was made with fumaric acid [193] which demonstrated the benefits of both the crystalline form in thermodynamic stability as well as the benefits of the amorphous form with improved solubility, rapid onset action and enhanced bioavailability. The cocrystal is now sold under the name Abilify®.

Co-crystal technology can also be used as a way of improving the solubility of classes of molecule that are notoriously insoluble making solid formulations particularly challenging. Kinase inhibitors are one such class, often neutral or weakly lipophilic compounds, exhibiting poor solubility and with low permeability through cell walls resulting in low bioavailability. An example is TAK-020 [194] which is used in the treatment of rheumatoid arthritis. Modification of the TAK-020 through cationic salt formation was unsuccessful providing no chemically stable solid form. During preclinical studies the cyclodextrin Capitsol® was used to enhance solubility, however despite being a accepted pharmaceutical excipient cyclodextrins typically have side effects including diarrhoea and cecal enlargement. Therefore, co-crystals were seen as a way of improving solubility without introducing additional side effects. A screen was conducted using grinding, slurrying and reactive methods with 29 different coformers, which successfully identified several co-crystals by XRPD, NMR and thermal analysis, confirmed by SXRD. Dissolution analysis showed two- to sixfold improvement in both FaSSIF and FeSSIF media compared to the free API. Furthermore, the IDR profiles of all co-crystals formed were increased compared to the free form, particularly co-crystals using gentisic acid, salicylic acid and maleic acid which showed a 47-, 52- and 54-fold increases respectively, providing three solid forms that could be used for future formulation development.

5.3 Bioavailability

Bioavailability is a measure of the rate and fraction of the initial dose of a drug that successfully reaches either the site of action or the bodily fluid domain from which the drug's intended targets have unobstructed access. The extent and rate of drug absorption are usually measured by the area under the blood or plasma concentration-time curve (AUC) and the maximum concentration (C_{max}), respectively [195]. As most APIs are crystalline solids at room temperature, and therefore commonly delivered as a solid oral dosage (tablet form), it is important that APIs have a good aqueous solubility profile. There are, however, several APIs that show exemplary therapeutic profiles and low toxicity but have low aqueous solubility. This often leads to poor absorption and consequently results in subtherapeutic concentrations at the intended target site [196]. By complexing APIs with nontoxic coformers it is possible to improve the solubility and subsequently the bioavailability of an API. This has been demonstrated with the molecule fluoxetine hydrochloride, which is the API in the antidepressant Prozac. After forming co-crystals of fluoxetine hydrochloride with succinic acid, fumaric acid and benzoic acid the IDR of each co-crystal was determined and compared to that of fluoxetine hydrochloride on its own. For the succinic acid co-crystal, an approximate threefold increase in the dissolution rate was observed, while for the benzoic acid co-crystal it was approximately half that of the pure API. The fumaric acid co-crystal had a dissolution rate that was similar to that of the pure API [197].

Wang et al. investigated the bioavailability of 6-mercaptopurine (6-MP), which is used to treat acute lymphoblastic leukaemia, and its co-crystal with isonicotinamide [119]. In a preliminary pharmacokinetic study it was found that the bioavailability of the 6-MP hydrated form was both low and variable; data showed a low plasma concentration (AUC = $131.3 \pm 60.0\, ng\, h\, mL^{-1}$), probably as a result of the API's poor solubility, as well as a long absorption process (T_{max} $1.6 \pm 1.4\, h$). The co-crystal, however, had a higher C_{max} and narrower T_{max} ($1.2 \pm 0.7\, h$) suggesting better dissolution in the gastrointestinal tract. The plasma concentration (AUC = $221.5 \pm 98.9\, ng\, h\, mL^{-1}$) for the co-crystal was also higher suggesting improved bioavailability of the co-crystal when compared to pure 6-MP·H_2O.

Apixaban is a highly potent, selective and useful inhibitor for blood coagulation, however, it has a poor oral bioavailability; approximately 50% for a single 10 mg dose. As the drug molecule has no ionisable moieties, increasing the solubility and bioavailability by salt formation would be challenging. Instead, co-crystal formation was attempted and a novel co-crystal with oxalic acid was identified [198]. The oxalic acid co-crystal showed favourable bioavailability properties with the measured plasma concentration in beagle dogs being 2.7 times higher than that of the pure API. The in vivo pharmacokinetic studies agreed with the results from the in vitro powder dissolution and IDR experiments, demonstrating that the oral bioavailability of apixaban can be improved via formation of the oxalic acid co-crystal.

In a more recent study by Pi et al., a nano-co-crystal of baicalein, an active flavonoid prepared by high pressure homogenization was produced and its pharmacokinetic profile studied [199]. Combining the benefits of co-crystal and nanocrystal technologies provided a novel approach to achieve improved dissolution rate and oral bioavailability [200,201]. The bioavailabilities of pure baicalein as a coarse powder and as a nanocrystal, as well as the baicalein-nicotinamide co-crystal and nano-co-crystal were evaluated in rats. The relative bioavailability enhancement of the co-crystal, pure nanocrystals, and nano-co-crystals were 2.87-, 3.32- and 6.02-fold, respectively, when compared to that of pure baicalein as a coarse powder [199].

5.4 Improving formulation

Acceptability of a drug is defined as the willingness for a patient to take the medication. The acceptability is a product of several factors including shape, texture, packaging, odour and taste. Particularly in children, adolescents and the elderly, taste has been highlighted as an important factor in the acceptance of oral drug products. The formation of co-crystals can offer a way of masking bad tasting APIs by using artificial sweeteners and flavours.

Arafa et al. [202] investigated the formulation of a disintegrating tablet of pure hydrochlorothiazide against its co-crystal with sucralose. Hydrochlorothiazide is a diuretic used in the treatment of high blood pressure and is formulated as a disintegrating tablet, removing the need for water or chewing. In this case, sucralose was selected as the coformer due to already being approved by the FDA. The co-crystal was generated by grinding in the presence of acetone giving three forms with final compositions of 1:1, 1:2 and 1:3 of API to coformer. Dissolution profile investigation showed that with pure hydrochlorothiazide only 30% of the drug had dissolved after 1 h. The co-crystal, however, showed dissolutions of 46%, 80% and 74% after 1 h for 1:1, 1:2 and 1:3 respectively. The inclusion of the sucralose molecule also resulted in an improved taste of the fast disintegration tablets.

Ogata et al. [203] investigated the formation of co-crystals with propiverine with the aim of improving its physiochemical properties. A co-crystal screen with 42 compounds was conducted using grinding methods and nine novel complexes were identified, all of which were successfully scaled up by either cooling or evaporative crystallisation techniques. All the co-crystals were subsequently shown to have increased melting points with respect to pure propiverine (~65°C), with the largest increase (to ~160°C) observed for the salicylic acid co-crystal. Dissolution studies indicated that the novel cocrystals tested all possessed a similar dissolution rate to propiverine hydrochloride, a salt form which is currently used in formulations. However, during taste analysis, it was found that the salicylic acid co crystal was less bitter than both pure propiverine and the propiverine hydrochloride.

Atipamula et al. [204] investigated the ability to improve the taste of theophylline by utilising co-crystal formation with artificial sweeteners for its use in suspension formulations. The artificial sweeteners investigated were saccharin and acesulfame which are well established taste maskers in pharmaceuticals. The co-crystals were generated by solvent drop grinding in a ball mill with methanol for the saccharin system and ethanol for the acesulfame system. The 1:1 theophylline-saccharin co-crystal was found to have an improved drug release profile within the formulation media compared to pure theophylline. The co-crystal was also shown to be stable for up to 9 weeks in suspension. The 1:2 theophylline-acesulfame co-crystal, however, exhibited a lower drug release profile compared with the pure form and was unstable at the four-week time point. In both cases, the sweetness was increased in taste sensor measurements, successfully masking the bitterness of the theophylline.

Co-crystal technology can offer a formulation where no stable API formulation is known, opening a way of salvaging molecules that might have been destined to go through a lengthy and expensive formulation process. An example is the work by Chen et al on a Merck proprietary compound [205]. The initial solid form used for formulation studies was an amorphous bis-HCl salt, however this was found to be hygroscopic and chemically unstable making it unsuitable for further development. Experiments to obtain a solid form of the free API were unsuccessful therefore a salt screen was conducted using evaporation and cooling with inorganic and organic acids to find a stable solid form, with the only crystalline form coming from experiments with phosphoric acid. Initial characterisation by ICP showed the presence of 2 mol eq of phosphate ions, SXRD showed one phosphate bonding as a salt and the other bonding as a co-crystal. Characterisation of the new solid form showed good physiochemical properties including high solubility, >250 mg/mL, and an excellent in vivo performance. From these superior properties this from was selected for further development.

5.5 Permeability

Dermal administration is an effective delivery method for some treatments, offering benefits such as enhanced patient compliance, less frequent dosing and more consistent plasma levels. However, for a molecule to be considered for dermal formulation it must have good skin membrane permeability, which is not always the case for APIs. Co-crystals have the potential to offer a route for improving the permeability of an API, opening different administration routes.

The treatment of cutaneous tuberculosis is conducted using a combination of oral drug therapies including isoniazid; however, this often results in damage to the liver. An alternative treatment route through dermal application of isoniazid could mitigate this side effect. The cutaneous permeation of isoniazid, however, is too high to be used in the treatment of cutaneous tuberculosis, resulting in no topical formulation being available at present. Rosa et al. [206] investigated co-crystal technology with an aim to modulate the permeability for a more localised treatment. An isoniazid-resveratrol co-crystal was produced by reactive co-crystallisation in methanol and the permeation was measured using a Franz cell with a pig's ear skin membrane. The permeation measurement of the co-crystal showed an 87% drug permeation decrease compared to pure isoniazid, while the drug retention in the skin was 5 times higher for the co-crystal.

Meloxicam is a nonsteroidal anti-inflammatory drug used to treat pain and inflammation, however, in oral formulations side effects include abdominal pain, ulceration and haemorrhage. Furthermore, meloxicam solubility has a very dependence on pH, limiting oral formulation development. Work by Machado et al. [207] looked to investigate the use of co-crystals for transdermal administration. A co-crystal with salicylic acid (1:1) was generated by slurrying both components in ethyl acetate. Again, the permeation of the co-crystal was tested using a Franz cell with a pig's ear membrane. During these experiments it was found that the co-crystal skin permeation rate was 1.6 times higher than pure meloxicam when in a suspension, opening the possibility for future topical application of meloxicam.

Dai et al. investigated, both computationally and through permeability studies, the influence co-crystals of 5-fluorouracil. The work initially focussed on benzoic acid coformers [208] and then moved onto lipophilic hydroxy-2-naphthoic acids [209]. 5-Fluorouracil is used in the treatment of skin cancer by topical application, however, its effectiveness is limited due to both poor permeation and skin penetration and is consequently classified as a BSC class III drug (high solubility, low permeability). The initial screen identified three co-crystals, 3-hydroxybenzoic acid, 4-aminobenzoic acid and cinnamic acid, by liquid assisted grinding and slurry screening methods. Permeability studies were conducted using a Franz cell with a silicone membrane and found that all three co-crystals showed an increase in permeability compared to pure 5-fluorouracil, as well as a steady penetration rate after 10 h. Analysis of both the structure of the coformer and the packing in the co-crystals through single crystal analysis revealed that the lipophilicity of the coformer was an important factor in improving the permeability of the co-crystal. This trend was subsequently investigated in further work using lipophilic coformers. Using liquid assisted grinding, two novel co-crystals were identified with 6-hydroxy-2-naphthoic acid and 3-hydroxy-2-naphthoic acid. These two co-crystals exhibited very different permeability behaviour that was shown to correlate to the lipophilicity of the coformer being investigated. When tested by a Franz cell through a silicon membrane, 6-hydroxy-2-naphthoic acid showed a significantly improved permeability compared to that of pure 5-fluorouracil due to the higher lipophilicity, whereas the co-crystal from the less lipophilic 3-hydroxy-2-naphthoic acid showed a reduced permeability.

5.6 Drug–drug co-crystals

5.6.1 Dual action

It may be considered a natural step in the development of co-crystals to replace the inactive coformer, that does not display any therapeutic benefit, with a second drug entity. As shown in previous pharmaceutical co-crystal examples, the ability to tune properties such as tabletability, permeability, solubility and drug release profiles have been demonstrated and one can foresee this may also be attainable in drug–drug co-crystals. In addition to these properties, drug–drug co-crystals can provide an alternative to the administration of fixed dose drug combinations (FDCs). FDCs are drug products that contain at least two drugs in combination within a single tablet dose. The nature of the drug product requires formulation strategies to account for differences in stability, solubility and incompatibility of the intended drugs. An example of a fixed drug dose would be a bilayer drug product where the separate drug components are formulated in a bottom-up approach.

The drug–drug co-crystal has the potential to overcome the formulation difficulties encountered with fixed dose combinations and may also impart some of the enhancing properties already discussed with co-crystal technology. Currently, there are three approved drug–drug co-crystal developed and marketed by Novartis: Entresto®. This is an ionic sodium sacubitril-disodium valsartan co-crystal. Entresto® is a complex co-crystal from a regulatory stand point due to its large unit cell consisting of 6 sacubitril anions, 6 valsartan anions, 18 sodium cations and 15 water molecules [210]. The marketed co-crystal results in improved bioavailability of valsartan, which is reflected by the reduced dosing in Entresto® vs Diovan® (valsartan monotherapy).

The other two drug–drug co-crystals are currently available on the market, one of which is dimenhydrinate, a co-crystal of diphenhydramine-8-chlorotheophylline. The diphenhydramine imparts therapeutic benefits against nausea and vomiting; however, it has a side effect of drowsiness. The 8-chlorotheophylline acts as a stimulant and counteracts the negative effects of the diphenhydramine [211].

The drug Depakote® is marketed as a co-crystal of valproic acid with the salt sodium valproate in a 1:1 stoichiometry, a further example of an ionic co-crystal. Valproic acid exists as a liquid in isolation and the formation of a salt can result in a solid form. The pure sodium salt, however, exhibits undesirable hygroscopic behaviour. The 1:1 stoichiometric co-crystal imparts stability benefits over the sodium salt and no significant weight gain on exposure to a humid atmosphere is observed when compared with the pure salt [50]. Further co-crystal forms have also been identified, one in which a co-crystal forms between three valproate ions with a single valproic acid. The charge balance is maintained by three sodium cations and an additional water molecule is also present in the crystal structure [212]. Further research identified a third form using a combination of XRPD and FT-IR analysis, although no structure has been determined.

The above examples are all the co-crystals currently available on the market, however, there are examples of other promising co-crystals such as the celecoxib-tramadol hydrochloride (CTC) co-crystal [213] which is currently in Phase III trials. The ionic co-crystal exists in a 1:1 stoichiometry, forming a 3D network mediated by a series of hydrogen bonds between the entities. The IDR of CTC was compared against the two isolated APIs. The authors observed a drop in the dissolution rate of tramadol in the CTC, whereas the IDR was increased for celecoxib. It was argued that the reduction in IDR of tramadol would support a delayed and sustained release and thereby reduce the side effects experienced with tramadol, while the increase in IDR of celecoxib would support increased bioavailability. This is an excellent example whereby the co-crystal has appropriately tuned the dissolution of bioavailability of both components appropriately to give an improvement over each of the monotherapies.

5.6.2 Non-commercial drug–drug co-crystals

Beyond the limited number of commercially available drug–drug co-crystals, a growing body of research investigating these co-crystals using different drug entities demonstrates the various benefits of drug–drug co-crystallisation. Additional examples of improvements in solubility, that often lead to an improvement in bioavailability, include a co-crystal of meloxicam and aspirin which resulted in an increase in the solubility of the meloxicam and a consequent fourfold increase in the bioavailability of the drug [214].

The stability of drug compounds is often a concern in manufacturing, which can again be overcome by the formation of co-crystals. For example, the co-crystal of tranilast-nicotinamide resulted in an improvement of the photo-stability of tranilast. In addition, an increase in the solubility of the tranilast was observed [215].

A ternary co-crystal consisting of drug-bridge-drug has been prepared between isoniazid and pyrazinamide overcoming previous difficulties encountered in the co-crystallisation of these two molecules directly [216]. Fumaric acid was used as the bridging coformer, utilising strong heterosynthon interactions with each drug entity and resulting in a 1:2:1 isoniazid:

fumaric acid: pyrazinamide ternary co-crystal. The co-crystal effectively modulated the solubility and IDR of both APIs as well as displaying increased permeability with relation to the individual APIs. The authors concluded that these properties resulted in the co-crystal displaying improved formulation capacity and in vitro/in vivo synergistic effects in comparison to the individual APIs.

The above are some of the myriad examples in using co-crystals to modulate the effects of two separate drugs to provide a single dual action platform. The growing interest in the area is reflected by a recent review [217] that identified all investigations to date. The authors further organised all studies into therapeutic area and list the known benefits and investigations completed for each drug–drug-co-crystal system.

5.7 Purification and chiral resolution

Liquid–liquid extraction, precipitation of salts, chromatography, impurity adsorption by charcoal or recrystallisation are the most common methods for impurity removal. However, crystallisation is the most economical way to purify chemical intermediates and APIs, given that the impurities are rejected during the crystal formation [218]. The principle behind purification by co-crystal formation is to precipitate the API from solution alongside a coformer, leaving the impurities in solution. The isolated solid can then be dissolved, dissociating the co-crystal, and the coformer can either be washed away after a liquid–liquid extraction, or a controlled crystallisation of the API in a specific solvent system can be conducted, leaving the coformer in solution.

Crystallisation from a very impure feedstock is sometimes challenging as the kinetics of nucleation can be significantly inhibited by the presence of impurities. Billot et al. [219] demonstrated the use of co-crystallisation for the purification of SAR1, an SRC kinase inhibitor developed by Sanofi (Fig. 23). The well-known aminopyrimidine synthon on the final product [220] was probed for co-crystal formation and a co-crystal with benzoic acid was successfully formed which displayed improved crystallisation kinetics. An optimisation of the co-crystal formation through crystallisation development provided robust conditions for the pilot lab. The resultant purification was effective; from a feedstock of 60% purity, a product with greater than 99% purity was obtained after cleavage of the co-crystal.

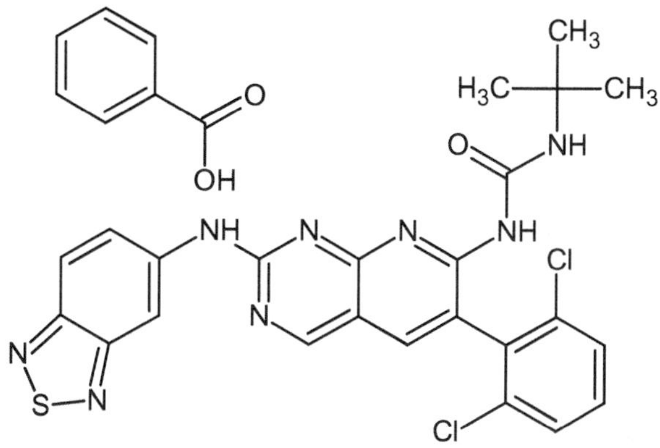

Fig. 23 Structure of co-crystal of SAR1 and benzoic acid.

5.7.1 Diastereomeric co-crystals

When a racemic mixture is ionisable, one of the most popular industrial methods used [221] for chiral resolution is the formation of diastereomeric salts [222]. The separation is based on the crystallisation of diastereomeric salts with a suitable resolving agent [223] (chiral acid or base) which will need to be identified by a screening strategy [224] for each system studied. The isolation of the desired diastereomeric salt can be optimised by a systematic solid state understanding of the crystalline phases involved [225]. In an optimal separation, the unwanted enantiomer remains in solution as the most soluble salt and the enantiomer of interest is isolated by filtration.

With the growth in knowledge in the co-crystal field, chiral resolution by co-crystal formation appears to be an option for the separation of racemic molecules not bearing an ionisable centre. Early work showed that some chiral alcohols could be successfully resolved [226–228] by the formation of diastereomeric co-crystals with chiral amines, (R,R)-tartaric acid or (−)-2,3-dibenzoyl-L-tartaric acid.

As it is not yet possible to predict if a chiral coformer will form a diastereomeric co-crystal phase, a co-crystal screen [229,230] is required to identify a suitable coformer to perform the chiral resolution. Liquid-assisted or neat grinding has been established as a practical way to rapidly generate hits [231,232]; the racemic mixture is typically ground with one equivalent of a chirally pure coformer, the resulting powder is analysed by XRPD and compared against the XRPD of the single components. If new phases are identified then further work will be pursued with this chiral resolving agent.

A solid-state characterisation of the phases involved coupled with measurement of the enantiomeric excess by chiral HPLC may help to develop a solution-based separation system.

Chiral resolution by co-crystal formation has now been exemplified for APIs. Leyssens et al. demonstrated that the co-crystallisation of racemic etiracetam with (S)-mandelic acid was stereospecific, resulting in the formation and isolation of a co-crystal of levetiracetam-(S)-mandelic acid with 70% yield [233]. Separation by co-crystal formation may differ from the separation by formation of diastereomeric salts as 85% of co-crystal formations seem stereospecific [234,235].

Chiral resolution by diastereomeric co-crystal formation was also demonstrated on a large scale with lamivudine, an antiretroviral drug which was separated from its enantiomer with (S)-BINOL on a 30 kg scale with a 67% yield. The API was easily liberated from the co-crystal after a liquid–liquid extraction with an enantiomeric excess of 99.9% [236].

By coupling with a racemisation agent, the chiral resolution can give over the maximum yield of 50%. A screen performed on 4,4-dimethyl-1-(4-fluorophenyl)-2-(1H-1,2,4-triazol-1-yl)-pentan-3-one, a compound with fungicide activity [237], with 40 chiral resolving agents identified (S)-3-phenylbutyric acid as a coformer for its chiral resolution [230]. This formation, when performed in a two reactor process in the presence of 1,8-diazabicyclo-[5.4.0]undec-7-ene (DBU) as a racemising base, has the potential for high yield when optimised (Fig. 24).

5.7.2 Preferential crystallisation of co-crystals

Chiral molecules can be separated by preferential crystallisation if a racemic mixture crystallises as a conglomerate [238]. In the case of a conglomerate, both of the enantiomers crystallise in separate crystallites as opposed to racemic compounds [239] where the enantiomers co-crystallise in the same crystal lattice. Preferential crystallisation, however, is possibly limited to 5%–10% of all compounds as most chemicals tend to form racemic compounds [221]. To identify novel conglomerates, a co-crystal screen using non-chiral coformers can be conducted. Once a suitable system is identified and the system and conditions are favourable, the enantiomers of a conglomerate can be separated by kinetic entrainment (Fig. 25). Typically, a supersaturated solution is seeded with one enantiomer which will induce crystallisation of that enantiomer before the second one spontaneously crystallises [240]. This can then be separated from the solution mixture by filtration.

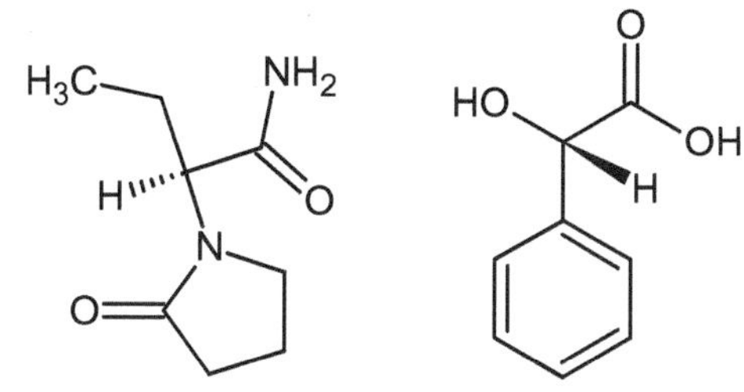

Co-crystal of levetiracetam and (S)-mandelic acid

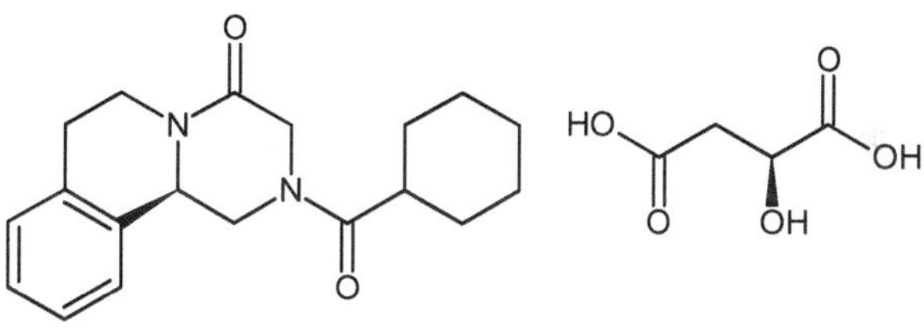

Co-crystal of R-praziquantel and L-malic acid

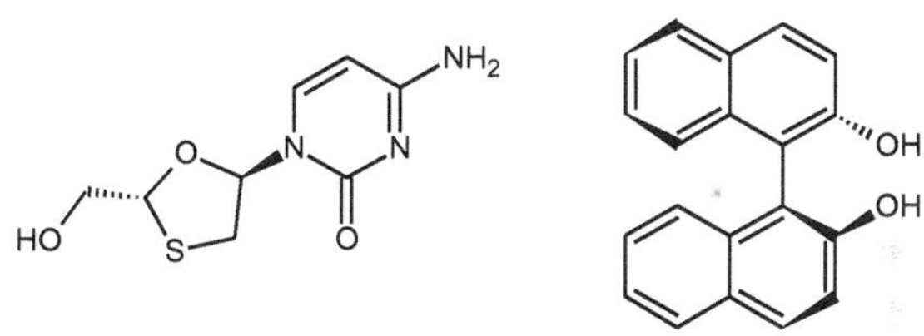

Co-crystal of lamivudine and (S)-BINOL

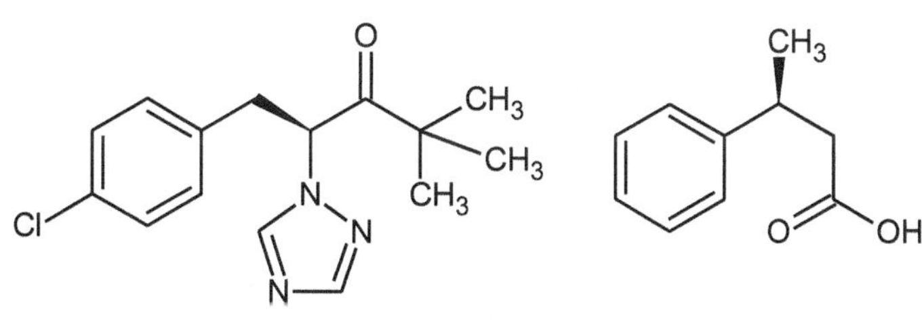

Co-crystal of 4,4-dimethyl-1-(4-fluorophenyl)-2-(1H-1,2,4-triazol-1-yl)-pentan-3-one and (S)-3-phenylbutyric acid

Fig. 24 Examples of chiral API resolved by co-crystal formation.

Harfouche et al. identified [241] a favourable conglomerate forming a monohydrate co-crystal salicylic acid-proxyphylline. The resolution by preferential crystallisation was exemplified in this system on a small scale [242].

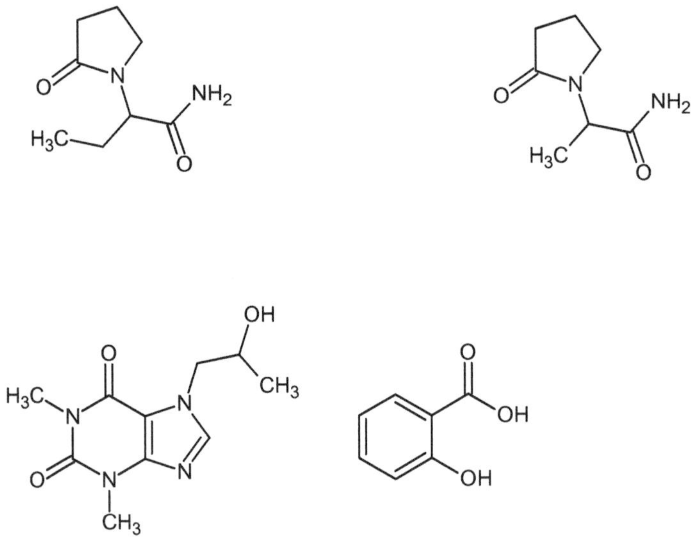

Fig. 25 Examples of co-crystals forming conglomerates.

6. Summary and looking forward

Co-crystal technology offers a unique way of not only altering and improving the properties of final API candidates, but also as a way of resolving potential issues with purity, both chiral and chemical. In recent years, the body of research investigating and demonstrating the developmental process as well as benefits of pharmaceutical co-crystals has expanded, however, despite this their acceptance is not as widespread as other solid forms such as salts and hydrates. The reluctance to adopt co-crystals can be attributed to their perceived ambiguity from a regulatory standpoint, as well as their added complexity in terms of scale-up and formulation; this is further compounded by fears of potential dissociation and stability issues. However, with many API's demonstrating poor solubility and subsequently falling into the BCS Class II and IV categories, scientists must maximise the number of tools at their disposal in order to overcome these issues and co-crystallisation has the potential to be a powerful technique.

This chapter has addressed these hesitations by presenting research and insights into the screening, manufacturing and formulation of co-crystals, as well as the benefits achievable with this technology. Furthermore, we have presented the regulatory landscape faced in the development of co-crystals from an innovator and generic standpoint.

Despite the abundance of academic literature on developing co-crystals there are still some outstanding aspects of co-crystal development where additional research could aid in the wider acceptance of co-crystals. Due to the limited number of co-crystals on the market, there are few examples of live successful formulations. Therefore, further exploration is needed investigating factors, such as excipients and processing techniques, to identify successful ways of formulating and stabilising co-crystals for manufacturing. Another area requiring additional research is the scale up of co-crystals that are only accessed through mechanochemistry techniques. While twin screw extrusion appears to be a promising solution, co-crystals that can only be accessed through mechanochemistry could require lengthy and expensive crystallisation development in order to be successful at an industrial scale.

Ultimately, it is the change in regulatory position in the last couple of years, particularly from the FDA, that has created an environment more hospitable for co-crystals in drug development. In addition, the retrospective analysis of the solid form for existing drug products has further demonstrated that some pharmaceutical co-crystals are already in use on the market. However, their regulatory approval was not granted with this knowledge. With the change in regulatory stance for co-crystals to be treated in the same way as polymorphs, it is anticipated that an increasing number of molecules will be developed as co-crystals. This will give rise to new opportunities to be effectively exploited as a strategy by the generics space being predicted. While the acceptance of co-crystals is in all probability inevitable; its uptake across the pharmaceutical sector is expected to be slow. The pharmaceutical industry is often conservative in the introduction of novel technologies; some could say understandably considering the end-use of the final product and even the nimblest of companies are hesitant of change.

By addressing the potential issues demonstrating the untapped opportunities that co-crystals can offer, we argue that they should be given equal consideration with other established solid forms when trying to resolve physicochemical problems and potential formulation concerns encountered in drug development.

References

[1] Chemburkar SR, Bauer J, Deming K, Spiwek H, Patel K, Morris J, et al. Dealing with the impact of ritonavir polymorphs on the late stages of bulk drug process development. Org Process Res Dev 2000;4(5):413–7.

[2] Elder DP, Holm R, de Diego HL. Use of pharmaceutical salts and cocrystals to address the issue of poor solubility. Int J Pharm 2013;453(1):88–100.

[3] Berry DJ, Steed JW. Pharmaceutical cocrystals, salts and multicomponent systems; intermolecular interactions and property based design. Adv Drug Deliv Rev 2017;117:3–24.
[4] Duggirala NK, Perry ML, Almarsson Ö, Zaworotko MJ. Pharmaceutical cocrystals: along the path to improved medicines. Chem Commun 2016;52(4):640–55.
[5] Kavanagh ON, Croker DM, Walker GM, Zaworotko MJ. Pharmaceutical cocrystals: from serendipity to design to application. Drug Discov Today 2019;24(3):796–804.
[6] Zimmermann B, Baranović G. Thermal analysis of paracetamol polymorphs by FT-IR spectroscopies. J Pharm Biomed Anal 2011;54(2):295–302.
[7] Di Martino P, Guyot-Hermann A-M, Conflant P, Drache M, Guyot J-C. A new pure paracetamol for direct compression: the orthorhombic form. Int J Pharm 1996; 128(1):1–8.
[8] Braun DE, Griesser UJ. Stoichiometric and nonstoichiometric hydrates of brucine. Cryst Growth Des 2016;16(10):6111–21.
[9] Aitipamula S, Banerjee R, Bansal AK, Biradha K, Cheney ML, Choudhury AR, et al. Polymorphs, salts, and cocrystals: what's in a name? Cryst Growth Des 2012; 12(5):2147–52.
[10] Almarsson Ö, Zaworotko MJ. Crystal engineering of the composition of pharmaceutical phases. Do pharmaceutical co-crystals represent a new path to improved medicines? Chem Commun 2004;(17):1889–96.
[11] Braga D, Grepioni F, Shemchuk O. Organic–inorganic ionic co-crystals: a new class of multipurpose compounds. CrstEngComm 2018;20(16):2212–20.
[12] Braga D, Grepioni F, Lampronti GI, Maini L, Turrina A. Ionic co-crystals of organic molecules with metal halides: a new prospect in the solid formulation of active pharmaceutical ingredients. Cryst Growth Des 2011;11(12):5621–7.
[13] Pepinsky R. Crystal engineering: a new concept in crystallography. Phys Rev 1955;100:952.
[14] Wohler F. Untersuchungen über das chinon. Justus Lieb Ann Chem 1844;51:153.
[15] Matsuda H, Osaki K, Nitta I. Crystal structure of quinhydrone. Bull Chem Soc Jpn 1958;31(5):611.
[16] Cohen MD, Schmidt GMJ, Sonntag FI. 384. Topochemistry. Part II. The photochemistry of trans-cinnamic acids. J Chem Soc 1964;2000–13.
[17] Schmidt GMJ. Photodimerization in the solid state. Pure Appl Chem 1971; 27(4):647–78.
[18] Schmidt GMJ. 385. Topochemistry. Part III. The crystal chemistry of some trans-cinnamic acids. J Chem Soc 1964;2014–21.
[19] Aakeroy CB, Seddon KR. The hydrogen bond and crystal engineering. Chem Soc Rev 1993;22(6):397–407.
[20] Pfeiffer P. Organische Molekülverbindungen; 1922.
[21] Lemmerer A, Bernstein J, Griesser UJ, Kahlenberg V, Többens DM, Lapidus SH, et al. A tale of two polymorphic pharmaceuticals: pyrithyldione and propyphenazone and their 1937 co-crystal patent. Chem A Eur J 2011;17(48):13445–60.
[22] Etter MC, MacDonald JC, Bernstein J. Graph-set analysis of hydrogen-bond patterns in organic crystals. Acta Crystallogr B Struct Sci 1990;46(2):256–62.
[23] Europa. Reflection paper on the use of cocrystals and other solid state forms of active substances in medicinal products. [Internet]. [cited 2020 Aug 13]. Available from https://www.ema.europa.eu/en/documents/scientific-guideline/reflection-paper-use-cocrystals-active-substances-medicinal-products_en.pdf; 2020.
[24] Childs SL, Stahly GP, Park A. The salt – cocrystal continuum: the influence of crystal structure on ionization state. Mol Pharm 2007;4(3):323–38.
[25] Bhogala BR, Basavoju S, Nangia A. Tape and layer structures in cocrystals of some di- and tricarboxylic acids with 4,4′-bipyridines and isonicotinamide. From binary to ternary cocrystals. CrstEngComm 2005;7(90):551–62.

[26] Grothe E, Meekes H, Vlieg E, ter Horst JH, de Gelder R. Solvates, salts, and cocrystals: a proposal for a feasible classification system. Cryst Growth Des 2016;16(6):3237–43.
[27] Johnson Matthey, n.d. Manuscipt in preperation.
[28] Wilson CC. Migration of the proton in the strong O—H···O hydrogen bond in urea–phosphoric acid (1/1). Acta Crystallogr B 2001;57(3):435–9.
[29] Jones CL, Skelton JM, Parker SC, Raithby PR, Walsh A, Wilson CC, et al. Living in the salt-cocrystal continuum: indecisive organic complexes with thermochromic behaviour. CrstEngComm 2019;21(10):1626–34.
[30] Rajput L, Banik M, Yarava JR, Joseph S, Pandey MK, Nishiyama Y, et al. Exploring the salt-cocrystal continuum with solid-state NMR using natural-abundance samples: implications for crystal engineering. IUCrJ 2017;4(Pt. 4):466–75.
[31] Aitipamula S, Chow PS, Tan RBH. Polymorphs and solvates of a cocrystal involving an analgesic drug, ethenzamide, and 3,5-dinitrobenzoic acid. Cryst Growth Des 2010;10(5):2229–38.
[32] Liu X, Tang CC, Boldyreva EV, Pulham CR. Influence of crystal packing on the mechanism of decomposition of the acetonitrile-solvated cocrystal of piroxicam and succinic acid. Cryst Growth Des 2019;19(12):7315–23.
[33] Muresan-Pop M, Braga D, Pop MM, Borodi G, Kacso I, Maini L. Crystal structure and physicochemical characterization of ambazone monohydrate, anhydrous, and acetate salt solvate. J Pharm Sci 2014;103(11):3594–601.
[34] Jiao L, Yang D, Zhang L, Yang S, Du G, Lu Y. Salt solvates of quinolones and oxicams: theoretical computation, structural characterization and dissolution studies. J Mol Struct 2021;1223, 128865.
[35] Wang T, Stevens JS, Vetter T, Whitehead GFS, Vitorica-Yrezabal IJ, Hao H, et al. Salts, cocrystals, and ionic cocrystals of a "simple" tautomeric compound. Cryst Growth Des 2018;18(11):6973–83.
[36] Lee M-J, Wang I-C, Kim M-J, Kim P, Song K-H, Chun N-H, et al. Controlling the polymorphism of carbamazepine-saccharin cocrystals formed during anti-solvent cocrystallization using kinetic parameters. Korean J Chem Eng 2015; 32(9):1910–7.
[37] Porter IIIWW, Elie SC, Matzger AJ. Polymorphism in carbamazepine cocrystals. Cryst Growth Des 2008;8(1):14–6.
[38] Ueto T, Takata N, Muroyama N, Nedu A, Sasaki A, Tanida S, et al. Polymorphs and a hydrate of furosemide–nicotinamide 1:1 cocrystal. Cryst Growth Des 2012; 12(1):485–94.
[39] Byrn SR, Pfeiffer RR, Stowell JG. Solid-state chemistry of drugs [Internet]. West Lafayette, Ind: SSCI, Inc; 1999. Available from http://books.google.com/books?id=zAZtAAAAMAAJ.
[40] Raheem Thayyil A, Juturu T, Nayak S, Kamath S. Pharmaceutical co-crystallization: regulatory aspects, design, characterization, and applications. Adv Pharm Bull 2020; 10(2):203–12. 2020/02/18 ed.
[41] FDA. Regulatory classification of pharmaceutical co-crystals guidance for industry. U.S. Department of Health and Human Services Food and Drug Administration Center for Drug Evaluation and Research (CDER); 2016. [Internet] [cited 2020 Aug 19]. Available from https://www.fda.gov/media/99917/download.
[42] Desiraju GR. Cryptic crystallography. Nat Mater 2002;1(2):77–9.
[43] Meade Edwin M. Sodium hydrogen divalproate oligomer. [Internet]. US; US 4988731 A. Available from https://lens.org/117-885-257-672-819; 1991.
[44] Boegesoe Klaus P, Jens P. Pharmaceutically useful (+)-1-(3-dimethylaminopropyl)-1-(4′-fluorophenyl)-1,3-dihydrosobenzofuran-5-carbonitrile and non-toxic acid addition salts thereof. [Internet]. US; US 4943590 A. Available from https://lens.org/003-034-852-408-036; 1990.

[45] Lili F, Erik GS, Allen SP, Mahavir P, Girgis Michael J, Bin H, et al. Compounds containing S-n-valeryl-n-{[2′-(1h-tetrazole-5-yl)-biphenyl-4-yl]-methyl}-valine and (2r,4s)-5-biphenyl-4-yl-4-(3-carboxy-propionylamino)-2-methyl-pentanoic acid ethyl ester moieties and cations. [Internet]. US; US 9388134 B2. Available from https://lens.org/083-337-018-634-832; 2016.
[46] Masakazu I, Keita N, Ryota S, Kenichi O, Daisuke S, Masamichi Y. Cocrystal of C-glycoside derivative and L-proline. [Internet]. US; US 8097592 B2. Available from https://lens.org/197-214-014-255-222; 2012.
[47] Vincent M, Micah CB. Dioxa-bicyclo[3.2.1.]Octane-2,3,4-triol derivatives. [Internet]. EP; EP 2334687 B1. Available from https://lens.org/128-284-146-221-025; 2012.
[48] James MG, Ludek R, Jindrich R, Ondrej D, Tomas C, Tomas P, et al. Pharmaceutically acceptable cocrystals of N-[2-(7-methoxy-1-naphthyl)]acetamide and methods of their preparation. [Internet]. EP; EP 2517700 B1. Available from https://lens.org/183-947-897-671-78X; 2013.
[49] Kumar A, Kumar S, Nanda AA. Review about regulatory status and recent patents of pharmaceutical co-crystals. Adv Pharm Bull 2018;8(3):355–63. 2018/08/29 ed.
[50] Anon. Solid substances comprising valproic acid and sodium valproate. [Internet]. US6077542A [cited 2019 Aug 18]. Available from https://patents.google.com/patent/US6077542/en28.
[51] Harrison WTA, Yathirajan HS, Bindya S, Anilkumar HG, Devaraju. Escitalopram oxalate: co-existence of oxalate dianions and oxalic acid molecules in the same crystal. Acta Crystallogr C 2007;63(2):o129–31.
[52] Odomzo. International non-proprietary name: sonidegib [Internet]. European Medicines Agency; 2015. Jun [cited 2019 Aug 18]. Report No.: EMA/472165/2015. Available from https://www.ema.europa.eu/en/documents/assessment-report/odomzo-epar-public-assessment-report_en.pdf.
[53] EMA. Reflection paper on the use of cocrystals of active substances in medicinal products. European Medicines Agency; 2015. [Internet] May [cited 2020 Aug 20]. Report no.: reflection on the current thinking. Available from https://www.ema.europa.eu/en/documents/scientific-guideline/reflection-paper-use-cocrystals-active-substances-medicinal-products_en.pdf.
[54] Fukte SR, Wagh MP, Rawat S. Coformer selection: an important tool in cocrystal formation. Int J Pharm Pharm Sci 2014;6(7):9–14. [Internet]. [cited 2020 Aug 20]. Available from https://innovareacademics.in/journals/index.php/ijpps/article/view/1733.
[55] Devogelaer J-J, Meekes H, Vlieg E, de Gelder R. Cocrystals in the Cambridge structural database: a network approach. Acta Crystallogr B 2019;75(3):371–83.
[56] Mohammad MA, Alhalaweh A, Velaga SP. Hansen solubility parameter as a tool to predict cocrystal formation. Int J Pharm 2011;407(1):63–71.
[57] FDA. Food and drugs chapter I—food and drug administration department of health and human services subchapter B—food for human consumption. U.S Food & Drug Administration; 2021. [Internet]. [cited 2020 Aug 20]. Available from https://www.accessdata.fda.gov/scripts/cdrh/cfdocs/cfcfr/CFRSearch.cfm?CFRPart=182.
[58] FDA. Substances added to food (formerly EAFUS). U.S Food & Drug Administration; 2021. [Internet] [cited 2020 Aug 20]. Available from https://www.cfsanappsexternal.fda.gov/scripts/fdcc/?set=FoodSubstances.
[59] Guillory JK. Handbook of pharmaceutical salts: properties, selection, and use edited by P. Heinrich Stahl and Camile G. Wermuth. VHCA, Verlag Helvetica Chimica Acta, Zürich, Switzerland, and Wiley-VCH, Weinheim, Germany. 2002. J Med Chem 2003;46(7):1277.

[60] Ringle S. Regulatory implications using API cocrystals for generic medicinal products within the EU and US. Rheinische Friedrich-Wilhelms unviersity; 2018. [Internet] [cited 2020 Aug 20]. Available from https://www.dgra.de/media/pdf/studium/masterthesis/master_ringle_sam_2018.pdf.
[61] Anand O, Yu LX, Conner DP, Davit BM. Dissolution testing for generic drugs: an FDA perspective. AAPS J 2011;13(3):328.
[62] FDA. Size, shape, and other physical attributes of generic tablets and capsules. U.S Department of Health and Human Services, Food and Drug Administration; 2015. [Internet] [cited 2020 Aug 20]. Available from https://www.fda.gov/media/87344/download.
[63] Aitipamula S, Chow PS, Tan RBH. Polymorphism in cocrystals: a review and assessment of its significance. CrstEngComm 2014;16(17):3451–65.
[64] Aakeröy CB, Sinha AS, editors. Co-crystals: preparation, characterization and applications. Monographs in supramolecular chemistry. The Royal Society of Chemistry; 2018. p. P001–342. [Internet]. Available from https://doi.org/10.1039/9781788012874.
[65] Etter MC. Hydrogen bonds as design elements in organic chemistry. J Phys Chem 1991;95(12):4601–10.
[66] Saha S, Desiraju GR. Acid···amide supramolecular synthon in cocrystals: from spectroscopic detection to property engineering. J Am Chem Soc 2018;140(20):6361–73.
[67] Desiraju GR. Crystal engineering: from molecule to crystal. J Am Chem Soc 2013;135(27):9952–67.
[68] Hutchins KM. Functional materials based on molecules with hydrogen-bonding ability: applications to drug co-crystals and polymer complexes. R Soc Open Sci 2018;5(6), 180564.
[69] Desiraju GR. Supramolecular synthons in crystal engineering—a new organic synthesis. Angew Chem Int Ed Engl 1995;34(21):2311–27.
[70] Corey EJ. Robert Robinson lecture. Retrosynthetic thinking? Essentials and examples. Chem Soc Rev 1988;17:111–33.
[71] Taylor CR, Day GM. Evaluating the energetic driving force for cocrystal formation. Cryst Growth Des 2018;18(2):892–904.
[72] Issa N, Karamertzanis PG, Welch GWA, Price SL. Can the formation of pharmaceutical cocrystals be computationally predicted? I. Comparison of lattice energies. Cryst Growth Des 2009;9(1):442–53.
[73] Fábián L. Cambridge structural database analysis of molecular complementarity in cocrystals. Cryst Growth Des 2009;9(3):1436–43.
[74] Delori A, Galek PTA, Pidcock E, Patni M, Jones W. Knowledge-based hydrogen bond prediction and the synthesis of salts and cocrystals of the anti-malarial drug pyrimethamine with various drug and GRAS molecules. CrstEngComm 2013;15(15):2916–28.
[75] Wood PA, Feeder N, Furlow M, Galek PTA, Groom CR, Pidcock E. Knowledge-based approaches to co-crystal design. CrstEngComm 2014;16(26):5839–48.
[76] Devogelaer J-J, Brugman SJT, Meekes H, Tinnemans P, Vlieg E, de Gelder R. Cocrystal design by network-based link prediction. CrstEngComm 2019;21(44):6875–85.
[77] Shattock TR, Arora KK, Vishweshwar P, Zaworotko MJ. Hierarchy of supramolecular synthons: persistent carboxylic acid···pyridine hydrogen bonds in cocrystals that also contain a hydroxyl moiety. Cryst Growth Des 2008;8(12):4533–45.
[78] Oswald IDH, Allan DR, McGregor PA, Motherwell WDS, Parsons S, Pulham CR. The formation of paracetamol (acetaminophen) adducts with hydrogen-bond acceptors. Acta Crystallogr B 2002;58(6):1057–66.

[79] Walsh RDB, Bradner MW, Fleischman S, Morales LA, Moulton B, Rodríguez-Hornedo N, et al. Crystal engineering of the composition of pharmaceutical phases. Chem Commun 2003;(2):186–7.
[80] Hasa D, Schneider Rauber G, Voinovich D, Jones W. Cocrystal formation through mechanochemistry: from neat and liquid-assisted grinding to polymer-assisted grinding. Angew Chem 2015;127(25):7479–83.
[81] Trask AV, Shan N, Motherwell WDS, Jones W, Feng S, Tan RBH, et al. Selective polymorph transformation via solvent-drop grinding. Chem Commun 2005;(7):880.
[82] Hasa D, Miniussi E, Jones W. Mechanochemical synthesis of multicomponent crystals: one liquid for one polymorph? A myth to dispel. Cryst Growth Des 2016; 16(8):4582–8.
[83] Fischer F, Scholz G, Benemann S, Rademann K, Emmerling F. Evaluation of the formation pathways of cocrystal polymorphs in liquid-assisted syntheses. CrstEngComm 2014;16(35):8272–8.
[84] Caira MR. Molecular complexes of sulfonamides. 2.1:1 complexes between drug molecules: sulfadimidine-acetylsalicylic acid and sulfadimidine-4-aminosalicylic acid. J Cryst Spectrosc 1992;22(2):193–200.
[85] Mukherjee A, Rogers RD, Myerson AS. Cocrystal formation by ionic liquid-assisted grinding: case study with cocrystals of caffeine. CrstEngComm 2018;20(27):3817–21.
[86] Hasa D, Carlino E, Jones W. Polymer-assisted grinding, a versatile method for polymorph control of cocrystallization. Cryst Growth Des 2016;16(3):1772–9.
[87] Friščić T, Halasz I, Beldon PJ, Belenguer AM, Adams F, Kimber SAJ, et al. Real-time and in situ monitoring of mechanochemical milling reactions. Nat Chem 2013;5 (1):66–73.
[88] Julien PA, Germann LS, Titi HM, Etter M, Dinnebier RE, Sharma L, et al. *In situ* monitoring of mechanochemical synthesis of calcium urea phosphate fertilizer cocrystal reveals highly effective water-based autocatalysis. Chem Sci 2020;11 (9):2350–5.
[89] Halasz I, Puškarić A, Kimber SAJ, Beldon PJ, Belenguer AM, Adams F, et al. Real-time in situ powder X-ray diffraction monitoring of mechanochemical synthesis of pharmaceutical cocrystals. Angew Chem Int Ed 2013;52(44):11538–41.
[90] Mullin JW, John W. Crystallization. 4th ed. Oxford: Oxford; 2001.
[91] Sheikh AY, Rahim SA, Hammond RB, Roberts KJ. Scalable solution cocrystallization: case of carbamazepine-nicotinamide I. CrstEngComm 2009;11(3):501–9.
[92] Bag PP, Patni M, Malla Reddy C. A kinetically controlled crystallization process for identifying new co-crystal forms: fast evaporation of solvent from solutions to dryness. CrstEngComm 2011;13(19):5650–2.
[93] Rahim SA, Hammond RB, Sheikh AY, Roberts KJ. A comparative assessment of the influence of different crystallization screening methodologies on the solid forms of carbamazepine co-crystals. CrstEngComm 2013;15(19):3862–73.
[94] Morissette SL, Almarsson Ö, Peterson ML, Remenar JF, Read MJ, Lemmo AV, et al. High-throughput crystallization: polymorphs, salts, co-crystals and solvates of pharmaceutical solids. Adv Drug Deliv Rev 2004;56(3):275–300.
[95] Morissette SL, Soukasene S, Levinson D, Cima MJ, Almarsson Ö. Elucidation of crystal form diversity of the HIV protease inhibitor ritonavir by high-throughput crystallization. Proc Natl Acad Sci U S A 2003;100(5):2180.
[96] Almarsson Ö, Hickey MB, Peterson ML, Morissette SL, Soukasene S, McNulty C, et al. High-throughput surveys of crystal form diversity of highly polymorphic pharmaceutical compounds. Cryst Growth Des 2003;3(6):927–33.
[97] Lee EH. A practical guide to pharmaceutical polymorph screening & selection. Asian J Pharm Sci 2014;9(4):163–75.

[98] Zhang GGZ, Henry RF, Borchardt TB, Lou X. Efficient co-crystal screening using solution-mediated phase transformation. J Pharm Sci 2007;96(5):990–5.
[99] Bučar D-K, Henry RF, Lou X, Borchardt TB, Zhang GGZ. A "hidden" co-crystal of caffeine and adipic acid. Chem Commun 2007;(5):525–7.
[100] Fleischman SG, Kuduva SS, McMahon JA, Moulton B, Bailey Walsh RD, Rodríguez-Hornedo N, et al. Crystal engineering of the composition of pharmaceutical phases: multiple-component crystalline solids involving carbamazepine. Cryst Growth Des 2003;3(6):909–19.
[101] Li Z, Yang B-S, Jiang M, Eriksson M, Spinelli E, Yee N, et al. A practical solid form screen approach to identify a pharmaceutical glutaric acid cocrystal for development. Org Process Res Dev 2009;13(6):1307–14.
[102] Zhao L, Hanrahan MP, Chakravarty P, DiPasquale AG, Sirois LE, Nagapudi K, et al. Characterization of pharmaceutical cocrystals and salts by dynamic nuclear polarization-enhanced solid-state NMR spectroscopy. Cryst Growth Des 2018; 18(4):2588–601.
[103] Aitipamula S, Vangala VR. X-ray crystallography and its role in understanding the physicochemical properties of pharmaceutical cocrystals. J Indian Inst Sci 2017; 97(2):227–43.
[104] Yamashita H, Hirakura Y, Yuda M, Teramura T, Terada K. Detection of cocrystal formation based on binary phase diagrams using thermal analysis. Pharm Res 2013;30(1):70–80.
[105] Datta S, Grant DJW. Crystal structures of drugs: advances in determination, prediction and engineering. Nat Rev Drug Discov 2004;3(1):42–57.
[106] Tan D, Loots L, Friščić T. Towards medicinal mechanochemistry: evolution of milling from pharmaceutical solid form screening to the synthesis of active pharmaceutical ingredients (APIs). Chem Commun 2016;52(50):7760–81.
[107] Braga D, Maini L, Grepioni F. Mechanochemical preparation of co-crystals. Chem Soc Rev 2013;42(18):7638–48.
[108] Shankland K, Spillman MJ, Kabova EA, Edgeley DS, Shankland N. The principles underlying the use of powder diffraction data in solving pharmaceutical crystal structures. Acta Crystallogr C 2013;69(11):1251–9.
[109] Harris KDM. Structure solution from powder X-ray diffraction data by genetic algorithm techniques, applied to organic materials generated as polycrystalline products from solid state processes. Mater Manuf Process 2009;24(3):293–302.
[110] Cheung EY, Kitchin SJ, Harris KDM, Imai Y, Tajima N, Kuroda R. Direct structure determination of a multicomponent molecular crystal prepared by a solid-state grinding procedure. J Am Chem Soc 2003;125(48):14658–9.
[111] Bolla G, Chernyshev V, Nangia A. Acemetacin cocrystal structures by powder X-ray diffraction. IUCrJ 2017;4(3):206–14.
[112] Tatton AS, Pham TN, Vogt FG, Iuga D, Edwards AJ, Brown SP. Probing hydrogen bonding in cocrystals and amorphous dispersions using 14N–1H HMQC solid-state NMR. Mol Pharm 2013;10(3):999–1007.
[113] Tatton AS, Pham TN, Vogt FG, Iuga D, Edwards AJ, Brown SP. Probing intermolecular interactions and nitrogen protonation in pharmaceuticals by novel 15N-edited and 2D 14N-1H solid-state NMR. CrstEngComm 2012;14(8):2654–9.
[114] Vogt FG, Clawson JS, Strohmeier M, Edwards AJ, Pham TN, Watson SA. Solid-state NMR analysis of organic cocrystals and complexes. Cryst Growth Des 2009;9 (2):921–37.
[115] Khan M, Enkelmann V, Brunklaus G. Heterosynthon mediated tailored synthesis of pharmaceutical complexes: a solid-state NMR approach. CrstEngComm 2011; 13(9):3213–23.

[116] Luedeker D, Gossmann R, Langer K, Brunklaus G. Crystal engineering of pharmaceutical co-crystals: "NMR crystallography" of niclosamide co-crystals. Cryst Growth Des 2016;16(6):3087–100.
[117] Mukherjee A, Tothadi S, Chakraborty S, Ganguly S, Desiraju GR. Synthon identification in co-crystals and polymorphs with IR spectroscopy. Primary amides as a case study. CrstEngComm 2013;15(23):4640–54.
[118] da Silva CC, Guimarães FF, Ribeiro L, Martins FT. Salt or cocrystal of salt? Probing the nature of multicomponent crystal forms with infrared spectroscopy. Spectrochim Acta A Mol Biomol Spectrosc 2016;167:89–95.
[119] Wang J-R, Yu X, Zhou C, Lin Y, Chen C, Pan G, et al. Improving the dissolution and bioavailability of 6-mercaptopurine via co-crystallization with isonicotinamide. Bioorg Med Chem Lett 2015;25(5):1036–9.
[120] Kojima T, Tsutsumi S, Yamamoto K, Ikeda Y, Moriwaki T. High-throughput cocrystal slurry screening by use of in situ Raman microscopy and multi-well plate. Int J Pharm 2010;399(1):52–9.
[121] Ahuja D, Svärd M, Rasmuson ÅC. Investigation of solid–liquid phase diagrams of the sulfamethazine–salicylic acid co-crystal. CrstEngComm 2019;21(18):2863–74.
[122] Zhang S, Chen H, Rasmuson ÅC. Thermodynamics and crystallization of a theophylline–salicylic acid cocrystal. CrstEngComm 2015;17(22):4125–35.
[123] DeVoe H. Thermodynamics and chemistry. First. University of Maryland; 2014.
[124] Atkins PW, Peter W. Elements of physical chemistry. 7th ed. Oxford University Press; 2017.
[125] Coquerel G. Crystallization of molecular systems from solution: phase diagrams, supersaturation and other basic concepts. Chem Soc Rev 2014;43(7):2286–300.
[126] Garside J, Davey R. From molecules to crystallizers. Oxford University Press; 2000.
[127] Zaworotko M, Bond A, Row TNG, Rodriguez-Hornedo N, Coquerel G, Hilfiker R, et al. Pharmaceutical salts and co-crystals [Internet]. Royal Society of Chemistry; 2011. (ISSN). Available from https://books.google.co.uk/books?id=3HIoDwAAQBAJ.
[128] Habgood M, Deij MA, Mazurek J, Price SL, ter Horst JH. Carbamazepine co-crystallization with pyridine carboxamides: rationalization by complementary phase diagrams and crystal energy landscapes. Cryst Growth Des 2010;10(2):903–12.
[129] Pal S, Roopa BN, Abu K, Manjunath SG, Nambiar S. Thermal studies of furosemide–caffeine binary system that forms a cocrystal. J Therm Anal Calorim 2014;115(3):2261–8.
[130] ter Horst JH, Deij MA, Cains PW. Discovering new co-crystals. Cryst Growth Des 2009;9(3):1531–7.
[131] Chiarella RA, Davey RJ, Peterson ML. Making co-crystals the utility of ternary phase diagrams. Cryst Growth Des 2007;7(7):1223–6.
[132] Jia Q, Zhang J, Zhang S, Lei D, Xu Y, Kou K. Investigation of the phase behavior of a HNIW·TNT cocrystal system and construction of ternary phase diagrams. Cryst Growth Des 2019;19(11):6370–6.
[133] Blagden N, Berry DJ, Parkin A, Javed H, Ibrahim A, Gavan PT, et al. Current directions in co-crystal growth. New J Chem 2008;32(10):1659–72.
[134] Ainouz A, Authelin J-R, Billot P, Lieberman H. Modeling and prediction of cocrystal phase diagrams. Int J Pharm 2009;374(1):82–9.
[135] Ahuja D, Svärd M, Lusi M, Rasmuson ÅC. Solution and calorimetric thermodynamic study of a new 1 : 1 sulfamethazine–3-methylsalicylic acid co-crystal. CrstEngComm 2020;22(20):3463–73.
[136] Steed JW. The role of co-crystals in pharmaceutical design. Trends Pharmacol Sci 2013;34(3):185–93.

[137] Reus MA, van der Heijden AEDM, ter Horst JH. Solubility determination from clear points upon solvent addition. Org Process Res Dev 2015;19(8):1004–11.
[138] Friščić T, Jones W. Recent advances in understanding the mechanism of cocrystal formation via grinding. Cryst Growth Des 2009;9(3):1621–37.
[139] Karimi-Jafari M, Padrela L, Walker GM, Croker DM. Creating cocrystals: a review of pharmaceutical cocrystal preparation routes and applications. Cryst Growth Des 2018;18(10):6370–87.
[140] Trask AV, Motherwell WDS, Jones W. Pharmaceutical cocrystallization: engineering a remedy for caffeine hydration. Cryst Growth Des 2005;5(3):1013–21.
[141] Ross SA, Lamprou DA, Douroumis D. Engineering and manufacturing of pharmaceutical co-crystals: a review of solvent-free manufacturing technologies. Chem Commun 2016;52(57):8772–86.
[142] Aitipamula S, Chow PS, Tan RBH. Trimorphs of a pharmaceutical cocrystal involving two active pharmaceutical ingredients: potential relevance to combination drugs. CrstEngComm 2009;11(9):1823–7.
[143] Simões MF, Pinto RMA, Simões S. Hot-melt extrusion in the pharmaceutical industry: toward filing a new drug application. Drug Discov Today 2019; 24(9):1749–68.
[144] Kesisoglou F, Hermans A, Neu C, Yee KL, Palcza J, Miller J. Development of in vitro–in vivo correlation for amorphous solid dispersion immediate-release suvorexant tablets and application to clinically relevant dissolution specifications and in-process controls. J Pharm Sci 2015;104(9):2913–22.
[145] Maniruzzaman M, Boateng JS, Chowdhry BZ, Snowden MJ, Douroumis D. A review on the taste masking of bitter APIs: hot-melt extrusion (HME) evaluation. Drug Dev Ind Pharm 2014;40(2):145–56.
[146] Daurio D, Nagapudi K, Li L, Quan P, Nunez F-A. Application of twin screw extrusion to the manufacture of cocrystals: scale-up of AMG 517–sorbic acid cocrystal production. Faraday Discuss 2014;170(0):235–49.
[147] Daurio D, Medina C, Saw R, Nagapudi K, Alvarez-Núñez F. Application of twin screw extrusion in the manufacture of cocrystals, part I: four case studies. Pharmaceutics 2011;3(3):582–600.
[148] Gajda M, Nartowski KP, Pluta J, Karolewicz B. The role of the polymer matrix in solvent-free hot melt extrusion continuous process for mechanochemical synthesis of pharmaceutical cocrystal. Eur J Pharm Biopharm 2018;131:48–59.
[149] Saerens L, Ghanam D, Raemdonck C, Francois K, Manz J, Krüger R, et al. In-line solid state prediction during pharmaceutical hot-melt extrusion in a 12mm twin screw extruder using Raman spectroscopy. Eur J Pharm Biopharm 2014;87(3):606–15.
[150] Kelly AL, Gough T, Dhumal RS, Halsey SA, Paradkar A. Monitoring ibuprofen–nicotinamide cocrystal formation during solvent free continuous cocrystallization (SFCC) using near infrared spectroscopy as a PAT tool. Int J Pharm 2012; 426(1):15–20.
[151] Gajda M, Nartowski K, Pluta J, Karolewicz B. Continuous, one-step synthesis of pharmaceutical cocrystals via hot melt extrusion from neat to matrix-assisted processing—state of the art. Int J Pharm 2019;558:426–40.
[152] Reitz E, Podhaisky H, Ely D, Thommes M. Residence time modeling of hot melt extrusion processes. Eur J Pharm Biopharm 2013;85(3, Part B):1200–5.
[153] Jerry S, Anderson Stephen R, Am Ende David J. A method to produce and scale-up cocrystals and salts via resonant acoustic mixing. [Internet]. EP; EP 2845852 A1. Available from https://lens.org/113-918-186-392-758; 2015.
[154] am Ende DJ, Anderson SR, Salan JS. Development and scale-up of cocrystals using resonant acoustic mixing. Org Process Res Dev 2014;18(2):331–41.

[155] Michalchuk AAL, Hope KS, Kennedy SR, Blanco MV, Boldyreva EV, Pulham CR. Ball-free mechanochemistry: in situ real-time monitoring of pharmaceutical co-crystal formation by resonant acoustic mixing. Chem Commun 2018;54(32):4033–6.
[156] Anderson SR, am Ende DJ, Salan JS, Samuels P. Preparation of an energetic-energetic cocrystal using resonant acoustic mixing. Propellants Explos Pyrotech 2014; 39(5):637–40.
[157] Nagapudi K, Umanzor EY, Masui C. High-throughput screening and scale-up of cocrystals using resonant acoustic mixing. Int J Pharm 2017;521(1):337–45.
[158] Parikh DM. Handbook of pharmaceutical granulation technology. 2nd ed; 2005. p. 1.
[159] Rehder S, Christensen NPA, Rantanen J, Rades T, Leopold CS. High-shear granulation as a manufacturing method for cocrystal granules. Eur J Pharm Biopharm 2013;85(3, Part B):1019–30.
[160] Ondrej D, Veronika S, Eliska S, Gregor S, Cocrystal A. Of ivabradine hydrochloride and (r)-mandelic acid and its pharmaceutical composition. [Internet]. WO; WO 2016/004911 A8. Available from https://lens.org/126-089-117-650-174; 2016.
[161] Sládková V, Dammer O, Sedmak G, Skořepová E, Kratochvíl B. Ivabradine hydrochloride (S)-mandelic acid co-crystal: in situ preparation during formulation. Crystals 2017;7:13.
[162] Tanaka R, Duggirala NK, Hattori Y, Otsuka M, Suryanarayanan R. Formation of indomethacin–saccharin cocrystals during wet granulation: role of polymeric excipients. Mol Pharm 2020;17(1):274–83.
[163] Tanaka K, Takiyama H. Effect of oiling-out during crystallization on purification of an intermediate compound. Org Process Res Dev 2019;23(9):2001–8.
[164] Alvarenga RN, Bernardo A, Pessoa Filho PA. Improvement of an industrial crystallization process: the production of Virginiamycin. Ind Eng Chem Res 2020; 59(16):7839–48.
[165] Horgan DE, Crowley LM, Stokes SP, Lawrence SE, Moynihan HA. Impurity exclusion and retention during crystallisation and recrystallisation—the phenacetin by ethylation of paracetamol process. In: Mastai Y, editor. Advanced topics in crystallization. Rijeka: IntechOpen; 2015. [Internet]. Available from https://doi.org/10.5772/59715.
[166] Beck R, Andreassen J-P. Influence of crystallization conditions on crystal morphology and size of CaCO3 and their effect on pressure filtration. AIChE J 2012;58(1):107–21.
[167] Shiau L-D. Modelling of the polymorph nucleation based on classical nucleation theory. Crystals 2019;9:69.
[168] Beckmann W. Seeding the desired polymorph: background, possibilities, limitations, and case studies. Org Process Res Dev 2000;4(5):372–83.
[169] Mohnicke GM, Ulrich J. Seeding techniques in the crystallization of metastable riboflavin polymorphs. VDI Ber 2005;617–22.
[170] Nicoud L, Licordari F, Myerson AS. Polymorph control in batch seeded crystallizers. A case study with paracetamol. CrstEngComm 2019;21(13):2105–18.
[171] Simone E, Klapwijk AR, Wilson CC, Nagy ZK. Investigation of the evolution of crystal size and shape during temperature cycling and in the presence of a polymeric additive using combined process analytical technologies. Cryst Growth Des 2017; 17(4):1695–706.
[172] Perini G, Salvatori F, Ochsenbein DR, Mazzotti M, Vetter T. Filterability prediction of needle-like crystals based on particle size and shape distribution data. Sep Purif Technol 2019;211:768–81.
[173] Leung DH, Lohani S, Ball RG, Canfield N, Wang Y, Rhodes T, et al. Two novel pharmaceutical cocrystals of a development compound—screening, scale-up, and characterization. Cryst Growth Des 2012;12(3):1254–62.

[174] Yu ZQ, Chow PS, Tan RBH, Ang WH. Supersaturation control in cooling polymorphic co-crystallization of caffeine and glutaric acid. Cryst Growth Des 2011; 11(10):4525–32.
[175] Wang I-C, Lee M-J, Sim S-J, Kim W-S, Chun N-H, Choi G. Anti-solvent co-crystallization of carbamazepine and saccharin. Int J Pharm 2013;450:311–22.
[176] Svoboda V, MacFhionnghaile P, McGinty J, Connor LE, Oswald IDH, Sefcik J. Continuous cocrystallization of benzoic acid and isonicotinamide by mixing-induced supersaturation: exploring opportunities between reactive and antisolvent crystallization concepts. Cryst Growth Des 2017;17(4):1902–9.
[177] Powell KA, Bartolini G, Wittering KE, Saleemi AN, Wilson CC, Rielly CD, et al. Toward continuous crystallization of urea-barbituric acid: a polymorphic co-crystal system. Cryst Growth Des 2015;15(10):4821–36.
[178] Duggirala NK, Vyas A, Krzyzaniak JF, Arora KK, Suryanarayanan R. Mechanistic insight into caffeine–oxalic cocrystal dissociation in formulations: role of excipients. Mol Pharm 2017;14(11):3879–87.
[179] Koranne S, Sahoo A, Krzyzaniak JF, Luthra S, Arora KK, Suryanarayanan R. Challenges in transitioning cocrystals from bench to bedside: dissociation in prototype drug product environment. Mol Pharm 2018;15(8):3297–307.
[180] Kale DP, Zode SS, Bansal AK. Challenges in translational development of pharmaceutical cocrystals. J Pharm Sci 2017;106(2):457–70.
[181] Wang C, Tong Q, Hou X, Hu S, Fang J, Sun CC. Enhancing bioavailability of dihydromyricetin through inhibiting precipitation of soluble cocrystals by a crystallization inhibitor. Cryst Growth Des 2016;16(9):5030–9.
[182] Nayak A, Pedireddi VR. Rational analysis of melting point behavior of co-crystals of 4-nitrophenol with some aza-compounds. Cryst Growth Des 2016;16(10):5966–75.
[183] Ahmed H, Shimpi MR, Velaga SP. Relationship between mechanical properties and crystal structure in cocrystals and salt of paracetamol. Drug Dev Ind Pharm 2017;43(1):89–97.
[184] Bhatt JA, Bahl D, Morris K, Stevens LL, Haware RV. Structure-mechanics and improved tableting performance of the drug-drug cocrystal metformin:salicylic acid. Eur J Pharm Biopharm 2020;153:23–35.
[185] Arabiani MR, K BR, Bhunia S, Teja PK, Lodagekar A, Chavan RB, et al. Brexpiprazole–catechol cocrystal: structure elucidation, excipient compatibility and stability. CrstEngComm 2019;21(44):6703–8.
[186] Bofill L, de Sande D, Barbas R, Prohens R. New cocrystal of ubiquinol with high stability to oxidation. Cryst Growth Des 2020;20(8):5583–8.
[187] Shinozaki T, Ono M, Higashi K, Moribe K. A novel drug-drug cocrystal of levofloxacin and metacetamol: reduced hygroscopicity and improved photostability of levofloxacin. J Pharm Sci 2019;108(7):2383–90.
[188] Thakuria R, Delori A, Jones W, Lipert MP, Roy L, Rodríguez-Hornedo N. Pharmaceutical cocrystals and poorly soluble drugs. Int J Pharm 2013; 453(1):101–25.
[189] Bavishi DD, Borkhataria CH. Spring and parachute: how cocrystals enhance solubility. Prog Cryst Growth Charact Mater 2016;62(3):1–8.
[190] Martin FA, Pop MM, Borodi G, Filip X, Kacso I. Ketoconazole salt and co-crystals with enhanced aqueous solubility. Cryst Growth Des 2013;13(10):4295–304.
[191] Aitipamula S, Cadden J, Chow PS. Cocrystals of zonisamide: physicochemical characterization and sustained release solid forms. CrstEngComm 2018;20(21):2923–31.
[192] Cho M-Y, Kim P, Kim G-Y, Lee J-Y, Song K-H, Lee M-J, et al. Preparation and characterization of aripiprazole cocrystals with coformers of multihydroxybenzene compounds. Cryst Growth Des 2017;17(12):6641–52.

[193] Narayana DS, Krishnamurthi V, Reddy BS, Reddy PP, Balaji R. Aripiprazole co-crystals. [Internet]. WO; WO 2007/092779 A2. Available from https://lens.org/098-229-989-322-934; 2007.
[194] Kimoto K, Yamamoto M, Karashima M, Hohokabe M, Takeda J, Yamamoto K, et al. Pharmaceutical cocrystal development of TAK-020 with enhanced oral absorption. Crystals 2020;10:211.
[195] Chow S-C. Bioavailability and bioequivalence in drug development. Wiley Interdiscip Rev Comput Stat 2014;6(4):304–12.
[196] Chadha K, Karan M, Bhalla Y, Chadha R, Khullar S, Mandal S, et al. Cocrystals of hesperetin: structural, pharmacokinetic, and pharmacodynamic evaluation. Cryst Growth Des 2017;17(5):2386–405.
[197] Childs SL, Chyall LJ, Dunlap JT, Smolenskaya VN, Stahly BC, Stahly GP. Crystal engineering approach to forming cocrystals of amine hydrochlorides with organic acids. Molecular complexes of fluoxetine hydrochloride with benzoic, succinic, and fumaric acids. J Am Chem Soc 2004;126(41):13335–42.
[198] Chen Y, Li L, Yao J, Ma Y-Y, Chen J-M, Lu T-B. Improving the solubility and bioavailability of apixaban via apixaban–oxalic acid cocrystal. Cryst Growth Des 2016; 16(5):2923–30.
[199] Pi J, Wang S, Li W, Kebebe D, Zhang Y, Zhang B, et al. A nano-cocrystal strategy to improve the dissolution rate and oral bioavailability of baicalein. Asian J Pharm Sci 2019;14(2):154–64.
[200] De Smet L, Saerens L, De Beer T, Carleer R, Adriaensens P, Van Bocxlaer J, et al. Formulation of itraconazole nanococrystals and evaluation of their bioavailability in dogs. Eur J Pharm Biopharm 2014;87(1):107–13.
[201] Huang Y, Li J-M, Lai Z-H, Wu J, Lu T-B, Chen J-M. Phenazopyridine-phthalimide nano-cocrystal: release rate and oral bioavailability enhancement. Eur J Pharm Sci 2017;109:581–6.
[202] Arafa MF, El-Gizawy SA, Osman MA, El Maghraby GM. Sucralose as co-crystal co-former for hydrochlorothiazide: development of oral disintegrating tablets. Drug Dev Ind Pharm 2016;42(8):1225–33.
[203] Ogata T, Tanaka D, Ozeki T. Enhancing the solubility and masking the bitter taste of propiverine using crystalline complex formation. Drug Dev Ind Pharm 2014; 40(8):1084–91.
[204] Aitipamula S, Wong ABH, Kanaujia P. Evaluating suspension formulations of theophylline cocrystals with artificial sweeteners. J Pharm Sci 2018;107(2):604–11.
[205] Chen AM, Ellison ME, Peresypkin A, Wenslow RM, Variankaval N, Savarin CG, et al. Development of a pharmaceutical cocrystal of a monophosphate salt with phosphoric acid. Chem Commun 2007;(4):419–21.
[206] Rosa J, Machado TC, da Silva AK, Kuminek G, Bortolluzzi AJ, Caon T, et al. Isoniazid-resveratrol cocrystal: a novel alternative for topical treatment of cutaneous tuberculosis. Cryst Growth Des 2019;19(9):5029–36.
[207] Machado TC, Gelain AB, Rosa J, Cardoso SG, Caon T. Cocrystallization as a novel approach to enhance the transdermal administration of meloxicam. Eur J Pharm Sci 2018;123:184–90.
[208] Dai X-L, Li S, Chen J-M, Lu T-B. Improving the membrane permeability of 5-fluorouracil via cocrystallization. Cryst Growth Des 2016;16(8):4430–8.
[209] Dai X-L, Voronin AP, Huang Y-L, Perlovich GL, Zhao X-H, Lu T-B, et al. 5-fluorouracil cocrystals with lipophilic hydroxy-2-naphthoic acids: crystal structures, theoretical computations, and permeation studies. Cryst Growth Des 2020; 20(2):923–33.
[210] Feng L, Karpinski PH, Sutton P, Liu Y, Hook DF, Hu B, et al. LCZ696: a dual-acting sodium supramolecular complex. Tetrahedron Lett 2012;53(3):275–6.

[211] Putra OD, Yoshida T, Umeda D, Higashi K, Uekusa H, Yonemochi E. Crystal structure determination of dimenhydrinate after more than 60 years: solving salt–cocrystal ambiguity via solid-state characterizations and solubility study. Cryst Growth Des 2016;16(9):5223–9.
[212] Petruševski G, Naumov P, Jovanovski G, Bogoeva-Gaceva G, Ng SW. Solid-state forms of sodium valproate, active component of the anticonvulsant drug epilim. ChemMedChem 2008;3(9):1377–86.
[213] Almansa C, Mercè R, Tesson N, Farran J, Tomàs J, Plata-Salamán CR. Co-crystal of tramadol hydrochloride–celecoxib (ctc): a novel API–API co-crystal for the treatment of pain. Cryst Growth Des 2017;17(4):1884–92.
[214] Cheney ML, Weyna DR, Shan N, Hanna M, Wojtas L, Zaworotko MJ. Coformer selection in pharmaceutical cocrystal development: a case study of a meloxicam aspirin cocrystal that exhibits enhanced solubility and pharmacokinetics. J Pharm Sci 2011;100(6):2172–81.
[215] Geng N, Chen J-M, Li Z-J, Jiang L, Lu T-B. Approach of cocrystallization to improve the solubility and photostability of tranilast. Cryst Growth Des 2013; 13(8):3546–53.
[216] Liu F, Song Y, Liu Y-N, Li Y-T, Wu Z-Y, Yan C-W. Drug-bridge-drug ternary cocrystallization strategy for antituberculosis drugs combination. Cryst Growth Des 2018;18(3):1283–6.
[217] Wang X, Du S, Zhang R, Jia X, Yang T, Zhang X. Drug-drug cocrystals: opportunities and challenges. Asian J Pharm Sci 2020. [Internet]. Available from http://www.sciencedirect.com/science/article/pii/S181808762030427X.
[218] Gao Z, Rohani S, Gong J, Wang J. Recent developments in the crystallization process: toward the pharmaceutical industry. Engineering 2017;3(3):343–53.
[219] Billot P, Hosek P, Perrin M-A. Efficient purification of an active pharmaceutical ingredient via cocrystallization: from thermodynamics to scale-up. Org Process Res Dev 2013;17(3):505–11.
[220] Garg U, Azim Y, Kar A, Pradeep CP. Cocrystals/salt of 1-naphthaleneacetic acid and utilizing Hirshfeld surface calculations for acid–aminopyrimidine synthons. CrstEngComm 2020;22(17):2978–89.
[221] Eliel EL, Wilen SH, Mander LN. Stereochemistry of organic compounds. New York: Wiley; 1994.
[222] Snatzke G. J. Jaques, A. Collet, and S. H. Wilen: enantiomers, racemates, and resolutions, J. Wiley & Sons, Inc., New York, Chichester, Brisbane, Toronto 1981. 447 Seiten, Preis: £ 38.75. Ber Bunsen Phys Chem 1982;86(11):1087.
[223] Prusov EV. Stereoselective synthesis of drugs and natural products. Edited by Vasyl Andrushko and Natalia Andrushko. Angew Chem Int Ed 2014;53(24):6037.
[224] Borghese A, Libert V, Zhang T, Alt CA. Efficient fast screening methodology for optical resolution agents: solvent effects are used to affect the efficiency of the resolution process. Org Process Res Dev 2004;8(3):532–4.
[225] Marchand P, Lefèbvre L, Querniard F, Cardinaël P, Perez G, Counioux J-J, et al. Diastereomeric resolution rationalized by phase diagrams under the actual conditions of the experimental process. Tetrahedron Asymmetry 2004;15(16):2455–65.
[226] Kawashima M, Hirayama A. Direct optical resolution of vicinal diols and an α-hydroxy oxime with a vicinal diamine. Chem Lett 1991;20(5):763–6.
[227] Molnár P, Thorey P, Bánsághi G, Székely E, Poppe L, Tomin A, et al. Resolution of racemic trans-1,2-cyclohexanediol with tartaric acid. Tetrahedron Asymmetry 2008;19(13):1587–92.
[228] Kassai C, Juvancz Z, Bálint J, Fogassy E, Kozma D. Optical resolution of racemic alcohols via diastereoisomeric supramolecular compound formation with O, O′-dibenzoyl-(2R,3R)-tartaric acid. Tetrahedron 2000;56(42):8355–9.

[229] Sánchez-Guadarrama O, Mendoza-Navarro F, Cedillo-Cruz A, Jung-Cook H, Arenas-García JI, Delgado-Díaz A, et al. Chiral resolution of RS-praziquantel via diastereomeric co-crystal pair formation with l-malic acid. Cryst Growth Des 2016; 16(1):307–14.
[230] Guillot M, de Meester J, Huynen S, Collard L, Robeyns K, Riant O, et al. Cocrystallization-induced spontaneous deracemization: a general thermodynamic approach to deracemization. Angew Chem Int Ed 2020;59(28):11303–6.
[231] Eddleston MD, Arhangelskis M, Friščić T, Jones W. Solid state grinding as a tool to aid enantiomeric resolution by cocrystallisation. Chem Commun 2012;48(92):11340–2.
[232] Springuel G, Norberg B, Robeyns K, Wouters J, Leyssens T. Advances in pharmaceutical co-crystal screening: effective co-crystal screening through structural resemblance. Cryst Growth Des 2012;12(1):475–84.
[233] Springuel G, Collard L, Leyssens T. Ternary and quaternary phase diagrams: key tools for chiral resolution through solution cocrystallization. CrstEngComm 2013;15 (39):7951–8.
[234] Springuel G, Robeyns K, Norberg B, Wouters J, Leyssens T. Cocrystal formation between chiral compounds: how cocrystals differ from salts. Cryst Growth Des 2014;14(8):3996–4004.
[235] Habgood M. Analysis of enantiospecific and diastereomeric cocrystal systems by crystal structure prediction. Cryst Growth Des 2013;13(10):4549–58.
[236] Roy BN, Singh GP, Srivastava D, Jadhav HS, Saini MB, Aher UP. A novel method for large-scale synthesis of lamivudine through cocrystal formation of racemic lamivudine with (S)-(−)-1,1′-bi(2-naphthol) [(S)-(BINOL)]. Org Process Res Dev 2009; 13(3):450–5.
[237] Sugavanam B, Shephard Margaret C. Fungicidal compounds. [Internet]. US; US 4243405 A. Available from https://lens.org/045-929-212-914-062; 1981.
[238] Sakamoto M, Mino T. Total resolution of racemates by dynamic preferential crystallization. In: Tamura R, Miyata M, editors. Advances in organic crystal chemistry: comprehensive reviews 2015. Tokyo: Springer Japan; 2015. p. 445–62. [Internet]. Available from https://doi.org/10.1007/978-4-431-55555-1_22.
[239] Srisanga S, ter Horst JH. Racemic compound, conglomerate, or solid solution: phase diagram screening of chiral compounds. Cryst Growth Des 2010;10(4):1808–12.
[240] Gonella S, Mahieux J, Sanselme M, Coquerel G. Spotting a conglomerate is just Halfway to achieving a preparative resolution by preferential crystallization. Org Process Res Dev 2012;16(2):286–93.
[241] Harfouche LC, Couvrat N, Sanselme M, Brandel C, Cartigny Y, Petit S, et al. Discovery of new proxyphylline-based chiral cocrystals: solid state landscape and dehydration mechanism. Cryst Growth Des 2020;20(6):3842–50.
[242] Harfouche LC, Brandel C, Cartigny Y, Petit S, Coquerel G. Resolution by preferential crystallization of proxyphylline by using its salicylic acid monohydrate co-crystal. Chem Eng Technol 2020;43(6):1093–8.

MIX
Papier | Fördert gute Waldnutzung
FSC® C083411

Kolibri 360 GmbH
Ferdinand-Jühlke-Straße 7
99095 Erfurt, Deutschland
produktsicherheit@kolibri360.de